W0232275

Metal-Containing
Polymeric Materials

Metal-Containing Polymeric Materials

Edited by

Charles U. Pittman, Jr.

Mississippi State University
Mississippi State, Mississippi

Charles E. Carraher, Jr.

Florida Atlantic University
Boca Raton, Florida

Martel Zeldin

City University of New York
New York, New York

John E. Sheats

Rider College
Lawrenceville, New Jersey

and

Bill M. Culbertson

Ohio State University
Columbus, Ohio

Plenum Press • New York and London

Library of Congress Cataloging-in-Publication Data

Metal-containing polymeric materials / edited by Charles U. Pittman,
Jr. ... [et al.].
        p.   cm.
    "Proceedings of the International Symposium on Metal-containing
Polymeric Materials, held August 21-25, 1994, at the 208th American
Chemical Society Meeting in Washington, D.C."--T.p. verso.
    Includes bibliographical references and index.
    ISBN 0-306-45295-2
    1. Polymeric composites--Congresses.. 2. Metallic composites-
-Congresses. 3. Silicon polymers--Congresses. I. Pittman, Charles
U. II. International Symposium on Metal-containing Polymeric
Materials (1994 : Washington, D.C.) III. American Chemical Society.
Meeting (208th : 1994 : Washington, D.C.)
    TA418.9.C6M459 1996
    620.1'92--dc20   96-1765
                                                            CIP

Proceedings of the International Symposium on Metal-Containing Polymeric Materials,
held August 21 – 25, 1994, at the 208th American Chemical Society Meeting in Washington D.C.

ISBN-13: 978-1-4613-8018-4   e-ISBN-13: 978-1-4613-0365-7
DOI: 10.1007/978-1-4613-0365-7

© 1996 Plenum Press, New York
Softcover reprint of the hardcover 1st edition 1996
A Division of Plenum Publishing Corporation
233 Spring Street, New York, N. Y. 10013

10 9 8 7 6 5 4 3 2 1

All rights reserved

No part of this book may be reproduced, stored in a retrieval system, or transmitted in any form or by any
means, electronic, mechanical, photocopying, microfilming, recording, or otherwise, without written
permission from the Publisher

PREFACE

Metal-containing polymeric materials now represent a major niche area within polymer science. This topic, once a curiosity among polymer scientists, owes its growth to the explosive development of organometallic chemistry since the early 1960s. The huge increase in knowledge concerning the chemistry of main group metals, transition metals and lanthanide/actinide metals has been followed by a variety of applications and mergers of these chemistries with polymer science. In the 1960s and early 1970s vinyl ferrocene and other polymerizable ferrocene derivatives were incorporated into polymers as a direct outgrowth of the rapidly developing field of organometallic chemistry of π-complexed transition metal systems. Likewise, the expansion of metal cluster chemistry was followed by incorporation of metal clusters into polymers. Silicon chemistry advances led directly to polysilanes and two-dimensional sheet systems involving silicon. Inorganic ceramic chemistry advances spurred research on metal-containing fiber development for preceramic applications. Current explosive growth in materials science, optical properties, catalysis, three-dimensional networks, etc. are fueling interesting mergers of organometallic, inorganic, and polymer chemistry. In order to follow and emphasize these trends, the editors organized the "International Symposium on Metal-Containing Polymeric Materials" at the 208th National American Chemical Society Meeting in Washington, DC, August 20-25, 1994. Following the symposium selected authors were asked to prepare chapters both introducing and summarizing their work for this book. This was the sixth such symposia which we organized. The previous symposia were in 1977, 1979, 1983, 1987 and 1989 four of which were subsequently published in book form. This and the earlier books provide a glimpse of the field's evolution.

As one would expect from a field of study involving polymer science, materials chemistry, organometallic and inorganic chemistry, catalysis, and biochemical concepts, metal-containing polymers has become a highly interdisciplinary topic. This becomes apparent simply looking at the chapter titles. Scientists from around the world have contributed discussions of topics ranging from structure and functioning of metalloenzymes, shish kabob polymers with metals along the spine, preceramic polymers, metal-cluster containing materials, mixed-metal polymers, polymer-bound Ziegler Natta catalysts, polymer precursors for conducting and ferromagnetic materials, synthesis of three-dimensional cage and channel networks, plant growth regulators, novel polyelectrolytes, metallized polyimides to novel photonic materials. Synthetic methodologies such as ring-opening polymerization of strained metallocenophanes, liquid crystalline transition metal-containing polymers, thermosetting systems, vinyl polymers, coordination polymers condensation polymerization and oxidative polymerizations have been applied to metal-containing polymers and all these topics represented herein.

There was no 'best' way in which to organize this text. The method selected was to follow the introductory overview chapter with six major thematic divisions. Synthesis and characterization of new systems was the first theme. Ten chapters fall into this general classification. This is followed by a section containing seven chapters on silicon-containing polymers, preceramic systems, 3-D cages and networks. Then eight chapters have been grouped under the theme: electrical, magnetic, photonic, and ion-exchange properties. This is followed by the sections: polyelectrolytes and ion-binding systems (four chapters) and one chapter on mass spectroscopy. The final section of the text covers biopolymers and their structure, function, and reactivity relationships (three chapters). One chapter is a major review by G. B. Jameson on transition metal-containing biopolymers and models.

In closing, I would like to acknowledge the special efforts of John E. Sheats of Rider University who is mainly responsible for the introductory chapter. I also want to thank Donna P. Murrah who played a major role in the clerical and technical aspects of bringing the text together. She also played a pivotal role in composing the subject index. Along with the other editors, I want to thank these individuals and all the authors for making this text possible.

Charles U. Pittman, Jr.

CONTENTS

SYNTHESIS AND CHARACTERIZATION OF NEW SYSTEMS

SILICON-CONTAINING POLYMERS, PRECERAMIC SYSTEMS, 3-D CAGES AND NETWORKS

CATALYTIC SYSTEMS

ELECTRICAL, MAGNETIC, PHOTONIC, AND
ION-EXCHANGE PROPERTIES

POLYELECTROLYTES AND ION-BINDING SYSTEMS

MASS SPECTROSCOPY

BIOLOGICAL SYSTEMS

Metal-Containing
Polymeric Materials

INORGANIC AND ORGANOMETALLIC POLYMERS - AN OVERVIEW

John E. Sheats,*[1] Charles E. Carraher, Jr.,[2] Charles U. Pittman, Jr.,[3] Martin Zeldin,[4] Bill M. Culbertson[5]

[1]Department of Chemistry, Rider University, Lawrenceville, NJ 08648

[2]College of Sciences, Florida Atlantic University, Boca Raton, FL 33431

[3]University/Industry Chemical Research Center, Department of Chemistry, Mississippi State University, Missisippi State, MS 39762

[4]CUNY - College of Staten Island, Staten Island, NY 10314

[5]College of Dentistry, Ohio State University, Columbus, Ohio 43210

INTRODUCTION

In the years since we organized our first symposium on inorganic and organometallic polymers in 1971[1] and five subsequent symposia in 1977,[2] 1979,[3] 1985,[4] 1989,[5] and 1994,[6] we have seen this field produce a wide variety of materials with diverse applications such as polymeric conductors and semiconductors, preceramic materials, polymer bound catalysts, shields for UV and other high energy radiation, biosensors, polymers with high thermal stability, flame retardancy, and a wide variety of applications. Indeed, what can be expected when you incorporate the entire periodic table into a field that formerly had concentrated primarily on C, H, N, O, S, Si and the halogens?

Other groups of researchers[7] have also endeavored to summarize developments in this field and a new journal, the *Journal of Inorganic and Organometallic Polymers*, arose from the 1989 symposium and is now in its fifth year, having weathered many of the storms and shoals that sink new journals.

To summarize all of the developments in this diverse field is beyond the scope of this short review. For a more detailed discussion, we refer the reader to the excellent reviews by Manners for the years 1991-94[8], to the works previously cited and to the *Journal of Inorganic and Organometallic Polymers*. What we will try to accomplish is to summarize developments in four major areas: Silanes and Siloxanes; Phosphazenes, Carbo, Thio and Thionylphosphazenes; Coordination Polymers and other polymers containing transition metals in their backbone and Metallocene Polymers.

The goal is not only to summarize what has happened in these fields, but also to predict directions future research may take. A distinction is made between organometallic polymers, which contain the metal atom as an intimate part of the polymeric backbone in near stoichiometric amounts and macromolecule-metal complexes, where metal ions or complexes are bound to or trapped within cavities in a macromolecule, but represent only a small part of the total structure. Macromolecule-metal complexes have been the subject of several recent symposia, some of which have appeared as special issues of journals[9] or as books.[10]

Siloxanes

Polysiloxanes are currently the most commercially significant class of inorganic polymers because of their high thermal stability, flexibility at low temperatures, hydrophobicity, and general chemical inertness. The basic procedures for preparing polysiloxanes were established in the 1940's and can be found in any current textbook on polymer chemistry. The significant new developments in this field involve formation of copolymers of siloxanes with other materials, grafting of siloxane oligomers to other polymer backbones, and attaching pendant functional groups to the normally inert polymer backbones.

An example of the first type of material is the synthesis of triblock siloxane polystyrene siloxane polymers **1** as shown in equation (1).[11] These materials show a microphase separation in the solid state to give a silicone phase with a T_g of -110°C and a glassy polystyrene phase, T_g 65-85°C.

$$Na^{+-}\underset{H}{\overset{C_6H_5}{C}}-CH_2-\left(\underset{H}{\overset{C_6H_5}{C}}-CH_2\right)_n-CH_2-\underset{H}{\overset{C_6H_5}{C}}{}^{-}\ Na^+\ +\left[Me_2SiO\right]_3$$

(1)

$$\longrightarrow \left(\underset{CH_3}{\overset{CH_3}{Si}}-O\right)_n-(polystyrene)-\left(\underset{CH_3}{\overset{CH_3}{Si}}-O\right)_n$$

1

Synthesis of siloxane-silica gel copolymers has been the subject of a recent review.[12] These materials vary in composition from silicone elastomers which have small particles of silica gel formed in the pores as fillers to polymer-modified ceramics. Three techniques [13-14] have been used to create these materials. The first technique involves swelling the pores of an elastomer, either a hydrocarbon rubber, a silicone, or a polyphosphazene with an organometallic species such as $Si(OEt)_4$ and hydrolyzing it to form silica particles (2).

$$Si(OEt)_4 + 2H_2O \longrightarrow SiO_2 + 4EtOH \qquad (2)$$

The second technique employs hydroxyl-terminated polysiloxanes and the organometallic species to produce polymer-crosslinked silica particles **2** (3).

$$HO(R_2SiO)_nOH + Si(OEt)_4 \longrightarrow -O(R_2SiO)_n O(SiO_2)n - \qquad (3)$$

2

The third technique blends $Si(OEt)_4$ with siloxane polymers containing terminal vinyl groups. After hydrolysis, the vinyl groups can be crosslinked by vinyl-silane coupling.

Sesquisiloxane-siloxane copolymers **3** have been prepared by Lichtenham et al.[15] from polyhedral dihydroxysilsequioxanes and R_2SiX_2 or $R_2Si(NH_2)_2$ or other siloxane oligomers. The resulting materials, with M_n of 15,000 to 200,000, are readily processable, but can be pyrolyzed to form ceramics containing SiO_2 and SiOC.

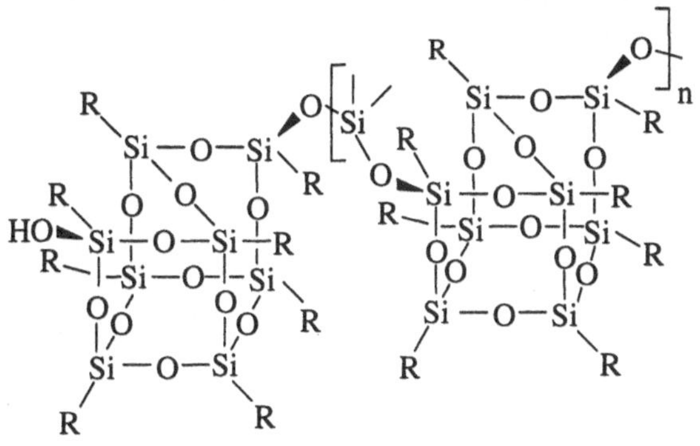

3

Ferrocenyl-sesquisiloxane networks **4** and hybrid silica gels **5** have been prepared by Corriu et al.[16] (4-5)

$$(MeO)_3SiC_5H_4FeC_5H_4Si(OMe)_3 \longrightarrow (O_{1.5}SiC_5H_4FeC_5H_4SiO_{1.5})_n \quad (4)$$

4

$$(5)$$

$$(MeO)_3SiC_5H_4FeC_5H_4Si(OMe)_3 + Si(OMe)_4 \xrightarrow{H_2O}$$

$$- O_{1.5}SiC_5H_4FeC_5H_4SiO_{1.5} (SiO_2)_n^{-}$$

5

Polyphenylmetallosiloxanes **6** were reported by Zhdanov[17] in our most recent symposium (6-7). These materials were obtained as crystalline solids and their 3-dimensional structure determined by X-ray crystallography. The crystals obtained when metals such as Fe, Cu, and Ni were employed were highly paramagnetic and could be pyrolyzed to form metal carbide or metal carbide-silicide and SiO_2 composites which were also paramagnetic.

$$(PhSiO_{1.5})_x + NaOH + Na \longrightarrow [PhSi(O)ONa]_n \quad (6)$$

$$[PhSi(O)ONa]_n + MCl_2 \longrightarrow [(PhSiO_{1.5})_{12} (MO)_6]_y \quad (7)$$

6

Soluble metallosiloxanes, **7** and **8**, suitable for preparing ceramic materials containing Ti or Zr, were prepared by Gunji and Abe[18] (8-9). The metal to Si ratio could be varied by use of oligomeric siloxanes instead of the monomers.

5

$$2 \text{ MeO}-\underset{\underset{\text{O}-\text{t-Bu}}{|}}{\overset{\overset{\text{O}-\text{t-Bu}}{|}}{\text{Si}}}-\text{OH} \quad + \quad \text{i-Pr}-\text{O}-\underset{\underset{\text{acac}}{\diagdown}}{\overset{\overset{\text{acac}}{\diagup}}{\text{M}}}-\text{O}-\text{i-Pr} \longrightarrow$$

M = Ti, Zr

$$\text{MeO}-\underset{\underset{\text{O}-\text{t-Bu}}{|}}{\overset{\overset{\text{O}-\text{t-Bu}}{|}}{\text{Si}}}-\text{O}-\underset{\underset{\text{acac}}{\diagdown}}{\overset{\overset{\text{acac}}{\diagup}}{\text{M}}}-\text{O}-\underset{\underset{\text{O}-\text{t-Bu}}{|}}{\overset{\overset{\text{O}-\text{t-Bu}}{|}}{\text{Si}}}-\text{OMe}$$

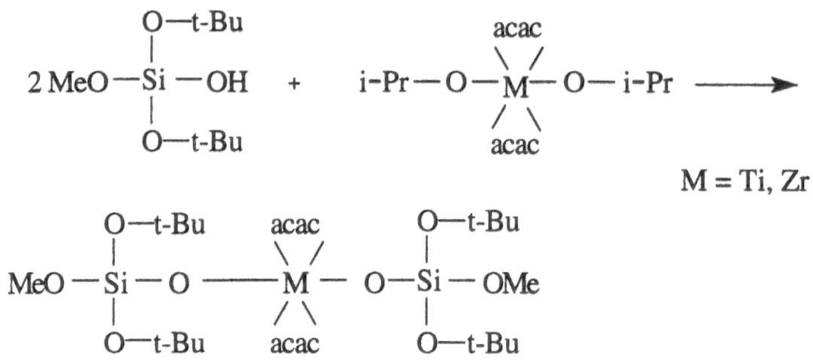

7

(9)

$$4 \text{ MeO}-\underset{\underset{\text{O}-\text{t-Bu}}{|}}{\overset{\overset{\text{O}-\text{t-Bu}}{|}}{\text{Si}}}-\text{OH} \quad + \quad \text{M(O-iPr)}_4 \longrightarrow \text{M}\left(\text{O}\,\underset{\underset{\text{O}-\text{t-Bu}}{|}}{\overset{\overset{\text{O}-\text{t-Bu}}{|}}{\text{Si}}}-\text{OMe}\right)_4$$

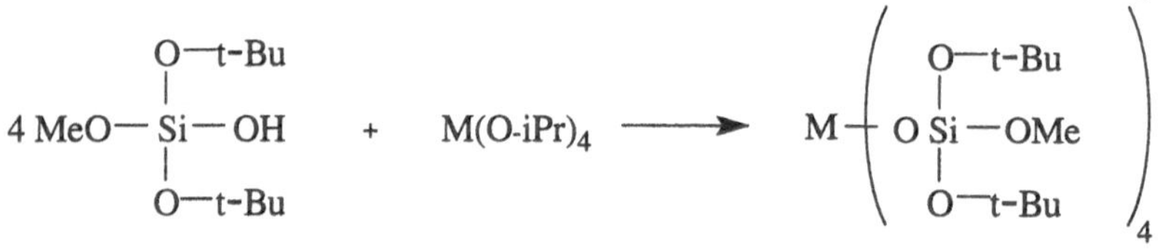

8

The same authors[19] reported a procedure for preparing a soluble, spinable polymer, **9**, which could be pyrolyzed to form continuous ZrO_2 fibers or, if $Y(acac)_3$ were added to the reaction, forming **9**, ZrO_2 - Y_2O_3 fibers (10-11).

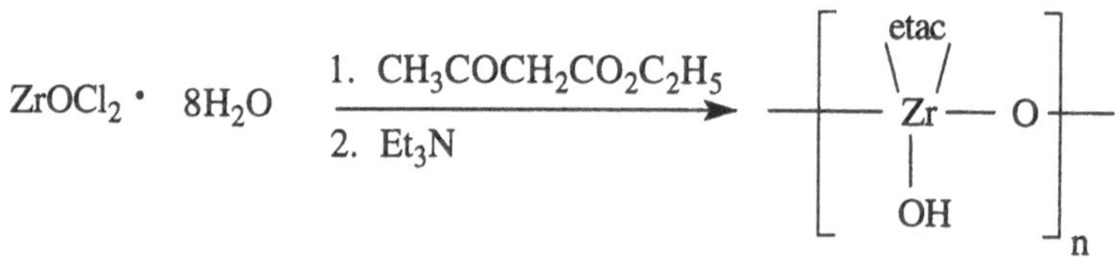

$$\text{ZrOCl}_2 \cdot 8\text{H}_2\text{O} \quad \xrightarrow[\text{2. Et}_3\text{N}]{\text{1. CH}_3\text{COCH}_2\text{CO}_2\text{C}_2\text{H}_5} \quad \left[\underset{\underset{\text{OH}}{|}}{\overset{\overset{\text{etac}}{\diagdown\diagup}}{\text{Zr}}}-\text{O}\right]_n$$

(10)

9

$$\textbf{9} \quad \xrightarrow[\text{2. Steam Treatment}]{\text{1. Dry Spinning}} \quad \begin{matrix} \text{ZrO}_2 \\ \text{or} \\ \text{ZrO}_2 \cdot \text{Y}_2\text{O}_3 \end{matrix} \quad \text{fibers}$$

Terpolymerization of 1-butene and SO_2 with vinyl-terminated polysiloxanes (10) produces a comb polymer, **10,** with a poly(butenesulfone) backbone and poly(dimethylsiloxane) side groups.[20] These materials can be used as electron beam resists and as polymers with resistance to oxygen ion etching.

$$SO_2 + CH_2 = CHC_2H_5 + CH_2 = CH\ (CH_2)_4(SiMe_2O)_nSiMe_2Bu \longrightarrow \qquad (12)$$

——— PBS

wwwwww PDMS

10

Comb polymers with a poly-methacrylate backbone were prepared by Puyenbroek et al.[21] from **11** employing either radical or anionic initiators. The resulting materials had M_n values of 2-9 x 10^4. Films of these materials exhibited spherical siloxane domains embedded in a methacrylate matrix.

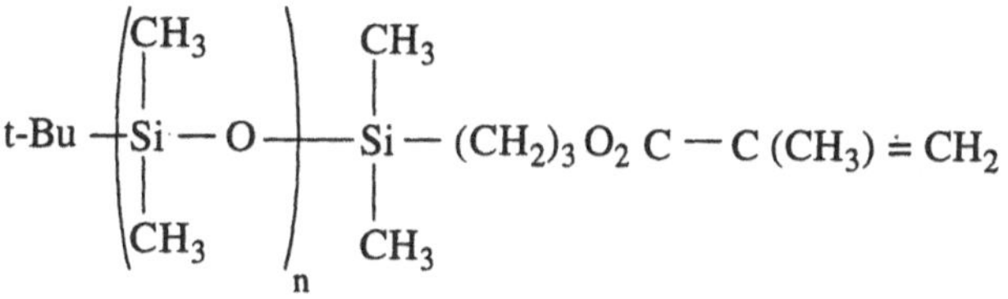

11

Functional groups can be attached to a siloxane backbone either by employing monomers with a reactive functional group or by hydrosilylation reactions of silicone polymers that contain a SiH bond with functionalized terminal olefins. Zeldin and coworkers [22-24] have prepared **12-14** by the first procedure.

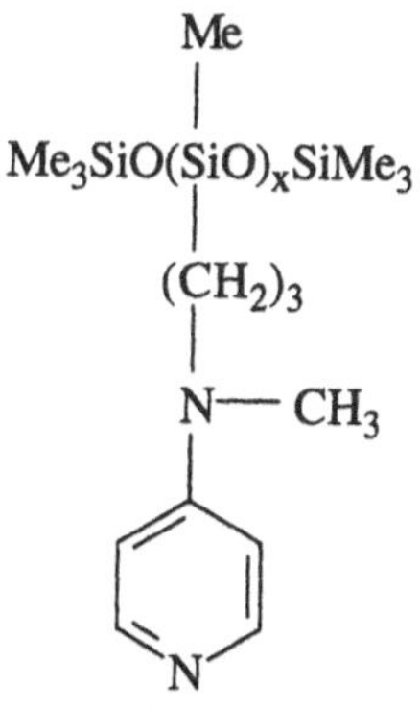

12

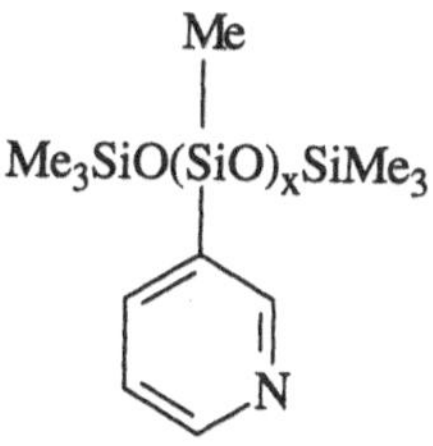

13

$$\underset{\textbf{14}}{\overset{\displaystyle \overset{\displaystyle Me}{|}}{Me_3SiO(SiO)_xSiMe_3}}$$

Polymers **12-14**, when partially protonated, are water soluble and exhibit a tendency to self-associate as micelle-like species. The pyridine and pyridine N–oxide side groups can act as powerful nucleophilic transacylation catalysts. The siloxane backbone is strongly lipophilic. This combination produces a water-soluble catalyst that readily binds and cleaves[22,24] organic carboxylates and phosphates.

Hydrosilylation of functionalized terminal olefins (9)[25] provides a route to a wide variety of functionalized silicones **15**, including glucose and galactose units,[25a] cinnamic acid derivatives,[26] methacrylate,[28] epoxy,[27] amino,[29] metal carbonyls,[30] and mesogenic groups which produce liquid crystallinity[31,32] Hydrosilylation, however, is far from being a clean, quantitative and well-understood reaction.[32] Side reactions include crosslinking, α-addition and decomposition of the unsaturated mesogen.

$$(13)$$

15

To avoid side reactions, it is essential to work under nitrogen in dry conditions and to insure the completion of the reaction by addition of a small molecule such as 1-octene to consume any remaining Si–H groups.

A number of other investigators have prepared siloxane polymers with mesogenic side groups.[33-42] Siloxane polymers which exhibit liquid crystallinity in the main backbone have also been described.[32]

Silanes

Carbon possesses a unique ability to form stable single, double, and triple bonds to itself. The field of organic chemistry is therefore filled with chains, rings, sheets and three-dimensional carbon structures, from which a variety of functional groups can be appended. Early attempts to form similar structures with silicon were largely unsuccessful. It was only in the early 1980's that soluble, well characterized long silylene chains were produced.[43-45] Possible applications of these materials in electronics and as precursors to SiC and other ceramic materials had led to extensive research in this field summarized in several recent reviews.[46-48] We will therefore only summarize the main points.

Four methods of preparing polysilylenes have been developed:

1. Reductive Coupling
$$n\ R_2SiCl_2 + 2n\ Na\ \text{or other metal} \longrightarrow 2n\ NaCl + (R_2Si)_n \qquad (14)$$

2. Dehydrogenative Coupling
$$n\ R_2SiH_2 + \text{metal} \longrightarrow n\ H_2 + (R_2Si)_n \qquad (15)$$

3. Polymerization of Masked Disilenes $\qquad\qquad\qquad (16)$

$$\longrightarrow \ -(R_2SiSiR_2)_n + n\ C_6H_5\text{-}C_6H_5$$

4. Anionic Ring-Opening Polymerization

$$\qquad (17)$$

Reductive coupling initially employed Na or other alkali metals. R groups therefore could not contain any reactive groups. The reaction is heterogeneous, occuring at the surface of tiny metal globules and appears to follow a chain mechanism rather than a step mechanism.[48] At low conversion, high molecular weight polymer is obtained, but in later stages only oligomers, many of which are cyclic Si_6 or Si_8 units, are formed. Thus, the final product has a bimodal distribution. In order to increase the surface area, ultrasonic radiation[49] is used to disperse the sodium into microspheres.

Ziegler et al.[50] offer strong evidence for a radical mechanism in which chain transfer to solvents such as toluene leads to termination, crosslinking, or incorporation of toluene into

9

the polymer. Radicals also caused chain scission. When $Br(C_6H_5SiMe)_6Br$ was coupled, no higher molecular weight polymer was obtained than when $C_6H_5SiCl_2CH_3$ was coupled. The longer monomer should produce longer chains if only ring-chain equilibrium is involved. Further evidence for radical induced chain cleavage was offered by West et al.,[47] who observed scrambling of SiR_2 units when disilanyl monomers $Cl((CH_3)_2Si)_2Cl$ and $Cl((C_6H_{13})_2Si)_2Cl$ were copolymerized. When benzene, a solvent which is incapable of chain transfer, is employed and an Na/K alloy which is molten at 80°C, the reflux temperature of benzene, is used as the reductant, high yields of polysilylenes are obtained.[47]

Thus, it appears that formation of silylyl chains is a considerably more complex process than formation of carbon chains. Because of the high C–C bond energy (80 k cal), once carbon chains are formed, they do not easily cleave, but Si-Si bonds ($\approx$ 30 k cal), are cleaved more readily, leading to a wide variety of products. The authors [46-50] appear to have gained considerable understanding of the process and how to control it, but much more development is needed before commercially useful material can be prepared reproducibly.

Coupling of dichlorodisilanes with dilithium salts of 1,2-diethynyl disilanes[51] (18) or dilithio polythiophenes[52] (18a) produce alternating copolymers, 16 and 17, which exhibit σ–π conjugation, earlier observed by Ishikawa et al.[53-54] for $(-SiR_2C_6H_4SiR_2-)_n$. Polymers 16 and 17 also undergo solid state transitions to form liquid crystalline mesophases.[51]

(18)

$$LiC\equiv C-SiR_2SiR_2C\equiv CLi + ClSiR_2'SiR_2'Cl \longrightarrow$$

$$+(SiR_2)_2\, C\equiv C\text{-}(SiR_2')_2\, C\equiv C\!\!+_n$$

16

(18a)

n = 1, 2 or 4

17

Dehydrogenative coupling of dialkyl and monoalkyl silanes (19) (Method 2) in the presence of a transition metal catalyst has provided an alternate route to polysilylenes.[55-62] Titanocene and zirconocene [55-61] dialkyls have served as effective catalysts for the coupling, but hafnocene catalysts are, surprisingly, inactive.[62] Organolanthanide catalysts such as $(C_5Me_5)_2$ LaR (R = H, $CH(SiMe_3)_2$) are also effective.[63]

A detailed mechanistic study of the reaction was also included. The same oxidative coupling mechanism can also be employed to synthesize disilanylarylene polymers[65] such as **18** (19) as an alternative to the procedure given in (18).

(19)

$$RH_2SiC_6H_4SiH_2R \xrightarrow{\text{cat}} \left[\begin{array}{c} R\ R \\ C_6H_4SiSi \\ H\ H \end{array} \right]_n$$

18

Dehydrogenative coupling (20) has also been employed by Tilley et al.[64] as a route to poly(stannanes) such as **19**. Soluble polymers with M_n up to 17,500 were obtained in 93% yield. The highest molecular weight fraction showed λ max at 382 nm, considerably red-shifted compared to silane and germane polymers and exhibited photosensitivity and photo blending typical of poly(silanes) and poly(germanes).

(20)

$$\text{n-Bu}_2\text{SnH}_2 \longrightarrow \text{(n-Bu)}_2\text{Sn)}_n$$

19

Polymerization of masked disilenes **20** (Method 3) proceeds readily as shown in (21-22)[65].

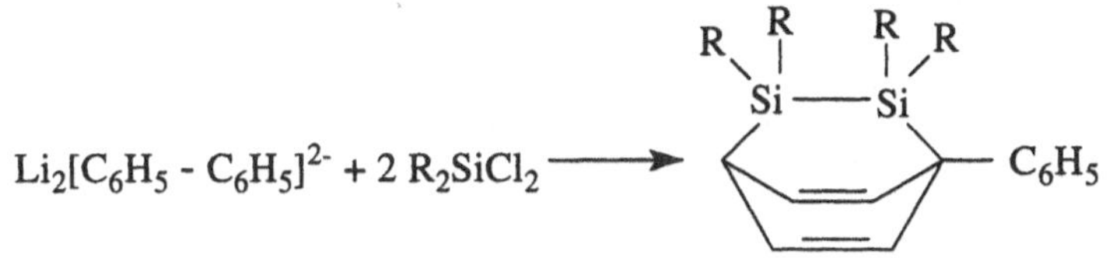

$$Li_2[C_6H_5 - C_6H_5]^{2-} + 2\ R_2SiCl_2 \longrightarrow$$

(21)

20

$$\text{20} + \text{BuLi} \longrightarrow (R_2Si)_n + n\ C_6H_5 - C_6H_5$$

(22)

Molecular weights of up to 5×10^4 were obtained. Addition of methyl methacrylate led to formation of a block silylene-methacrylate copolymer, **21** (23). This method appears quite versatile and will be investigated in greater detail.

$$Bu(R_2Si)_n^- + m\ CH_2 = C(CH_3)CO_2CH_3 \longrightarrow$$

(23)

$$Bu(R_2Si)_n(CH_2 - C\ (CH_3)CO_2CH_3)_m H$$

21

Ring opening polymerization of strained cyclic silanes (Method 4) has proven useful for preparing a variety of polysilylenes.[66] This method had been described in two recent reviews.[47,67] Cyclotetrasilanes, **22**, and other cyclosilanes can be prepared from 1,2-dichloro disilanes by reductive coupling (24). Steric strain arises more from repulsion of substitutents than from angle strain.

$$\text{ClR}_2\text{SiSiR}_2\text{Cl} + 2\text{ Na} \longrightarrow$$

$$
\begin{array}{ccc}
 & R & R \\
 & | & | \\
R - & Si - & Si - R \\
 & | & | \\
R - & Si - & Si - R \\
 & | & | \\
 & R & R
\end{array}
$$

22

Thus polymerization is favored only when nonbonded repulsions can be decreased. Since **22** can possess a variety of stereoisomers if all the R groups are not equivalent, separation of the stereoisomers before polymerization can lead to stereoregular polymers.[68] In the absence of steric strain, cyclic penta- and hexa-silanes can be coupled to form a great variety of structures not involving ring opening.[68]

The great interest in silane polymers arises because they show $\sigma-$ delocalization and $\sigma-\pi$ delocalization when they are conjugated with arenes or acetylenes. This property is greatly enhanced when holes are created by doping, by illumination, or by an electric field. Haarer et al.[69] describe a light emitting diode, LED, which uses poly(methylphenylsilane) as the hole transporting material. The LED's electroluminesced at a threshold voltage of 40V in a continuous DC mode under forward bias. A bright, easily seen yellow light was produced. The silane polymer shows a higher effective mobility of holes than polyphenylene vinylene which had been used previously in LED fabrication. The lack of visible absorption for polymethylphenylsilane permits LED's which can emit light in any wavelength in the visible range, since absorption of the emitted light does not occur. Similar luminescence was observed for the thiophene block polymers, **17**. The thiophene chain appears to be the source of the luminescence. The wavelength emitted can be tuned from red to blue by adjusting the length of the thiophene block.[52] LED's fabricated from **17** appear to have a much longer lifetime than those from poly(methylphenylsilane). Doping of poly(methylphenylsilane) with fullerene greatly enhances its photoconductivity.[70] Applications of these materials in electronic devices are only in their infancy, but vigorous research is continuing, much of it behind closed doors.

Polyphosphazenes

Next to polysiloxanes, the polyphosphazenes are the inorganic polymers which have achieved the greatest commercial success. The first step to formation of polyphosphazenes, the ring-opening polymerization of $P_3N_3Cl_6$ has been known for many years, but Allcock and his coworkers[71-75] in 1965 were the first to demonstrate that a wide variety of useful, hydrolytically stable materials could be obtained by replacing the Cl with alkoxy, aryloxy, amino, or alkyl groups. By the end of 1990, roughly 300 types of polyphosphazenes had been prepared, 2,000 publications and patents had appeared, and 200 or more new works are appearing each year.[75-76]

$$P_3N_3Cl_6 \xrightarrow[\text{2 days}]{200 - 250^o} \left(\begin{array}{c} Cl \\ | \\ P=N \\ | \\ Cl \end{array} \right)_n \quad \mathbf{23} \qquad (25)$$

$$\xrightarrow{ArO^-} \left(\begin{array}{c} OAr \\ | \\ P=N \\ | \\ OAr \end{array} \right)_n \quad \text{flame retardants} \qquad (26)$$

24

$$\mathbf{23} \xrightarrow{\begin{array}{c} C_3F_7CH_2OH \\ CF_3CH_2OH \end{array}} \left(\begin{array}{c} OCH_2CF_3 \\ | \\ P=N \\ | \\ OCH_2C_3F_7 \end{array} \right)_n \quad \text{elastomers}$$

25

$$\xrightarrow{RNH_2} \left(\begin{array}{c} HNR \\ | \\ P=N \\ | \\ HNR \end{array} \right)_n \quad \text{water soluble film-forming}$$

26

$$\xrightarrow{H_2NCH_2CO_2H} \left(\begin{array}{c} HN-CH_2CO_2H \\ | \\ P=N \\ | \\ HN-CH_2CO_2H \end{array} \right)_n \quad \text{biodegradable polymers}$$

27

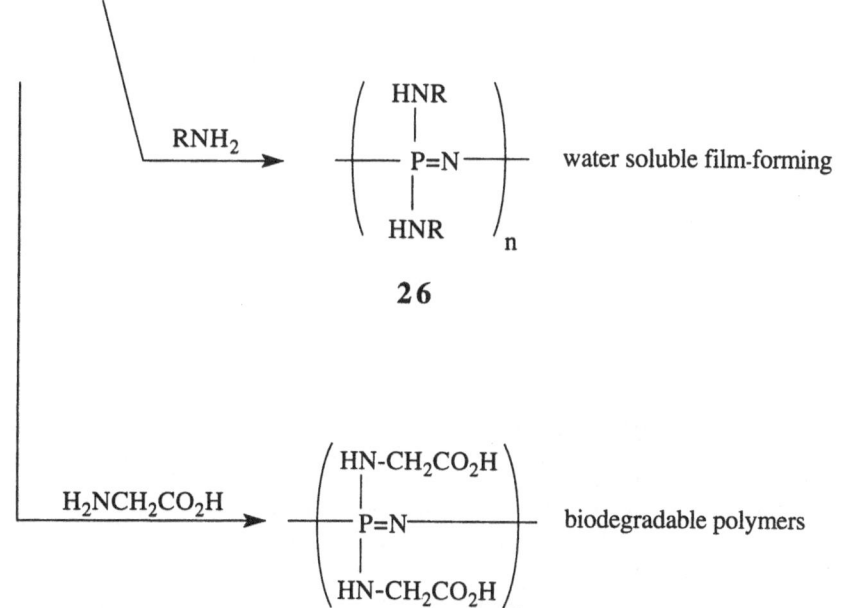

The biggest difficulty appears to be the control of the pattern of substitution so that well defined structures can be formed.[77] The method of synthesis of polyphosphazenes and efforts to control their structure will therefore be discussed.

There are three basic methods of synthesis of polyphosphazenes: ring opening polymerization of $P_3N_3Cl_6$, followed by substitution of the polyphosphonitrilic chloride **23** (Equation 25); substitution of the cyclic trimer, followed by ring-opening polymerization, and formation of the polyphosphazene chain by condensation reactions not involving the cyclic trimer. Ring opening polymerization of $P_3N_3Cl_6$ (25), the first method employed in synthesis of polyphosphazenes, is still the most widely employed, but it has several major disadvantages. Temperatures > 200° and reaction times of several days are required. Polymers with a broad range of molecular weights are obtained. The reaction must be terminated at 60-80% conversion to prevent crosslinking or other undesirable side reactions. Substitution along the chain is random in many cases; although two different substitutents can be placed on the same phosphorous, if the first substitutent reduces the reactivity of the P-Cl bond significantly (27). Formation of block or alternating copolymers by this method is not possible.

(27)

$$\left(\begin{array}{c} Cl \\ | \\ \text{---P} = \text{N---} \\ | \\ Cl \end{array}\right)_n \xrightarrow{RO^-} \left[\begin{array}{c} OR \\ | \\ \text{---P} = \text{N---} \\ | \\ Cl \end{array}\right]_n$$

$$\xrightarrow{RNH_2} \left(\begin{array}{c} OR \\ | \\ \text{---P} = \text{N---} \\ | \\ NHR \end{array}\right)_n$$

28

The second method, substitution of the cyclic trimer prior to polymerization, offers the possibility of better control of polymer structure. Monomers with a definite structure and stereochemistry can be prepared, purified, and polymerized. Two problems, however, arise in this procedure. The first problem is that more highly substituted trimers do not polymerize readily. The second problem with this method is that initiation of polymerization can occur at any one of the three phosphorus atoms in the trimer so that the structure becomes random. One solution to the first problem is to use partially substituted trimers and complete the substitution once polymerization has taken place. The other solution is to introduce strain into the ring by use of bridging ligands such as ferrocenyl[78-79] (28) and other transannular bridging ligands.[80]

14

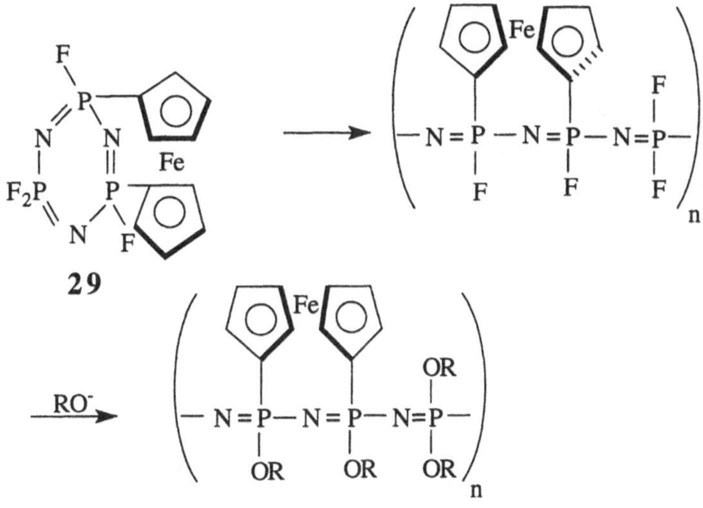

29

30

X-ray crystallography indicates that the phosphazene ring of **29** shows significant nonplanarity induced by stress,[79] which can be relieved in the polymer.

The third method of synthesis, preparation of phosphazenes by condensation routes, has received much attention in recent years as the best route to well defined structures. Polymerization of **31** at 200° produces gel-free **32**[81]. The molecular weight of **32** can be controlled more readily than that of **23** obtained from polymerization of $P_3N_3Cl_6$.

(29)

$$n \ Cl_3P = N - P(O)Cl_2 \xrightarrow{200°} Cl_3P = N -\!\!\!(PCl_2 = N)_{\overline{n\text{-}1}}P(O)Cl_2$$

$$+ \ n\text{-}1 \ POCl_3$$

31   **32**

Polymerization of phosphoranimines **33** and **35** (30) and (31) leads to soluble, tractable polymers **34** and **36** at moderate temperatures without the need for a subsequent substitution step.[82] If two different phosphoranes are included in the same pot, random copolymers can be formed.[83]

$$\text{Ph(CF}_3\text{CH}_2\text{O})_2 \text{ P} + \text{Me}_3\text{SiN}_3 \xrightarrow{\ 70^{\circ}\ }$$

33

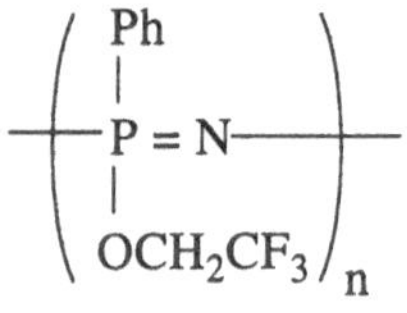

$$\left(\begin{array}{c} \text{Ph} \\ | \\ \text{—P = N—} \\ | \\ \text{OCH}_2\text{CF}_3 \end{array} \right)_n \qquad (30)$$

34

$$\text{(RO)}_3 \text{ P = N - SiMe}_3 \xrightarrow{\ \text{Bu}_4\text{NF}\ } \left(\begin{array}{c} \text{OR} \\ | \\ \text{—P = N—} \\ | \\ \text{OR} \end{array} \right)_n \quad \left(\begin{array}{c} \text{OR}' \\ | \\ \text{—P = N—} \\ | \\ \text{OR}' \end{array} \right)_n \qquad (31)$$

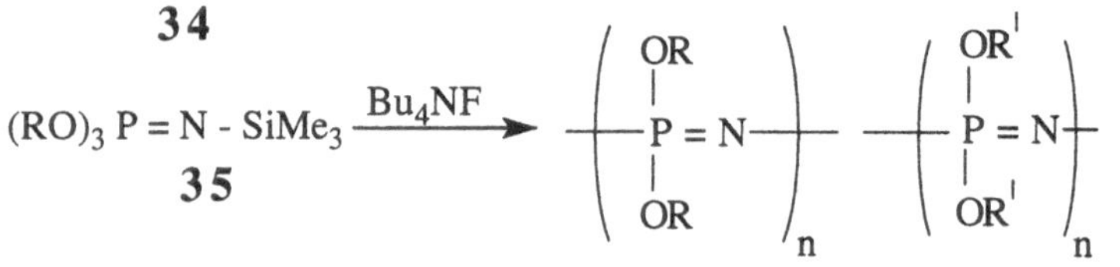

35

36

Branched chain phosphazenes can be formed by polymerization of **37** (32).[84] The activation energy for this polymerization is 46 kcal/mole[-1], which is significantly lower than that for $P_3N_3Cl_6$. The PCl_3 branches on **38** undergo substitution more readily than the main chain P-Cl bonds, creating the possibility for a wide range of functional derivatives. It may also be possible to use a condensation reaction to grow phosphazene chains from the branches.

$$\qquad (32)$$

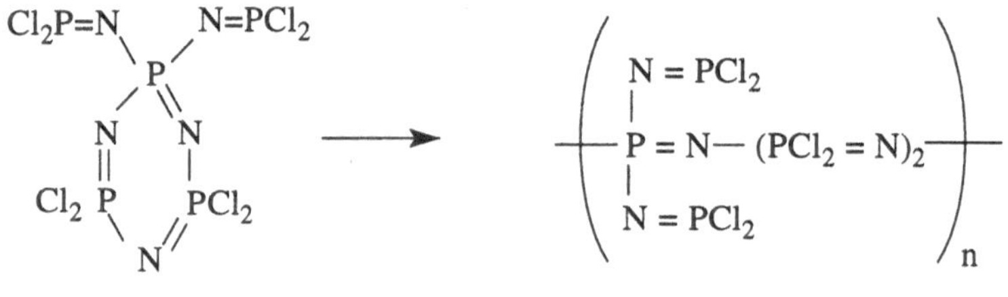

$$\left(\begin{array}{c} \text{N = PCl}_2 \\ | \\ \text{—P = N— (PCl}_2 \text{ = N)}_2\text{—} \\ | \\ \text{N = PCl}_2 \end{array} \right)_n$$

37

$$\longrightarrow \left(\begin{array}{c} \text{N = P(OR)}_2 \\ | \\ \text{—P = N — (PCl}_2 \text{ = N)}_2\text{—} \\ | \\ \text{N = P(OR)}_2 \end{array} \right)_n$$

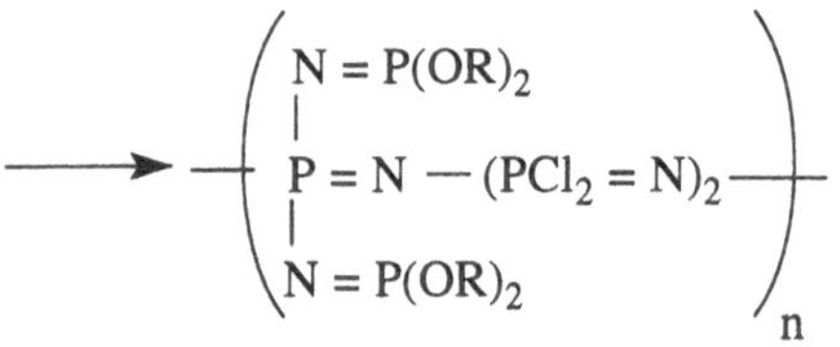

38

Certain properties of the polyphosphazenes, high skeletal flexibility, thermo-oxidative and photo-oxidative stability[76] and the broad range of electromagnetic transmission (near

IR to far UV) are characteristic of the phosphazene backbone. The diversity of properties obtainable is dependent on the wide variety of side groups which can be appended to the backbone. Indeed, substitution on the phosphazene backbone appears to be easier to accomplish than for either siloxane, silane, or hydrocarbon polymers. We shall summarize a few of the interesting recent developments, but cannot hope to be complete because of the rapid development in the field.

The transparency and photostability of the polyphosphazene chain makes it an ideal backbone to anchor and stabilize photochemically active species.[85] Dyes such as Rose Bengal, which can be used as a photosensitizer for production of singlet oxygen, are significantly more stable and have longer lifetimes when anchored to polyphosphazene than to polystyrene.[86] Spiropyrans and other photochromic side chains[87-88] and side chains possessing nonlinear optical properties[89-91] or liquid crystalline properties[88] can also be anchored to the phosphazene backbone.

Organometallic species such as ferrocene,[92] $C_6H_6Cr(CO)_3$ [93] and $C_5H_5Fe(CO)_2$ [94] have been attached to phosphazene backbones.

(33)

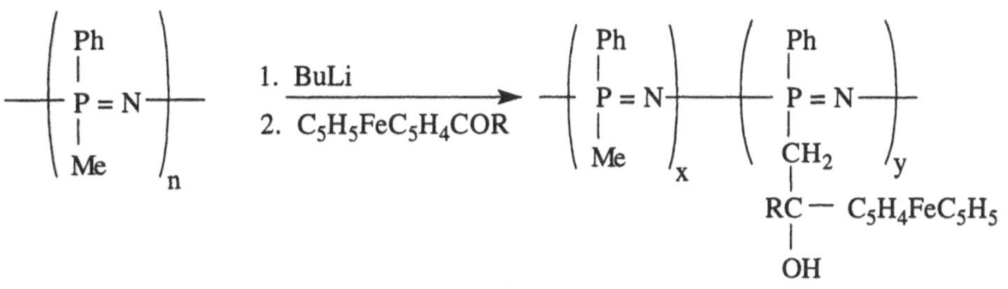

39

The ferrocene polymers, **39**, formed flexible films suitable for coating electrodes. The ferrocenyl groups could be oxidized reversibly and electron-hopping between ferrocenyl centers was observed. In general, electrical conductivity improved as the ferrocenyl content increased. With further development, these materials may serve as an alternative to the vinylferrocene polymers widely employed in biosensors.

Electrical conductivity in the solid state can be achieved by use of polyphosphazenes with ether side chains, **40**, impregnated with salts such as lithium triflate.

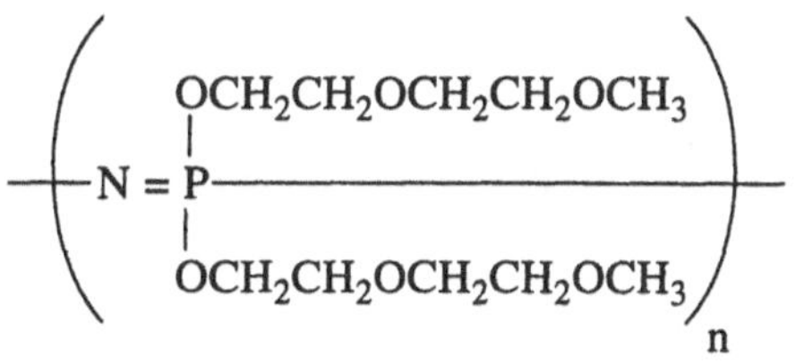

40

These materials can be used as the solid electrolyte in lightweight rechargable lithium batteries.[76,94] The materials, however, have poor mechanical stability and are difficult to cast as thin films. Chen-Yang et al.[95] have prepared a series of polyaminophosphazes, **41**, with varying content of $AgNO_3$. The Ag^+ ions appear to complex to both the side chain and the backbone N atoms and to move down the chain when an electric field is applied. Clear films with mechanical properties better than **40** were obtained, but stability decreased as the AgNO$_3$ content increased.

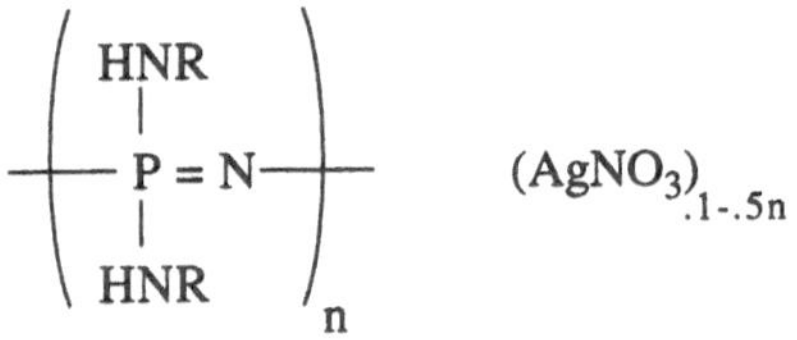

$$\left(\begin{array}{c} \text{HNR} \\ | \\ -\text{P} = \text{N}- \\ | \\ \text{HNR} \end{array}\right)_n \qquad (\text{AgNO}_3)_{.1-.5n}$$

$$R = \text{n-C}_3\text{H}_7, \text{ n-C}_4\text{H}_9$$

41

Polyphosphazenes containing C[96,98], S(IV)[97-99] and S(VI)[100-102] in the polymer backbone have been prepared by the reaction sequences shown below.

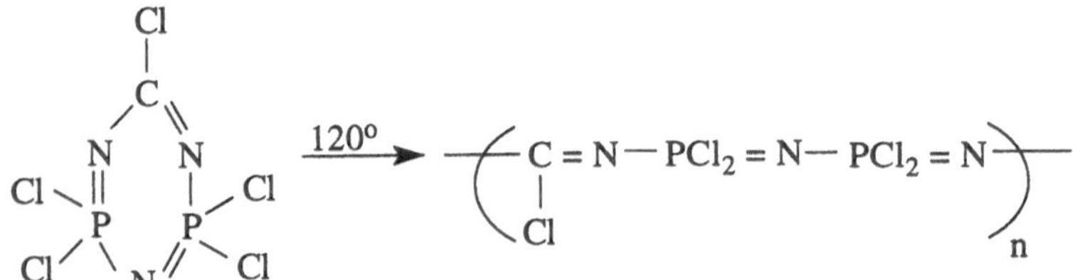

(34)

42

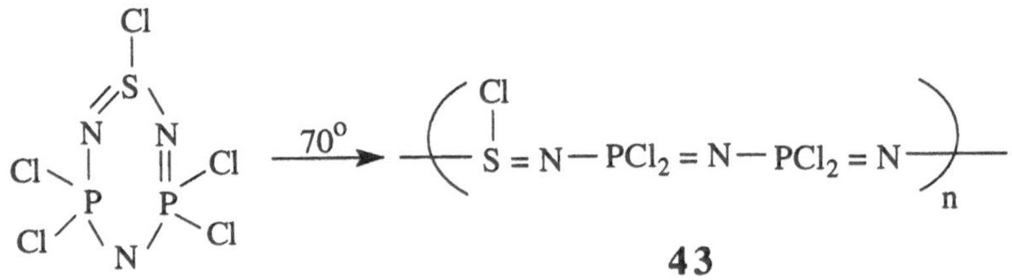

(35)

43

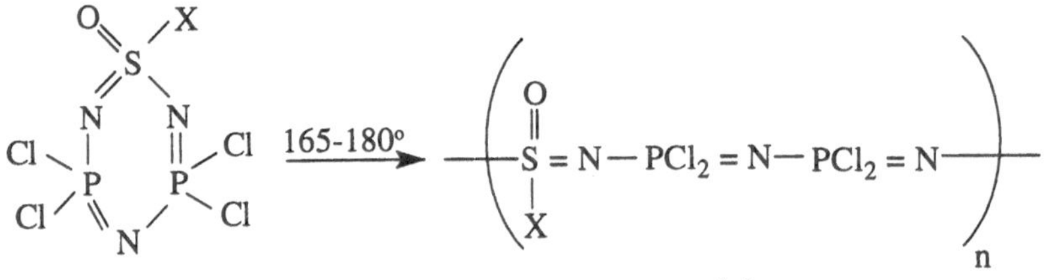

(36)

44

X = Cl, F

In general, incorporation of either C or S into the backbone increases the rigidity and raises T$_g$ substantially.[104,105] In the case of carbon, the C=N bond is pπ-pπ rather than dπ-p-π and does not undergo rotation readily. In **42**, both C-Cl and P-Cl undergo substitution readily. In **43**, the S-Cl bond undergoes substitution more readily than the P-Cl bond. The polymers hydrolyze readily and can be stabilized only with bulky substituents.[97] The reverse order of reactivity was observed with the S(VI) species, **44**. P-Cl bonds react much more readily than the S-Cl or S-F bonds. Ring opening and chain propagation occur preferentially at the sulfonyl–N bond.[103] Therefore, the possibility exists for creation of well defined structures with different substituents on the S and P.

By analogy with the condensation route to polyphosphazenes, **32**, Roy et al.[106-107] have prepared poly(oxothiazenes) **46** (37). Polymerization is catalyzed by both Lewis acids and bases.

(37)

$$
\begin{array}{c}
\quad\quad O \\
\quad\quad \| \\
R - S = N - SiMe_3 \\
\quad\quad | \\
\quad\quad OR \\
\\
\quad\quad 45
\end{array}
\quad\longrightarrow\quad
\left(
\begin{array}{c}
O \\
\| \\
S = N \\
| \\
R
\end{array}
\right)_n
\quad\quad 46
$$

R = Me or Ph

The polymers have a higher T_g than the corresponding polyphosphazenes. A helical conformation of the polymer chain was predicted by molecular orbital calculations.

In summary, the field of heteropolyphosphazenes has broken wide open in the last five years.[104] By the time this review appears, there will be at least ten to twenty new papers in print describing more diverse materials with unusual properties.

Polymers Containing Transition Metals in the Polymer Backbone

Stable, processable, high-molecular-weight polymers containing transition metals as a regular structural unit in the polymer backbone have long been sought. Transition metals contain unpaired electrons, in contrast to the neat pairing in compounds of the main group elements. They possess variable oxidation states which can be fine-tuned by changing the ligands on the metal. They are often highly colored, but absorption of visible or UV light can be fine-tuned by changing the ligands. Because electrons are unpaired, paramagnetism or ferromagnetism is often observed. The heavy nuclei can absorb neutrons, X-rays or other types of radiation, either to protect underlying material from radiation or to promote selective cleavage and degradation of the polymer. Thus, a wealth of potential applications exists for transition metal-containing polymers. Unfortunately, most attempts to prepare these materials have led to low molecular weight oligomers or to "brick dust" - insoluble intractible materials of poorly defined composition. Two techniques for preparing transition metal-containing polymers which have led to tractable materials will be discussed: metallocene polymers and coordination polymers.

Polymers Containing Metallocenes

Within a few years of the discovery of ferrocene in 1951,[108] attempts were made to prepare polymers containing ferrocene. The driving force for this research was the aerospace industry, which needed polymers with high thermal stability for lubricants and gaskets for jet engines, radiation shields, ablative materials, and combustion regulators for solid-state rocket fuel. Early research in this field was summarized by Neuse and Rosenberg in 1970.[109] Three major routes to metallocene polymers were developed: vinyl polymers, ferrocenyl methylene polymers, and poly-ferrocenylenes.

Vinylic organometallic monomers, **49**, have been prepared by a variety of techniques.[110] The most commonly used technique involves acetylation, reduction, and dehydration (38). This is successful for derivatives of ferrocene,[110] ruthenocene,[111] osmocene,[112] $C_5H_5Mn(CO)_3$[113] and $C_5H_5Re(CO)_3$[114] which can undergo electrophilic substitution. Other vinylic monomers may be prepared from vinylcyclopentadienyl lithium by procedures perfected by Rausch and his coworkers.[115]

(38)

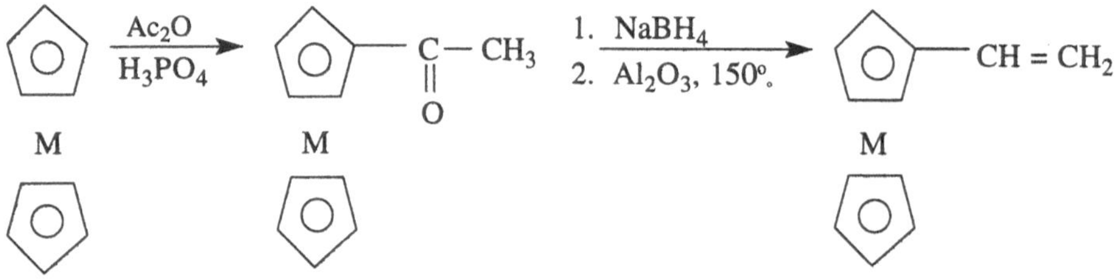

M

47   48   49

M = Fe, Ru, Os

Polymerization of vinylferrocene can be accomplished by either radical or cationic initiation.[109,116] The resulting polymers generally have molecular weight less than 10,000 (DP < 50) and often are highly branched. A unique chain transfer mechanism exists (39) for vinylferrocene in which an electron is transferred from Fe to the radical site on the α-carbon, effectively quenching polymerization.[117-118] This process occurs to a much lesser extent in ruthenocene and osmocene which are less readily oxidized.[111,112]

(39)

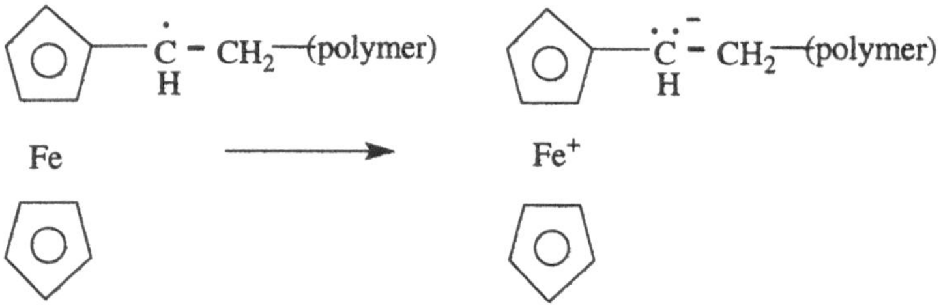

Fe $\longrightarrow$ Fe$^+$

Higher molecular weight polymers can be obtained by anionic polymerization of ferrocenylmethyl and ferrocenylethyl methacrylates, **50 a-b**.[116] Removal of the ferrocenyl group from the backbone effectively eliminates electron transfer.

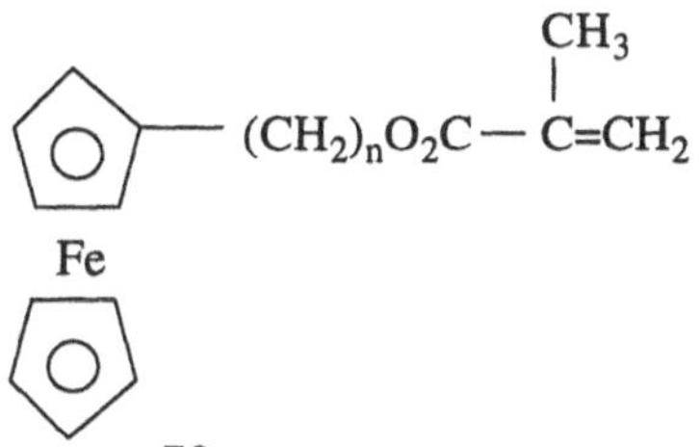

$$CH_3$$

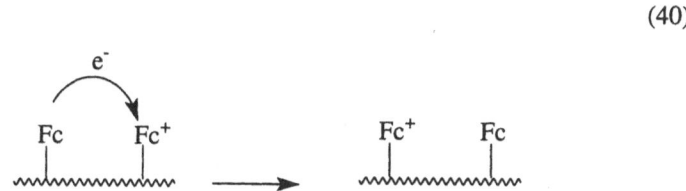

$$—(CH_2)_nO_2C—C{=}CH_2$$

Fe

50 a $n = 1$
b $n = 2$

Partial oxidation of vinylic ferrocene polymers produces semiconducting materials[113] in which current can be conducted by "electron - hopping" between ferrocene and ferrocenium centers.

(40)

In the 25 years since the pioneering studies were performed, a wide variety of conducting ferrocenyl polymer films have been prepared which are now widely employed as biosensors.[119-120] Several examples of silicon and phosphazene polymers with pendant ferrocenyl groups have already been mentioned in this review. A computer search of the chemical literature from 1990-95 produced 277 citations related to the electrochemistry of ferrocene derivatives. Further discussion of these materials is therefore beyond the scope of this review.

Ferrocenylmethanol, **51a**, ferrocenylethanol, **51b**, and other ferrocenyl carbinols readily undergo cationic polymerization in the presence of Lewis acid or protic acid catalysts.[121] Polymers **52-54** can also be prepared by acid-catalyzed condensation of aldehydes with ferrocene. The resulting materials usually have molecular weights below 10,000 and possess a mixture of 1, 1'; **52 a-b**; 1, 2; **53 a-b**, and 1, 3 linkages, **54 a-b**. Crosslinking to form insoluble material and partial oxidation occur frequently. The polymer could be further crosslinked to form thermoset resins which charred and ablated when exposed to high temperatures, but showed limited practical applications otherwise. As is often the case with cationic polymerizations in the presence of aromatic rings which can undergo electrophilic substitution, pure materials with high molecular weight and definite structure cannot be obtained.

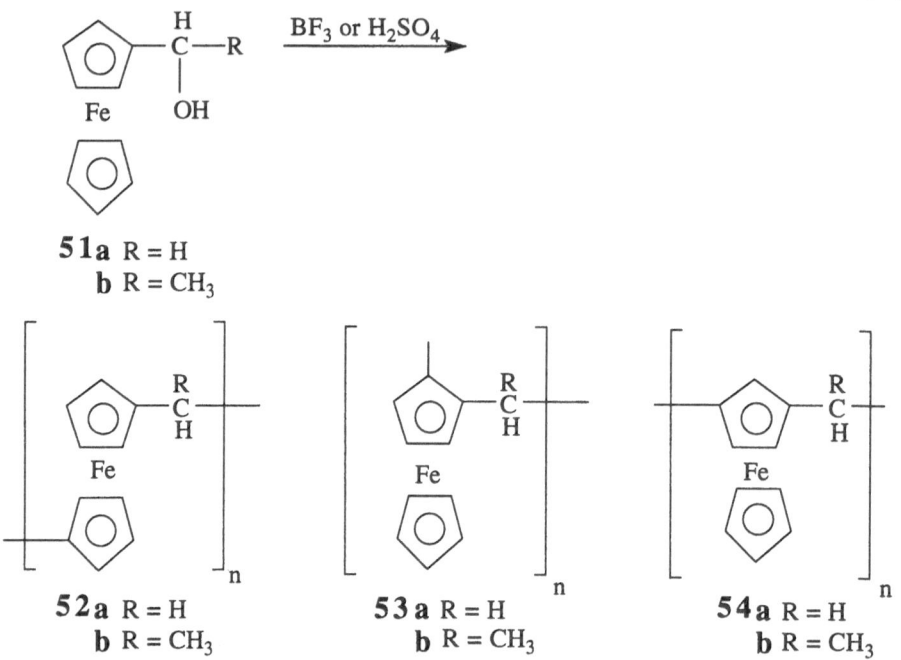

51a R = H
b R = CH₃

52a R = H
b R = CH₃

53a R = H
b R = CH₃

54a R = H
b R = CH₃

Poly 1,1' - ferrocenylenes, **55**, have been prepared directly from ferrocene or by a variety of coupling reactions (42).

(42)

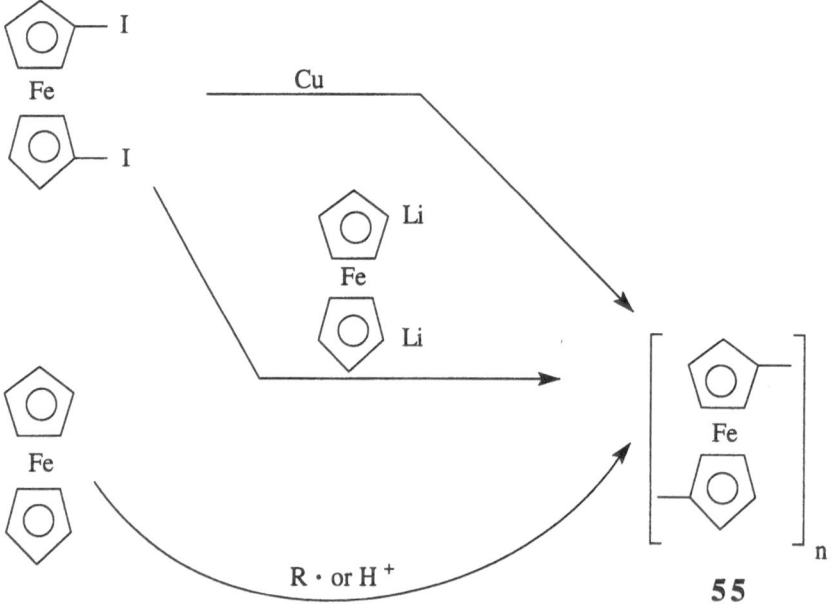

55

Generally, **55** prepared by any of these routes has molecular weight below 5,000. Solubility decreases as the number of ferrocenyl units increases. Branching or incorporation of solvent or other organic species such as dislodged cyclopentadienyl rings into the polymer chain often occurs.[121] Neuse[122] has more recently characterized a series of linear highly purified polyferrocenylenes containing up to 20 ferrocenyl units, but well defined high polymers have not been obtained. The polymer **55** can be partially oxidized to form a semiconductor, but conductivity is not much better than for polyvinyl ferrocene. The polymer adopts a helical conformation so that the ferrocenyl units do not lie coplanar. The charge is therefore confined on the individual ferrocenyl units and conductivity must take place largely by "electron hopping" as described in (40).

Thus, as of 1992, no tractable high molecular weight polymers with ferrocene in their backbone were available. This changed rapidly when Manners et al. discovered that silicon-bridged 1-ferrocenophanes previously prepared by a number of investigators[124-127] underwent ring-opening polymerization at temperatures of 150-180°.[128] Processable polymers with molecular weights in the range 1×10^5 to 1×10^6 were obtained. Extensive research on ring opening polymerization of metallocenophanes has followed, which has been summarized in an extensive review.[129]

(43)

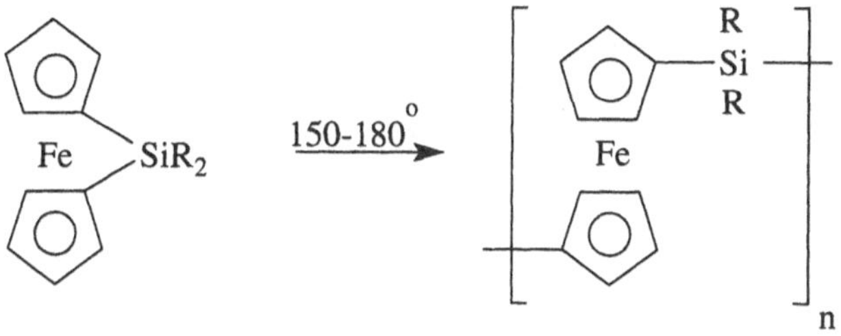

56 a R = H
   b R = CH$_3$
   c R = C$_4$H$_9$
   d R = C$_6$H$_{13}$
   e R = C$_{18}$H$_{37}$
   f R = C$_6$H$_5$
   g R = CH$_3$, Fc

57 a-g

The ring opening polymerization reaction has been explored for a variety of substituted derivatives of **56**, including both symmetrical and unsymmetrically substituted compounds. In all cases, high polymers, **57**, are obtained. Copolymers can be prepared from mixtures of **56 a - f** and block copolymers **59** can be prepared from **58** and **56b**.[130]

56 b   **58**

59

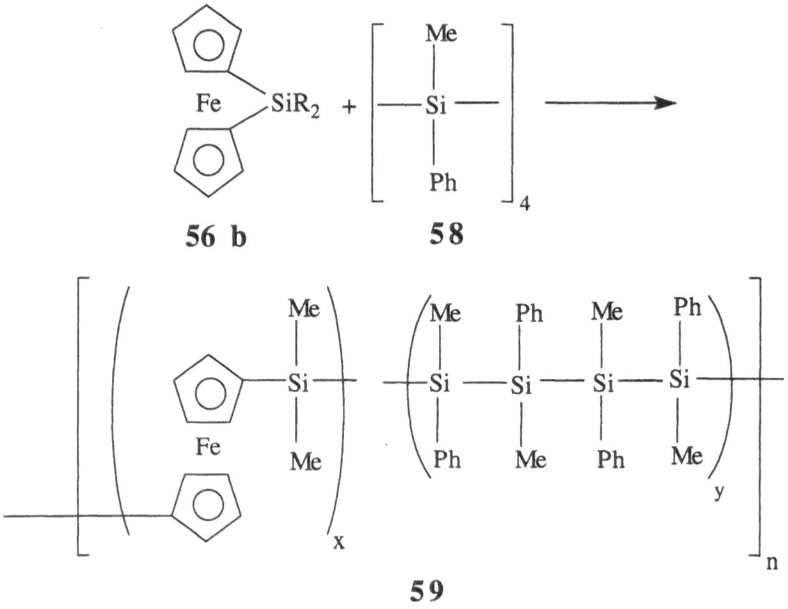

Compound **56b** also undergoes anionic-initiated polymerization at room temperature[131] in the presence of FcLi or other anionic initiators (45). This technique has been used primarily to prepare oligomers, **60**, (n = 3-30) for structural studies, but polymers with higher molecular weights have also been prepared.

(45)

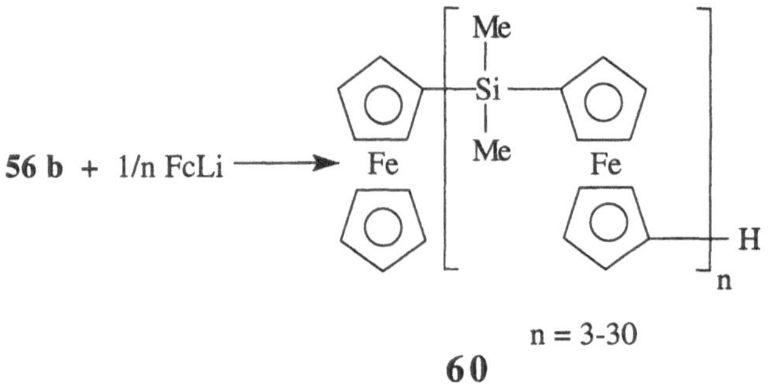

56 b + 1/n FcLi

n = 3-30

60

The driving force for polymerization in all cases appears to be relief of steric strain of 20-25 kcal mole^{-1}.[128] Oligosilane-bridged ferrocenes which do not possess steric strain do not undergo ring-opening polymerization. The ease with which anionic polymerization occurs provides evidence for an ionic mechanism for the thermal polymerization also, possibly catalyzed by traces of nucleophiles. Indeed, a polymer film can be coated on a glass surface which contains traces of basic species. Further mechanistic studies of this reaction are currently under way.

Poly(ferrocenylsilanes) prepared as described above are readily soluble in organic solvents. The polymers with short side chains, **57 a-b**, are more soluble in aromatic or chlorinated solvents, but polymers with longer side chains, **57 c-e**, are soluble in hexane. The phenylated polymer **57f** has greatly reduced solubility. All of the polymers can form films, but those with longer side chains, **57 c-e**, tend to be gummy materials. The T_g values tend to be 20-30° higher than the corresponding silane polymers. This indicates that incorporation of the ferrocenyl moiety into the backbone[133-134] decreases the conformational flexibility. Unsymmetrically substituted polymers tended to be amorphous[135], but the C_{18} side chains of **57e** tend to crystallize.

Solutions of poly(ferrocenylsilanes) exhibit two reversible oxidation waves of roughly equal intensity when studied by cyclic voltammetry. The separation of the two peaks varies from 0.21V for **57b** to 0.29V for **57 c-d**.[136] These results are consistent with initial oxidation of alternating ferrocenyl sites, followed by a more difficult oxidation of a ferrocenyl group which lies between two charged sites. Polymer **57g**, which possesses both backbone and pendent ferrocenyl groups, displays a complex redox behavior in which both backbone and pendant ferrocenyl groups can be oxidized.[134] The polymer films are electrochromic, being amber in the reduced form and blue when oxidized.[129] The partially oxidized polymers are weak semiconductors (10^{-7} -10^{-8} S/cm). The charges are therefore essentially localized on the individual ferrocenyl units, in contrast to polysilanes which can exhibit σ-delocalization.[129]

1-Germaferrocenophanes, **61 a-d**, have been synthesized[138] and undergo thermal polymerization at 90°, about 30° lower than **56 a-f**.

(46)

61 **a** R = Me
    b R = Et
    c R = n-Bu
    d R = Ph

62 **a-d**

The resulting polymers, **62 a-d**, have molecular weights above 1×10^5, are soluble in organic solvents, and exhibit two reversible oxidation waves in CH_2Cl_2 similar to **57 a-f**. Attempts to prepare 1-stanna- or 1-plumbaferrocenophanes have failed, probably because the strain energy is comparable to the bond energy of the C-Sn or C-Pb bonds.[129]

1-Phosphaferrocenophanes also undergo both thermal[137-139] and anionic-initiated[134] polymerization. Preliminary studies indicate that the polymers possess similar redox properties to the Si and Ge analogs. The presence of phosphine groups in the polymer backbone offers the possibility of binding other transition metal species such as metal carbonyls to the polymer backbone.[129]

2-Ferrocenophane **63 a-b**, which is also highly strained, undergoes thermal ring-opening polymerization at 300°.[140] The resulting polymer, **64a**, is insoluble, but **64b** is readily soluble in organic solvents.

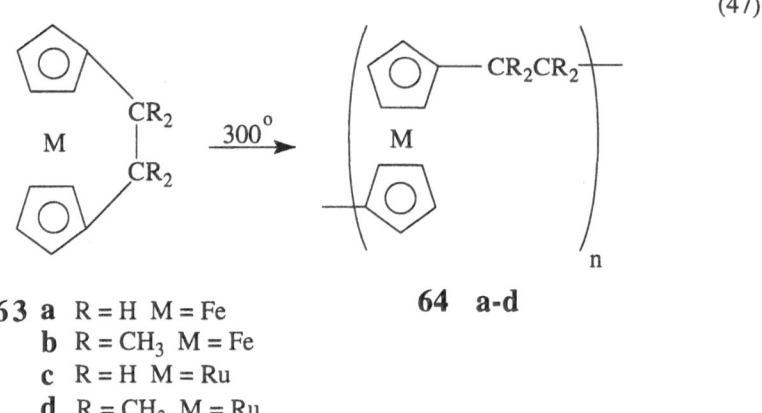

63 a R = H M = Fe   64 a-d
   b R = CH₃ M = Fe
   c R = H M = Ru
   d R = CH₃ M = Ru

The ferrocenyl groups are sufficiently well separated that cyclic voltammetry shows only one reversible redox wave. The corresponding Ru derivatives, **63 c-d**, are more highly strained and undergo polymerization at 220°.[141] Cyclic voltammetry of **64 c-d** shows only irreversible oxidation, which is typically observed for ruthenocene derivatives.[111]

Ferrocene derivatives with three membered bridges normally show little steric strain, so ring-opening polymerization should not take place. Rauchenfuss et al.[142] in 1992, however, prepared polyferrocenyl persulfides, **66 a-b**, by a novel S abstraction reaction (48).[142-144]

(48)

65 a R = H   66 a-c
   b R = n-Bu
   c R = t-Bu

Polymers **66 a-c** can be cleaved by Li[BEt₃H] to form the dimercaptides **67 a-c** and regenerated by oxidation with I₂.

$$\text{66 a-c} \quad \xrightleftharpoons[I_2]{Li[BEt_3H]} \quad \text{67 a-c}$$

Polymers **66 a-c** can be prepared with molecular weights in excess of 25,000. n-Butyl or t-butyl groups are necessary to maintain solubility in organic solvents. If ferrocenes with two or more trisulfide bridges are employed, network polymers with molecular weights up to 1×10^6 can be obtained.[144] The polymers are air stable, but photosensitive, undergoing cleavage at the S-S bonds. Cyclic voltammetry shows two reversible waves, indicating oxidation at alternate sites as is observed in other ferrocenyl polymers. Extensive research on these materials is currently under way.

In conclusion, since 1992 there has been an explosion of research on soluble, tractable, photosensitive, and electrochemically active ferrocene polymers. After years of goos and brick dust, we should see some commercially useful materials shortly.

Coordination Polymers

Coordination polymers may be defined as macromolecules composed of extended systems of repeating units of metal centers bridged by σ-bonded organic ligands. The metal ion is an essential part of the polymer backbone rather than being appended to it. The first attempts to prepare such polymers were summarized in a seminal review by Bailar in 1964[145] and have been reviewed more recently by Archer [146] and Cronin.[147]

The simplest way to produce such polymers would be to combine a soluble transition metal salt such as a chloride or sulfate with a bis-monodentate, bidentate, tridentate, or tetradentate ligand which cannot chelate. The resulting polymer should be less soluble in polar solvents and should precipitate from solution. It does, usually before more than a few units have been coupled. A strong tendency toward crosslinking between metal sites which are not coordinatively saturated or strong Van der Waals interactions between planar aromatic bridging ligands leads to intermolecular forces which cause precipitation. The resulting "brick dust" has limited usefulness.

A further complication is that metals which form labile complexes couple rapidly, but the chains can be cleaved readily. Metals which form inert complexes often react with the ligand too slowly to form long chains.

The goal of research in this field is to form high polymers that combine both the flexibility of an organic species and the thermal stability of inert coordination complexes. These materials must be either soluble or fusible so they can be cast as films, spun as fibers, or blended with other organic polymers. So far, only a few polymers of this type have been prepared. We will therefore summarize the most successful examples of these materials.

Numerous stacked or "shish kabob" polymers have been prepared, which are one-dimensional rigid-rod polymers formed from metalloporphyrins, metallophthalocyanines[148] or other metallomacrocycles bridged by single atoms such as O, S, Se, or Te; small ions such as CN^- or heterocycles such as pyrazine, pyrimidine, or 4,4'- bipyridine. These materials often function as conductors or semiconductors if the distance between the macrocyclic ligands is not too great. Extensive research has been performed on these compounds in recent years. Most of these materials are not soluble in organic solvents so they fall outside the domain of this review. The reader is referred to a recent review by Chen and Sustich, [149] and to recent papers by Thompson et al.[150-152]

Soluble rigid rod polyyne polymers, **68**, were first prepared by Hagihara et al. in 1980-82.[153-154] The bulky Bu groups promote solubility by hindering close approach of the individual chains. Materials of this type, with Pt as the central metal, possess a high

$$
\left[\begin{array}{c} PBu_3 \\ | \\ -Pt - C\equiv C - X - C\equiv C \\ | \\ PBu_3 \end{array} \right]_n
$$

68   X = p-C$_6$H$_4$, p-C$_6$H$_4$-C$_6$H$_4$

degree of non-linear optical activity.[146] As such, they attracted the attention of Lewis, Marder et al.[155-156] who have studied similar materials containing conjugated acetylenic groups. Extended Hückel molecular orbital calculations indicated that the degree of optical nonlinearity and also the electrical conductivity was dependent on the length of the conjugated chain between the metal centers, the coordination number of the metal, and also the nature of the metal itself.[157] The best electrical conductivity appears to arise from a d^8 square planar configuration (Pt or Pd) which has a larger band width for the conduction band than a d^6 octahedral transition metal (Fe or Ru).

In order to test this hypothesis, Ru polymers, **69**, were prepared.[158] Soluble,

$$
\left[\begin{array}{c} CH_2 \\ Ph_2P \quad PPh_2 \\ -Ru - C\equiv C - R - C\equiv C \\ Ph_2P \quad PPh_2 \\ CH_2 \end{array} \right]
$$

69

processable polymers with molecular weights of 5-6 x 10^4 were obtained. The studies have been extended to include monomeric Os acetylides,[159] but no Os polymers have been reported yet. Investigations of the nonlinear optical properties of these materials are currently under way.

Carraher et al.[160-161] have investigated coordination polymers, **70**, formed by condensation mechanisms for the past 30 years (49), employing a vast array of metals which can act as Lewis acids and Lewis bases. These have been prepared by interfacial polymerization with the Lewis acid in the organic phase and the Lewis base in the aqueous phase.

$$ML_nX_2 \quad + \quad BRB \longrightarrow \left[ML_n-B-R-B \right]_m$$

70

n = 2, 3, 4

M = Ge, Sn, Pb, As, Sb, Bi, UO_2^{2+}

B = NH_2, NR_2, O^-, S^-

Most of these materials are insoluble, but their lack of solubility combined with ease of degradation can sometimes be an advantage, i.e. controlled release of drugs[162] or growth hormones.[163]

Two of the more soluble condensation polymers, **71** and **72**, are shown below.[164-165]

(50)

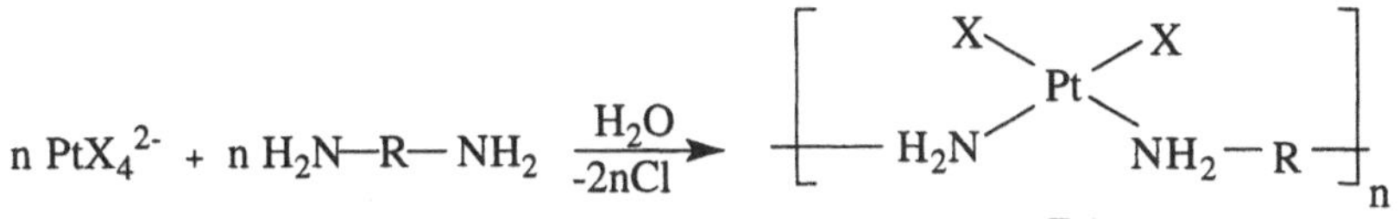

$$n\ PtX_4^{2-} + n\ H_2N-R-NH_2 \xrightarrow[\text{-2nCl}]{H_2O} \left[H_2N \overset{\displaystyle X}{\underset{\displaystyle NH_2}{\diagdown Pt \diagup}} \overset{X}{\diagup} NH_2-R \right]_n$$

71

(51)

$$n\ HS-R-SH + n\ \underline{cis}\text{-Ru(bpy)}_2\ Cl_2 \longrightarrow$$

$$\left[S-R-S-\underline{cis}\text{-Ru(bpy)}_2 \right]_n n\ HCl$$

72 \quad R = n-C_8H_{16}, $(CH_2)_2\ O(CH_2)_2$

In both cases, solubility is enhanced by the <u>cis</u>-coordination at the metal sites, which places kinks in the polymer chain, and by the flexibility of the aliphatic linkages, R.

Archer, et al.[166-171] have investigated polymers containing 8-coordinate Zr [166,168,170] for the past 16 years. 8-Coordination, rather than 6-coordination, is commonly observed for the lanthanides and actinides as well as second and third row transition metals such as Zr and W. The 8-coordinate state is stereochemically non-rigid, with low energy barriers between the trigonal-faced dodecahedron (D_2d symmetry), the square antiprism (D_4d symmetry), and the bicapped trigonal prism (C_2v symmetry). The Zr and Ce polymers can be prepared as shown below.

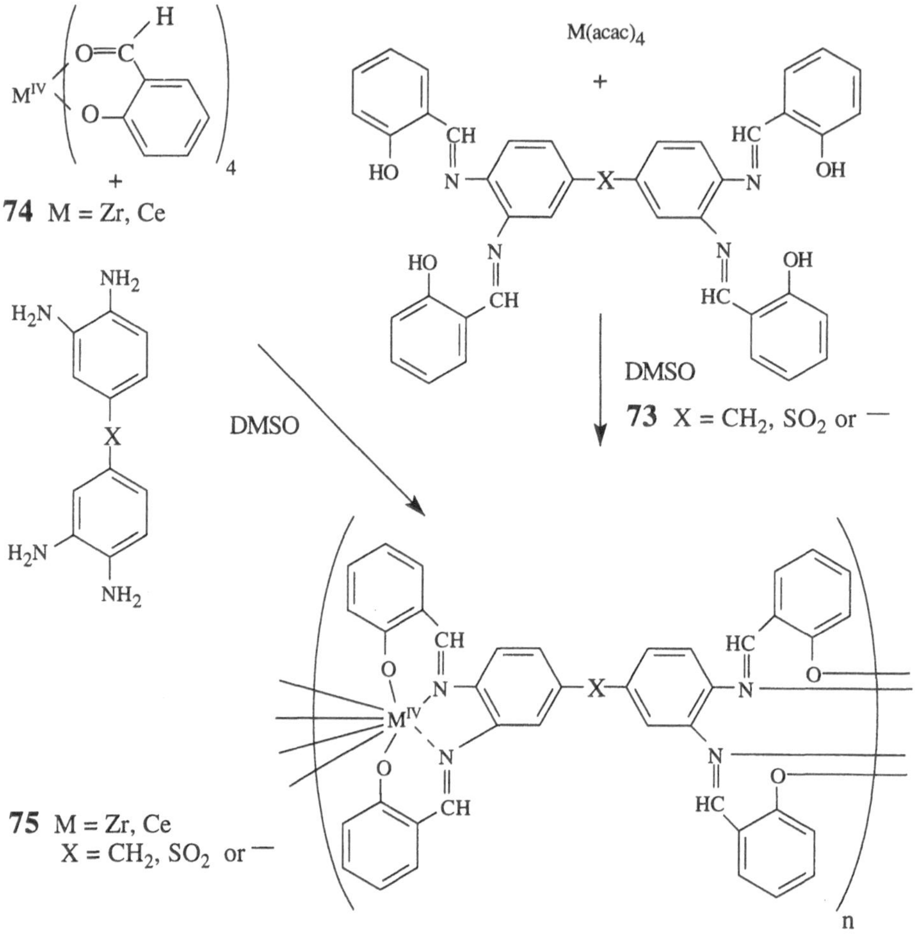

M(acac)$_4$

+

74 M = Zr, Ce

NH$_2$

H$_2$N

X

H$_2$N

NH$_2$

DMSO

DMSO

HO

CH

N

HO

N

CH

HC

N

OH

N

HC

OH

73 X = CH$_2$, SO$_2$ or —

75 M = Zr, Ce
    X = CH$_2$, SO$_2$ or —

CH

N

M^{IV}

N

CH

X

HC

N

N

HC

n

Because of the relatively low solubility of the tetradentate ligands,73, they were initially formed in situ from tetrasalicaldehydato Zr or Ce, 74, and the appropriate aromatic tetramine. Subsequent studies showed that the preformed ligand could be used if the reaction temperature were raised to 100°.[171] The polymers, 75, formed have molecular weight of 2-5 x10^4 and are soluble in organic solvents. The T$_g$ for Zr polymers is in the range 80-100°C; whereas the Ce polymers usually lie 80-100° higher.[170] The T$_g$ can be lowered by inserting spacers such as SO$_2$ or CH$_2$ between the phenyl rings to increase flexibility. The difference in the Zr and Ce materials appears to be that the Zr polymers adopt a flexible trigonal-faced dodecahedral configuration while the Ce polymers adopt a less flexible square antiprism configuration.

The polymers may be terminated either by coordinatively unsaturated metal centers or by excess amine groups. Polymers terminated by coordinatively-unsaturated Zr bind readily to silica or to the thin film of Al$_2$O$_3$ covering Al which has been exposed to air.[172] The resulting polymer films can be coated readily with polyethylene or other polymeric materials. Thus, small amounts of Zr polymer greatly improve the adhesion of polymers to glass. This may lead to important commercial applications.

Amine-terminated Zr or Ce polymers may be joined with carboxy- or epoxy- terminated organic species to form block copolymers[146] or other modified materials. Most of the Zr and Ce polymers used for molecular weight determinations have been end-capped to prevent adhesion to glass.

Trivalent lanthanide ions (La^{+3}, Gd^{+3}, and Yb^{+3}) and Y^{+3} can be incorporated into polymers by a procedure analogous to the synthesis of the Zr^{+4} and Ce^{+4} polymers.[173] Because the tetradentate ligands carry a -4 charge, a polyelectrolyte is formed with -1 net charge per metal ion. When Na^+ is employed as the counterion, the polymers are soluble in DMSO. Molecular weights of 2×10^4 have been obtained. The polymers are thermally stable up to 500°C. Use of other alkali metals or tetraalkylammonium ions as counterions produces insoluble materials. Further studies of these materials are in progress.

In conclusion, if conformational flexibility at the metal site and flexibility in the polymer chain can be varied, it is possible to prepare a variety of stable, soluble, processable coordination polymers. Many more materials of this type will be prepared in the next few years and extensive investigation of their properties will take place.

Conclusion

Exciting things are happening in the field of organometallic polymers. It is impossible to discuss all of the developments. We have not discussed polymers which contain organometallic groups pendant to the chain, polymers which are doped with metal ions, polymer-bound catalysts or biopolymers, all of which are discussed in other chapters of this book. We wait with excitement to see what the future will bring in this field.

Acknowledgements

The assistance of Ian Manners, who provided copies of several of his most recent publications and whose four review articles formed the framework for this chapter and the assistance of Jon Cronin who provided a rough draft of his thesis including a thorough discussion of coordination polymers are gratefully acknowledged. The assistance of Mary Kildea of Rider University who typed this manuscript and drew all of the chemical structures and Janet Weiss of the Rider University Library who conducted many of the literature searches is also gratefully acknowledged.

References

1. The proceedings of the First Symposium on Inorganic Polymers are included in A.C.S. *Org. Coatings & Plastics Chem.* **1971**, *31* (2).
2. Carraher, C. E., Jr.; Sheats, J. E.; Pittman, C. U., Jr. *Organometallic Polymers* **1978**, Academic Press, New York.
3. Carraher, C. E., Jr.; Sheats, J. E.; Pittman, C. U., Jr. *Advances in Organometallic and Inorganic Polymer Science* **1982**, Marcel Dekker, New York.
4. Sheats, J. E.; Carraher, C. E., Jr.; Pittman, C. U., Jr. *Metal Containing Polymeric Species* **1985**, Plenum Press, New York.
5. Sheats, J. E.; Carraher, C. E., Jr.; Pittman, C. U., Jr.; Zeldin, M.; Currell, B. *Inorganic and Metal-Containing Polymeric Materials* **1990**, Plenum Press, New York.
6. Pittman, C. U., Jr. et al. *Metal-Containing Polymeric Materials* **1996**, Plenum Press, New York.
7. Zeldin, M.; Wynne, K. J.; Allcock, H. R. *Inorganic & Organometallic Polymers* **1988**, A.C.S. Symposium Series #360, Washington, D.C.
8. Manners, I. *Ann. Rep. Prog. Chem., Sect. A. Inorg. Chem.* **1991**, *88*, 77-92; **1992**, *89*, 93-105; **1993**, *90*, 103-118.
9. "Macromolecule-Metal Complexes III," Selected Papers from Seminar of July 23-28, 1989, *J. Macromol. Sci. Chem.* **1990**, *A27 (9-11)*, 1109-1446.
10. Tsuchida, E. *Macromolecular Complexes - Dynamic Interactions and Electronic Processes* **1991**, VCH Publishers, New York.
11. Dems, A.; Strobin, G. *Makromol. Chem.* **1991**, *192*, 2521.
12. Mark, J. E. *J. Inorg. Organometal. Polym.* **1991**, *1*, 431-48.
13. Mark, J. E.; Erman, B. *Rubberlike Elasticity. A Molecular Primer* **1988**, Wiley Interscience, New York.
14. Mark, J. E. *Chemtech* **1990**, *19*, 230.
15. Lichtenhan, J. D.; Vu, N.Q.; Carter, J. A.; Gilman, J. W.; Feher, F. J. *Macromolecules* **1993**, *26*, 2141.
16. Cerveau, G.; Corriu, R. J. P.; Costa, N. *J. Non-Cryst. Sol.* **1993**, *163,* 226.
17. Zhdanov, A. A.; Shchegolikhina, O. I.; Blagodatskikh, I. V. *PMSE* **1994**, *71*, 321-2.
18. Gunji, T.; Abe, Y. *PMSE* **1994**, *71*, 314.
19. Gunji, T.; Abe, Y. *PMSE* **1994**, *71*, 393.
20. DeSimone, J. M.; York, G. A. ; McGrath, J. E.; Gozdz, A. S.; Bowden, M. J. *Macromolecules* **1991**, *24*, 5330.
21. Puyenbroek, R.; Werkman, P.; Jansema, J. H.; van de Grampel, J. C.; Rousseuw, B. A. C.; van der Drift, W. E. J. M. *Polym. Prepr.* **1993**, *34 (1)*, 238.
22. Rubinsztajn, S.; Zeldin, M.; Fife, W. K. *Macromolecules* **1991**, *24*, 2682.
23. Zeldin, M.; Rubinsztajn, S.; Fife, W. K. *J. Inorg. Organomet. Polym.* **1992**, *2*, 319.
24. Fife, W. K.; Rubinsztajn, S.; Zeldin, M. *J. Am. Chem. Soc.* **1991** *113*, 8535.
25. Stein, J.; Lewis, L. N.; Smith, K. A.; Lettko, K. X. *J. Inorg. Organomet. Polym.* **1991**, *1*, 325.
25a. Jonas, G.; Stadtler, R. *Makromol. Chem. Rapid Commun.* **1991**, *12*, 625.
26. Barley, S. H.; Gilbert, A.; Mitchell, G. R. *Makromol. Chem.* **1991**, *192*, 2801.
27. Duplock, S. K.; Matisons, J. G.; Swincer, A. G.; Warren, R. F. O. *J. Inorg. Organometal. Polym.* **1991**, *1*, 361.
28. Crivello J. V.; Fan, M. *Polymer Prep.* (Am. Chem. Soc., Div. Polym. Chem.) **1991**, *32(3)*, 171.

29. Babu, J. R.; Sinai-Zingde, G.; Rifle, J. S. *Ibid.* **1991**, *32(1)*, 152.
30. McCormick, F. B.; Wright, B. B.; Williams, J. W. *PMSE* **1994**, *71*, 729.
31. Stauffer, G.; Latterman, G. *Makromol. Chem.* **1991**, *192*, 2421.
32. Boileau, D; Teyssie, D. *J. Inorg. Organomet. Polym.* **1991**, *1*, 247.
33. Sauer, T. *Macromolecules* **1993**, *26*, 2057.
34. Chiellini, E.; Galli, G.; Dossi, E.; Cioni, F.; Gallot, B. *Macromolecules* **1993**, *26*, 849.
35. Shenouda, I. G.; Chien, L. C. *Macromolecules* **1993**, *26*, 5020.
36. Hsu, C. S.; Shih, L. J.; Hsiue, G. H. *Macromolecules* **1993**, *26*, 3161.
37. Fu, R.; Jing, P.; Gu, J.; Huang, Z.; Chen, Y. *Anal. Chem.* **1993**, *65*, 2141.
38. Kumar, U.; Kato, T.; Frechet, J. M. J. *J. Am. Chem. Soc.* **1992**, *114*, 6630.
39. Kumar, U.; Frechet, J. M. J.; Kato, T.; Ujiie, S.; Timura, K. ; *Angew. Chem. Int. Ed. Engl.* **1992**, *31*, 1531.
40. Poths, H.; Andersson, G.; Sharp, K.; Zentel, R. *Adv. Mater.* **1992**, *4*, 792.
41. Kapitza, H.; Zentel, R.; Twieg, R. J.; Nguyen, C.; Vallerien, S. U.; Kremer, F.; Willson, C. G. *Adv. Mater.* **1990**, *2*, 539.
42. Sledzinska, I; Soltysiak, J.; Stanczyk, W. *J. Inorg. Organometal. Polym.* **1994**, *4*, 199.
43. West, R.; David, L. D.; Djurovich, P. I.; Stearley, K. S.; Srinivasan, K. S. V.; Yu, H. *J. Am. Chem. Soc.* **1981**, *103*, 7352.
44. Wesson, J. P.; Williams, T. C. *J. Polym. Sci., Polym. Chem. Ed.* **1979**, *17*, 2833.
45. Trujillo, R. E. *J. Organometal. Chem.* **1980**, *198*, C27.
46. Matyjaszewski, K. *J. Inorg. Organometal. Polym.* **1991**, *1*, 463-85.
47. West, R.; Menescal, R.; Asuke, T.; Eveland, J. *J. Inorg. Organometal. Polym.*, **1992**, *2*, 29-45.
48. Hengge, E. F. *J. Inorg. Organometal. Polym.* **1993**, *3*, 287-303.
49. Price, G. J. *J. Chem. Soc., Chem. Commun.* **1992**, 1209.
50. Ziegler, J. M.; McLaughlin, L. I.; Perry, R. J. *J. Inorg. Organometal. Polym.* **1991**, *1*, 531.
51. West, R.; Hayase, S.; Iwahara, T. *J. Inorg. Organometal. Polym.* **1991**, *1*, 545.
52. Wildeman, J.; Herrema, J. K.; Hadziioannou, G.; Schomaker, E. *J. Inorg. Organometal. Polym.* **1991**, *1*, 567.
53. Shizuka, H.; Sato, Y.; Ueki, Y.; Ishikawa, M.; Kumada, M. *J. Chem. Soc. Faraday Trans.*, 1, **1984**, *80*, 341.
54. Shizuka, H.; Obuchi, H.; Ishikawa, M.; Kumada, M. *Ibid.* **1984**, *80*, 383.
55. Aitken, C.; Harrod, J. F.; Samuel, E. *J. Organomet. Chem.* **1985**, *279*, C11.
56. Aitken, C.; Harrod, J. F.; Samuel, E. *J. Am. Chem. Soc.* **1986**, *108*, 4059.
57. Harrod, J. F.; *A.C.S. Symposium Ser.* **1988**, *360*, 89.
58. Woo, H. G.; Tilley, T. D. *J. Am. Chem. Soc.* **1989**, *111*, 3757.
59. Tilley, T. D. *Acc. Chem. Res.* **1993**, *26*, 22.
60. Banovetz, J. P.; Stein, K. M.; Waymouth, R. M. *Organometallics* **1991**, *10*, 3430.
61. Corey, J. Y.; Zhu, X. H.; Bedard, T. C.; Lange, L. D. *Organometallics* **1991**, *10*, 924.
62. Li, H.; Gauvin, F.; Harrod, J. F. *Organometallics* **1993**, *12*, 575.
63. Forsyth, C. M.; Nolan, S. P.; Marks, T. J. *Organometallics* **1991**, *10*, 2543.
64. Imori, T.; Woo, H. G.; Walzer, J. F.; Tilley, T. D. *Chem. Mater.* **1993**, *5*, 1487.
65. Sakamoto, K.; Obata, K.; Hirata, H.; Nakajima, M.; Sakurai, H. *J. Am. Chem. Soc.* **1989**, *111*, 7641.
66. Cypryk, M.; Gupta, Y.; Matyjaszewski, K. *J. Am. Chem. Soc.* **1991**, *113*, 1046.
67. Matyjaszewski, K. *J. Inorg. Organomet. Polym.* **1992**, *2*, 5-27.
68. Hengge, E. F. *J. Inorg. Organometal. Polym.* **1993**, *3*, 287-303.

69. Suzuki, H.; Meyer, H.; Simmerer, J.; Yang, J.; Haarer, D. *Adv. Mater.* **1993**, *5*, 743.
70. Wang, Y.; West, R.; Yuan, C. H. *J. Am. Chem. Soc.* **1993**, *115*, 3844.
71. Allcock, H. R.; Kugel, R. L. *J. Am. Chem. Soc.* **1965**, *87*, 4216.
72. Allcock, H. R.; Kugel, R. L.; Valan, K. J. *Inorg. Chem.* **1966**, *5*, 1709.
73. Allcock, H. R.; Kugel, R. L. *Inorg. Chem.* **1966**, *5*, 1716.
74. Allcock, H. R.; Cook, W. J.; Mack, D. P. *Inorg. Chem.* **1972**, *11*, 2584.
75. Allcock, H. R. *J. Inorg. Organometal. Polym.* **1992**, *2*, 197-211.
76. Bortolus, P.; Gleria, M. *J. Inorg. Organometal. Polym.* **1994**, *4*, 205-236.
77. Matyjaszewski, K. *J. Inorg. Organometal. Chem.* **1992**, *2*, 5-28.
78. Manners, I.; Riding, G. H.; Dodge, J. A.; Allcock, H. R. *J. Am. Chem. Soc.* **1989**, *111*, 3067 and **1991**, *113*, 9596.
79. Allcock, H. R.; Dodge, J. A.; Manners, I.; Parvez, M.; Riding, G. H.; Visscher, K. B. *Organometallics* **1991**, *10*, 3098.
80. Allcock, H. R.; Turner, M. L. *Macromolecules* **1993**, *26*, 3.
81. D'Halluin, G.; DeJaeger, R.; Chambrette, J. P.; Potin, P. *Macromolecules* **1992**, *25*, 1254.
82. Cypryk, M.; Matyjaszewski, K.; Kojima, M.; Magill, J. H. *Makromol. Chem. Rapid Commun.* **1992** *13*, 39.
83. Matyjaszewski, K.; Lindenberg, M. S.; Moore, M. K.; White, M. L.; Kojima, M. *J. Inorg. Organometal. Polym.* **1993**, *3*, 317.
84. Allcock, H. R.; Ngo, D. C. *Macromolecules* **1992**, *25*, 2802.
85. Bortolus, P.; Gleria, M. *J. Inorg. Organometal. Polym.* **1994**, *4*, 95-142.
86. Facchin, G.; Minto, F.; Gleria, M.; Bertani, R.; Bortolus, P. *Ibid.* **1991**, *1*, 389.
87. Allcock, H. R.; Kim, C. *Macromolecules* **1991**, *24*, 2846.
88. Allcock, H. R.; Kim, C. *Macromolecules* **1989**, *22*, 2596.
89. Snigler, R.; Willingham, R. A.; Noel, C.; Friedrich, C.; Bosio, L.; Atkins, A. D. T.; Lenz, R. W. in *Liquid Crystalline Polymers* **1990**, *435*, 185, Weiss, R. A.; Ober, C. K., Eds., A.C.S. Symposium Series, Washington, DC.
90. Dembek, A.A.; Kim, C.; Allcock, H. R.; Devine, R. L. S.; Steier, W. H.; Spangler, C. W. *Chem. Mater.* **1990**, *2*, 97.
91. Allcock, H. R.; Dembek, A. A.; Kim, C.; Devine, R. L. S.; Shi, Y.; Steier, W. H.; Spangler, C. W. *Macromolecules* **1991**, *24*, 1000 .
92. Crumbliss, A. L.; Cooke, D.; Castillo, J.; Wisian-Neilson, P. *Inorg. Chem.* **1993**, *32*, 6088.
93. Allcock, H. R.; Dembek, A. A.; Klingenberg, E. H. *Macromolecules* **1991**, *24*, 5208.
94. Bennett, J. L.; Dembek, A. A.; Allcock, H. R.; Heyen, B. J.; Shriver, D. F. *Chem. Mater.* **1989**, *1*, 14 and earlier papers cited therein.
95. Chen-Yang, Y. W.; Hwang, J. J.; Kau, J. Y. *PMSE* **1994**, *71*, 727.
96. Manners, I.; Allcock, H. R.; Renner, G.; Nuyken, D. *J. Am. Chem. Soc.* **1989**, *111*, 5478.
97. Dodge, J. A.; Manners, I.; Allcock, H. R.; Renner, D.; Nuyken, D. J. *J. Am. Chem. Soc.* **1990**, *112*, 1268.
98. Allcock, H. R.; Coley, S. M.; Manners, I.; Visscher, K. B.; Parvez, M.; Nuyken, O.; Renner, G. *Inorg. Chem.* **1993**, *32*, 5088.
99. Allcock, H. R.; Dodge, J. A.; Manners, I. *Macromolecules* **1993**, *26*, 11.
100. Ni, Y.; Stammer, A.; Liang, M.; Massey, J.; Vancso, G. J.; Manners, I. *Macromolecules* **1992**, *25*, 7119.
101. Manners, I. *Polymer News* **1993**, *18*, 133.
102. Edwards, M.; Ni, Y.; Liang, M.; Stammer, A.; Massey, J.; Manners, I. *Polymer Prepr.*, (Am. Chem. Soc., Div. Polym. Chem.), **1993**, *34*, 324 and 326.

103. Liang, M.; Waddling, C.; Honeyman, C.; Foucher, D.; Manners, I. *Phosphorus, Sulfur, Silicon and Rel. Elem.* **1992,** *64,* 113.

104. Manners, I. *Coord. Chem. Rev.* **1994,** *137,* 109-129.

105. Jaeger, R.; Lagowski, J. B.; Manners, I.; Vancso, G. J. *Macromolecules* **1995,** *28,* 539.

106. Roy, A. K. *J. Am. Chem. Soc.* **1993,** 2598.

107. Roy, A. K.; Burns, G. T.; Lie, G. C.; Grigoras, S. *J. Am. Chem. Soc.* **1993,** *115,* 2604.

108. Kealy, T. J.; Pauson, P. L. *Nature* **1951** *168,* 1039.

109. Neuse, E. W. ; Rosenberg, H. *Metallocene Polymers* **1970,** Marcel Dekker, New York.

110. Pittman, C. U., Jr. *Organometallic Reactions* **1977,** *6,* 1-61.

111. Willis, T. C.; Sheats, J. E. *J. Polymer. Sci., Polym. Chem. Ed.* **1984,** *22,* 1077.

112. Sheats, J. E.; Hessel, F.; Tsarouhas, L.; Podejko, K. G.; Porter, T.; Kool, L. B.; Nolen, R. L., Jr. in B. Culbertson, *New and Unusual Monomers & Polymers* **1983,** Plenum Press, New York.

113. Pittman, C. U. Jr. *J. Paint Technol.* **1971,** *43,* 29.

114. Sheats, J. E. Unpublished results.

115. Rausch, M. D.; Macomber, D. W.; Gonsalves, K.; Fang, F. G.; Lin, Z. R.; Pittman, C. U., Jr. in *Metal-Containing Polymeric Systems* **1986,** 43-57, Sheats, J. E.; Carraher, C. E., Jr.; Pittman, C. U., Jr., Eds., Plenum Press, New York.

116. Pittman, C. U., Jr.; Lai, J. C.; Vanderpool, D. P.; Good, M.; Prado, R. *Macromolecules* **1970,** *3,* 746.

117. George, M. H.; Hayes, G. F. *J. Polym. Sci., Polym. Lett. Ed.* **1973,** *11,* 471.

118. George, M. H.; Hayes, G. F. *J. Polym. Sci., Polym. Chem. Ed.* **1976,** *14,* 475.

119. Hale, P. D.; Inagaki, T.; Karan, H. I.; Okamoto, Y.; Skotheim, T. A. *J. Am. Chem. Soc.* **1989,** *111,* 3482.

120. Kittlesen, G. P.; White, H. S.; Wrighton, M. S. *J. Am. Chem. Soc.* **1985,** *107,* 7373.

121. Ref. 106, pp 45-63.

122. Neuse, E.W. *J. Macrol. Sci., Chem.* **1981,** *A16(1),* 3-72.

123. Ref. 106, pp 29-42.

124. Osburne, A. G.; Whiteley, R. H.; *J. Organometal. Chem.* **1975,** *101,* C27.

125. Fischer, A. B.; Kinney, J. B.; Staley, R. H.; Wrighton, M. S. *J. Am. Chem. Soc.* **1979,** *101,* 6501.

126. Osborne, A. G.; Whiteley, R. H.; Meads, R. E.; *J. Organomet. Chem.* **1980,** *193,* 345.

127. Butler, I. R.; Cullen, W. R.; Rettig, S. J. *Can. J. Chem.* **1987,** *65,* 1452.

128. Foucher, D. A.; Tang, B. Z.; Manners, I. *J. Am. Chem. Soc.* **1992,** *114,* 6246.

129. Manners, I. *Adv. Organometal. Chem.* **1995,** *37,* 131-68.

130. Fossum, E.; Matyjaszewski, K.; Rulkens, R.; Manners, I. *Macromolecules* **1995,** *28,* 401.

131. Rulkens, R.; Lough, A. J.; Manners, I. *J. Am. Chem. Soc.* **1994,** *116,* 797.

132. Foucher, D. A.; Ziembinski, R.; Tang, B. Z.; Macdonald, P. M.; Massey, J.; Jaeger, C. R.; Vancso, G. J.; Manners, I. *Macromolecules* **1993,** *26,* 2878.

133. Manners, I. *J. Inorg. Organometal. Polym.* **1993,** *3,* 185.

134. Foucher, D.; Ziembinski R.; Petersen, R.; Pudelski, J.; Edwards, M.; Ni, Y.; Massey, J.; Jaeger, C. R.; Vansco, G. J.; Manners, I. *Macromolecules* **1994,** *27,* 3992.

135. Pudelski, J. K.; Foucher, D. A.; Manners, I. *PMSE* **1994,** *71,* 312.

136. Foucher, D. A.; Honeyman, C. H.; Nelson, J. M.; Tang, B. Z.; Manners, I. *Angew. Chem. Int. Ed. Engl.* **1993,** *32,* 1709.

137. Foucher, D. A.; Ziembinski, R.; Rulkens, R.; Nelson, J.; Manners, I. in *Inorganic and Organometallic Polymers* **1994**, *572*, Allcock, H. R.; Wynne, K.; Wisnian-Neilson, P., Eds., A.C.S. Symposium Ser., American Chemical Society, Washington, D.C.

138. Foucher, D. A.; Manners, I. *Makromol. Chem. Rapid. Commun.* **1993**, *14*, 63.

139. Withers, H. P.; Seyferth, D.; Fellman, J. D.; Garrou, P. E.; Martin, S. *Organometallics* **1982**, *1*, 1275-1288.

140. Nelson, J. M.; Rengel, H.; Manners, I. *J. Am. Chem. Soc.* **1993**, *115*, 7035.

141. Nelson, J. M.; Lough, A. J. ; Manners, I. *Angew. Chem. Int. Ed. Engl.* **1994**, *33*, 989.

142. Brandt, P. F.; Rauchfuss, T. B. *J. Am. Chem. Soc.* **1992**, *114*, 1926.

143. Compton, D. L.; Rauchfuss, T. B. *Organometallics* **1994**, *13*, 4367.

144. Galloway, C. P.; Rauchfuss, T. B. *Angew. Chem. Int. Ed. Engl.* **1993**, *32*, 1319.

145. Bailar, J. E. in *Preparative Inorganic Reactions* **1964**, *1*, 1-27, W. Jolly, Ed., Interscience, New York.

146. Archer, R. D. *Coordination. Chem. Rev.* **1993**, *128*, 49-68.

147. Cronin, J. A. **1995** Ph.D. Thesis in Chemistry, University of Massachusetts.

148. Wynne, K. J. *J. Inorg. Organometal. Polym.* **1992**, *2*, 79.

149. Chen, C. T.; Suslick, K. S. *Coord. Chem. Rev.* **1993**, *128*, 293-322.

150. Chiang, W.; Thompson, M. E.; Van Engen, D. in *Organic Materials for Nonlinear Optics II*, Royal Society of Chemistry, **1991**, pp 211-16, Hann, R. A.; Bloor, D., Eds., London.

151. Thompson, M. E.; Chiang, W.; Myers, L. K.; Langhoff, C. *Proc. SPIE Int. Soc. Opt. Eng.* **1991**, *1497*, 423.

152. Chiang, W.; Ho, D. M.; Van Engen, D.; Thompson, M. E. *Inorg. Chem.* **1993**, *32*, 2886.

153. Hagihara, N.; Sonogashira, K.; Takahoshi, S. *Adv. Polym. Sci.* **1981**, *41*, 149.

154. Takahashi, S.; Morimoto, E.; Murata, E.; Kataoka, S.; Sonogashira, K.; Hagihara, N. *J. Polym. Sci.., Polym. Chem. Ed.*, **1982**, *20*, 565.

155. Lewis, J.; Khan, M. S.; Kakkar, A. K.; Johnson, B. F. G.; Marder, T. B.; Fyfe, H. B.; Wittman, F.; Friend, R. H.; Dray, A. E. *J. Organometal. Chem.* **1992**, *425*, 165.

156. Khan, M. S.; Kakkar, A. K.; Long, N. J.; Lewis, J.; Raithby, P.; Nguyen, P.; Marder, T. B.; Wittman, F.; Friend, R. H. *J. Mater. Chem.* **1994**, *4*, 1227.

157. Frapper, G.; Kertesz, M. *Inorg. Chem.* **1993**, *32*, 732.

158. Faulkner, C. W.; Ingham, S. L.; Khan, M. S.; Lewis, J.; Long, N.J.; Raithby, P. R. *J. Organometal. Chem.* **1994**, *482*, 139.

159. Hodge, A. J.; Ingham, S. L.; Kakkar, A. K.; Khan, M. S.; Lewis, J.; Long, N. J.; Parker, D. G.; Raithby, P. R. *J. Organometal. Chem.* **1995**, *488*, 205.

160. Carraher, C. E., Jr., Ref. 2, pp 79-86.

161. Carraher, C. E., Jr.; Louda, J. W.; Sterling, D.; Rivalta, A.; Zhang, Q.; Baker, E. *PMSE* **1994**, *71*, 386.

162. Mbonyana, C. W.; Neuse, E. W.; Perlwitz, A. G. *Appl. Organomet. Chem.* **1993**, *7*, 279.

163. Carraher, C. E., Jr., Stewart, H. H.; Soldani, W. J. II; DeLaTorre, J.; Pandya, B.; Reckleben, L.; Hibiscus, F. *PMSE* **1994**, *71*, 783.

164. Carraher, C. E., Jr.; Scott, W. J.; Schroeder, J. A. *J. Macromol. Sci. Chem.* **1981**, *A15*, 625.

165. Carraher, C. E., Jr.; Zhang, Q.; Párkányi, C. *PMSE* **1994**, 71, 505.

166. Archer, R. D.; Illingsworth, M. L.; Rau, D. N.; Hardiman, C. J. *Macromolecules* **1985**, *18*, 1371.

167. Archer, R. D.; Batschelet, W. H.; Illingsworth, M. L. *J. Macromol. Sci. Chem.* **1981**. A16. 261.

168. Archer, R. D.; Wang, B. *Inorg. Chem.* **1990**, *29*, 39-43.
169. Archer, R. D.; Tramontana, V. J.; Ochaya, V. O.; West, P. V.; Cummings, W. E. in *Inorganic and Metal-Containing Polymeric Materials* **1991**, Carraher, C. E., Jr.; Pittman, C. U., Jr.; Sheats, J. E.; Zeldin, M.; Currell, B., Eds., Plenum Press, New York.
170. Chen, H.; Cronin, J. A.; Archer, R. D. *Macromolecules* **1994**, *27*, 2174.
171. Archer, R. D.; Chen, H.; Cronin, J.A. *PMSE* **1994**, *71*, 316.
172. Wang, B.; Archer, R. D. *PMSE* **1988**, *59*, 120 and **1988**, *60*, 710.
173. Chen, H.; Archer, R.D. *Macromolecules* **1995**, *28*, 1609.

RING-OPENING POLYMERIZATION (ROP) OF STRAINED METALLOCENOPHANES

John K. Pudelski, Daniel A. Foucher and Ian Manners[*]

Department of Chemistry, University of Toronto, 80 St. George Street, Toronto M5S 1A1, Ontario, Canada

INTRODUCTION

Synthetic organic polymers are predominantly based on chains of carbon atoms and have tremendous technological utility. The presence of transition metal elements in the main chain of a polymer is expected to lead to macromolecules which combine processability with novel physical or catalytic properties. To date the preparation of transition metal-based polymers in which the metal atoms are held in close proximity so as to promote interactions has been dominated by synthetic problems and examples of soluble, well-defined, and well-characterized materials of reasonable molecular weight ($M_n > 10^4$) are very rare.[1]

As part of our studies of new classes of polymers based on transition metal elements, in 1992 we reported the discovery of a novel, ring-opening polymerization (ROP) route which provided access to the first examples of high molecular weight poly(ferrocenylsilanes).[2] This involved the thermal ROP (TROP) of [1]ferrocenophanes **1** containing a silicon atom in the bridge and the resulting macromolecules **2** possess an unusual main chain comprising ferrocene units and silicon atoms. A general and comprehensive overview of research in this area up to the Spring of 1994 has been recently published.[3] In this Chapter we review some very recent developments and also discuss the extension of the ROP route to prepare other classes of transition metal-based macromolecules.

Metal-Containing Polymeric Materials
Edited by C.U. Pittman, Jr., *et al.*, Plenum Press, New York, 1996

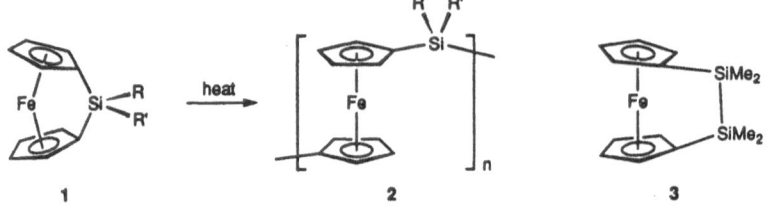

Figure 1. TROP of silicon-bridged [1]ferrocenophanes.

RESULTS AND DISCUSSION

Synthesis of Poly(ferrocenylsilanes) via Thermal Ring-Opening Polymerization

Silicon-bridged [1]ferrocenophanes were first reported in 1975 by Osborne and co-workers who prepared the diphenyl species **1a**.[4] Later Wrighton and co-workers reported the dimethyl analog **1b**.[5] These species were readily obtained as red crystalline solids via the reactions of the N,N,N,N-tetramethylethylenediamine (TMEDA) complex of 1,1-dilithioferrocene (η-$C_5H_4Li)_2Fe \cdot$TMEDA with dichlorosilanes. Crystallographic studies suggested that these [1]ferrocenophanes were highly strained,[6] i.e see Figure 2. Whereas in ferrocene the cyclopentadienyl (Cp) rings are parallel, the Cp rings in **1b** were found to be tilted relative to each other by 20.8(5)°.[3] Additionally, the Cp ring planes in **1b** were found to make an angle of ca 37.0(6)° with the exocyclic Si-Cp_{ipso} bond resulting in distortion at the ipso carbon atoms. X-ray crystallographic analysis revealed similar distortion in the structure of diphenyl analog **1a**. In addition to X-ray crystallography, NMR studies, UV/Vis studies and early reactivity studies provided evidence for the strained nature of silicon-bridged [1]ferrocenophanes.[3]

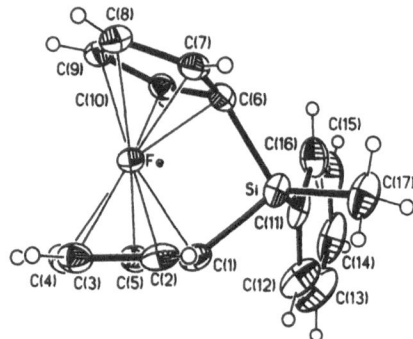

Figure 2. ORTEP perspective of strained silicon-bridged [1]ferrocenophane **1l**.

Although the strained nature of silicon-bridged ferrocenophanes had long been recognized, the first thermal ROP reactions of these species were not discovered until 1992.[2] We found that when a sample of **1b** was sealed in a Pyrex tube and heated at 130 °C, the sample became molten, then viscous and finally immobile over the course of 10 minutes. The soluble poly(ferrocenylsilane) **2b** was then recovered in good yield (Figure 1). The solubility of **2b** in common organic solvents (eg THF) allowed for extensive characterization by ^{1}H, ^{13}C and ^{29}Si NMR and UV/Vis spectroscopies which confirmed the assigned polymeric structure. GPC analysis vs. polystyrene standards found the material to be of high molecular weight. Molecular weight estimates for **2b** are included in Table 1. Similarly, heating **1a** at 230 °C afforded **2a** although this material suffered poor solubility in common organic solvents.

The strained nature of **1a** and **1b** was further probed by following the TROP reactions of these ferrocenophanes via differential scanning calorimetry (DSC).[2] DSC heating traces of **1a** and **1b** show sharp melting endotherms followed by more broad polymerization exotherms at higher temperature. Integration of the exotherms provided strain energy estimates of ca 60 and ca 80 kJ/mol for **1a** and **1b**, respectively. In comparison, the strain energies of cyclobutane, cyclopentane and norbornene are 113, 30, and 114 kJ/mol, respectively.[7]

We discovered that a strained structure was a prerequisite for successful TROP of ferrocenophanes.[3] When disilane-bridged [2]ferrocenophane **3** was heated under similar conditions as **1a** and **1b**, only unreacted starting material was recovered. Likewise, attempts to polymerize **3** via heating at temperatures as high as 340 °C for several days and via heating at 200 °C in the presence of anionic initiator K[OSiMe$_3$] proved unsuccessful. The failure of **3** to undergo TROP suggested that this compound was considerably less strained than the [1]ferrocenophane analogs. Indeed, X-ray crystallography, ^{13}C NMR and UV/Vis spectroscopy provided evidence that the degree of strain in the former compound is small. For example, X-ray crystal structure analysis found a small tilt angle of 4.19(2)° in **3** and an angle of only 10.8(3)° between the Cp ring planes and the Cp$_{ipso}$-Si bonds for this compound.

Armed with the knowledge that strain was critical for ferrocenophane polymerization we proceeded to probe the scope of the TROP reaction by preparing and studying the TROP reactions of an additional series of silicon-bridged [1]ferrocenophanes symmetrically substituted at silicon with alkyl groups.[3] Thus, diethyl, di *n*-propyl, di *n*-butyl, di *n*-pentyl, and di *n*-hexyl substituted ferrocenophanes **1c** - **1g** were prepared from reactions of (η-C$_5$H$_4$Li)$_2$Fe•TMEDA with the corresponding dialkyldichlorosilanes. TROP of **1c** - **1g** at ca. 130 °C afforded poly(ferrocenylsilanes) **2c** - **2g** in high yields (Figure 1). These materials, as expected, were highly soluble allowing for thorough characterization and study. As the GPC estimates in Table 1 indicate, the poly(ferrocenylsilanes) were also of high molecular weight.

We then further probed the scope of the TROP reaction by preparing and studying the TROP reactions of a series of similar [1]ferrocenophane monomers which were unsymmetrically substituted at silicon.[8] Thus, silicon-bridged [1]ferrocenophanes **1h** - **1n** were prepared in modest to high yields via reactions of (η-C$_5$H$_4$Li)$_2$Fe•TMEDA with the corresponding unsymmetrical diorganodichlorosilanes. Notably, the organic substituents at Si in these ferrocenophanes contained a variety of functionality previously unexplored in terms of compatiblity with the TROP reaction. Additionally, the groups at Si varied greatly in structure from small (H) to long and floppy (*n*-C$_{18}$H$_{37}$) to bulky and rigid (5-norbornyl, (η-C$_5$H$_4$)Fe(η-C$_5$H$_5$)). All of the sidegroup structure and functionality contained in **1h** - **1n** proved compatible with the polymerization reaction as TROP of these monomers afforded unsymmetrically substituted poly(ferrocenylsilanes) **2h** - **2n** in high yields (See Table 1 for R, R' assignments). These materials were all found to be highly soluble, which was particularly notable in the case of 2l with methyl and phenyl substituents at silicon since the diphenyl analog 2a was found to be poorly soluble in common organic solvents. Accordingly, the poly(ferrocenylsilanes) could be fully characterized and studied. The GPC data contained in Table 1 indicate that high molecular weight materials were obtained in all cases. Successful polymerization of 1h - 1n demonstrated that TROP of silicon-bridged [1]ferrocenophanes provides an attractive method of incorporating a variety of side chain structure and functionality into poly(ferrocenylsilane) materials. Since electronic and morphological properties of these polymers were found to vary greatly with side chain identity (see below), this work represented an important step toward control of the poly(ferrocenylsilane) properties in a rational manner.

Properties of Poly(ferrocenylsilanes) Prepared via TROP

Poly(ferrocenylsilanes) display a range of interesting properties and because of their relative ease of synthesis from readily available starting materials, ferrocene and organodichlorosilanes, they have attracted significant industrial interest. These polymers are isolated as amber fibrous solids, powders, or in some cases, as gums. In general they are soluble in organic solvents and most cases readily form solvent-cast flexible films. The more crystalline examples can be drawn into fibers from the melt. Some of the properties which have been studied to date are discussed below.

Thermal Transition Behavior and Morphological Properties. DSC and dynamic mechanical analysis (DMA) studies have been used to identify the glass transitions (T_gs) of poly(ferrocenylsilanes). Table 1 lists T_g values for both symmetrically and unsymmetrically substituted polymers. In the symmetrical series, T_g values generally decrease as the length of the organic side group at silicon increases.[9] This is a well-established trend for many polymer systems and can be attributed to increased free volume as the polymer main chains are pushed further apart. In the case of unsymmetrical poly(ferrocenylsilanes), in general, the T_g values increase with increasing steric bulk and rigidity of the non-methyl substituent at silicon.[8] The flexible n-$C_{18}H_{37}$ side chain of 2k presumably pushes the polymer chains apart thereby increasing the the free volume and lowering the T_g relative to the other polymers.

The polymers with fairly short n-alkyl chains at silicon are also capable of crystallizing, as shown by wide-angle X-ray diffraction which shows the presence of sharp lines for certain samples and DSC which shows melting transitions. Indeed, a recent detailed study of films of the dimethylated poly(ferrocenylsilane) 2b using techniques such as atomic force microscopy has uncovered a spherulitic morphology.[10] With the exception of 2h and 2k, DSC analysis of the unsymmetrical poly(ferrocenylsilanes) showed no evidence for melting transitions. Notably, DSC analysis of 2k found a large melting endotherm at 16 °C with the corresponding exotherm detected upon cooling. We rationalize this transition as crystallization of the n-$C_{18}H_{37}$ side chain of 2k below this temperature.

Table 1. GPC molecular weight estimates, T_g values and voltammetric $\Delta E_{1/2}$ values for poly(ferrocenylsilanes) prepared via TROP.

Polymer	R	R'	$M_w{}^a$	$M_n{}^a$	$T_g{}^b$ (°C)	$\Delta E_{1/2}{}^c$ (V)
2a	Ph	Ph	5.1×10^4	3.2×10^4		
2b	Me	Me	5.2×10^5	3.4×10^5	33	0.21
2c	Et	Et	7.4×10^5	4.8×10^5	22^d	0.27
2d	n-Pr	n-Pr	2.3×10^5	8.5×10^4		
2e	n-Bu	n-Bu	8.9×10^5	3.4×10^5	3^d	0.29
2f	n-Pen	n-Pen	4.9×10^5	3.0×10^5		
2g	n-Hex	n-Hex	1.2×10^5	7.6×10^4	-26	0.29
2h	Me	H	8.6×10^5	4.2×10^5	23	0.18
2i	Me	$CH_2CH_2CF_3$	2.7×10^6	8.1×10^5	59	0.16
2j	Me	$CH=CH_2$	1.6×10^5	7.7×10^4	28	0.20
2k	Me	$n\text{-}C_{18}H_{37}$	1.4×10^6	5.6×10^5	1	0.26
2l	Me	Ph	3.0×10^5	1.5×10^5	54	0.22
2m	Me	Fc^e	1.6×10^5	7.1×10^4	99	0.20^f
2n	Me	5-norbornyl	1.6×10^5	1.1×10^5	81	0.25

^{a} Obtained from analysis of THF polymer solutions which contained 1% [Bu$_4$N][Br] and estimated vs. polystyrene standards. ^{b} Obtained by DSC analysis unless otherwise noted. ^{c} $\Delta E_{1/2} = {}^2E_{1/2} - {}^1E_{1/2}$. ^{d} T_g values obtained by DMA analysis ^{e} Fc = $(\eta\text{-}C_5H_4)Fe(\eta\text{-}C_5H_5)$. ^{f} A third oxidation wave was observed for this polymer at $^3E_{1/2}$ corresponding to a ΔE value of 0.23 V between the second and third waves.

Electrochemical Characteristics and Properties of Oxidized Materials. One of the first characteristic features of poly(ferrocenylsilanes) to be noted was their intriguing electrochemical behavior.[2] The cyclic voltammograms of poly(ferrocenylsilanes) show two reversible oxidation waves in a ca. 1:1 intensity ratio.[2,3] This response has been attributed to interactions between the iron centers which lead to initial oxidation of alternating iron centers (at $E = {}^1E_{1/2}$) followed by subsequent oxidation, at significantly higher potential ($E = {}^2E_{1/2}$), of the iron centers in between, an explanation that has been fully supported by cyclic voltammetry studies of well-defined oligo(ferrocenyldimethylsilanes).[11] The peak separation ΔE ($\Delta E = {}^2E_{1/2} - {}^1E_{1/2}$), which gives a useful measure of the interaction between the iron centers, varies significantly with the substituents at silicon which may be a consequence of electronic or conformational effects. Table 1 lists poly(ferrocenylsilane) ΔE values. As expected, the cyclic voltammetric response of 2m was more complex than observed for the other polymers due to the presence of backbone and side chain iron centers in this material. The presence of interactions between the iron centers is in contrast to the situation with polymers containing ferrocenyl groups in the side chain structure, such as poly(vinylferrocene), which exhibit only a single reversible oxidation wave. This response indicates that all of the iron centers oxidize at the same potential.

Partially oxidized samples of poly(ferrocenylsilanes) possess electronic conductivities of ca 10^{-6} - 10^{-7} Scm^{-1} which is significantly greater than that that of the insulating, neutral precursor which is indicative of significant hole mobility in the former.[3] However, all of the high polymer samples studied to date are localized on the Mossbauer timescale. The magnetic properties of a polydisperse mixture of low molecular weight oligo(ferrocenylsilanes) ($M_n < 3,000$) generated by condensation routes have been examined by Garnier et al.[12] Evidence for the presence of ferromagnetic properties at low temperatures in tcne-oxidized materials was found and evidence for electron delocalization on the Mössbauer timescale was provided for a partially oxidized oligo(ferrocenyldihexylsilane).

Thermal Stability and Pyrolysis Behavior. Because of their processing advantages, polymers containing inorganic elements are of considerable interest as pyrolytic precursors to ceramics. However, the use of polymers as precursors to transition metal containing solids, which are known to possess a wide range of interesting electrical, magnetic, and optical properties, is virtually unexplored. Studies of the thermal stability of the poly(ferrocenylsilanes) **2a** and **2b** by Thermogravimetric Analysis (TGA) have indicated that the polymers undergo significant weight loss at 350 - 500 °C to yield ceramic residues in <u>ca</u> 35 - 40% yield with no further weight loss up to 1000 °C.[3] The lustrous ceramic products formed when **2a** or **2b** are heated at 500°C under a slow flow of dinitrogen have been characterized as amorphous iron silicon carbide materials by techniques such as X-ray photoelectron spectroscopy (XPS) and energy-dispersive X-ray (EDX) microanalysis. Mössbauer spectra and magnetic characterization (eg the presence of hysteresis in the B vs H plot) are consistent with a ferromagnetic nature of these materials. In addition, orange-yellow small molecule depolymerization products have been identified and in some cases characterized.

Poly(ferrocenylsilanes) via the Anionic Ring-Opening Polymerization of Silicon-Bridged [1]Ferrocenophanes

In order to investigate whether the ROP of silicon-bridged [1]ferrocenophanes could be achieved under more mild conditions than the elevated temperatures necessary for TROP we have studied anionic ROP in solution. Anionic ROP is known for a variety of silicon containing monomers including cyclotetrasilanes and silacyclopentenes. In addition, Seyferth et al have shown that phosphorus-bridged [1]ferrocenophanes will give short chain oligomers with up to 4 repeat units in the presence of anionic reagents but attempts to generate polymers using low concentrations of anionic initiator were unsuccessful.[13]

We found that **1b** undergoes anionic ring-opening oligomerization when treated with 0.5 - 1.0 equivalents of ferrocenyllithium, FcLi (Fc = Fe(η-C_5H_4)(η-C_5H_5)), in THF to afford linear oligo(ferrocenylsilanes) which function as excellent models for the corresponding high polymers with respect to conformational and electrochemical properties (Figure 3).[11] W e have also demonstrated that if a low concentration of anionic initiator is used poly(ferrocenylsilanes) are formed. Furthermore, very recent studies have demonstrated that under certain conditions the living anionic ROP of **1b** can be achieved.[14] This permits molecular weight control, end-group control, and the formation of novel transition metal-containing block copolymers.

Reaction of **1b** with FcLi, PhLi, or *n*-BuLi in THF at 25°C for 15 min followed by quenching of the living polymer **4** with either H_2O or $SiMe_3Cl$ yielded the H- or $SiMe_3$-capped poly(ferrocenylsilanes) **5** (Figure 3). By varying the initiator:monomer ratio from 1 :

20 to 1 : 99 controllable molecular weights from $M_n = 4.0 \times 10^3$ - 3.4×10^4 and narrow polydispersities ($M_w/M_n = 1.02$ - 1.26) were achieved. Treatment of the intermediate polymer **4** with additional monomer **1** before quenching with H_2O led to polymers **5** which showed the expected increase in molecular weight characteristic of a living process.

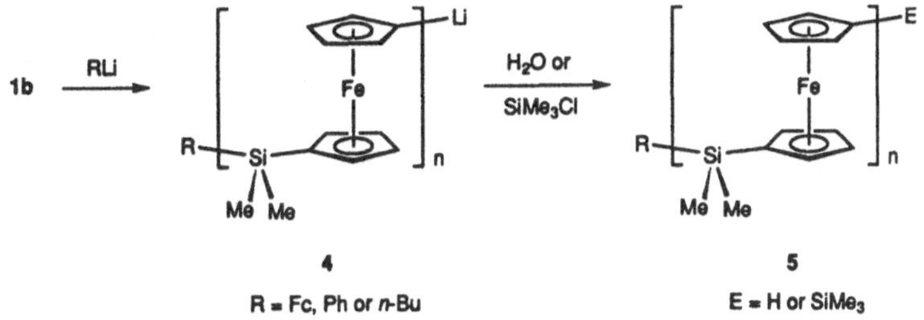

Figure 3. Anionic ROP of silicon-bridged [1]ferrocenophane **1b**.

Novel, well-defined poly(ferrocenylsilane)-poly(dimethylsiloxane) block copolymers **6** were prepared by the reaction of the living polymer **4** (R = *n*-Bu) with the strained cyclotrisiloxane [Me$_2$SiO]$_3$ in THF followed by the addition of Me$_3$SiCl (Figure 4). These yellow copolymers were, in contrast to the poly(ferrocenylsilane) **5**, found to be soluble in hexanes in which they appear to form micelles.

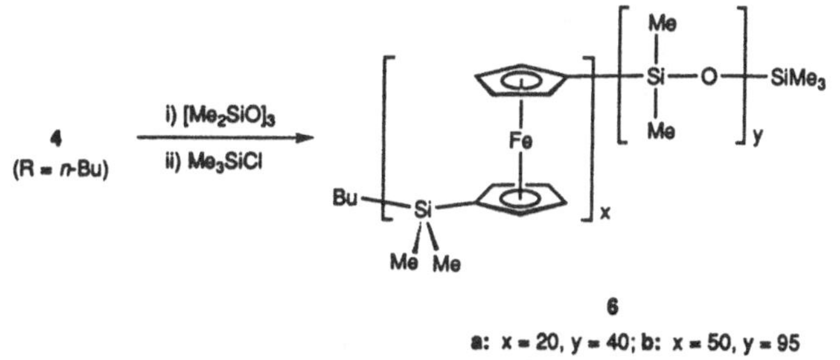

Figure 4. Preparation of poly(ferrocenylsilane)-poly(dimethylsiloxane) copolymers.

Thermal Ring-Opening Polymerization of other [n]Metallocenophanes

Since the initial discovery of the thermal ROP of **1a** and **1b** we have been interested in extending the ROP chemistry to other transition metal-containing rings. Germanium-bridged [1]ferrocenophanes can be prepared via the same route as their silicon-bridged analogs. Thus the reactions of (η-C$_5$H$_4$Li)$_2$Fe·TMEDA with the appropriate diorganodichlorogermane afford germanium-bridged [1]ferrocenophanes **7a** - **c** (Figure 5) in yields ranging from 30 - 35%.[15] Significantly, X-ray crystallographic analysis of the dimethyl- and diphenylgermanium-bridged [1]ferrocenophanes **7a** and **7c** found features characteristic of strained structures. For example, a tilt angle of 19.0 (9)° was found between the Cp ring planes of **7a**. This angle is slightly less than that found for silicon-bridged analog **1b**, consistent with the slightly larger covalent radius of germanium (1.22 Å) relative to silicon (1.17Å). Distortion at the ipso Cp carbon atoms of **7a** was indicated by angles between the the Cp ring planes and the C$_{ipso}$-germanium bonds of 37.5 (5)° and 35.9 (5)°.

Analysis of the germanium-bridged [1]ferrocenophanes by ^{1}H and ^{13}C NMR provided additional evidence for their strained nature.

Germanium-bridged [1]ferrocenophanes 7a - c were indeed found to undergo facile TROP reactions when heated in the melt (Figure 5).[3,15] For example, the poly(ferrocenylgermane) 8a resulting from TROP of 7a was isolated as a golden yellow powder in 80% yield. Similarly, poly(ferrocenylgermanes) 8b and 8c were recovered in high yield after heating the respective ferrocenophane monomers at 120 °C and 230 °C. Table 2 gives molecular weight data for the poly(ferrocenylgermanes) obtained by GPC analysis vs. polystyrene standards. While 8a and 8b were found to be highly soluble in common organic solvents such as THF, 8c was found to be less soluble.

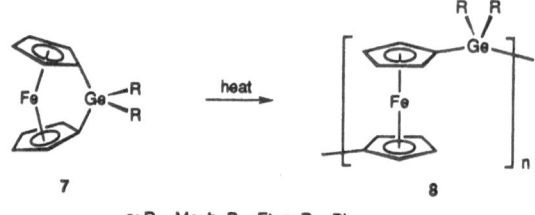

a: R = Me; b: R = Et; c: R = Ph

Figure 5. TROP of germanium-bridged [1]ferrocenophanes.

The thermal behavior of the poly(ferrocenylgermanes) was investigated by DSC which showed weak, yet identifiable glass transitions for 8a (28 °C) and 8b (12 °C). However, no evidence for melting transitions was detected, even after substantial annealing periods at 85 °C. Comparison indicates that T_g values of the poly(ferrocenylgermanes) are slightly lower than the values of their poly(ferrocenylsilane) analogs. The lower T_g values are possibly a consequence of the longer $Ge-C_{ipso}$ (≈ 1.95 Å) bonds vs $Si-C_{ipso}$ (≈ 1.86 Å) bonds, which give the poly(ferrocenylgermane) chains a greater degree of flexibility.

Table 2. GPC Molecular Weight Estimates for Poly(ferrocenylgermanes) and Poly(metallocenylethylenes) Prepared via TROP.[a]

Polymer	R	M_w	M_n
8a	Me	2.0×10^6	8.5×10^5
8b	Et	1.0×10^6	8.1×10^5
8c	Ph	1.0×10^6	8.2×10^5
10b	Me		
Fraction 1		8.1×10^4	6.6×10^4
Fraction 2		4.8×10^3	3.5×10^3
10d	Me		
Fraction 1		4.3×10^4	1.2×10^4
Fraction 2		1.3×10^4	4.4×10^3

[a] Obtained from analysis of THF polymer solutions which contained 1% [Bu$_4$N][Br] and estimated vs. polystyrene standards.

Poly(ferrocenylgermane) 8a displayed electrochemical behavior similar to the analogous poly(ferrocenylsilane) 2b. Thus, the cyclic voltammogram of 8a showed reversible oxidation waves in an intensity ratio of ca. 1:1 at $^1E_{1/2}$ = -0.01 V and $^2E_{1/2}$ = 0.19 V vs. the ferrocene/ferrocenium ion couple. The resulting ΔE value of 0.20 V suggests a similar extent of electrochemical interaction between Fe centers in the poly(ferrocenylgermane) and the poly(ferrocenylsilane) chains.

Preliminary work has shown that phosphorus-bridged [1]ferrocenophanes also undergo thermally-induced ROP.[3] The polymers formed, poly(ferrocenylphosphines), are spectroscopically similar to those previously prepared by condensation routes. Electrochemical studies have also indicated that the iron centers interact with one another.

We felt that TROP of [2]ferrocenophanes with hydrocarbon bridging groups represented an attractive route to poly(ferrocenylethylenes) because, unlike analogous disilane-bridged species, the former compounds are known to be considerably strained. Hydrocarbon-bridged [2]ferrocenophanes such as 9a (Figure 6) can be prepared in modest yields of ca. 20 - 30% via the the reaction of FeCl$_2$ with Li$_2$[C$_5$H$_4$CH$_2$]$_2$.[3] The strained nature of [2]ferrocenophanes with hydrocarbon bridges is undoubtedly related to the smaller size of the carbon bridging atoms in these species relative to the silicon bridging atoms in the unstrained disilane-bridged analog 3. X-ray crystallographic analyses of 9a found that the Cp rings of this compound were tilted relative to each other by an angle of 21.6°, compared to 4.19° in 3.

When a sample of 9a was heated in a sealed Pyrex tube at 300 °C the sample became molten, then viscous and, after one hour, immobile. Analysis of the product was impossible due to its near complete insolubility. In order to prepare soluble poly(ferrocenylethylenes) the synthesis and TROP reaction of methylated, hydrocarbon-bridged [2]ferrocenophane 9b was undertaken.[16] When this species, generated as an inseparable mixture of isomers, was treated under the same TROP conditions as 9a, a completely immobile material was again produced. The resulting poly(ferrocenylethylene) 10b (Figure 6) is isolated as a yellow fibrous material, readily soluble in THF. GPC analysis vs. polystyrene standards revealed that 10b posessed a bimodal molecular weight distribution consisting of high molecular weight and essentially oligomeric fractions. GPC molecular weight estimates for 10b are included in Table 2. It is possible that generation of bimodal poly(ferrocenylethylene) results from the operation of two TROP mechanisms in this system.

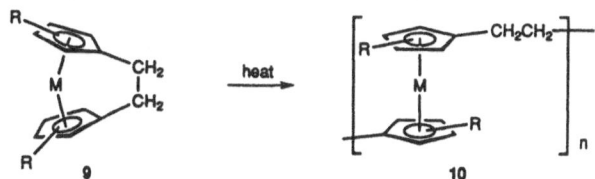

a: R = H, M = Fe; b: R = Me, M = Fe; c: R = H, M = Ru; d: R = Me, M = Ru

Figure 6. TROP of hydrocarbon-bridged [2]metallocenophanes.

In contrast to the cyclic voltammetric response of poly(ferrocenylsilanes) and poly(ferrocenylgermanes), the cyclic voltammogram of poly(ferrocenylethylene) 10b shows a single wave with a partially resolved shoulder at $E_{1/2}$ = -0.27 V vs. the ferrocene/ferrocenium ion couple. This behavior is indicative of minimal interaction between the iron centers in the poly(ferrocenylethylene) backbone and suggests that the ethylene groups which bridge the ferrocenyl units are electronically insulating.[17]

Very recently we reported that hydrocarbon-bridged [2]ruthenocenophanes could be readily prepared in a manner similar to the preparation of their ferrocenophane analogs.[18] Thermal ROP of these species yields poly(ruthenocenylethylenes) **10c/d** (Figure 6). GPC molecular weight estimates for **10d** are included in Table 2.

SUMMARY

Thermal ROP of strained metallocenophanes provides a new and general route to high molecular weight metal-containing polymeric materials.[3] Strained monomer structures are critical for ROP reactivity since unstrained metallocenophanes have been found to be unreactive. In the case of silicon-bridged [1]ferrocenophanes we have shown that anionic ROP in solution also occurs. Moreover, living polymerizations can be performed and this allows molecular weight control and the formation of novel block copolymers.[14] Further developments in the area of well-defined transition metal-based polymers with controlled architectures can be expected in the future. In addition, efforts are now also targeted at gaining a complete understanding of the mechanism(s) of the thermal ROP reactions of strained metallocenophanes.

REFERENCES

(1)   Sheats, J. E.; Carraher, C. E.; Pittman, C. U. *Metal Containing Polymer Systems*; Plenum: New York, 1985.

(2)   Foucher, D. A.; Tang, B.-Z.; Manners, I. *J. Am. Chem. Soc.* **1992**, *114*, 6246-6248.

(3)   Manners, I. *Adv. Organomet. Chem.* **1995**, *37*, 131.

(4)   Osborne, A. G.; Whiteley, R. H. *J. Organomet. Chem.* **1975**, *101*, C27.

(5)   Fischer, A. B.; Kinney, J. B.; Staley, R. H.; Wrighton, M. S. *J. Am. Chem. Soc.* **1979**, *101*, 6501-6506.

(6)   Stoeckli-Evans, H.; Osborne, A. G.; Whitely, R. H. *Helv. Chim. Acta* **1976**, *59*, 2402-2406.

(7)   Schleyer, P. v. R.; Williams, J. E.; Blanchard, K. R. *J. Am. Chem. Soc.* **1970**, *92*, 2377-2386.

(8)   Foucher, D.; Ziembinski, R.; Petersen, R.; Pudelski, J.; Edwards, M.; Ni, Y.; Massey, J.; Jaeger, C. R.; Vancso, G. J.; Manners, I. *Macromolecules* **1994**, *27*, 3992-3999.

(9)   Foucher, D. A.; Ziembinski, R.; Tang, B.-Z.; Macdonald, P. M.; Massey, J.; Jaeger, C. R.; Vancso, G. J.; Manners, I. *Macromolecules* **1993**, *26*, 2878-2884.

(10)   Rasburn, J.; Petersen, R. J.; Jahr, T.; Rulkens, R.; Manners, I.; Vancso, G. J., unpublished results.

(11)   Rulkens, R.; Lough, A. J.; Manners, I. *J. Am. Chem. Soc.* **1994**, *116*, 797-798.

(12)   Hmyene, M.; Yassar, A.; Escorne, M.; Percheron-Guegan, A.; Garnier, F. *Adv. Mater.* **1994**, *6*, 564.

(13)   Withers, H. P.; Seyferth, D.; Fellmann, J. D.; Garrou, P. E.; Martin, S. *Organometallics* **1982**, *1*, 1283-1288.

(14)   Rulkens, R.; Ni, Y.; Manners, I. *J. Am. Chem. Soc.* **1994**, *116*, 12121.

(15)   Foucher, D. A.; Edwards, M.; Burrow, R. A.; Lough, A. J.; Manners, I. *Organometallics* **1994**, *13*, 4959.

(16)   Nelson, J. M.; Rengel, H.; Manners, I. *J. Am. Chem. Soc.* **1993**, *115*, 7035-7036.

(17)   Foucher, D. A.; Honeyman, C. H.; Nelson, J. M.; Tang, B.-Z.; Manners, I. *Angew. Chem. Int. Ed. Engl.* **1993**, *32*, 1709-1711.

(18)   Nelson, J. M.; Lough, A. J.; Manners, I. *Angew. Chem. Int. Ed. Engl.* **1994**, *33*, 989-991.

CLUSTER-CONTAINING METAL (CO)POLYMERS: PRODUCTION, STRUCTURE AND THERMAL PROPERTIES

N. M. Bravaya,[1] A. D. Pomogailo,[1] V. A. Maksakov,[2] V. P. Kirin[2]

[1]Institute of Chemical Physics in Chernogolovka Russian Academy of Sciences, Chernogolovka 142432, Moscow Region, Russia;
[2]Institute of Inorganic Chemistry Siberian Department Russian Academy of Sciences, Novosibirsk 630090, Pr. Lavrent'eva 5, Russia

Introduction

The increasing interest in cluster-containing polymers is nowadays based mainly on two aspects:

(i) It is reasonable to expect that clusters or cluster complexes incorporated into the polymers would considerably modify polymer materials properties (adhesive, magnetic, thermal, etc.).

(ii) Cluster-containing polymers are of great interest as catalysts for different reactions,[1] combining the advantages both of homogeneous catalysts (due to solubility or swelling in organic solvents) and heterogeneous ones (due to the increase of cluster stability by their bonding to the polymer which prevents cluster fragmentation or aggregation to large metal particles).

Many investigators have been studied metal cluster formation or incorporation in polymers. All the ways for making cluster-containing polymer systems can be divided into two main groups. The first group yields polymer-incorporated clusters or cluster complexes of indefinite or nonuniform structure. These systems are characterized with the average values of the cluster particle sizes (usually metal dispersions). These are formed by reduction (chemical, photochemical, or electrochemical) of metal salts or mononuclear complexes in polymer medium, decomposition of volatile metal salts, metal carbonyls,[2] organometallic derivatives, spraying of metal atoms (at low temperature and pressure) on thin polymer films,[3] mixing of clusters with polymers (microcupsulation),[4] etc. The main product of the reactions are small (10-35 Å) cluster-type metal particles stabilized within polymer matrixes.

Other synthetic methods employ rather well or partially characterized cluster precursors and give rise to cluster-type moieties. These methods assemble polynuclear complexes from mononuclear species by several chemical reactions with the participation of polymer functional groups, cluster immobilization, etc. This approach generates complexes with different structures before and after the polymer-analogous reaction.

The formation of polymer-immobilized cluster complexes during the preparation of a polymer matrix is an attractive approach for simplifying the synthesis of polymer cluster complexes or dispersions. However, it is rather difficult to prepare stable metal particles on a cluster scale in a medium containing monomers. The more promising approach consists in handling stable cluster complexes to preserve their nuclearity and chemical structure after incorporation into polymer matrix. Recently, the preparation of uniform molybdenum chloride cluster dispersions ($\sim$6 Å) in polyvinylimidazole matrix was reported.[5] The authors have used vinylimidazole as a solvent. Six molecules provided strong coordination with the cluster complex. Radical initiated polymerization of the vinylimidazole yielded the polymer containing cluster complexes in cross-links. It seems probable that the ability to reproduce properties of these cluster-containing polymers would depend on a variety of reaction parameters. It is rather difficult on our opinion to control the behaviour of all six coordinated vinyl groups and consequently the properties of obtained cluster-containing polymers.

We have succeeded in the synthesis of cluster-type monomers capable of radical poly- and/or copolymerization with several traditional monomers. The choice of comonomers depends on which desirable properties are wanted (solubility, thermal stability, the presence of functional groups, etc.). The copolymerizations of cluster-containing monomers with styrene, acrylonitrile, and methylmethacrylate were studied. The presence of cluster units in these copolymers should affect their properties. Therefore, some thermal properties of new cluster-containing polymer materials are reported.

Synthesis of Triosmium Carbonyl Cluster Monomers.

We have synthesized a number of Os_3- and Ru_3-cluster monomers[6] possessing a double bond capable of addition polymerization. The double-bond was in a μ-coordinated ligand. The synthesis of Os_3-containing monomers is illustrated below by following a general scheme:

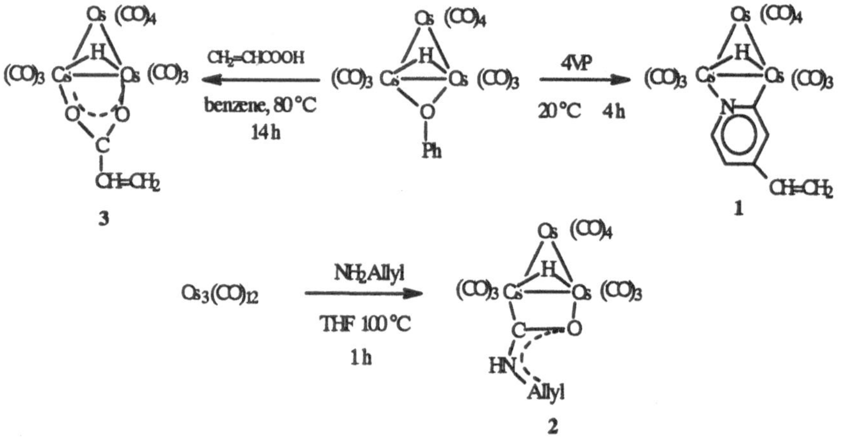

Scheme 1

The details of synthesis have been omitted but IR- and NMR-spectra are listed.

(μ-H)Os₃(CO)₁₀(μ-NC₅H₃CH=CH₂) (1) IR-spectrum: (cyclohexene); ν(CO), cm⁻¹); 2101 m, 2061 s, 2051 s, 2020 s, 2008 s, 2000 m, 1988 m, 1973 w; (CH₂Cl₂ ν(C=C), cm⁻¹); 1597 w. NMR-spectra (CDCl₃; δ, ppm): 8.01 (d, 1H, H heterocycle), 7.25 (m, 2H, –CH= + H heterocycle), 5.72 (m 2 H, =CH₂), –14.85 (s, 1 H, Os₂(μ-H)).

(μ-H)Os₃(CO)₁₀(μ-CONHCH₂CH=CH₂) (2) IR-spectrum (hexane); ν(CO), cm⁻¹); 2106 w, 2067 s, 2056 s, 2022 s, 2013 vs, 2010 sh, 1993 s, 1985 m, 1976 m, 1967 w, 1951 w; (CH₂Cl₂; ν(C=C), cm⁻¹); 1670 w; (CH₂Cl₂; ν(NC=O), cm⁻¹): 1504 w; (CH₂Cl₂; ν(NH), cm⁻¹): 3423 w, 3330 w. NMR-spectra (CDCl₃; δ, ppm): 7.17 (broad, 1H, NH), 6.6 (m, 1H, -CH=CH₂), 1.59 (m, 3 H, -CH₂- + -CH=CH₂), –14.21 (s, 1 H, Os₂(μ-H)).

(μ-H)Os₃(CO)₁₀(μ-COOCH=CH₂) (3) IR-spectrum (hexane); ν(CO), cm⁻¹); 2013 w, 2073 vs, 2062 s, 2027 vs, 2016 vs, 2010 s, 1988 m, 1984 m, 1722 (μ-CO₂); (CH₂Cl₂; ν(C=C), cm⁻¹); 1645 w. NMR-spectra (CDCl₃; δ, ppm): 5.95—5.16 (m, 3 H, -CH=CH₂), –10.42 (s, 1 H, Os₂(μ-H)).

Attempts to prepare a single crystal of complex **1** for X-ray diffraction study failed; a fine powder precipitated from various solvents. However the replacement of one CO group with PPh₃ allowed the preparation of a single crystal of phosphine derivative (**4**). Figure 1 shows the structure of **4**. The vinylpyridine ligand is bridge-coordinated by C and N atoms of the pyridine ring and occupies a semiaxial position (the angle between the Os₃ and Os₂CN planes is 101.4°). The Os atoms bonded with it are located almost in the plain of aromatic system (the deflections do not exceed 0.03 Å). The vinyl group does not participate in the coordination. The bond length and angles in the vinylpyridine, PPh₃, and CO ligands have normal values. Neither hydrogen bonds nor other short intermolecular contacts are observed in the structure under consideration.

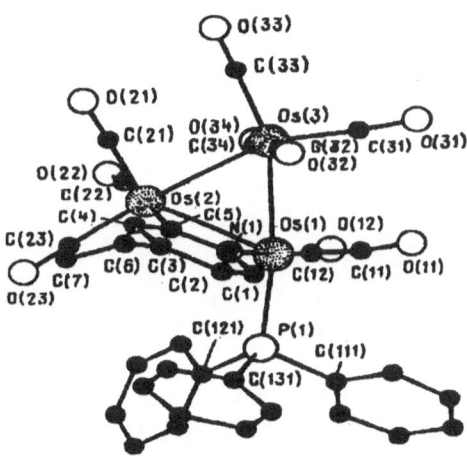

Figure 1. The projection of molecule **4** onto the plain of metal framework. Main bond length, Å: Os(1)-Os(2) 2.932(1), Os(1)-Os(3) 2.866(1), Os(2)-Os(3) 2.875(1); Os(1)-P(1) 2.369(6); Os(1)-N(1) 2.13(2); Os(2)-C(5) 2.12(2); N(1)-C(5) 1.31(3); N(1)-C(1) 1.38(3); C(1)-C(2) 1.32(4), C(2)-C(3) 1.39(4), C(3)-C(4) 1.38(4), C(4)-C(5) 1.46(3), C(3)-C(6) 1.47(5), C(6)-C(7) 1.25(7);Os-C(CO) 1.83-1.93(3) (the average meaning 1.89(3)); C-O(CO) 1.12-1.22(4) (the average meaning 1.16(4)).

Copolymerization of Cluster-Containing Monomers 1-3 with Styrene and Acrylonitrile

Three cluster-containing monomers **1-3**, differing in the nature of the double bond (4-vinylpyridine, allyl amine, and acrylic acid) have been selected for the study.

There were some principal questions to be answered. Are cluster-type monomers 1-3 capable of addition to growing macroradical if they have a bulky substituent in the vicinity of their double bond? Do cluster substituents affect the reactivity of double bond due to possible redistributions of electron density? Does the nature of double bond (4-vinylpyridine, allyl-type, and acrylic type) effect copolymerization with styrene and acrylonitrile as it does for their nonmetallic analogs?

The triosmium carbonyl cluster framework interacts with the CN-group of AIBN[7] which was used as initiator of radical copolymerization. However, both IR- and PMR spectra of the mixture of these reagents were superpositions of the spectra of the individual compounds.

Homopolymerizations of 1-3 are hindered, probably due to steric reasons. Only oligomeric products composed of 5-6 monomeric units (GPC data), have been obtained (1-2% yields) during radical initiated polymerizations of 1 in toluene.

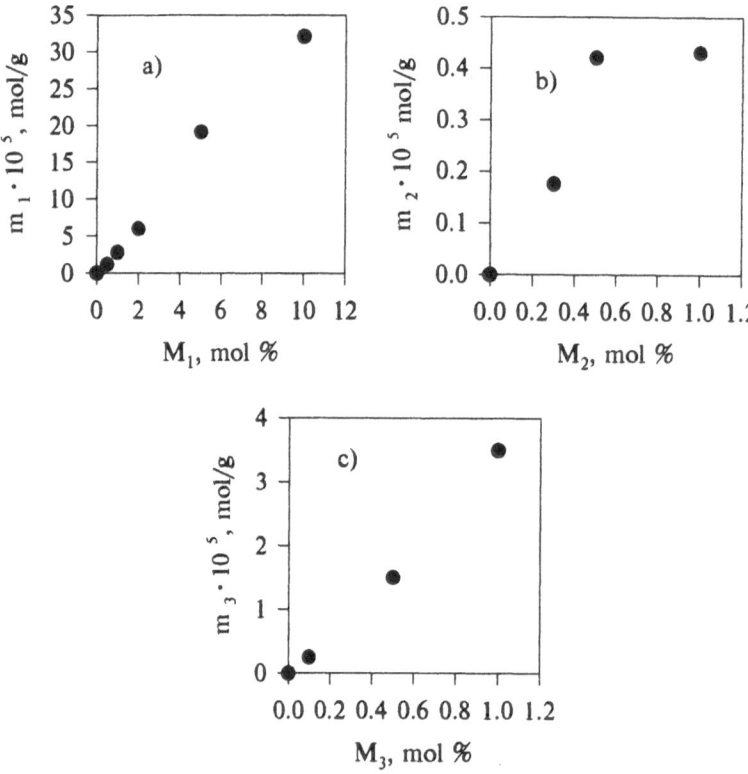

Figure 2. The dependence of cluster units content (m_i) in copolymers of styrene with cluster monomers 1 (a), 2 (b), and 3 (c) on the cluster comonomer content (M_i) in the mixture of comonomers.

Copolymerization of 1-3 with both styrene and acrylonitrile did not change either the ligand surroundings (IR-spectra) or the structure of cluster monomer framework (UV visible spectra).

The influence of the cluster monomer on the kinetics of styrene copolymerization with 1 (microcalorimetric measurements of liberated heat) as well on molecular weight characteristics of product copolymers (GPC) has been studied. Increasing the feed content of 1 from 0.5 to 10 mol % is accompanied with an increase of cluster units content in copolymers (Fig.2 a). However, the yields of precipitated copolymer decrease from 65 to 5 %, and sharp decrease of molecular weight from 37000 to 5000 occurs (Table 1). The comparative analysis of the molecular weights of the copolymers, the elemental analysis data for osmium content, and the analysis of double bond content in copolymers (ozonolysis) (Table 1) show that each 1-styrene copolymer chain contains in average one cluster unit and one probably terminal double bond. From this data one concludes that cluster monomer addition restricts chain propagation. However, copolymerization kinetic studies showed that the presence of the cluster comonomer does not decrease the rate of polymerization (Fig.4). These observations suggest that cluster monomer addition to growing polymer chains terminates propagation by fast chain transfer:

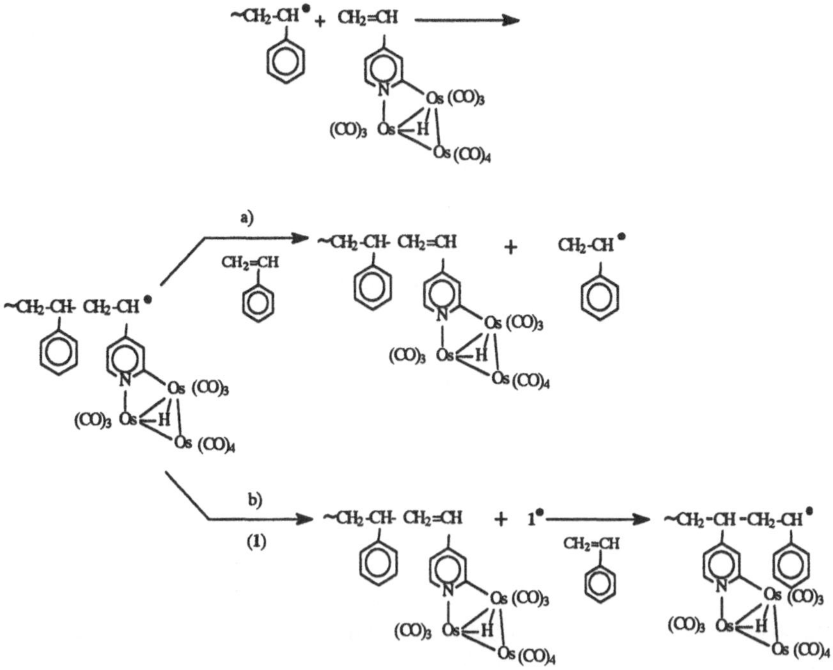

Scheme 2

Both styrene (the route a) and cluster monomer (the route b) can act as chain transfer agents. The last reaction is however less probable due to steric reasons which also hinder homopolymerization of 1. So the probability of this last reaction is rather small. Thus, styrene is the most probable chain transfer agent.

The rate constants for chain transfer in homopolymerizations of both styrene and 4-vinylpyridine are known[8] to be 4 orders of magnitude higher than chain propagation rate constants. It is also known that the copolymerization reactivity ratios for styrene with 4-vinylpyridine are $r_1=0.5-0.7$, $r_2=0.5-0.7$.[9] So, the addition of 4-vinylpyridine to styrene macroradicals is twice as fast as the styrene-to-styrene addition ($\sim 2 \cdot 10^2$ l/(mol·s)) at 60 °C),[10] i.e. about 400. This value will be the highest for 1 adding to the styrene macroradical; the presence of bulky substituents should probably decrease this value. The

55

lowest value of the rate constant for addition of **1** to the styrene macroradical can be estimated on the basis of copolymer molecular weights by the Mayo equation:

$$1/P_n = 1/P_0 + C_sS/M$$

where P_n is number average degree of polymerization; P_0 is the degree of polymerization in the absence of chain termination agent; $C_s = K_s/K_p$ is the relative rate constant for chain propagation in the presence of chain termination agent; K_s is the absolute value for this process; S is chain terminator concentration; M is monomer concentration.

Thus estimated value of C_s (Fig.3) is equal to ~0.2; K_s thus will be within the range 40-500 at 60 °C.

If one takes the highest value of K_s to estimate the influence of the cluster monomer on copolymerization kinetics, the presence of cluster-containing monomer leads to a decrease of copolymer molecular weights. Therefore, cluster units do not terminate the chain. Instead, they cause chain transfer, most probably to styrene molecules. This conclusion is also confirmed by the correspondence between the cluster content and the specific number of double bonds in obtained copolymers. So, the copolymers produced are macromolecules of polystyrene which are terminated with a cluster unit.

Table 1. Copolymers of styrene with cluster-containing monomers **1-3** and the corresponding nonmetallic monomers, 4-vinylpyridine (4-VP), allyl amine (AA), and acrylic acid (Aac). Comparison of their number average molecular weight, M_n, average number of polymer chains per gram of polymer, average number of double bonds $(mol/g \cdot 10^5)$, and cluster units content $(mol/g \cdot 10^5)$.

Molar ratio styrene:CM	M_n	[q]	[C=C]	[CM]
Polystyrene[*]	31000	3.2	2.8	
St:4-VP=0.005[*]	44000	2.3	2.7	
St:4-VP=0.01[*]	23000	4.3	2.6	
St:ÀÀ=0.005[*]	30000	3.3	2.5	
St:ÀÀ=0.01[*]	30000	3.3	3.0	
St:AAc=0.005[*]	28000	3.6	4.7	
St:AAc=0.01[*]	28000	3.6	5.0	
St:Os$_3$(CO)$_{12}$=0.01[*]	30000	3.3	2.7	
St:1 0.005[*]	31000	3.2	2.2	1.2
St:1 0.01[*]	37000	2.7	3.1	2.8
St:1 0.02[*]	35000	2.9	6.5	6.0
St:1 0.05[**]	6000	11.6	12.3	19.2
St:1 0.1[**]	5000	19.1	17.6	32.1
St:2 0.005[*]	29000	3.4	4.2	0.42
St:2 0.01[*]	21000	4.8	5.0	0.43
St:3 0.005[*]	15000	6.7	3.2	1.5
St:3 0.01[*]	12000	8.3	4.0	3.5

[*] Polymerization in mass; 60 °C; [AIBN]=0.5 mol %.
[**] Polymerization in toluene solution; 60 °C; [AIBN]=0.5 mol %.

It is interesting to note that bulky cluster substituents in these cluster monomers did not affect the nearest ligand surrounding of double bond. The correlation between reactivities of vinyl-groups, allyl-groups, and acrylate-groups on cluster monomers with respect to styrene macroradical is in accordance with those for nonmetallic monomers ($1/r_1$=1.6, 0.01-0.03, 0.4). The content of cluster **2** in the copolymers is an order of magnitude lower than that for **1** and **3** (see Fig.2 and Table1). However, the discrepancy between the average number of polymer chains for copolymers of styrene with **3** and sharp drop of M_n values for these copolymers remains unclear. Both the kinetics of heat release in styrene-cluster copolymerization and the values of reactivity ratios for acrylic acid are the same as for **1** and 4-vinylpyridine.

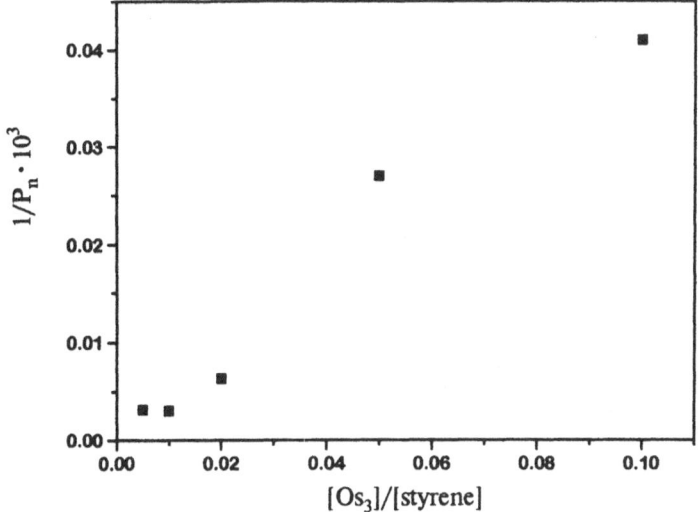

Figure 3. The dependence of reciprocal value of number average polymerization degree on the molar ratio [Os$_3$]-to-[styrene].

Copolymerization of cluster monomers 1-3 with acrylonitrile

Copolymerization of cluster monomers **1-3** with acrylonitrile is complicated by the heterogeneity of the process. However the specific features of copolymerization seem to be similar to those in copolymerization with styrene. The cluster content for the copolymers (see Table 2) also correlates with reactivity ratios of the corresponding nonmetallic monomers with acrylonitrile radicals: 9 (r_1=0.11), 0.2-0.3 (r_1 3-5), and 0.9 (r_1=1.1), respectively.[9] The heterogeneity of the process also leads to the fact that polymers obtained are mixtures of copolymer and the homopolymer of acrylonitrile. We have also observed a decrease in copolymer yield from 70 to 48 % by increasing the content of **3** from 0.5 to 1.0 mol % in the mixture of comonomers. The number averaged molecular weights of these products also drops. The effect of monomer **3** on these copolymer characteristics is unclear.

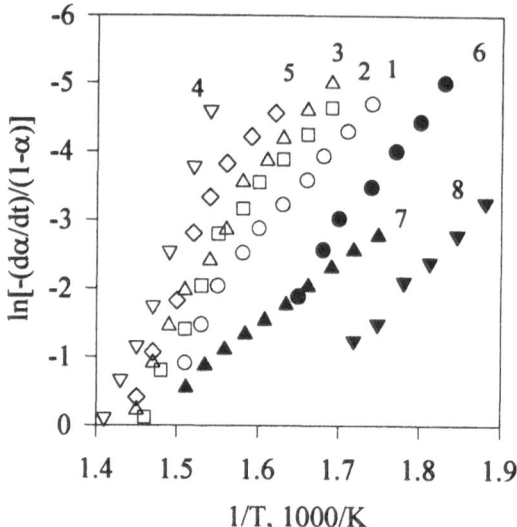

Figure 4. The kinetics of specific heat release in radical copolymerization of styrene (1) and styrene copolymerization with cluster monomer 1 at cluster content 0.5 mol % (2) and 1.0 mol % (3), as well copolymerization of styrene with 3 at 1.0 mol % (4).

Table 2. Copolymers of acrylonitrile with cluster comonomers **1-3**. Analysis of the number average molecular weights, cluster content (mol/g•10^5) and percentage of cluster-containing polymer chains

Polymer	M_n	[CM] in copolymer	%
PAN	18000		
AN:1=0.005	15000	0.61	9
AN:1=0.01	15000	1.12	17
AN:2=0.005	17000	0.41	7
AN:2=0.01	15000	0.61	10
AN:3=0.005	12000	0.62	7
AN:3=0.01	8000	0.64	5

Thermooxidative Degradation of Styrene and Acrylonitrile Copolymers with 1-3

The thermal gravymetric analysis (TGA) of both styrene and acrylonitrile copolymers with **1-3** reveals a strong mutual influence of their components on thermal decomposition.[11] Degradation of carbonyl cluster complexes is known to begin at rather low temperatures (~200-250 °C).[7] Cluster immobilization on metal oxide surfaces does not lead to their thermal stabilization.[12] DTA curves show a significant shift of the strong exothermic peak (T_{cl} corresponds to peak maximum) to higher temperatures. This reflects cluster decomposition processes (Table 3). This temperature is about 150 °C which is higher then that for unattached clusters (individual clusters and those incorporated into polymer mixtures mechanically). The thermal oxidative decomposition is accompanied with a number of transformations in the cluster skeleton and in the metal's coordination sphere. The role of polymer chain consists of the dissipation of thermal energy by free rotational-oscillatory movements of the cluster into the polymer backbone, increasing the thermal stability of cluster units.

Analysis of TGA- (mass loss) and DTA-curves (temperature effects) for styrene copolymers shows that the organic monomers, such as: 4-vinylpyridine (4VP), allylamine (AA), acrylic acid (AAc) in comparable amounts do not change significantly the thermal stability of copolymers (Table 3). The temperatures of decomposition beginning (T_0) are essentially the same as those for polystyrene. Copolymers containing clusters (1-3) exhibit an onset of decay that is significantly higher (~50-100 °C when the cluster content is 0.5-1 mol.%).

Table 3. Some characteristic temperatures of thermal oxidative decomposition of cluster monomers, cluster-containing styrene and acrylonitrile copolymers and those for copolymers without metal

Sample	$T_0{}^a$, °C	$T_{cl}{}^b$, °C	$T_1{}^c$, °C	decomposition degree
1	222	295		100
PS	303			97
St-4VP (1.0)^d	336			98
St-AA (1.0)	308			99
St-AAc(1.0)	310			98
St-1 (0.5)	378	450		98
St-1 (1.0)	392	450		98
St-2 (1.0)	355	444		98
St-3 (0.1)	363	280 sh, 426		98
St-3 (1.0)	277	278 sh, 413		99
PAN	273			48
AN-1 (1.0)	267	405	267	96
AN-2 (1.0)	264	457	264	99
AN-3 (1.0)	258	275 sh, 484	258	98
PS-+Os$_3$(CO)$_{12}$	360	280, 450 sh		99
PAN+1	243	280	243	95

[a] The temperature of polymer decay outset.
[b] The temperature corresponding to the maximum of exothermic peak of cluster decay on DTA curve.
[c] The temperature corresponding to the appearence of sharp exthermic peak on acrylonitrile decay DTA curves.
[d] In brackets the content of second comonomer is shown.

To estimate the activation energy values for decomposition of styrene copolymers the following approach was used:

$$[PS] \xrightarrow{k} gas\ products$$

$$\frac{d[PS]}{dt} = -k[PS] \quad where$$

$$k = k_0 e^{-E_d/RT}$$

Substituting the [PS] with the portion of disintegrated polymer, α, to operate with dimentionless values we have:

$$\ln[-\frac{d\alpha/dt}{(1-\alpha)}] = \ln k_0 - E_{eff}/RT$$

This dependence for polystyrene, styrene copolymers with monomers without metal cluster units, and styrene copolymers having a small amount of cluster monomers consists of two linear plots (curves 1, 2, 3, 5 on Fig.5). This effect can be explained by the existence of two competitive thermal oxidative decomposition processes within these polymers. First, an oxidative destruction occurs which has a smaller activation energy. This predominates at lower temperatures over the depolymerization processes with higher activation energies (Table 4). The role of the cluster complexes in this process consists in the suppression of oxidative copolymer decay. The low temperature decay onset for styrene-3 copolymers at decomposition is probably connected with the intrinsic nature of 3 (monomer of acrylate type). The lower value of the specific polymerization heat compared with that of the vinyl type monomers can lead to both partial decomposition of the polymer during polymerization (the lower values of M_n for these copolymers, see Table 1) and lower thermal stability of these copolymers. The acrylate polymers are mainly decomposed via depolymerization. This process is secondary for styrene copolymers and it can be hindered by cross-linking of polymer chains. The copolymer of methyl methacrylate with 1 decomposes by a depolymerization mechanism only. Cluster monomer units do not influence significantly the polymer decomposition process (curves 7,8 Fig.2). Thus, cluster additives are potential inhibitors of the oxidative decomposition in the copolymers studied.

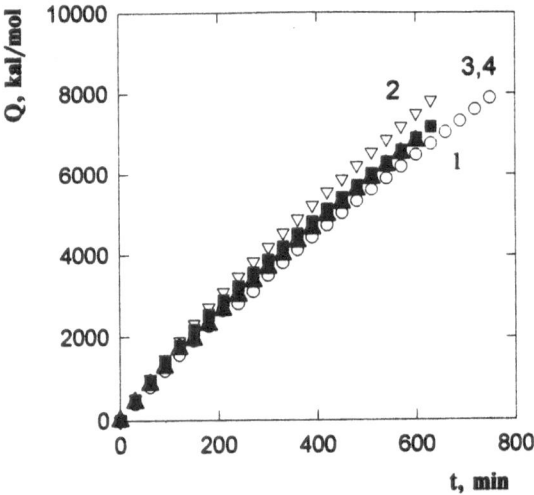

Figure 5. The dependence of ln of specific rate decomposition on the 1/T for polymers: 1 - polystyrene; copolymers: 2 - styrene and 4-VP; 3 - styrene with 1 (cluster content $1.2 \cdot 10^{-5}$ mol/g), 4 - styrene with 1 (cluster content $2.8 \cdot 10^{-5}$ mol/g), 5 - styrene with 2 (cluster content $0.43 \cdot 10^{-5}$ mol/g), 6 - styrene with 3 (cluster content $3.5 \cdot 10^{-5}$ mol/g); 7 - PMMA; 8 - copolymer of MMA with 1 (cluster content $2.3 \cdot 10^{-5}$ mol/g).

Table 4. The effective values of activation energies for the decomposition of polystyrene and its copolymers both with 4VP, AA, AAc and **1-3**.

Polymer	$[CM] \cdot 10^{-5}$ mol/g	E_{a1}	E_{a2},
PS		27	56
St-4-VP (1.0)		26	57
St-AA (1.0)		25	66
St-AAc (1.0)		25	67
St-1 (1.0)	2.8	-	69
St-1-(2.0)	6.0	-	64
St-2-(0.5)	0.42	45	72
St-2 (1.0)	0.43	34	65
St-3 (0.1)	1.5	33	-
St-3-(1.0)	3.5	30	-

The influence of **1-3** on AN copolymer decay is quite different. The cluster units do not promote the increase of copolymers thermal stability, but rather (even at small amounts) lead to the complete polymer decay.

The decomposition of polyacrylonitrile (PAN) starts via the formation of conjugated nitrile structure[13]. This process is accompanied by the appearance of a sharp exothermic peak in its DTA curves (T_1 in Table 3). The presence of the cluster units does not affect the above process. The other specific feature of PAN decomposition is the formation of nonvolatile naphtydine-like compounds as the final product (~50 w.%). Only cluster monomers that are connected with a polymer chain prevent the formation of these nonvolatiles. Decomposition of these copolymers proceeds sharply and completely (Table 3). The details of the process are now under study.

AKNOWLEDGEMENTS

The authors would like to thank the International Science Foundation (grant BNG000) for their support.

REFERENCES

1. Metal Clusters in Catalysis Eds. B.C.Gates, L.Guczi and H.Knözinger, Elsevier, Amsterdam-Oxford-New York-Tokyo (1986).
2. For example: R.Tannenbaum, C.L.Flenniken, E.P.Goldberg, Magnetic metal-polymer composites: thermal and oxidative decomposition of $Fe(CO)_5$ and $Co_2(CO)_8$ in a poly(vinylidene fluoride) matrix, .Polym.Sci., Part B: Polym. Phys., 28:2421 (1990).
3. A.Yu.Vasil'kov, P.V.Pribytkov, E.A.Fedorovskaya, A.A.Slinkin, A.S.Kogan, V.A.Sergeev, The specific features of magnetical behaviour of nickel- and cobalt-containing polyethylenes obtained by cryochemical methods, Dokl.Rus.Akad. Nauk, 331:179 (1993).
4. S.Y.Lee, R.Aris, The distribution of active ingredients in supported catalysts prepared by impregnation, Catal.Rev.-Sci.Eng., 27:207 (1985).

5. J.H.Golden, H.Deng, F.J.DiSalvo, and J.M.J.Fréchet, Poly(N-vinylimidazole) crosslinked by [Mo_6Cl_8(N-vinylimidazole)$_6$(triflate)$_4$ clusters, *Proc. ACS: Div. Polym. Mater.: Sci.-Eng,* 7:308 (1994).

6. V.A.Maksakov, V.P.Kirin, S.N.Konchenko, N.M.Bravaya, A.D.Pomogailo, et al., Preparation and reactivity of metal-containing monomers 31. Synthesis and properties of osmium- and ruthenium-based trinuclear cluster mnomers. The crystal structure of (μ-H)Os_3(μ-4-Vpy)(PPh_3)$(CO)_9$, *Russ. Chem. Bull.,* 42:1236, 1993.

7. S.P.Gubin, Khimiya Klasterov, Nauka, Moscow, (1987).

8. E.T.Denisov, Kinetika Gomogennykh Khimicheskikh Reaktsii, Vysshaya Shkola, Moskow (1988).

9. Copolymerization, Ed. G.E.Ham, Interscience Publishers, Div. John Wiley & Sons, New York, London, Sydney (1967).

10. J.Brandrup, E.H.Immergut, Polymer Handbook, Interscience Publishers (1966).

11. N.M.Bravaya, A.D.Pomogailo, V.A.Maksakov, V.P.Kirin, G.P.Belov, T.I.Solovieva, Preparation and reactivity of metal-containing monomers 38. Thermodestruction of polystyrene and polyacrylonitrile modified with triosmiumcarbonyl cluster monomers, *Izv. Acakad. Nauk, Ser. Khim.,* N3:423 (1994).

12. L.N.Arzamaskova, Yu.I.Ermakov, Polynuclear complexes on the surface of supports, *Zh. Vsesouyz. Khim Obshch. im.Mendeleeva,* 32:75 (1987).

13. N.Grassi, G.Scott, Polymer Degradation and Stabilization, Cambridge University Press, Cambridge, London, New York, New Rochelle, Melburne, Sydney (1985).

VARIABILITY OF MIXED-UNIT CHAINS IN METAL-CONTAINING POLYMERS

G. I. Dzhardimalieva and A. D. Pomogailo

Institute of Chemical Physics in Chernogolovka
Russian Academy of Sciences,
Chernogolovka 142432, Moscow Region, Russia

INTRODUCTION

Metal-containing polymers are presently interesting not only as individual compounds but as ingredients of polymer composites such as stabilizers, apprets, adhesive, coloring and dying agents, etc. A novel approach to the synthesis of structurally uniform metal-containing polymers is based on the homo- and copolymerization of metal-containing monomers (MCM): transition metal salts of unsaturated carboxylic acids, mono- and polynuclear complexes and monomers of cluster and chelate type. The dual nature of metal-containing polymers (MCP) (metal complexes on one side, and polymers with pendant metal substituents on the other) predetermines their properties and applications.

Polymerization and copolymerization of metal-containing monomers is a direct way for MCP preparation. The advantages of the method are: simplicity (one step synthesis), unreacted functional groups of the polymer are excluded, the possibility to control the metal ion distribution along the polymer chain.

The idea of mixed-linked polymers[1] may be successfully applied to the study of high-molecular compounds including macromolecular metal complexes. Chemical heterogeneity ("defects") in MCP structure may be caused by the appearance of the units with the structure or geometry different from those for initial ones. Therefore, metal-containing polymer molecules cannot always be considered as a sequence of similar units; all possible anomalies in the units should be taken into consideration. The effects which arise from chain variability are more probable if such polymers are produced by (co)polymerization of metal-containing monomers with metal center MX_n or $M''Y_n$ (where M is transition metal) in a pendant unit connected with a backbone by some Z group:

$$(p+q)CH_2=CH \xrightarrow{\ R^\bullet\ } (\text{-}CH_2\text{-}CH\text{-})p\text{——}(\text{-}CH_2\text{-}CH\text{-})q$$

$$\underset{\underset{MX_n}{|}}{\overset{|}{Z}} \qquad\qquad \underset{\underset{MX_n}{|}}{\overset{|}{Z}} \qquad\qquad \underset{\underset{M'Y_m}{|}}{\overset{|}{Z}}$$

This chapter reports the synthesis and polymerization of MCM based on unsaturated metal carboxylates. The main types of mixed-unit chains of metallopolymers and the properties of products synthesized are also discussed.

RESULTS

Synthesis of metal acrylates

The neutralization of metals oxides, hydroxides or carbonates with the calculated amounts of unsaturated acids is a simple and convenient method of metal carboxylate[2] synthesis.:

$M_2O_n + 2RCOOH \rightarrow M(OOCR)_n + nH_2O$

$M(OH)_n + nRCOOH \rightarrow M(OOCR)_n + nH_2O$

$MCO_3 + nRCOOH \rightarrow M(OOCR)_n + nCO_2 + nH_2O$

These reactions are straight forward and the side-products (water and carbon dioxide) are easily removed. Using this technique we prepared a number of metal acrylates (MAcr$_n$):

$Ni(OCOCH=CH_2)_2 \cdot H_2O$, (NiAcr$_2$), $Co(OCOCH=CH_2)_2 \cdot H_2O$, (CoAcr$_2$),

$Cu(OCOCH=CH_2)_2$,[3] (CuAcr$_2$), $[Fe_3O(OCOCH=CH_6) \cdot 3H_2O]OH$, (FeAcr$_3$),[3,4]

$Cr_3O(OCOCH=CH_6) \cdot 3H_2O)]$, (CrAcr$_3$),[5] $V_3O(OCOCH=CH_6) \cdot 3H_2O$, (VAcr$_3$), etc.

The elemental analyses and infrared spectra of the monomers synthesized are given in Table 1.

One general synthetic approach is to use excess acid both in the absence of solvent[6] and in methanol or ethanol solutions, and in hydrocarbon suspensions.[7] The latter process is distinguished by a high yield (over 95%) of the product containing more than 94% of the double bonds with respect to the calculated value. The preparation of metal acrylates in the absence of a solvent is very convenient, but the process is often accompanied by heating of reaction mixture and side-reactions (complexation, polymerization, etc.). The application of a threefold excess of acrylic acid (AA) results in the same MCM yield that occurs when a tenfold excess is used.[6] The problem of the isolating of the reaction product is essentially simplified using a threefold excess of acrylic acid because total deposition is achieved with significantly less solvent precipitant. We used diethyl ether as a selective precipitant becuase AA is quite soluble in this solvent.

Table 1. The elemental analyses and key infrared absorptions for $MAcr_n$ synthesized in this work.

$MAcr_n$	Found/Calcd, %			IR spectral characteristic band frequencies, cm^{-1}				
	C	H	M	ν, >C=C<	ν, =CH- C	ν_{as}, COO	ν_s, COO	ν, M-O
$CuAcr_2$	34.5/ 34.95	2.94/2 .91	29.9/ 31.06	1645	1065	1580, 1560	1370	315
$CoAcr_2$	32.6/32 .88	3.71/3. 65	27.5/ 26.94	1640	1070	1595, 1585	1370	280
$NiAcr_2$	32.4/32 .88	3.70/3. 65	26.6/26 .94	1640	1065	1560	1360	290
$FeAcr_3$	31.7/31 .98	3.67/3. 05	24.6/24 .06	1635	1065	1575, 1515	1435, 1370	525
$CrAcr_3$	32.0/ 32.29	3.91/3. 73	24.6/ 23.32	1635	1065	1575, 1525	1440, 1370	540
$VAcr_3$	32.1/ 32.43	3.43/3. 75	21.5/ 22.97	1635	1065	1590, 1527	1444, 1375	

Structure and physical-chemical properties of metal acrylates

It should be mentioned that the carboxylate group (RCOO⁻) is multifunctional.[8] It can act as a mono-, bi-, tri-, and even tetradentate ligand (a,b,c,d - respectively) depending on the coordination with the metal ion:

a   b1   b2

C1   C2

$$[R-C \underset{O}{\overset{O}{\cdots}} \begin{matrix} \cdots M^{n+} \\ -M^{n+} \\ -M^{n+} \\ M^{n+} \end{matrix}]_{4n} \qquad [R-C \underset{O}{\overset{O}{\cdots}} \begin{matrix} \cdots M^{n+} \\ M^{n+} \\ M^{n+} \end{matrix}]_{3n}$$

d1   d2

However, most frequently $RCOO^-$ performs as a mono (a) and a bidentate (b) ligand. The coordination type can be proved both by X-ray analysis and infrared (IR) spectroscopy: for the structures (b) $vCOO^-$ appears at 1560 to 1580 cm^{-1} [9] and at 1700 cm^{-1}, indicating the absorptions of the coordinated and the undissociated group, respectively.

IR-spectra of metal acrylates exhibit absorption bands at 1530-1570 cm^{-1} and 1370 cm^{-1} associated with the antisymmetric and symmetric stretching vibrations ($v_{as}(COO)$ and $v_s(COO)$, respectively) of the carboxyl group which have bridging coordination (type b1, scheme 1).[10] In addition, for $FeAcr_3$, $CrAcr_3$, and $VAcr_3$, very intense bands are observed at 1515-1527 cm^{-1} and 1435-1444 cm^{-1}. These can be attributed to the antisymmetric and symmetric stretching vibrations of carboxyl groups possessing bidentate coordination (type b2).[10] The absorption bands at 1635-1640 cm^{-1} ($v(C=C)$), 330 cm^{-1} ($\rho_w(CH_2)$), 910 cm^{-1} ($\pi(=CH_2)$), 1665-1070 cm^{-1} ($v(=CH-C)$, 1635-1645 cm^{-1} ($v(>C=C<)$) are also observed.[10,11] The absorption in 200-500 cm^{-1} range can be attributed to the valence vibrations of M-O bonds![10]

The synthesis of some metal acrylates is accompanied by an increase in metal ion nuclearity. The metal acrylates synthesized in this work have mono- ($NiAcr_2$, $CoAcr_2$), bi- ($CuAcr_2$) and tri- ($FeAcr_3$, $CrAcr_3$, $VAcr_n$) nuclear structures.

The molecular structure of $CuAcr_2$. The binuclear structure of $CuAcr_2$ was determined by X-ray analysis.[3,12] Blue-green single crystals of $[Cu_2(OCOCH=CH_2)_4(ROH)_2]ROH$, where R = CH_3(I) or C_2H_5 (II) were isolated from ethanol or methanol solutions. The crystals of I are monoclinic, with space group $P2_{1/b}$ and a = 7,593(2), b = 15.061(2), c = 9.032(3) Å, γ = 112.48(2)°, Z = 2, d_{calc} = 1.66 g · cm^{-1} (Fig.1).

Its structure consists of the formation of a binuclear cluster with Cu-Cu distance equal to 2.617 Å. In this case the copper atoms are interconnected via carboxylate bridges to form a complex complemented with two alcohol molecules. The studies of the nearest atomic surroundings of copper indicate that its coordination is best described as a square-pyramidal one. The bond distances and valence angles have typical sizes: Cu-$O_{carboxyl}$ - 1.962 Å, C-O - 1.26 Å, C-C -1.50 Å, C=C - 1,34 Å.

Polynuclear metal carboxylates. Trinuclear transition metal μ_3-oxocarboxylates containing $[M_3O(OCOR)_6]^+$ (R = H, CH_3, C_6H_5 etc.) cluster cation are well known.[13-16] They have a cluster structure in which the metal atoms form an equilateral triangle with the oxygen atom in at the center. The carboxylic groups form bridges between the metal atoms. The total formula for these compounds may be represented as $[M_3O(RCOO)_6(H_2O)_3]^+X^- \cdot nH_2O$, where X = OH, Cl, Br etc.[14,15] However, there is no information in the literature on polynuclear carboxylates with unsaturated carboxyl ligands. We have obtained the first trinuclear chromium, iron and vanadium acrylates.[4,5] Taking into account that the bands

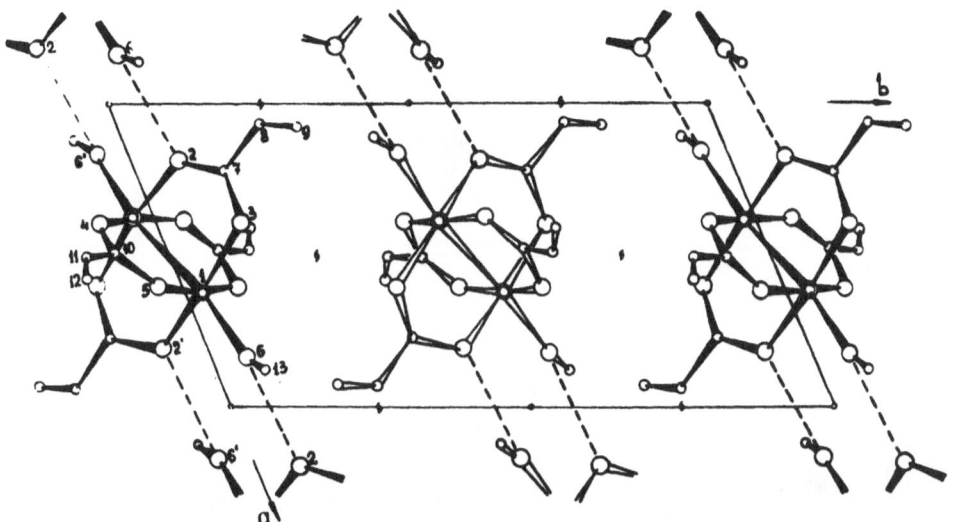

Figure 1. Molecular structure of copper acrylate [Cu$_2$(OCOCH=CH$_2$)$_4$(CH$_3$OH)$_2$] CH$_3$OH

associated with both types of carboxyl group coordination have approximately equal intensities (IR data), the structure of the trinuclear carboxylates may be represented as follows:

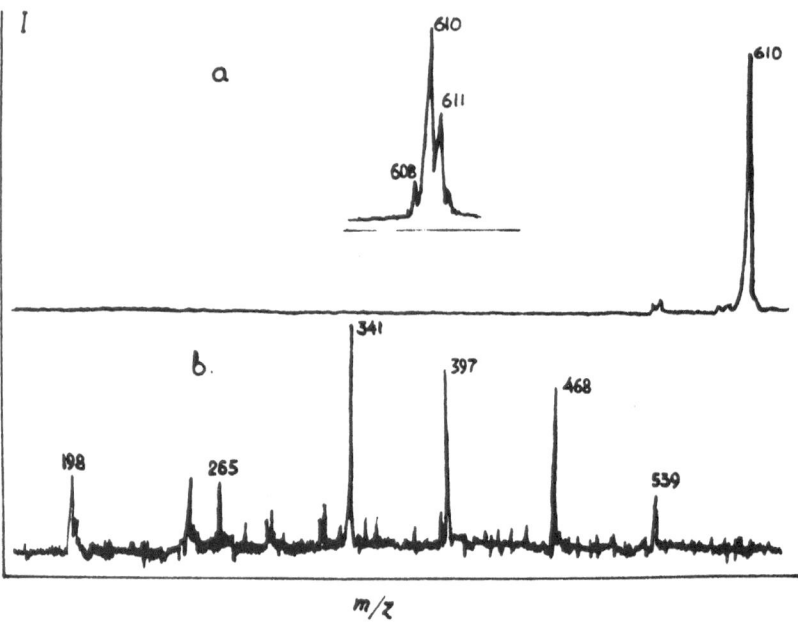

Figure 2. Mass spectra of FeAcr₃ (positive ions) extracted from an alcohol solution recorded at U = 200 V (a) and 400 V (b).

Keeping in mind that the cluster of the proposed structure is charged, one can identify its structure by time-of-slight mass spectrometry with the extraction of dissolved ions. Figure 2 shows the mass spectrum (U=200 V and 400V) of the positive ions extracted from a solution of the iron acrylates. The main peaks in the mass spectrum (m/z = 598 and 610) agree with the value calculated for the $[Cr_3O(Acr)_6]^+$ and $[Fe_3O(Acr)_6]^+$, respectively. Increasing the electric field intensity to U = 400 V makes it possible to control dissociative fragmentation of ions by their collision. For example, the ions with m/z = 539, 468 and 397 correspond to the loss of one, two, and three acrylate residues, respectively, from the $[Fe_3O(Acr)_6]^+$ molecular ion. The ion with m/z = 341 corresponds to the loss of the $Fe(CH_2=CHCOO)_3$ molecule (Table 2). These facts confirm the cluster structure of the compounds under study.

Table 2. Electrospray mass spectrometric data from $[M_3O(Acr)_6]^+$ fragmentation in the gas phase (U = 370 V)

The way of decay	M=Fe		M=Cr	
	ion mass m/z	intensity%	ion mass, m/z	intensity %
$[M_3O(Acr)_5]^+ + AA$	539	5	527	20
$[M_3O(Acr)_4]^+ + 2AA$	468	14	456	1
$[M_3O(Acr)_3]^+ + 3AA$	397	17	385	0
$[M_2O(Acr)_4]^+ + M(Acr)_2$	412	0	404	17
$[M_2O(Acr)_3]^+ + M(Acr)_3$	341	17	333	10
$[M(Acr)_3]^+ + [M_2O(Acr)_3]$	265	6	261	0
$[M(Acr)_2]^+ + [M_2O(Acr)_4]$	198	6	194	3

Magnetic properties. Antiferromagnetic interaction plays an important role in a clearer understanding metal-metal interactions in metal carboxylates. The magnetic moments both calculated and found from magnetic susceptibility measurements are given in Table 3 ($\mu_{ef} = 2.84 \sqrt{\chi_{at} \cdot T}$, χ_{at} - the measured paramagnetic susceptibility per metal 1 g-at, T - temperature, K).

Table 3. The magnetic properties of MAcr$_n$

MAcr$_n$	μ_{ef}, μB		μ_s, μB	Valence state of metal	Configuration	Antiferromagnetic exchange
	292K	77K				
CoAcr$_2$	4.00	4.52	3.87	Co^{2+}	$3d^7$	no echange
NiAcr$_2$	3.60	3.47	2.83	Ni^{2+}	$3d^8$	no echange
CuAcr$_2$	1.40	0.22	1.73	Cu^{2+}	$3d^9$	strong echange
FeAcr$_3$	3.36	2.31	5.92	Fe^{3+}	$3d^5$	echange
CrAcr$_3$	3.48	2.84	3.87	Cr^{3+}	$3d^3$	echange
VAcr$_3$	2.87	2.60	2.83	V^{3+}	$3d^2$	echange

Antiferromagnetic interactions are not observed for mononuclear metal acrylates (CoAcr$_2$, NiAcr$_2$) and magnetic moments correspond to the octahedral coordination of MO_6^{2+}. At the same time the cluster structures of polynuclear carboxylates with bridge and bidentate carboxyl ligands allow the strong antiferromagnetic interactions.

Radical homopolymerization and copolymerization of MAcr$_n$.

Homopolymerization of transition metal acrylates. The rate of transition metal acrylate polymerization (AIBN initiation, in ethanol) decreases as follows:

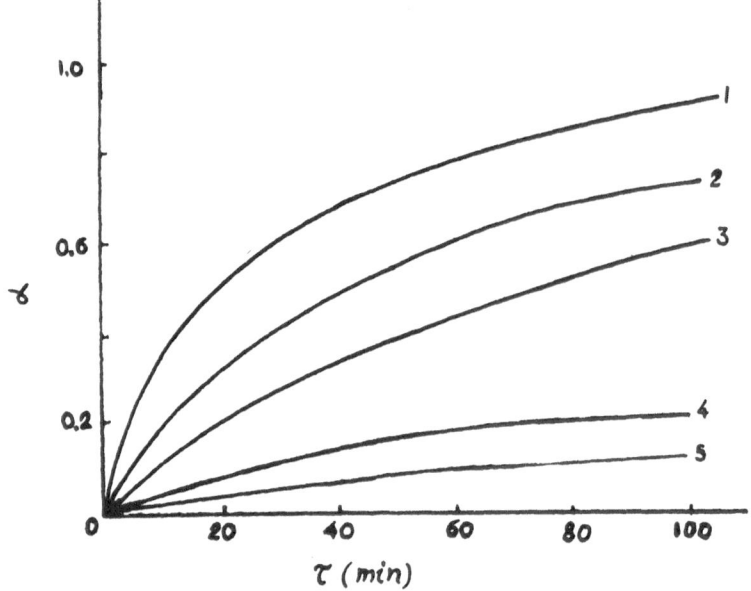

Figure 3. Kinetic curves of homopolymerization yield for AA (1), CoAcr₂ (2), NiAcr₂, (3), FeAcr₃ (4), and CuAcr₂ (5). [MCM] = 0.9 mol/l, [AIBN] = 2.5 · 10⁻² mol/l, ethanol, 78°C.

Co(II)>Ni(II)>Fe(III)>Cu(II) and in all cases it is lower than that of AA polymerization (Fig.3). Kinetic analysis of transition metal acrylate homopolymerization indicates that this process is characterized by the same elemental steps encountered with ordinary vinyl-type monomers. However, it is sometimes complicated by a purely individual effect of the transition metal.[17] The canonical radical initiated polymerization scheme can also be used to describe the polymerization of these MCMs:

$$I \xrightarrow{k_i} 2R^{\bullet}$$

$$R \xrightarrow{k_p''} RM^{\bullet}$$

$$RM^{\bullet} + M \xrightarrow{k_p} R_j^{\bullet}$$

$$R_j^{\bullet} + R_s^{\bullet} \xrightarrow{k_t} P$$

where $R^{\bullet}$, $RM^{\bullet}$, $R_j^{\bullet}$ and $R_s^{\bullet}$ are the initiating, primary, and propagating polymeric radicals, with the corresponding number of monomeric units, respectively; P is the resulting product.

Considering the initial $[M]_0$ and the current monomer $[M]$ concentrations to be related to the conversion α by the relationship $[M] = [M]_0 (1-\alpha)$ and a quasi steady state approximation with respect to the macroradicals to be valid, the polymerization rate is expressed as follows:

$$d\alpha / dt = k_p (k_i / k_t)^{1/2} I^{1/2} (1 - \alpha)$$

Provided that $I = I_o e^{-k_i t}$, its solution yields the following equation

$$\ln\left\{\ln(1-\alpha) + 2k_p[I_o / k_i k_t]^{1/2}\right\} = \ln 2k_p[I_o(k_i k_t)]^{1/2} - 1/2k_i t,$$

which provides a good description of the kinetics of polymer formation. Also, such an analysis characterizes the observed dependence of the limiting conversion on the initial initiator concentration:

$$\ln(1-\alpha_\infty) = 2k_p[I_o / (k_p k_t)]^{1/2}.$$

The value of k_i is found from these equations to be $3.16 \cdot 10^{-2}$ min which is close to the published data.[18]

A correlation between the rate of transition metal acrylate polymerization and metal electronegativity was established. With an increase of electronegativity the double bond electron density drops. As a result, the polymerization rate decreases. However, the effects produced by various transition metals on the polymerization kinetics are more profound and, to a large extent, individual. For example, a low polymerization activity of Cu(II) and Fe(III) - containing MCMs, is likely to result from an intramolecular chain termination:

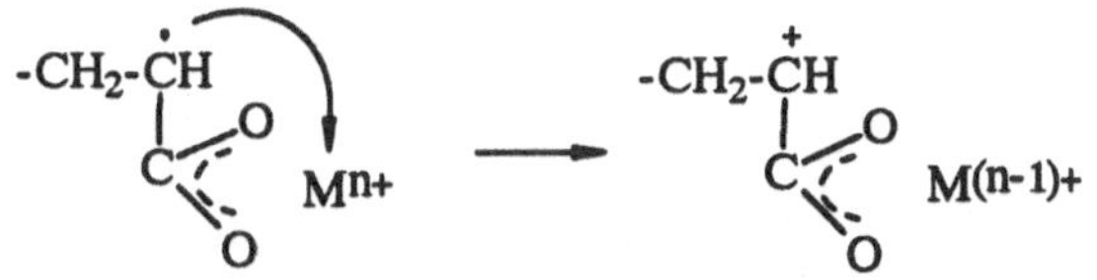

This occurs via electron transfer from the propagating macroradical to the metal ion thereby reducing its degree of oxidation. This process is favored by the fairly high standard electrode potentials of copper and iron ions:

$$E^o_{Cu^{2+} \to Cu^{1+}} = 0.15 \text{ V.} \quad E^o_{Fe^{3+} \to Fe^{2+}} = 0.77 \text{ V}$$

Table 4. Distribution of monomeric units in the chain of styrene copolymers (M_1) with metal acrylates

Unit	Content of MAcr$_a$ (mole fraction)	
	Ni(II)	Co(II)
-M_1-	0.25	0.11
-$M_1 M_1$-	0.10	0.13
-$M_1 M_1 M_1$-	0.03	0.11
Remaining M_1 units	0.02	0.06
Total number of M_1 units	0.41	0.41
-M_2-	0.25	0.17
-$M_2 M_2$-	0.15	0.11
-$M_2 M_2 M_2$-	0.06	0.06
Remaining M_2 units	0.13	0.07
Total number of M_2 units	0.59	0.59

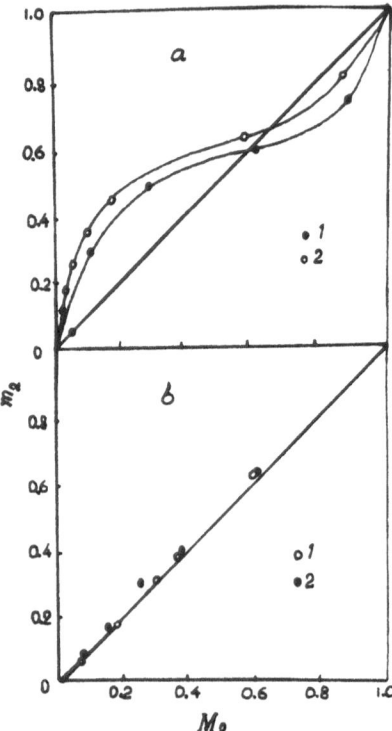

Figure 4. The mole composition curves of copolymers: a - NiAcr$_2$ (1), CoAcr$_2$ (2) with styrene; b - NiAcr$_2$ (1), CuAcr$_2$ (2) with Cp$_2$TiMAcr$_2$.

Copolymerization of metal acrylates. Copolymerization of a MCM with both conventional vinyl monomers and with another MCM permits one to obtain new polymers and also to modify well-known polymers.

The rate of radical initiated metal acrylate copolymerization with styrene depends upon the cation's type and correlates with the reactivity of the acrylate double bond in homopolymerization.[19] Diagrams of the feed versus copolymer compositions for styrene/CoAcr$_2$ or NiAcr$_2$ copolymers indicate a tendency to alternation (Fig.4): r_1 = 0.43-0.55, r_2 = 0.30-0.32 (M$_1$ - styrene, M$_2$ - CoAcr$_2$, NiAcr$_2$). The monomer sequence distribution in the macromolecular chain is important in defining the properties of the copolymer. It is evident from Table 4 that regularly alternating structures are responsible for 46% of the acrylate units in the Ni(II) acrylate - styrene copolymer.

The copolymerization of heterometallic MCMs is of great interest for producing bifunctional metal complex catalysts. Thus, by using dicyclopentadienyltitanium dimethacrylate (Cp$_2$TiMAcr$_2$) and either NiAcr$_2$ or CuAcr$_2$, the following copolymers (Scheme 2, Table 5) were obtained:[20]

$CH_2=C(CH_3)$ $CH_2=CH$ $(-CH_2-C(CH_3)-)_{n}- (-CH_2-CH-)_m$

(chemical reaction scheme: two monomers combining to give copolymer with Ti and M metal centers)

Table 5. Copolymerization parameters for transition metal acrylates (M_1) with $(C_5H_5)_2Ti(OCOC(CH_3)=CH_2)_2$ (M_2).

M_1	r_1	r_2	r_1r_2	Q_1	Q_2	e_1	e_2
NiAcr$_2$	0.95±0.02	0.65±0.03	0.62	0.53	0.56	-2.63	-1.96
CuAcr$_2$	1.09±0.03	0.89±0.04	0.97	0.89	0.56	-2.13	-1.96

It is important that such processes can proceed not only in solution, but in the solid state, under mechanochemical initiation or under a combination of high pressure and shear strain.

The thermal copolymerization of FeAcr$_3$ with CoAcr$_2$ (taken in molar ratios of 1:1 and 2:1) occurs in solid phase at 200-250°C. During thermal transformations of cocrystallyzed monomers various radicals, which can initiate polymerization, appear.[21] The thermal decomposition of resulting products leads to the formation of nano-sized particles stabilized in a polymer matrix. The mechanism and kinetics of metal acrylate thermal transformations are discussed in detail by Pomogailo and coworkers in another Chapter presented in this book.

DISCUSSION AND CONCLUSION

The structural features of metal acrylates and their specific behavior during polymerization affects the structure of resulting metal-containing polymers.

The main types of units varieties in the metallopolymer chain may be assumed as follows:

1. Mixed-unit metal-containing polymers are obtained by copolymerization of metal acrylates containing different metals. It was mentioned above that copolymerization of the heterometallic MCM yields metallopolymers with a variety of metals involved in their structure (Ti(IV) - Ni(II), Ti(IV) - Cu(II), Fe(III) - Co(II) etc.). M and M_1 units are randomly distributed over the chain.

2. The formation of mixed-unit metallopolymers containing different valence states of the same metal may be also expected. For example, Cu^{2+} acrylate polymers contain both Cu^{2+} and Cu^{1+} units:

$$(n+m)CH_2=CH \quad \xrightarrow{R^\cdot} \quad (-CH_2-CH-)_n \quad (-CH_2-CH-)_m$$

The Cu2p$_{3/2}$ and C1s regions of the XPS spectra for CuAcr$_2$ and its polymer are shown in Fig.5. Polymerization shifts the main peak Cu2p$_{3/2}$ by 1 eV to lower binding energies and the relative intensity of the satellite, located on the higher binding energy side of the main peak, is diminished from 0.38 to 0.20 (Fig.5). The effective charge on carbon atoms and relaxation energy may vary as a result of the change in the oxidation state of Cu into which the ligand is attached. Cu^{2+} ions constitute 30-50% of the overall Cu content in the polymerization product. This conclusion results from the relations [Cu^{2+}] ~ μ_{ef}^2 and $I_s/I_m = k[Cu^{2+}]$ (where I_s/I_m is a relative intensity of the satellite).[22] Reaction of the initiator with the monomer cannot account for the high Cu^{1+} content in the polymer (the initiator concentration was not more than 2% to that of the monomer). Probably, the autocatalytic reduction Cu^{2+} → Cu^{1+} occurred analogous to that known for the Rh^{3+} → Rh^{1+} system.[23]

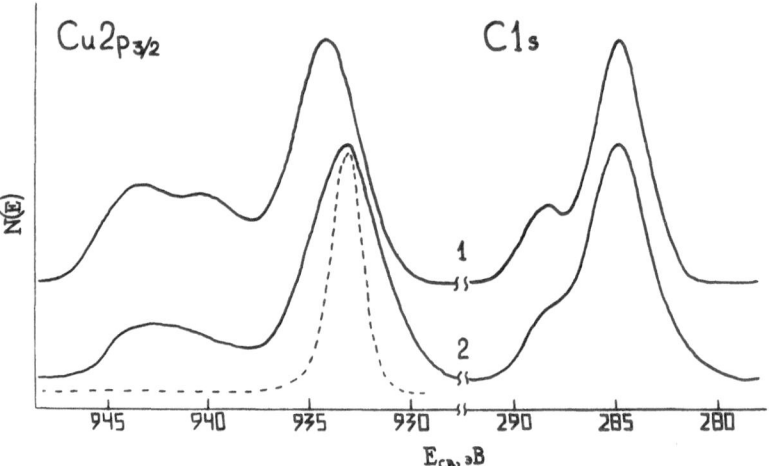

Figure .5. X-ray photoelectron spectra of Cu(II) acrylate (1) and Cu-containing polyacrylate (2). The spectra of Cu$_2$O - dashed line.

The combined data suggests that the polymeric chain consists of dimers of mono- and bivalent copper atoms with a bridged ligand structure (IR data). Their formation can be represented as follows:

$$4n \ CuL_2 \longrightarrow \left[-L'-L'- \begin{matrix} Cu_2L \\ | \\ Cu_2L_3 \end{matrix} - \right] + \ 2n \ L^\bullet$$

$$\left[-L'-L'- \begin{matrix} Cu_2L \\ | \\ Cu_2L_3 \end{matrix} - \right] = \left[\begin{matrix} CH_2=CH \\ Cu \cdots O-C-O \cdots Cu \\ O-C=O \\ -CH-CH_2-CH-CH_2- \\ O-C=O \\ Cu\cdots L\cdots Cu \\ O-C=O \\ CH_2=CH \end{matrix} \right]$$

where L' is a ligand in the polymer and L* is an ion-radical arising as a result of reduction.

3. The formation of mixed-unit metallopolymers might be caused by an increase in metal ion nuclearity. Besides the polynuclear acrylates described above we have obtained heterometallic μ_3 - oxocarboxylates, $[M_3'M_2''O(CH_2=CHCOO)_6]OH$ (M' = Ru(III), M'' = Cr(III), Co(II)). The investigation of their properties are in progress.

It is important whether or not the polynuclear structure of $MAcr_n$ remains unchanged during polymerization. In the case of $CuAcr_2$ the dimeric structure of copper remains unchanged upon polymerization. This is not clear for polynuclear μ_3 -oxoacrylates. Further studies are needed.

4. Coordination of similar (e.g., mono-, bi-, bridged coordinated) carboxylate groups and the formation of chelating units also cause the appearance of mixed-unit chains in metallopolymers.

The polymerization causes some transformations in the coordination sphere because the frequencies v_{as} (COO) for bidentate and bridged groups merge forming a broad band at 1550 cm^{-1} (for $FeAcr_3$). The greatest shift is observed for the vibration bands of carboxylate groups with bidentate coordination, indicating that these ligands are the most transformed during polymerization. In the case of $CrAcr_3$, monodentate units appear in the resulting polymer (1700 cm^{-1}).[5] During polymerization of $CuAcr_2$ the bridged coordination of its carboxyl groups remains unchanged. The electronic spectrum shows that the electronic structure of Cu ions incorporated in the polymer changes substantially. These spectral variations (hypsochromic shift) can be described[24] as a change in the symmetry of their nearest environment for the ions that retain their valence state during polymerization.

The electronic and Mössbauer spectra of $FeAcr_3$ and its polymer indicate that the inner sphere environment of all Fe^{3+} ions undergoes substantial changes upon polymerization. The symmetry of the inner spheres in the polymer is lower than that of the monomer. The value of Δ in the Mössbauer spectrum of the polymer increases versus those of $FeAcr_3$ (Table 6) and d-d transition band is displaced in the electronic spectrum (bathochromic shift).[4]

Sample	S,%	δ	Δ	G
			mm/s	
Monomer	85.5	0.30	0.52	0.37
Polymer	100	0.29	0.78	0.44

5. Mixed-unit chains formation in metal-containing polymers may also be defined by stereoregularity (more often by sindiotactic addition).

The polymerization of bifunctional acrylic monomers consists of two stages. The first stage is characterized by the formation of a linear polymer. Its stereoregularity depends both on the classical factors and the metal ion's nature. This is the stage in which the stereoregular fraction is formed. A spatial net-like structure is formed in the second stage where chain growth in the presence of strong steric factors leads to the formation of an atactic structure.

Direct evidence for internal stresses in a cross-linked metal-containing polymer is provided by the analysis of its far-IR spectra (Fig.6). Comparing the spectra of the metal-containing monomer and polymer reveals significant differences in the region corresponding to the vibrations of the O-M-O fragment. Instead of two sharp bands characteristic of the monomer (300 cm^{-1} and 410 cm^{-1}) a single broad band with a maximum at 340 cm^{-1} was observed for the polymer. This can be explained by a distortion of the geometry of the bridged groups under internal stresses present in the cross-linked structure.

[*] S - the square; δ is isomeric shift; Δ is the quadrupole splitting; G is the line width measured at the half-height.

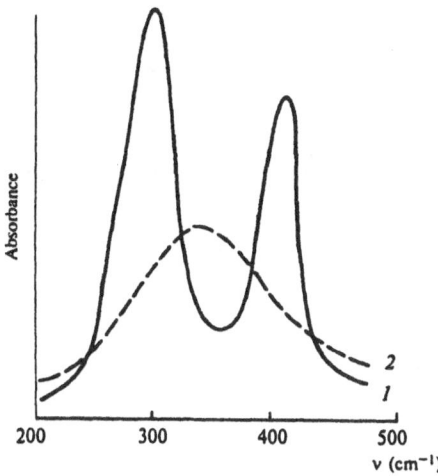

Figure 6. IR spectra of ZnAcr$_2$ (1) and of a polymer network obtained therefrom (2).

The metal acrylate carboxyl groups were shown[17] to promote the inversion of the configuration of the chain-growth center in each propagation step. This results in the formation of a predominantly syndiotactic polymer. Lower temperatures should increase the degree of tacticity in the macromolecules. Samples of polyacrylic acid were fractionated to confirm this hypothesis[25] (Table 7).

Table 7. The stereoregular composition of polyacrylic acid produced by hydrolysis of metal polyacrylates[26]

Polyacrylate	The fraction yield, %		Molecular weight
	atactic	syndiotactic	
Zn* polyacrylate	20	80	238000
Zn** polyacrylate	58	42	73000
Polyacrylic acid**	59	41	21800
Ba* polyacrylate	26	74	211000
Pb* polyacrylate	-	-	197000

* Polymerization at 9 °C; ** Polymerization at 70 °C.

6. The formation of mixed-units in metallopolymers may also occur due to macromolecular branching, chain cross-linking, metal ion participation during interchain coordination, etc. Divinyl monomers are capable of polymerization through one or both multiple bonds, producing either linear or cross-linked polymers:

Metal-containing polymers have a cross-linked structure with up to 50% of the double bonds remaining.[4,5,22] They may have practical importance in the production of network polymers of high strength and elasticity at high softening temperatures due to of vacant unsaturated bonds (inner binding agents).

7. The presence of stable isotopes in the organic portions of the polymer (^{13}C, ^{2}H, ^{18}O, etc.) and of the metal ions (^{54}Fe, ^{56}Fe, ^{234}U, etc.) may also result in the formation of mixed-unit MCP. For this reason metal ions were deliberately enriched with their isotopes (^{57}Fe) in some cases:

8. The formation of MCP with mixed-unit chains can also be provoked by the metal's ligand environment. Thus, the effect depends on such factors as the amount of groups capable of polymerization, keto-enol tautomerism of the ligands, D- and L-optical isomers, and the participation of ligand groups in some side reactions (hydrolysis, etherization, salt formation, residual unsaturation and the like).

Finally complex effects of several different factors may occur during mixed-unit metal-containing polymer formation. Mechanical, chemical and catalytic properties of such polymers are strongly affected by thesebove effects.

ACKNOWLEDGMENTS

We acknowledge the International Science Foundation (NJB000) for the Support of this reseach.

REFERENCES

1. V.V. Korshak. "Mixed-Unit Chains in Polymers," Nauka, Moscow (1977) (in Russian)
2. A.D.Pomogailo, V.S.Savostyanov. "Synthesis and Polymerization of Metal-Containing Monomers," CRC, Boca Raton (1994).
3. G.I.Dzhardimalieva, A.D.Pomogailo, V.I.Ponomarev, L.O.Atovmyan, Yu.M.Shulga, and A.G.Starikov, Izv. Akad. Nauk SSSR, Ser. Khim. 1525 (1988) (Bull. Acad. Sci. USSR, Div. Chem. Sci. 37:1352 (1988) (Engl. Transl.))
4. Yu.M.Shulga, O.S.Roshchupkina, G.I.Dzhardimalieva, I.V.Chernushchevich, A.F.Dodonov, Yu.V.Baldokhin, P.Ya.Kolotyrkin, A.S.Rozenberg, and A.D.Pomogailo, Izv. Akad. Nauk, Ser. Khim. 1739 (1993) (Russ. Chem. Bull., 42:1661 (1993) (Engl. Transl.)
5. Yu.M.Shulga, I.V.Chernushchevich, G.I.Dzhardimalieva, O.S.Roshchupkina, A.F.Dodonov, and A.D.Pomogailo, Izv. Akad. Nauk, Ser. Khim. 1047 (1994)
6. S.Besecke, G.Schroeder, W.Ude, and E.Baumgartner. German Patent 3, 224, 927 (1984); Chem. Abst. 100:104028q (1984)
7. Z.Wojtczak and A.Gronowski, Polimery 27:471 (1984).
8. M.A Poraj-Koshits, in: "Crystallochemistry. Advances in Science and Technology, VINITI, Moscow 15:3 (1981).
9. J.Zurahowska-Orszagh, J.Skupinska, and F.Suchan, Polimery 30:185 (1985).
10. K.Nakamoto. "Infrared and Raman Spectra of Inorganic and Coordination Conpounds," Wiley, New York (1986).
11. K. Nakanishi. "Infrared Absorption Spectroscopy," Holden-Day, Inc., San Francisco (1962).
12. V.I.Ponomarev, L.O.Atovmyan, G.I.Dzhardimalieva, A.D.Pomogailo, and I.N.Ivleva, Koord. Khimiya 14:1537 (1988) (Sov. J. Coord. Chem., 14 (1988) (Engl. Transl.))
13. B.N.Figgis, and G.B.Robertson, Nature 205:694 (1965).
14. G.J.Long, W.N.Robinson, W.P.Tappmeyer, and D.L.Bridges, J.Chem.Soc., Dalton Trans. 573 (1973).
15. R.C.Mehrotra, and R.Bohra. "Metal Carboxylates," Academic Press, London New York (1983).
16. A.S. Batsanov, Yu.T.Struchkov, N.V.Gerbeleu, G.A.Timko, and O.S.Manole, Koord. Khim. 20:833 (1994).
17. G.I.Dzhardimalieva, A.D.Pomogailo, S.P.Davtyan, and V.I.Ponomarev, Izv. Akad. Nauk SSSR, Ser. Khim., 1531 (1988) (Bull. Acad. Sci. USSR, Div. Chem. Sci., 37:1352 (1988).
18. Kh.S.Bagdasar'yan. "Theory of Radical Polymerization," Nauka, Moscow (1966).
19. G.I.Dzhardimalieva, and A.D.Pomogailo, Izv. Akad. Nauk SSSR, Ser. Khim., 352 (1991).
20. G.I.Dzhardimalieva, V.A.Zhorin, A.D.Pomogailo, and N.S.Enikolopyan, Dokl. Akad. Nauk SSSR, 287:654 (1986).
21. A.S.Rozenberg, G.I.Dzhardimalieva, and A.D.Pomogailo, Izv. Akad. Nauk, Ser. Khim. (1995) (in press).
22. Yu.M.Shul'ga, O.S.Roschupkina, G.I.Dzhardimalieva, and A.D.Pomogailo, Russ. Chem. Bull. 42:1498 (1993) (English Transl.).
23. A.D.Pomogailo. "Immobilized Polymeric Metallo-Complex Catalysts," Nauka, Moscow (1988) (in Russian).

24. F.Cotton, and G.Wilkinson. "Advanced Inorganic Chemistry," John Wiley and Sons, New York (1967).

25. A.Shapiro, D.Goedfeeld-Freilich, and J.Perichon, Europ. Polym. J., 11:515 (1975).

26. B.S.Selenova, G.I.Dzhardimalieva, M.V.Tsikalova, S.V.Kurmaz, V.P.Roschupkin, I.Ya.Levitin, A.D.Pomogailo, and M.E.Vol'pin, Russ. Chem. Bull., 42:453 (1993).

METAL COORDINATION POLYMERS: EIGHT-COORDINATE CERIUM(IV) AND ZIRCONIUM(IV) POLYMERS WITH VARIED FLEXIBILITY, CONJUGATION, AND STABILITY THROUGH LIGAND VARIATION

Ronald D. Archer, Huiyong Chen, Jon A. Cronin & Sharon M. Palmer[†]

Departments of Chemistry
University of Massachusetts and Smith College[†]
Amherst, MA 01003 and Northampton, MA 01060

INTRODUCTION

Tractable (soluble) linear transition metal coordination polymers in which the metal ions are a necessary part of the backbone (and without which the polymer will stay intact) have proven difficult to synthesize.[1-6] Bis(tetradentate) Schiff-base ligands coupled with zirconium(IV) ions large enough for eight-coordination can be used to overcome this insolubility,[7-8] and recently an analogous soluble cerium(IV) Schiff-base polymer has been prepared.[9] Although a number of reports had appeared regarding polymeric systems in which lanthanide metal ions are attached to a branch of a polymer chain[10-14] or solid state species that either remain insoluble in all solvents or fall apart into small molecules when dissolved,[15-17] soluble linear polymers with lanthanide ions in their backbone had been lacking prior to our study of *catena*-poly[cerium(IV)-μ-N,N',N",N"'-tetrasalicylidene(3,3'-diaminobenzidinato)-O,-N,N',O':O",N",N"',O"'], [Ce(tsdb)]$_n$.[9]

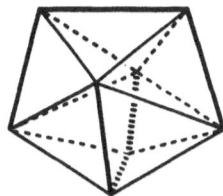

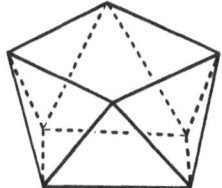

D$_{2d}$ Dodecahedron   D$_{4d}$ Square Antiprism

Figure 1. Ligand dispositions expected for eight-coordinate metal centers.

Whereas [Ce(tsdb)]$_n$ is stoichiometrically similar to the analogous zirconium species, [Zr(tsdb)]$_n$, that we reported previously,[8] the cerium polymer is thought to have a different stereochemical configuration. That is, the zirconium centers are thought to be eight-coordinate trigonal-faced dodecahedra (Figure 1) based on the analogous Zr(dsp)$_2$ structure,[18] where dsp^{2-} = the anion of N,N'-disalicylidene-1,2-phenylenediamine and the cerium species are thought to be eight-coordinate Archimedian or square antiprisms based on the analogous Ce(dsp)$_2$ structure.[19] Although 3,3'-diaminobenzidine (i.e., 3,4,3',4'-tetraaminodiphenyl) is available commercially, the two new ligands that are reported herein for the first time have been synthesized as shown in Figures 2 and 3. Whereas our earlier studies[7,8] had involved the step-growth condensation of the Schiff-base ligands during the polymerization reaction (cSB method), we have found that by preheating the ligands in dimethyl sulfoxide (DMSO) at 100 °C, sufficient ligand dissolves to allow chelate polymerization by the displacement of an alkoxide (dOR method), a β-diketone (dβ method) or salicylaldehyde (dsal method) by the ligand on a compound or complex of the appropriate metal as shown schematically in Figure 4. Note that the cSB method is effectively a reaction between tetrafunctional monomers since the phenolic oxygens of the salicylaldehydes stay intact, whereas the chelate polymerizations are essentially octafunctional monomer reactions. In either case, in order to obtain polymers, rather than oligomers, such step growth polymerizations require that the reactions go essentially to completion (p = extent of reaction ≥ 0.99) and the stoichiometric ratio must also be almost exactly 1:1; i.e., the degree of polymerization DP = $(N_M + N_L)/(N_M + N_L - 2p)$ where N_M and N_L are the moles of the two reactants.

RESULTS

The bridging ligand syntheses were thoroughly investigated. The 3,4,3',4'-tetraaminodiphenylmethane (TAPM) molecule was synthesized with a 49 % yield overall from 4,4'-diaminodiphenylmethane, and the N,N',N'',N'''-tetra-salicylidene-3,4,3'4'-tetraaminodiphenylmethane (H$_4$tstm) bridging ligand was synthesized from TAPM with an 87 % yield. [The nitration reaction must be kept at or below 15 °C in order to avoid polynitration (more than one nitro group per ring with concentrated nitric acid, which must be used to obtain even single nitration on the protected ring), and the Schiff-base ligand preparation must be conducted in neat salicylaldehyde in order to obtain a high yield of the tetrasalicylidene product. Species with less than four salicylidene groups precipitate in a variety of organic solvents.][20]

Ligand Structures

1 tsdb^{4-} Q = ——
2 tstm^{4-} Q = CH$_2$
3 tsts^{4-} Q = SO$_2$

4 dsp^{2-} R = H
5 bsp^{2-} R = t-Bu

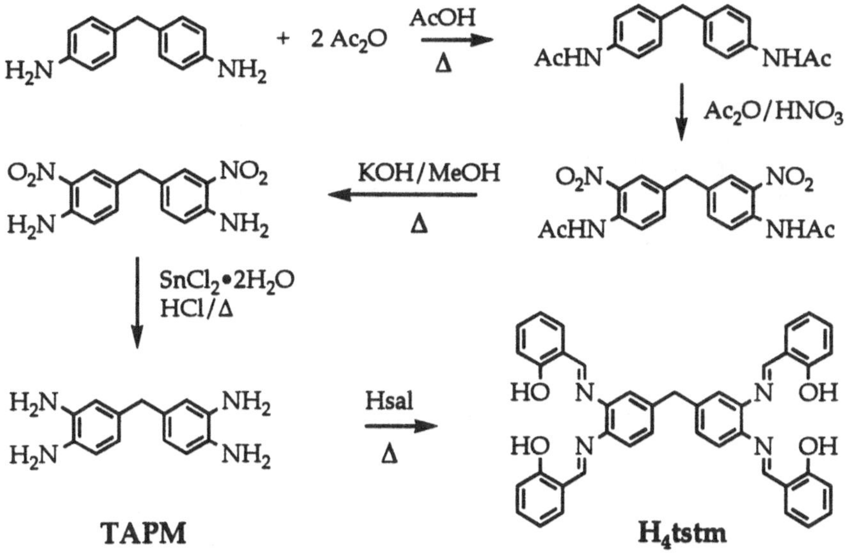

Figure 2. The synthesis of 3,4,3',4'-tetraaminodiphenylmethane (TAPM) and N,N',N'',N'''-tetrasalicylidene-3,4,3'4'-tetraaminodiphenylmethane (H₄tstm).

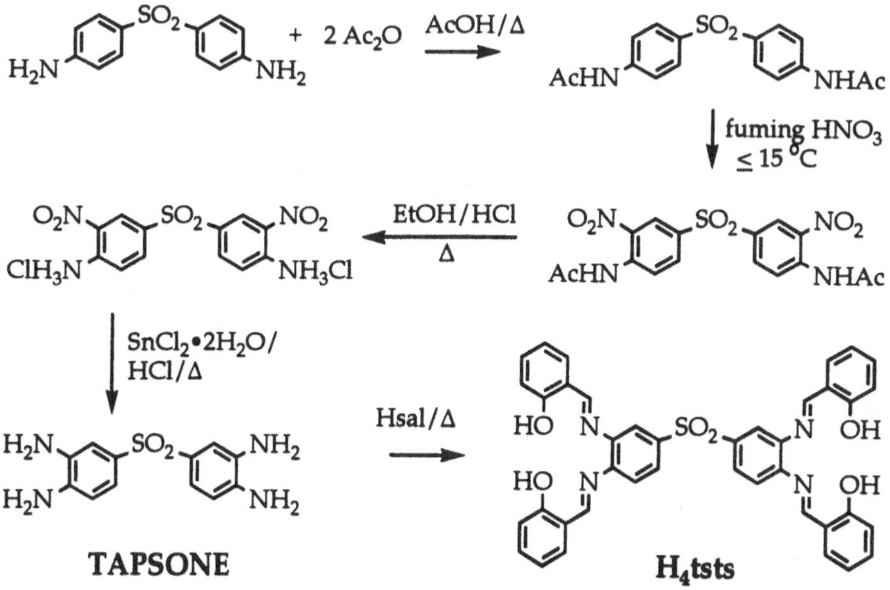

Figure 3. The synthesis of 3,4,3',4'-tetraaminodiphenylsulfone (TAPSONE) and N,N',N'',N'''-tetrasalicylidene-3,4,3'4'-tetraaminodiphenylsulfone (H₄tsts).

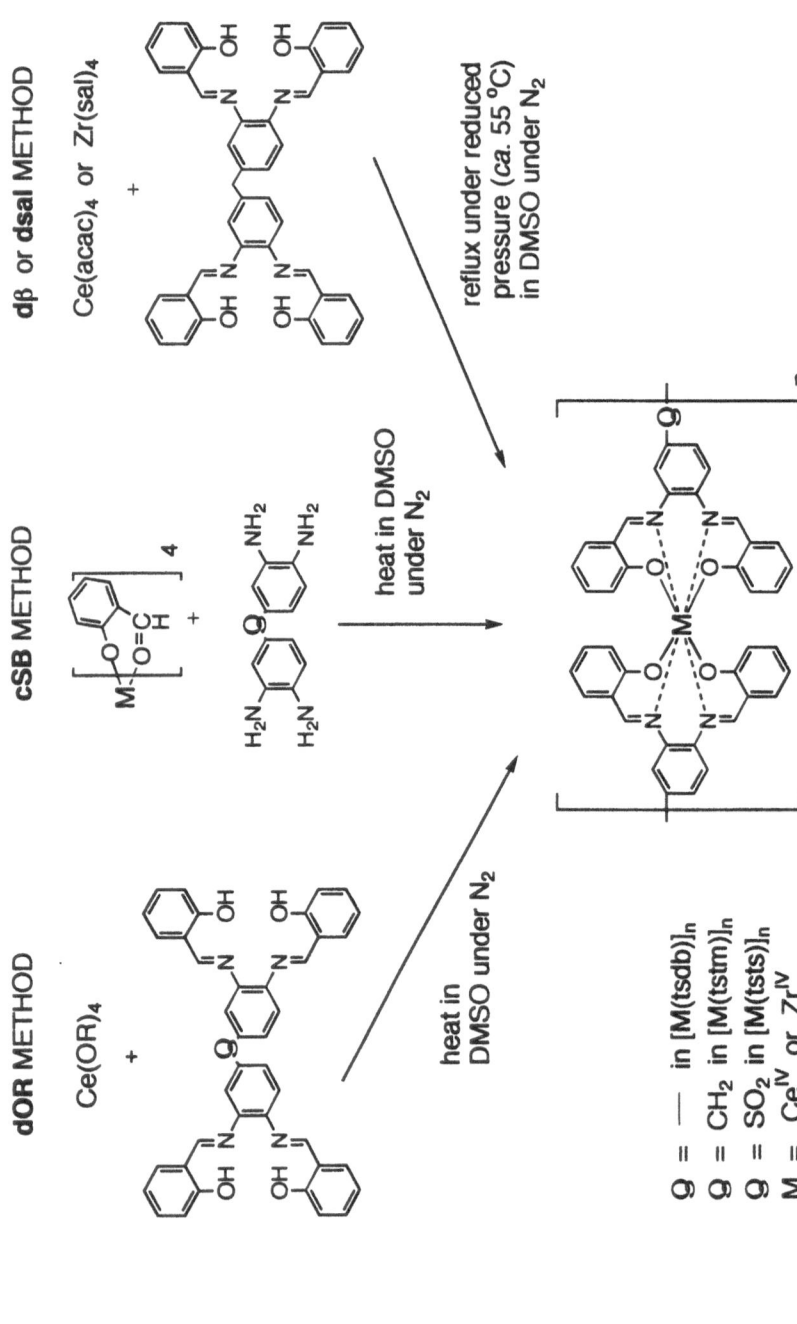

Figure 4. Methods for the successful synthesis of 8-coordinate polymers of Ce^{IV} and Zr^{IV}. **dOR**, **dβ**, and **dsal** are displacement reactions of alkoxides (OR^-), β-diketones and salicylaldehydes, respectively; and **cSB** is a condensation reaction of a Schiff-Base.

The 3,4,3',4'-tetraaminodiphenylsulfone (TAPSONE) molecule was synthesized with a 59 % yield overall from the corresponding diamine, and the N,-N',N",N"'-tetrasalicylidene-3,4,3'4'-tetraaminodiphenylsulfone (H4tsts) bridging ligand was synthesized from TAPSONE with an 85 % yield.

Two new linear cerium(IV) Schiff-base coordination polymers, *catena*-poly[cerium(IV)-μ-N,N',N",N"'-tetrasalicylidene-3,3',4,4'-tetraaminobiphenyl-methanato-O,N,N',O',O",N",N"',O"'], [Ce(tstm)]ₙ, and *catena*-poly[cerium(IV)-μ-N,N',N",N"'-tetrasalicylidene-3,3',4,4'-tetraaminodiphenylsulfonato-O,N,-N',O',O",N",N"',O"'], [Ce(tsts)]ₙ, have been successfully synthesized using step growth chelation polymerizations between either cerium(IV) octoxide (dOR method) or tetrakis(acetylacetonato)cerium(IV) (dβ method) and the new, premade Schiff-base bridging ligands, H4tstm and H4tsts. [The successful preparation of the cerium(IV) octoxide from ceric ammonium nitrate as shown in eq. 1 to 3, requires pure sodium methoxide, e.g., freshly cut Na metal dissolved in dry methanol. The methanol produced by reaction of the cerium methoxide intermediate with n-octanol (eq. 2) must be completely removed (overnight distillation) to drive the next reaction (eq. 3) to completion. Pure product from this stoichiometrically exact synthesis can only be obtained after all the methoxide is replaced by octoxide. Otherwise, the residual n-octanol, which has a high boiling point, is difficult to separate from the product, which is an orange oil. The complete removal of toluene from the product can be accomplished by vacuum distillation at about 40 °C. and is confirmed by ^{1}H-NMR.][20]

$$[Ce(NO_3)_6]^{2-} + 4\,CH_3O^- \rightarrow Ce(OCH_3)_4 + 6\,NO_3^- \tag{1}$$

$$2\,NH_4^+ + 2\,CH_3O^- \rightarrow 2\,CH_3OH + 2\,NH_3\uparrow \tag{2}$$

$$Ce(OCH_3)_4 + 4\,C_8H_{17}OH \rightarrow Ce(OC_8H_{17})_4 + 4\,CH_3OH \tag{3}$$

Two new linear zirconium(IV) Schiff-base coordination polymers, *catena*-poly[zirconium(IV)-μ-N,N',N",N"'-tetrasalicylidene-3,3',4,4'tetraaminobiphen-ylmethanato-O,N,N',O',O",N",N"',O"'], [Zr(tstm)]ₙ, and *catena*-poly[zirconium(IV)-μ-N,N',N",N"'-tetrasalicylidene-3,3',4,4'-tetraaminodiphenylsulfonato-O,N,N',O',O",N",N"',O"'], [Zr(tsts)]ₙ, have also been successfully synthesized. Whereas step growth chelation polymerizations between zirconium(IV) salicylaldehyde and H4tstm provides a reasonable polymer with a number-average molecular weight, $\overline{M}_n$, about 2×10^4, the analogous sulfone Schiff-base H4tsts undergoes decomposition under the reaction conditions (65 °C and 12 hours in DMSO); therefore, the condensation reaction (cSB method) between TAPSONE and zirconium(IV) salicylaldehyde was necesssary. The results obtained for each polymer using the polymerization reactions depicted in Figure 4 are shown in Table 1. The method listed for each polymer is the method that has provided the highest $\overline{M}_n$ value for each of the polymers. Note that the the T_g values for the cerium polymers are appreciably higher than those of the zirconium polymers. Also, the methylene bridged tstm^{4-} polymers show enhanced solubility over the sulfone bridged tsts^{4-}polymers and the directly linked (diphenyl) tsdb^{4-} polymers in DMSO and NMP. Furthermore, the tstm^{4-} polymers with moderate molecular weights are somewhat soluble in methylene chloride and tetrahydrofuran, whereas the others are all insoluble in low-polarity solvents .

Viscosity measurements in NMP on fractionated cerium polymers (Figure 5) provide Mark-Houwink constants for these new bridging groups relative to the the tsdb^{4-} polymer reported earlier.[9] The plots for [Ce(tstm)]ₙ give $a = 0.79$

Table 1. Characterization of Eight-Coordinate Schiff-Base Polymers

Polymer	(Method)[1]	Intrinsic Viscosity[2]	GPC $\overline{M_n}$[3]	^{1}H-NMR $\overline{M_n}$[4]	T_g(K)[5]	T_{dec}(K)[6]
[Ce(tstm)]$_n$	(dOR)	17	23,500	28,000	430	613
[Ce(tsts)]$_n$	(dOR)	19	26,500	31,500	454	623
[Ce(tsdb)]$_n$[7]	(dβ)	17	28,000	30,000	457	673
[Zr(tstm)]$_n$	(dsal)	11	15,000	26,500	--[8]	793; 718[9]
[Zr(tsts)]$_n$	(cSB)		13,400	17,000	401	714[9]
[Zr(tsdb)]$_n$[10]	(cSB)	20	30,000	33,000	347	773
[Zr(tsb)]$_n$[11]	(cSB)	18	49,000		360[10]	

[1]See Figure 4 for synthetic scheme. [2]Intrinsic viscosity in cm^3/g in NMP at 30 °C. [3]Polystyrene equivalent number-average molecular weights in NMP at room temperature. [4]$\overline{M_n}$ based on relative intensities of the bridging ligand protons to the end-capping reagent protons and assuming an average of one bsp^{2-} (or dsdt^{2-}) ligand per chain. [5]Glass transition temperature. [6]Decomposition temperature under N$_2$, except as noted otherwise. [7]Reference 9. [8]Not observed, even when the initial temperature was lowered below 300 K. [9]In air. [10]Reference 8. [11]Reference 7.

and $K = 5.84 \times 10^{-3}$ cm^3/g for the Mark-Houwink equation ($\eta_{int} = K\overline{M}^a$) with a linear regression with R^2 = 0.995. For [Ce(tsts)]$_n$, $a = 0.70$ and $K = 1.25 \times 10^{-2}$ cm^3/g with a linear regression with R^2 = 0.987. [Ce(tsdb)]$_n$ has $a = 0.76$ and $K = 7.36 \times 10^{-3}$ cm^3/g values in NMP.[9] The a value decreases from [Ce(tstm)]$_n$ to [Ce(tsdb)]$_n$ to [Ce(tsts)]$_n$. The high a value of [Ce(tstm)]$_n$ in NMP is consistent with NMP being a good solvent for this polymer.

The electron conductivity of the polymers has also been investigated. The polymers appear to be semiconductors with a sizable increase in conductivity possible through exposure to iodine vapor as shown in Figure 6 and Table 2. The native polymers have conductivities of the order of 10^{-7} S/cm. The iodine doped polymers have conductivities about four orders of magnitude greater than the undoped polymers.

The spectral properties of these polymers are very similar to those that we have reported previously. For example, the phenyl-oxygen stretching frequency (that is at 1278 cm^{-1} in each of the ligands) is shifted to above 1300 cm^{-1}

Table 2. Conductivity of cerium and zirconium coordination polymer films

Cerium Polymer	Conductivity (S/cm)		Zirconium Polymer	Conductivity (S/cm)	
	Native[1]	I$_2$ Doped[2]		Native[1]	I$_2$ Doped[2]
[Ce(tstm)]$_n$	4 x 10^{-7}	1 x 10^{-3}	[Zr(tstm)]$_n$	2 x 10^{-7}	5 x 10^{-3}
[Ce(tsts)]$_n$	1 x 10^{-7}	2 x 10^{-3}	[Zr(tsts)]$_n$	4 x 10^{-7}	3 x 10^{-3}
[Ce(tsdb)]$_n$	4 x 10^{-7}	1 x 10^{-3}	[Zr(tsdb)]$_n$	5 x 10^{-7}	5 x 10^{-4}

[1]Polymer films measured at room temperature (298 K). [2]Doped with iodine vapor in closed container.

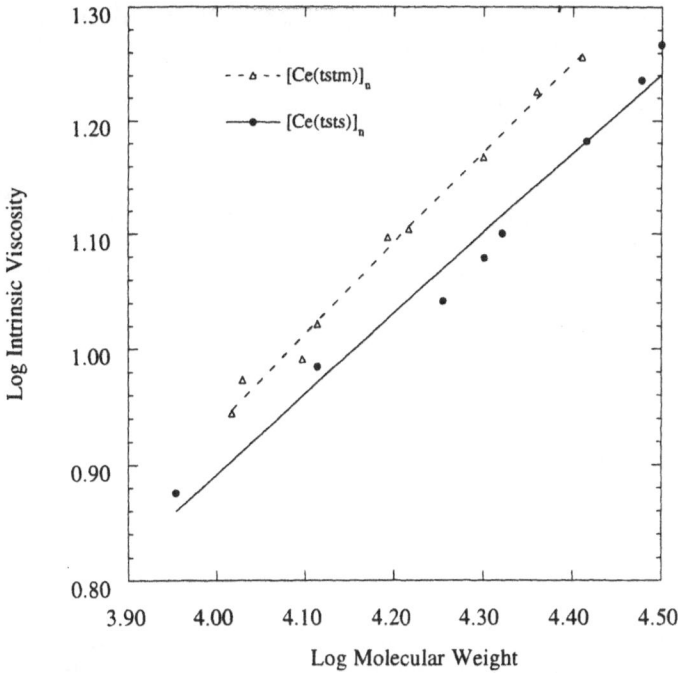

Figure 5. A composite Mark-Houwink plot for the $[Ce(tstm)]_n$ and $[Ce(tsts)]_n$ polymers.

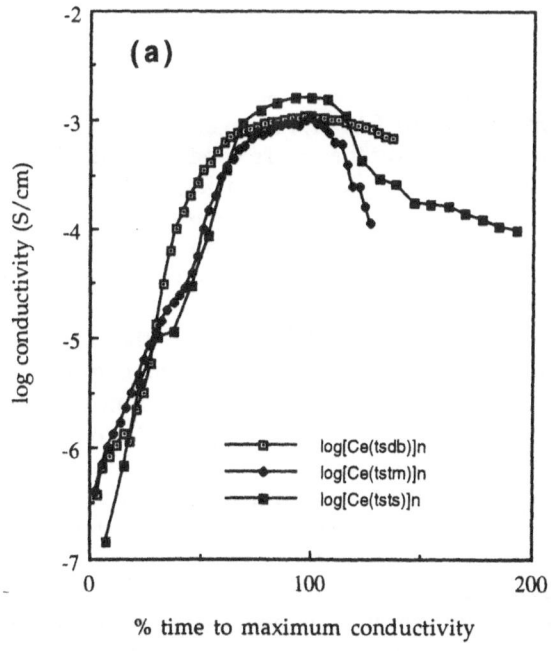

Figure 6. The conductivity of (a) cerium polymer films and (b) zirconium polymer films during iodine vapor doping experiments.

in all of the polymers. The methylene protons of the tstm^{4-} bridging ligand are found at 4.1 to 4.2 ppm downfield from tetramethylsilane, and the aromatic protons range from about 5.6 - 8.2 ppm and the aldimine protons just below that. To determine the molecular weights, the polymers were end-capped with the sodium salt of N,N'-bis(5-*tert*-butyl-2-hydoxybenzylidene)-1,2-diaminoben-zene (Na$_2$bsp), which displaces the two salicylaldehydato ligands from zirconium and cerium or the two acetylacetonato or alkoxide ligands from cerium as appropriate. The bsp^{2-} ligand, with its 18 *tert*-butyl protons, provides good sensitivity and accuracy for molecular weight determination by ^{1}H-NMR analysis. The polymer chains should have on average one end of this type. Schematically:

$$M(L-L)_{n-1}ML'_2 + Na_2bsp \rightleftarrows M(L-L)_{n-1}M(bsp) + 2\ NaL' \qquad (4)$$

DISCUSSION

Soluble, thermally stable, linear coordination polymers can be prepared between both cerium(IV) and zirconium(IV) with a variety of bis-tetradentate Schiff-base ligands. The preferred synthetic method is both metal and ligand specific.

The thermal stability is more metal than ligand specific. The zirconium polymers are more thermally stable in air than the cerium polymers are under nitrogen. Even so, the cerium polymers are stable to 340 - 400 °C under nitrogen and the zirconium polymers are stable to at least 500 °C under these conditions. The parent ligands are even less thermally stable; thus, the lower stability of the cerium polymers is probably related to the larger size of the cerium(IV) ion (111 pm)[21] relative to the zirconium(IV) ion (98 pm). Another factor in the lower thermal stability of the cerium(IV) polymers may be the accessibility of the cerium(III) state, which can help provide a redox decomposition path that is less likely for zirconium(IV). Redox involving the cerium(IV) was noted during the condensation reaction (cSB method) between Ce(sal)$_4$ and TAPM in particular. In fact, some crosslinked material always occurs during attempted synthesis by this method with TAPM; however, TAPSONE is less vulnerable to this problem, even when the syntheses are conducted at 65 °C in DMSO.

On the other hand, the glass transition temperatures of the cerium polymers are higher than those of the zirconium polymers. As noted in the introduction, the zirconium centers are thought to be eight-coordinate trigonal-faced dodecahedra, whereas the cerium species are thought to be eight-coordinate Archimedian or square antiprisms. The latter apparently expose more of the phenyl rings to interchain attractions that contribute to the sizable increases in the glass transition temperatures of the cerium polymers.

Separating the phenyl rings of our more classical tsdb^{4-} ligand with a methylene group enhances both the ligand and the polymer solubilities and lowers the glass transition temperatures of the polymers. The higher Mark-Houwink *a* value observed for the cerium tstm^{4-} polymer is apparently related to the lower polymer-polymer interactions in this system relative to the other two cerium polymers.

The conductivities of the polymers are typical for unsaturated systems and can be enhanced by iodine doping. However, the bulk conductivities of either the native or iodine doped films are not a function of the conjugation of the individual polymer molecules; thus, interchain resistance predominates, and our ability to provide tractable (soluble) polymers because of their lower polymer-polymer interchain attractions is a hindrance in conductivity studies. This is similar to the observations of the d^8 phosphine complexes reported recently by Wang and Fox,[22] where the large phosphines do not enhance conductivity either. Similar trends have also been noted by Hanack and his coworkers in their extensive studies with phthalocyanine and other macro-cycles.[23]

Furthermore, the successful synthesis of these tractable coordination polymers of both cerium and zirconium has led us to investigate the synthesis of Schiff-base polyelectrolytes of trivalent yttrium and several lanthanides.[24] A series of linear coordination polyelectrolytes formed by lanthanide(III) ions ($Ln^{3+} = La^{3+}$, Gd^{3+}, Y^{3+} or Yb^{3+}) with the anion of H_4tsdb have been synthesized with the resultant formula $[MLn(tsdb)]_n$ (where M^+ is an alkali metal ion). Some of the polyelectrolytes can dissolve in polar organic solvents with high dielectric constants, and the polymers exhibit the typical properties of polyelectrolytes and are very thermally stable ($\geq$ 773 K). The number-average molecular weights of $[NaY(tsdb)]_n$ have been estimated by NMR end-group analysis to be as high as 18,500. We are now in the process of completing similar studies with luminescent polyelectrolytes of europium(III).

Finally, we must note that the non-end-capped zirconium polymers adhere strongly to both silica and aluminum (actually to the alumina on the aluminum).[25] Thus, the basic research studies discussed herein are leading to a wide variety of coordination polymers, some of which may possess useful properties.

ACKNOWLEDGMENTS

The authors wish to acknowledge the financial support of the Petroleum Research Fund administered by the American Chemical Society.

REFERENCES

1. J.E. Sheats, C.E. Carraher, Jr., M. Zeldin, B. Currell, and C.U. Pittman, Jr. (Eds.), *Inorganic and Metal-Containing Polymeric Materials*, Plenum Publishing Co., New York, 1991, especially Chapter 1.
2. C.U. Pittman, Jr., C.E. Carraher, Jr. and J.R. Reynolds, *in*: "Encyclopedia of Polymer Science and Engineering," J.I. Kroschwitz, ed. 2nd edn., Vol. 10, Wiley, New York (1987) pp. 541-94.
3. B.M. Foxman and S.W. Gersten, *ibid.* Vol. 4, 1985, pp. 175-91.
4. N. Hagihar, K. Sonogashira and S. Takahashi, *Adv. Polym. Sci.*, 41: 149 (1981).
5. J. E. Sheats, in M. Grayson and D. Eckroth (Eds.), *Kirk-Othmer Encycl. Chem. Technol.* 3rd edn., Vol. 15, Wiley, New York, 1981, pp. 184-220.
6. R. D. Archer, *Coord. Chem. Revs.*, 128: 49 (1993).
7. R.D. Archer, M.L. Illingsworth, D.N. Rau, & C.J. Hardiman, A soluble linear Schiff-base coordination polymer containing eight-coordinate zirconium(IV), *Macromolecules*, 18: 1371 (1985).

8. R.D. Archer and B. Wang, Synthesis and characterization of the thermally stable copolymer of tetrakis(salicylaldehydato-O,O')zirconium(IV) and 3,3'-diaminobenzidine, *Inorg. Chem.*, 29: 39 (1990).

9. H. Chen, J.A. Cronin, and R.D. Archer, The synthesis and characterization of linear cerium(IV) Schiff-base coordination polymers, *Macromolecules*, 27: 2174 (1994).

10. Y. Yokamoto, H. Lujan, and M.D. Cho, Lanthanide ion containing polymers: Investigation of the ion binding properties of tacticity poly(methacrylic acid) and degradation of polymers by γ-irradiation using Tb^{3+} ion as a fluorescence probe, *Polym. Mater. Sci. Eng.*, 71: 723 (1994).

11. Y. Okamoto, Synthesis, characterization of polymers containing lanthanide metals, *J. Macromol. Sci.-Chem. A* 24: 455 (1987).

12. H. Nishide, T. Izusshi, N. Yoshioka, and E. Tsuchida, Complexation of europium ions with poly(methacrylic acid)s and fluorescent properties of the complexes, *Polym. Bull.* 14, 387 (1985).

13. Y. Okamoto, Y. Ueba, I. Nagata, E. Banks, B.A. Garetz, and J.M. Khosrofian, Rare earth metal containing polymers: energy transfer from uranyl to europium ions in ionomers, *Contemp. Top. Polym. Sci.* 4: 387 (1984).

14. Y. Ueba, K.J. Zhu, E. Banks, and Y. Okamoto, Rare earth metal containing polymers. 5. Synthesis, characterization and fluorescence properties of Eu^{3+} polymer complexes, *J.Polym. Sci.: Polym. Chem. Ed.* 20: 1278 (1982) *and earlier refs.*

15. D.K. Dwivedi, B.K. Shukla, and R.K. Shukla, Coordination polymers of La(III) acetate with terephthalaldehyde bis isonicotinic acid hydrazide, *Oriental J. Chem.*, 7: 99 (1991).

16. D.K. Dwivedi, B.K. Shukla, and R.K. Shukla, R. K., Coordination polymers of La(III) acetate with isatin-oxalyldihydrazone, *Acta Ciencia Indica,*, 17C: 383 (1991).

17. E.R. Birnbaum, Carboxylates, *in: Gmelin Handbook of Inorganic Chemistry, 8th Edn, Sc, Y, La-lu, Part D5"*; T. Moeller, ed., Gmelin, New York (1984), especially pp 112ff.

18. R.D. Archer, R.O. Day, and M.L. Illingsworth, Transition-metal eight-coordination. 13. Synthesis, Characterization, and crystal and molecular structure of the schiff-base chelate; Bis(N,N'-disalicylidene-1,2-phenylenediamino)zirconium(IV) benzene solvate, *Inorg. Chem.*, 18: 2908 (1979).

19. A. Terzis, D. Mentzafos, H.A. Tajmir-Riahi, Eight-coordination. synthesis and structure of the Schiff-base chelate bis(N,N'-disalicylidene-1,2-phenylenediamino)cerium(IV), *Inorg. Chim. Acta*, 84: 187 (1984).

20. H. Chen, J.A. Cronin, and R.D. Archer, The synthesis and characterization of two new Schiff-bases and their soluble linear cerium(IV) polymers, submitted for publication.

21. R.D. Shannon, *Acta Crystallogr.* A32: 751 (1976).

22. P.-W. Wang & M.A. Fox, Metal-metal interactions in tetrakis(diphenylphosphino)benzene bridged dimetallic complexes and their related coordination polymers, *Inorg. Chem.*, 33: 2938 (1994).

23. M. Hanack & J. Pohmer, Bridged macrocyclic transition metal oligomers synthesis and electrical properties, *Polym. Mater. Sci. Eng.*, 71: 391 (1994).

24. H. Chen and R.D. Archer, The synthesis and characterization of N,N',N",-N'"-tetrasalicylidene-3,3'-diaminobenzidine Schiff-base coordination polyelectrolytes of yttrium(III), lanthanum(III), gadolinium(III), and ytterbium(III), *Macromolecules*, in press.
25. R.D. Archer and B. Wang, The adhesion of the tetrakis(salicylidene)diaminobenzidinezirconium coordination polymer to silica and alumina, *Chem. Mater.*, 5: 317 (1993).

STRUCTURAL CHARACTERIZATION AND EFFECTS OF GIBBERELLIC ACID-CONTAINING ORGANOMETALLIC POLYMERS AS PLANT GROWTH REGULATORS

Herbert H. Stewart[a], Charles E. Carraher, Jr.[b,] Winn J. Soldani[c], Lisa Reckleben[b], Jose de la Torre[a], and Shi Li Miao[d].

Departments of Biological Sciences[a]
and Chemistry[b]
Florida Atlantic University
Boca Raton, Florida   33431

Fancy Hibiscus[c]
1142 S.W. First Avenue
Pompano Beach, Florida   33060

South Florida Water Management District[d]
West Palm Beach, Florida

INTRODUCTION

Gibberellins are cyclic diterpenes with the ability to induce a number of plant responses including cell elongation and cell division. Widespread in nature, the major commercially employed gibberellin is GA_3. GA_3 was condensed with bis (cyclopentadienyl) titanium IV dichloride and analyzed for structural characterization. Different concentrations of both GA_3 and GA_3 polymer were used with varieties of common bean (Phaseolus vulgaris L.) and Hibiscus rosa sinensis var. Albo Lacinatus, a commonly utilized rooting material employed in the commercial production of grafted hybrid hibiscus. Treatment with both GA_3 and the GA polymer showed great increase in height as compared with the untreated control. With respect to flowering, in general, there was an inverse relationship between GA concentrations and flower formation. With the hibiscus root stock, while roots per se were inhibited, the formation of adventitious roots was accelerated.

In the early 1900's Asian farmers found that rice (Oryza sativa L.) plants would grow tall and pale (chlorotic) and fail to produce seeds. Frequently the stem would be so fragile that the whole plant would fall over. Rice, a member of the grass (Graminaceae) family is a major food crop. Understandably, the Asian farmers were concerned. In Japan, they called the condition "foolish seedling" disease.

Plant pathologists investigating the disease linked it to an Ascomycete fungus. The perfect (sexual) stage of the fungus is called Gibberella fujikuroi and the imperfect (asexual) stage is Fusarium moniliforme. In the 1930's Japanese scientists isolated a compound with growth-promoting qualities which they named gibberellin A. Not until the mid-1950's did English and American research groups purify a substance from fungal culture filtrates which caused plant stem elongation. They named it gibberellic acid. Subsequently, in Japan three gibberellins were isolated from the original gibberellin A and named: gibberellin A_1, gibberellin A_2, and gibberellin A_3. Gibberellin A_3 and gibberellic acid proved to be identical.

The fascinating story of the discovery of the GAs as plant growth regulators (PGRs) is told by Stodola (1) in the Source Book on Gibberellins 1928-1957. For a discussion of terminology including plant hormone, plant growth substance, and plant growth regulator, see Stewart, et al (2). More recently, Phinney, 1983 (3) has provided a personal account of the history leading up to the discovery of GAs in higher plants.

One other fungus has been found to produce GAs. It is the pathogenic fungus Sphaceloma manihoticola which belongs to the Ascomycete family and produces what is called "super-elongation disease" of cassava, according to Rademacher and Graebe (4).

With the availability of gibberellic acid, testing began on a wide variety of plants. Exogeneous applications resulted in spectacular elongation of dwarf and rosette plants, especially with genetically dwarf peas (Pisum sativum) and dwarf maize (Zea mays). Mendel's tall/dwarf factors (or alleles) in peas are, in fact, genes that control gibberellin metabolism.

If exogenous applications of gibberellic acid could increase the height of dwarf plants, it seemed logical to wonder whether plants contain their own gibberellins. Subsequent research has revealed 84 different naturally occurring gibberellins, (Graebe 5), (Sponsel 6), and (Takahashi, et al 7). Seventy three have been found in plants, 25 in the Gibberella fungus, and 14 in both. Seeds of the common bean (Phaseolus vulgaris) contain at least 16 different gibberellins. According to Bottini (8) and Atzorn (9) gibberellins have been found in anthophytes, gymnosperms, ferns, mosses, algae, fungi and two species of bacteria.

As more and more gibberellins were discovered, first from fungal sources, and later from plants, a system (gibberellin A_x or GA_x) was adopted in which the gibberellins were numbered in order of their discovery. The numbers are strictly for convenience in classification and in no way indicate chemical relationships or biological activity.

While they are widespread in nature, most GAs are not biologically active. The formation of inactive species from active species may represent a mechanism by which excess biologically active GAs are removed. In addition, glucose may be attached to a GA via a carboxyl group giving a GA glycoside or a hydroxyl group giving a GA glycosyl ether. When GAs are applied exogenously to a plant, some become glycosylated, presumably being inactivated. But in some cases they are metabolized back to free GAs and as such represent a storage form, according to Schneider and Schmidt (10).

Phinney (11) indicates that GA_{53}, GA_{44}, GA_{19} and GA_1 are all biologically active but only GA_1 has not been found to be converted to other biologically active gibberellins. It is believed that GA_1 is directly active in normal cell elongation. GA_3 is active and differs from GA_1 only in having one additional double bond (Fig. 1).

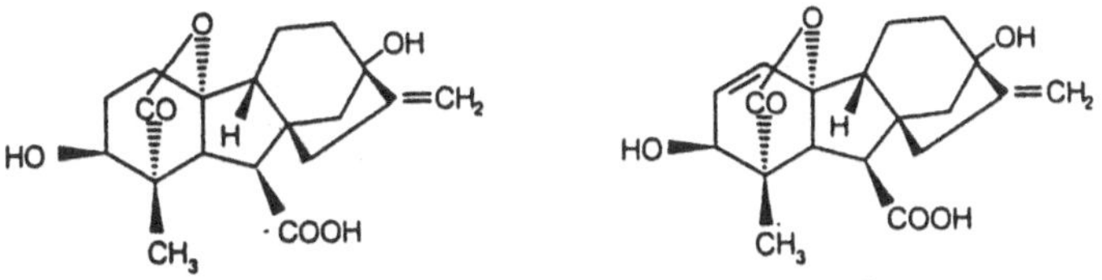

(a) Gibberellin A₁ (GA_1)   (b) Gibberellic acid (GA_3)

Fig. 1

Low levels of GAs occur in vegetative tissue of plants, particularly in young leaves and buds. Roots, too, may contain GAs. The beginning steps of GA synthesis from acetyl CoA may occur in one tissue and be metabolized to an inactive GA in another tissue (See "Genesis of Gibberellins" for the metabolic pathways).

GENESIS OF THE GIBBERELLINS

Joined head to tail, five-carbon (isoprene) units (Fig. 2) are the building blocks of a terpenoid.

(isoprene) building blocks $(-CH_2-\overset{\overset{\displaystyle CH}{|}}{C}=CH-CH_2-)$

Fig. 2

Gibberellins are terpenoids consisting of 20 carbons from four isoprene units. In tracing the synthesis of gibberellins, acetyl CoA is the starting compound. This is not surprising since Acetyl CoA is an important intermediate in cell respiration and the precursor of a whole series of compounds (See Fig. 3 for these biochemical pathways).

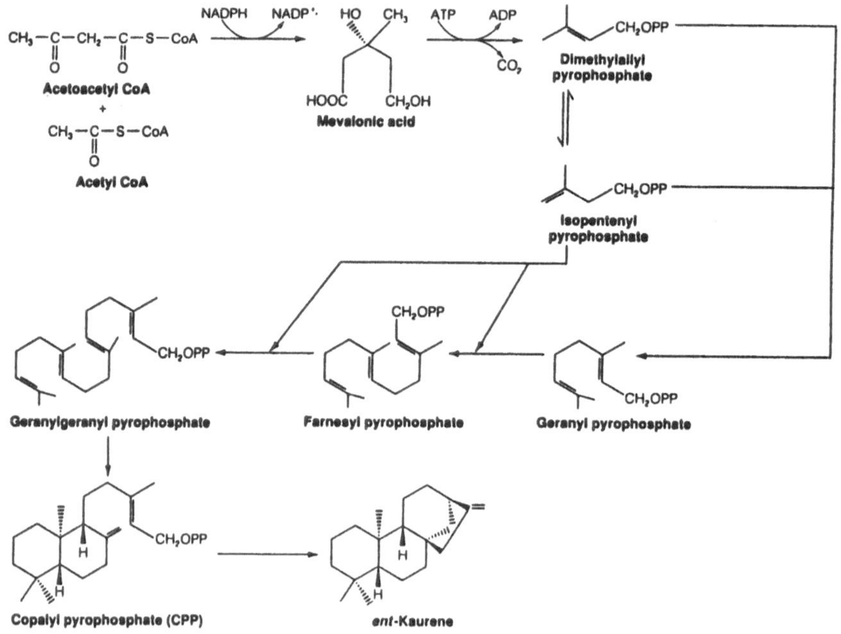

Fig. 3

Mevalonic acid, a six carbon compound, is synthesized from Acetyl CoA and in turn is phosphorylated by ATP, then decarboxylated to form isopentenyl pyrophosphate which is the first isoprenoid compound on the pathway leading to the synthesis of gibberellins. Following the synthesis of compounds in the pathway which is shared by all terpenes including essential plant oils, carotenoids, and steroids, the enantiomeric form of (+) -kaurene is formed.

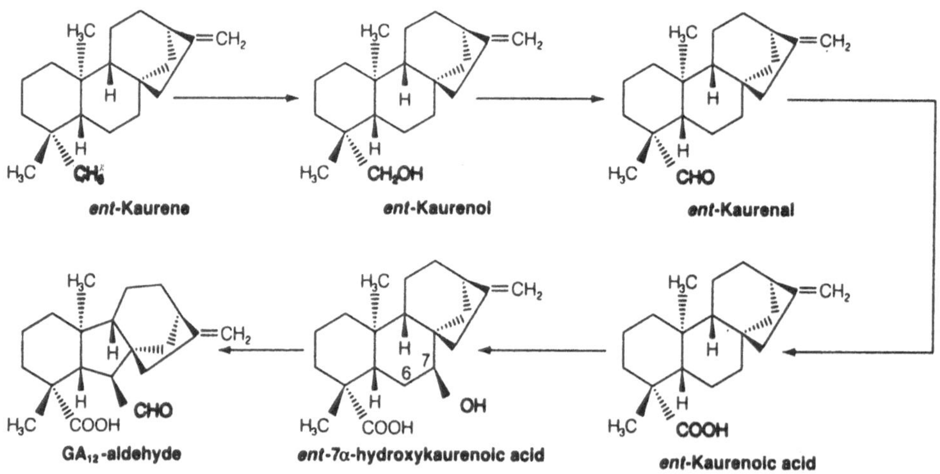

Fig. 4

In turn, (Fig. 4) 19 of ent-kaurene is oxidized to carboxylic acid, and the B ring is contracted from a six to five-carbon ring, giving GA_{12}-aldehyde which is the first gibberellin in plants and the precursor of all other gibberellins.

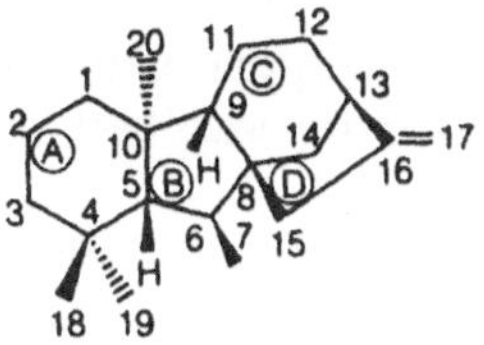

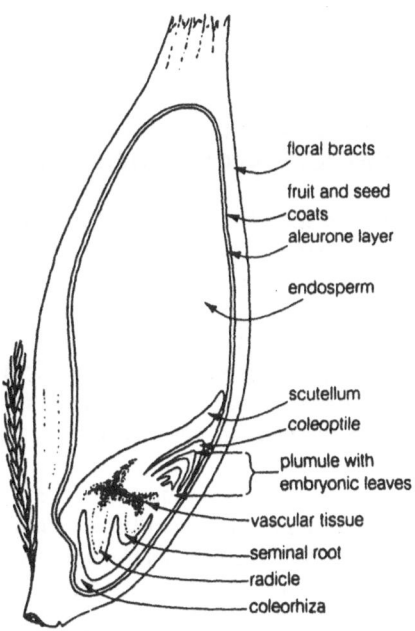

ent-Gibberellane skeleton

Fig. 5

Based on the ent-gibberellane skeleton (Fig. 5), some gibberellins have the full complement of 20 carbons while others have lost one carbon in metabolism, leaving only 19. The location of hydroxyl groups and their stereochemistry can have a strong bearing on their biological activity. For instance, hydroxylation in the beta configuration (in front of, as viewed on a page) of carbon 2 eliminates any biological activity.

A MODEL FOR GA_s IN SEED GERMINATION

The highest concentrations of GAs are found in seeds. It is believed that they are synthesized on site. The role of GAs in the germination of barley seeds has been studied extensively and the results serve as the basis of a widely-used model.

floral bracts

fruit and seed
coats
aleurone layer

endosperm

scutellum
coleoptile
plumule with
embryonic leaves
vascular tissue
seminal root
radicle
coleorhiza

Fig. 6

GAs are synthesized by the coleoptile and scutellum (cotyledon) of the embryo. In Graminaceae, grasses, the coleoptile is a protective organ around the seed leaves and the scutellum is a specially absorptive organ. The GAs diffuse from the embryo into the starchy endosperm tissue and from there into the aleurone layer where they induce the synthesis of a-amylase and b-amylase and other hydrolases which are secreted into the endosperm. Stored food in the endosperm is broken down into soluble sugars, amino acids, and other products and are transported back to a rapidly growing embryo.

The Florida Everglades is the largest freshwater subtropical wetland in the world (12) but it is undergoing change at an alarmingly rapid rate partly as a result of extensive drainage, agricultural runoff, and urban development. Sawgrass (Cladium jamaicense Krantz), which is not a grass but a sedge, is the dominant plant in the Everglades but is on the decline, while cattails (Typha domingensis) and (Typha latifolia) which are invasive plants, are on the increase. Since the freshwater supply for an exploding population in South Florida is dependent on the Everglades, any plans for restoration must include a better understanding of how sawgrass germinates. Surprisingly, very little research has been done on the physiology or seed germination of saw grass. Alexander (13) represents one of the few published sources.

While some seeds germinate readily, others are notoriously difficult. An example is found in an ongoing study of the germination of sawgrass and cattails in the Everglades. Current unpublished research (14) shows that cattail seeds germinate readily with high viability and within a short period of three of four days. Sawgrass, by contrast, requires at least a month to germinate, and then with very low viability. Through the use of GA_3 at a concentration of 1000 ppm, time to germination has been reduced from a month or more to five days. The per cent of seeds germinating, however, remains low.

GIBBERELLINS AND GENES

GAs act primarily by inducing the expression of the gene for a-amylase. A-amylase cDNA clones were isolated and used to make ^{32}P-labeled cDNA probes. According to Chandler (15), by hybridizing these radioactively labeled cDNA probes to Northern (RNA) blots, the level of a-amylase mRNA is seen to be greatly increased by GAs.

The transcription of information from DNA to RNA, which is turn translates into making the protein, a-amylase, is thought to be mediated by DNA-binding proteins, similar to regulatory proteins in operons of prokaryotic cells.

Using the model of an operon in a prokaryotic cell, it is believed that GAs increase the level of protein that switches on the production of a-amylase mRNA by binding to an upstream regulatory sequence of the a-amylase gene. While this is a proposed model, it is not known how GAs increase the level of this protein.

For gibberellins to be detected, there must be receptors, presumably proteins. Fence and Caruso (16) using antibodies that recognize GA_3 antibodies found that the GA_3 antibodies wound up on the surfaces of wild oat aluerone protoplasts. Additional research indicates that GA receptors may be located on the plasma membranes of aleurone cells.

Rapid stem elongation results from the exogenous application of GA_3. As part of a wider experiment involving the polymerization of GA_3, Carraher et al (17) reported that when common pole beans (Phaseolus vulgaris L.) were treated with 1000 ppm of GA_3, the stems elongated so rapidly that after two weeks they became thin and watery in appearance and could no longer support the growth. The plants fell over and subsequently died.

Cell elongation as well as cell division is thought to be involved in GA-induced stem elongation. In the case of cell elongation, plant cell walls must loosen, but how does this occur? Rayle and Cleland (18) have shown that when auxin (IBA) is applied, protons are extruded into the cell wall, and the decrease in pH activates wall-loosening enzymes that promote the breakage of key cross-linkages, increasing wall extensibility. This acid growth theory of auxin-stimulated plant cell elongation, however, has been ruled out as an explanation for GA-induced elongation. Stuart and Jones (19) have shown that GA_3 does not promote proton release form hypocotyl tissue under a wide range of incubation conditions.

Another explanation involves the proteins and inorganic ions which are present, as well as complex carbohydrates, in plant cell walls. The Ca^{+2} ion has long been known to cause a decrease in extensibility in dicots (But not monocots). Moll and Jones (20) worked with lettuce hypocotyls and found that GA_3 promoted the uptake of Ca^{+2} into the cytoplasm. In this manner, GAs could lower the calcium concentration in the walls by promoting the movement of calcium ions.

Still another hypothesis suggests that GAs prevent reactions that would result in cell wall stiffening. Fry (21) has proposed that cell wall stiffening is related to cross-linking of phenolic cell wall components such as ferulic acid residues, under the influence of the enzymes peroxidase. GAs inhibit cell wall peroxidase activity, and in this manner, may result in temporarily more flexible cell walls which are amenable to elongation.

In addition to rapid cell elongation, the application of GA_3 is believed to result in rapid cell division. Liu and Loy (22) have shown that (growth) GAs promote cell division because they stimulate cells in the G_1 phase to enter the S phase and because they also shorten the S phase. An increase in cell numbers then results in rapid stem growth. Not only is stem elongation promoted by GAs, but the growth of the whole plant is also promoted.

OTHER EFFECTS OF GAs

In addition to resulting in cell elongation and cell division, active GAs, whether occurring endogenously or applied exogenously, can induce a whole series of reactions.

GAs function to control the balance between stem elongation between nodes and leaf development. In many plants leaf development is profuse and internodal growth is retarded creating a pattern called a rosette. Just before the reproductive stage involving flowering, great internodal elongation of the stem occurs. This process is called bolting and appears to be separate from the flowering process which follows immediately. Some plants require a number of hours of daylength before bolting and are called long-day plants. Other plants require a cold treatment. The application of GA_3 to these plants can overcome the need for a longday or for cold treatment.

As indicated elsewhere, the application of GA_3 can overcome the effects of genetic dwarfism, resulting in plants of normal size. This is true for most but not all "Dwarf" plants. Inhibitors of gibberellin such as cycocel are sprayed on plants such as poinsettias to retard their stem growth, making them more visually appealing.

Application of GAs have been found to be very reliable in producing fruit parthenocarpically, i.e., without fertilization of the flower. Pome and stone fruits have proven to be unresponsive to auxin treatments, but quite responsive to the exogenous application of GA_3.

In terms of commercial applications, GA_3 is used as a spray on Thompson seedless grapes to increase grape and cluster size. GAs are also used to increase the amount of a-amylase in germinating barley seeds which are used to produce malt for the brewing industry.

Other uses include their application to sugarcane, which results in stalk elongation and an enhancement of sugar content. Their use with celery has not been successful, resulting in watery stalks (leaf petiole) not of commercial quality.

ISOLATION AND ANALYSIS OF GIBBERELLINS

Plants are sensitive to small concentrations of GAs. Dwarf plants are sensitive to as little as one nanogram of GA_3. Reeve and Crozier (23) report that dwarf rice is sensitive to as little as 3.5 picograms of GA_3.

Initially bioassays played an important role in the early research with GAs:

Dwarf corn was used. A GA solution was applied to the ligule of the first leaf of corn seedlings, causing a marked stimulation of the next node. This test required a period of 10 days and was sensitive down to 10 nanograms of GA_3.

Lettuce seedlings were placed whole in a GA_3 solution for two days and the elongation of the hypocotyl was used as a measure of GA_3 concentration. This test was sensitive down to 0.1 nanogram.

Half seeds of barley not containing the embryo were incubated in a solution of GA_3 for 1 day. If GAs were present they stimulated amylase activity, disappearance of starch, and buildup of reducing sugars. This test was sensitive down to 0.2 nanogram of GA_3.

A combination of gas chromatography and mass spectrometry (GC-MS) is used today to provide more accurate data on GAs. Following solvent extraction of plant tissue and some preliminary purification, the plant extract is injected into a HPLC-MS system (24).

EXPERIMENTAL

Titanocene dichloride (Aldrich, Milwaukee, WI) and GA$_3$ (Sigma, Saint Louis, MO) were used as received. The polymer was produced using the classical interfacial technique. Briefly, titanocene dichloride (1.00 mmole) in 50 ml of chloroform is added to a rapidly stirred (18,000 rpm, no load) aqueous solution (50 ml of water containing 1.00 mmole of GA$_3$ and 2.00 mmole sodium hydroxide) at room temperature (ca 25 C) for 30 seconds. A yellow/brown product (0.138 gram; 26% yield) precipitated from the reaction mixture. The product was collected by suction filtration and washed to remove unreacted material and then allowed to dry under room conditions.

FT-IR were recorded using KBr pellets employing a Mattson Alpha-Centauri spectrophotometer. High resolution electron impact positive ion mass spectral studies were carried out at the Midwest Center for Mass Spectroscopy, Lincoln, Neb. A Kratos MS-50 Mass Spectrometer, operating at 8KV accelerating voltage and a ten second per decade scan rate was employed for all measurements.

Rooting experiments were conducted employing sterilized coarse "Terra-Lite" perlite and Hibiscus rosa-sinensis, variety Albo Lacinatus (Anderson's Crepe Pink) wooden canes about ten to twelve inches in length and 0.25 to 0.75 inches in diameter. Fifty canes were employed for each test group. The stocks were dipped (0.037 gram average of talc mix or about 0.00007 grams of test material for a 2% sample) and then placed into the perlite. The samples were watered with a fine mist spray several times a day as needed and exposed to partial to full sun.

Bean seeds (Phaseolus vulgaris L.; Kentucky Wonder, Rust Resistant Pole; Ferry-Morse, 1990, Lot F-1) were soaked in water for one hour and planted in moist vermiculite in a greenhouse flat. They were germinated in an environmental chamber at 25 C (77 F). Twelve days after planting, the seedlings were harvested. A distance of eight cm was measured downward on each seedling from the primary foliage leaves and at that point the stem was cut, removing the roots. A distance of two cm was measured upward on the stem and marked. Subsequently, each plant was dipped in water to the two cm mark, dipped in the appropiate talc mix, tapped three time to remove excess powder and planted, one to a pot, with each plant residing in moist vermiculite up to the two cm mark. Subsequent measurements were made upwards from the two primary foliage leaves on a daily basis beginning on day 13 from the original planting. Measurements were continued to day 20.

A second series of experiments were carried out using pole bean seeds (Phaseolus vulgaris L.; Red Kidney) except that the roots were removed at a distance of ten cm from the primary leaves. Measurements were conducted for the next month.

RESULTS AND DISCUSSION

A more in depth discussion describing the structural characterization of the GA$_3$-containing polymers is contained in C. Carraher, et al. (25). Briefly, the products are crosslinked because of the trifunctionality of GA$_3$. Products were derived from a variety of organotin dichlorides and from titanocene dichloride and dipyridine manganese (II) dichloride. The mass spectra, infrared spectra and other structural characterization results are consistent with a structure as indicated in Fig. 7. For the present study, the product from the reaction of titanocene dichloride and GA_3 was used. Mass spectral analysis of the product showed the usual ion fragments from the titanocene moiety with ion fragments found at (all masses reported in m/z=1) 129 (CpTiO), 144 ($CpTiO_2$), 157 ($CpTiCO_2$), 241 (Cp_2TiOCO_2) and 265 ($Cp_2Ti(CO_2)_2$). Ion fragments were also found derived from the GA_3 moiety such as 303 (GA_3 minus CO_2), 331 (G-CH_3), 268 (G-CO_2, OH, CH_3) and 260 (G-$2CO_2$). In addition, there are numerous ion fragments containing one or more repeat units of the polymer. For instance, ion fragments are found at 525 (Unit), 482 (Unit minus CO_2), 491 (Unit minus OH and methyl) and 506 (Unit minus methyl).

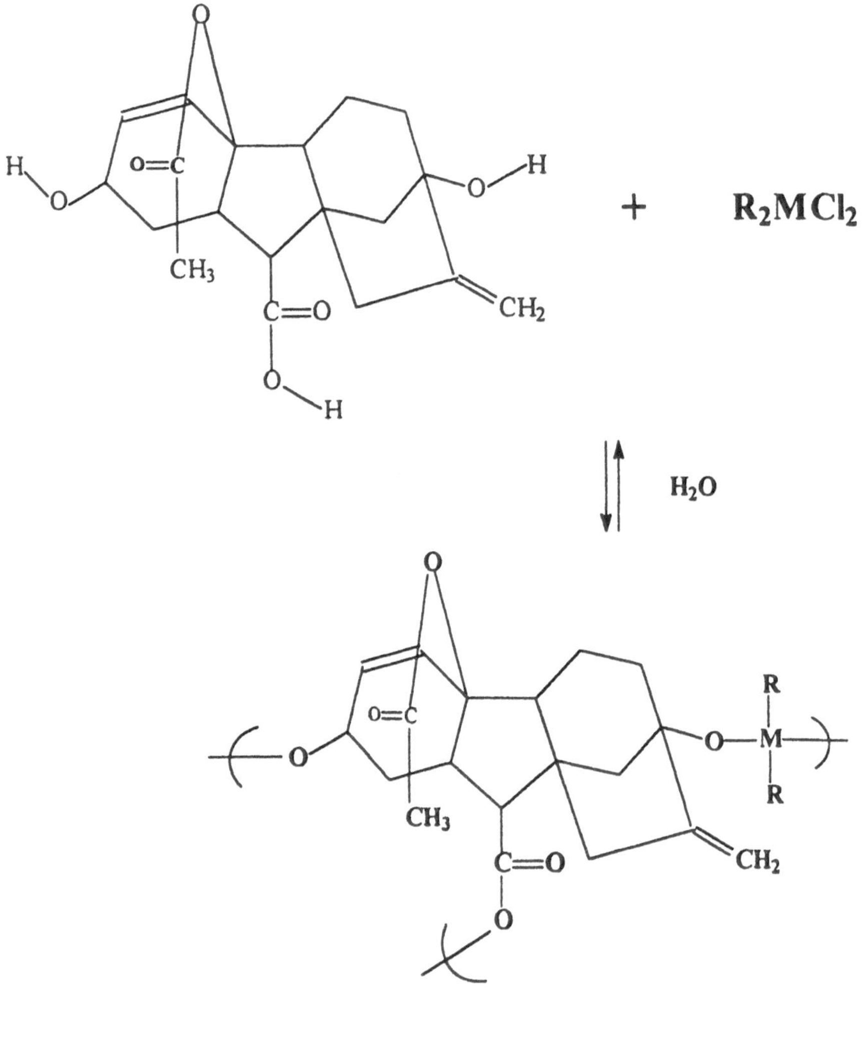

Fig. 7

Rooting experiments were conducted using Anderson's Crepe Pink, a common rooting stock employed as the "root" portion of grafted plants. Thus, it has a strong tendency to grow roots. Briefly, those stocks treated with both talc mixtures of GA_3 and the polymers containing the GA_3 moiety formed substantially fewer roots than the stock simply treated with talc alone and without any treatment. Of interest is the number of stocks that formed roots along the stem, above the treated portion at sites where small branches have been removed from the stocks.

Selected results evaluating growth and flowering of pole beans appear in Figures 8-9. The results from the Kentucky Wonder and Red Kidney pole beans were similar. In both cases, the bean plants were allowed to "stand" on their on. In all cases, those plants treated with GA_3 and polymer (containing GA_3) grew at a much faster rate with the rate increasing with increased concentration of GA_3 or polymer (Figures 7-8). For the Kentucky Wonder beans treated with 1000 ppm of the GA_3 or polymer, the plants grew so fast that after about two weeks, the stem could not support the growth and the plant "fell over" and subsequently died. By comparison, plants treated with the 100 ppm polymer sample showed more rapid growth and height through out the test period of four weeks (Figure 7).

Red Kidney bean stocks treated with GA_3 or polymer both showed enhanced growing rates through out the test period (about three weeks; Figure 9). Figure 9 contains the difference between the (mean) average height at 20 days subtracted with the initial (mean) height. The Y-axis is given in cm's. Thus, while the GA_3 treated beans exhibit the greatest growth, the polymer treated beans also exhibit dramatic growth increases. Thus, for the beans treated with 1000 ppm polymer the increase in height is about 50% greater than for the untreated beans.

With respect to flowering, the results are less clear. In general, as the GA_3 concentration is increased, the rate and number of flowers decreases. By comparison, the rate of flowering is most rapid for the two lowest concentrations of polymer and equal to that of the beans treated with 10 ppm GA_3.

In summary, the polymers containing GA_3 appear to act similar to GA_3 itself affecting growth, flowering and rooting changes.

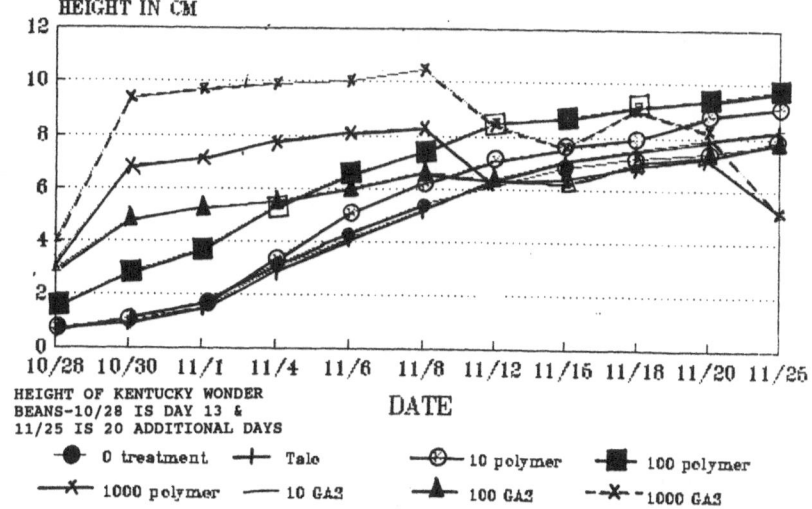

Fig. 8

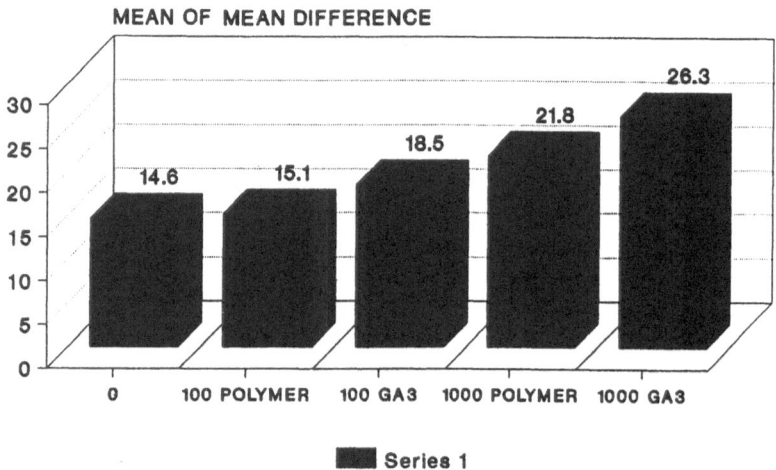

TREATMENTS

ppm

MEAN OF MEAN DIFFERENCE

Fig. 9

REFERENCES

1. Stodola, F.H., "Source Book on Gibberellins 1928-1957, "Agricultural Research Service, U.S. Department of Agriculture, 1958.

2. Stewart, Herbert H., Charles E. Carraher, Jr., Winn J. Soldani II, and Lisa Reckleben, "Polymeric Auxin Plant Growth Hormones Based on the Condensation Products of Indole-3-butyric Acid with Bis (Cyclopentadieny) Titanium IV Dichloride and Dypyridine Manganese II Dichloride" in "Inorganic and Metal-containing Polymeric Materials (John E. Sheats, Charles E. Carraher, Jr., Charles U. Pitman, Jr., Charles U. Pitman, Jr., Martel Zeldin and Brian Correll, eds.), Plenum Press, New York, 1990.

3. Phinney, B.O., "The History of the Gibberellins" in " The Biochemistry and Physiology of Gibberellins," vol. 1 (A. Crozier, ed.), Praeger Press, New York, 1983.

4. Rademacher, W., and J.E. Graebe, "Gibbereliin A, Produced by Sphaceloma manihoticola, the Carrier of the Hyperelongation Disease of Cassava (Marihot esculentum), Biochem. Biophys. Res. Commun. 91, 35-40, 1979.

5. Graebe, J.E., and H.J. Ropers, "Gibberellins" in "Phytohormones and Related Compounds - A Comprehensive Treatise," Vol. 1 (D.S. Letham, P.B. Goodwin and T.J.B. Higgins, eds.), Elsevier/North Holland Biomedical Press, Amsterdam, 1978.

6. Sponsel, V.M., "Gibberellin Biosynthesis and Metabolism" in "Plant Hormones and Their Role in Plant Growth and Development," (P.J. Davies, ed.) Kluwer, Boston, 1987.

7. Takahashi, N., B.O. Phinney and J. MacMillan, eds., "Gibberellins," Springer-Verlag, Berlin, 1990.

8. Bottini, R., M. Fulchieri, D. Pearce, and R.P. Pharis, "Identification of Gibberellins A_1, A_3 and iso-A_3 in Cultures of Azospirillum lipoferum, "Plant Physiology (90) 45-47, 1989.

9. Atzorn, D.P., A. Crozier, C. T. Wheeler and G. Sandberg "Production of Gibberellins and Indole-3-acetic Acid by Rhizobium phaseolis in Relation to Nodulation of Phaseolus vulgaris roots, "Planta (175) 532-538, 1988.

10. Schneider, G., and J. Schmidt, "Conjugation of Gibberellins in Zea mays L. "Plant Growth Substances (Pharis, R.P. and S.B. Rood, eds.)," 300-306, Springer-Verlag, Heidelberg, 1990.

11. Phinney, B.O., "Gibberellin A_1, Dwarfism and Shoot Elongation in Higher Plants, "Biol. Plant. (27) 172-179, 1985.

12. Loveless, C.M., "A Study of the Vegetation in the Florida Everglades, "Ecology (40) 1-9, 1959.

13. Alexander, T.R., "Sawgrass Biology Related to the Future of the Everglades Ecosystem," Soil and Crop Sci. Soc. of Florida, Proceedins, (31) 72-74, 1971.

14. Stewart, Herbert H., Charles Carraher, Jr., Gabriela Antunez, Adriana Antunez and Marsha Colbert, "Preliminary Data on Conditions of Seed Germination of Cattail and Sawgrass," unpublished report to South Florida Water Management District, 1994.

15. Chandler, P.M., J.A. Zwar, J. B. Jacobsen, T. Higgins and A.S. Inglis, "The Effects of Gibberellic Acid And Absciscic Acid on A-amylase mRNA levels in Barley Aleurone Layers. Studies Using an a-amylase cDNA Clone, "Plant Mol. Biol. (3) 407-418, 1984.

16. Pence, V.C. and J.L. Caruro, "Immunoassay Methods of Plant Hormone Analysis," in "Plant Hormones and Their Role in Plant Growth and Development (P.J. Davies, ed.), 240-256, Kluwer, Boston, 1987.

17. Carraher, Charles E., Jr., Herbert H. Stewart, Winn J. Soldani II, Jose De La Torre, Bhoomin Pandya and Lisa Reckelben, "Effects of Gibberellic Acid-Containing Organometallic Polymers as Plant Growth Hormones, "Polymeric Materials Science and Engineering," (71) 783-784, American Chemical Society, 1994.

18. Rayle, D.L. and R. Cleland, "Control of Plant Cell Enlargement by Hydrogen Ions," in "Current Topics in Developmental Biology," (11) 187-214, 1979.

19. Stuart, D.A. and R.L. Jones, "The Role of Acidification in Gibberellic Acid and Fusiococci-induced Elongation Growth of Lettuce Hypocotyl Sections" Planta (142) 135-145, 1978.

20. Moll, C. and R.L. Jones, "Calcium and Gibbereliin-induced Elongation of Lettuce Hypocotyl Sections" Planta (150) 450-456, 1981.

21. Fry, S.C., "Phenolic Components of the Primary Cell Walls and Their Possible Role in the Hormonal Regulation of Growth," Planta (146) 343-351, 1979.

22. Liu, P.B.W. and J. B. Loy, "Action of Gibberellic Acid on Cell Proliferation in the Subapical Shoot Meristem of Watermelon Seedlings, American Journal of Botany (63) 700-704, 1976.

23. Reeve, D.R. and A. Crozier, "Qualitative Analysis of Plant Hormones" in "Hormonal Regulation of Development. I Molecular Aspects of Plant Hormones Regulation." Encyclopedia of Plants Physiology, New Series (J. MacMillan, ed.) (9) 203-280, Springer-Verlag, New York, 1980.

24. Horgan, R., "Instrumental Methods of Plant Hormone Analysis," in "Plant Hormones and Their Role in Plant Growth and Development," (P.J. Davies, ed.) 223-239, Kluwer, Boston, 1987.

25. Carraher, Charles C., Jr., Herbert H. Stewart, Winn Soldani II, Lisa Reckleben and Bhoomin Pandya, Polymeric Materials: Science and Engineering (67) 270-271, 1992.

THE USE OF RUTHENIUM-CONTAINING POLYTHIOLS FOR SOLAR ENERGY CONVERSION

Charles E. Carraher, Jr. and Qingmao Zhang
Florida Atlantic University
Boca Raton, FL. 33431

INTRODUCTION

Solar energy conversion is an attractive and important area of study. Work has increased as the cost of fuels has increased and the availability of non-(rapidly) renewable fuels has decreased. This coincides with an increased awareness and need for "cleaner" fuels. Efforts have varied from simply capturing thermal solar energy in liquid holding tanks to the use of complex chemically-based systems. Many of these chemical efforts have involved the "splitting" of water into hydrogen and oxygen gasses and the subsequently reuniting of these gasses at a later time, subsequently capturing the expelled energy.

Two main processes have been considered in the chemical conversion of solar energy. These are first, photoinduced charge separation occurring within a very short excitation time and second, separated reactions occurring at a catalyst or at electrodes giving either products or "electrical" energy. In these processes, unidirectional transportation of the charge is necessary to prevent the energy-consuming reverse reaction to occur. It is very difficult to achieve anisotropic electron flow in a homogenous solution system. To achieve anisotropic electron flow, heterogenous reaction systems have become a focus of study with the emphases on the use of polymeric materials and molecular assemblies.

The present research effort focuses on the use of chemical intermediates to "capture" solar energy and with these "energy holding materials" then releasing energy appropriate to convert water into its elemental gasses-hydrogen and oxygen.

One major emphasis for the two step conversion of solar energy to effect the hydrolysis of water involves the use of ruthenium(II) bis(2,2'-bipyridine), RuBBPy, complexes. Here, the photochemical conversion model (Figure 1) requires that at the photoreaction converter, (P2-P1=P), the excitation energy be smaller than 3.5 ev in order to utilize visible light. T1 and T2 (acceptor and donor, respectively), are the electron mediators allowing charge separation and C1 and C2 are catalysts to facilitate production of reduction and oxidation products, respectively and P is ruthenium(II) tri-bipyridine plus two. Electrolytes or electricity can be formed. Thus, water (R2) is oxidized at C2 giving oxygen gas and H+ (R1) is reduced at C1 giving hydrogen gas. A free energy of 57 kcal/mol is obtained in this reaction (E(oxygen/water)=0.82 ev (vs NHE at pH=7) and E(H+/HH)= -0.41 ev). For the photolysis of water to occur, a potential of at least 1.23 V must be acquired (corresponding to a wavelength of 1000 nm).

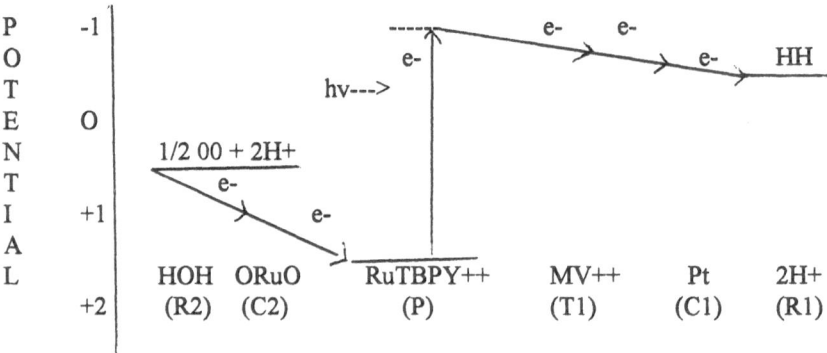

Figure 1.   Model for the ruthenium-associated photolysis of water.

Ruthenium bipyridine complexes are considered as promising candidates for the photoreaction center. These complexes have a moderate electrode potential as shown below for the tri-bipyridine complex.

RuTBPy+++ + e- -------> RuTBPy++   E=1.27 V (vs NHE at pH=7)

This redox reaction involving ruthenium bipyridines is reversible. The moderate electrode potential corresponds to 450 nm within the UV-Visible light range with its ground and excited states suitable for the redox hydrolysis of water (1-7).

Ruthenium(II) bis(2,2'-bipyridine) complexes were initially synthesized in 1955 (8). Since then nearly 100 cis-disubstituted complexes have been reported containing a wide variety of ligands, including pseudohalides, thiolates, thioethers, amines and organometallic donors (9-11). The reasons for their synthesis have been varied but here we will focus on their possible use in solar energy conversion. Like photosynthesis, where the conversion of carbon dioxide and water by sunlight to for carbohydrates is based on a metal-containing polymer-chlorophyll-there are also a number of other metal-containing critical enzymes such as hemoglobin, which also contains a metal as the key site for activity. Ruthenium-containing complexes have been studied as enzyme "mimic" models for chlorophyll and hemoglobin.

Several ruthenium bipyridine-containing polymers have been prepared for the purpose of solar energy conversion (12-15). One of the initial photoreaction centers studied was composed of polystyrene containing pendant RuTBPy++ groups. A similar product was prepared from poly(4-vinyl pyridine). Unfortunately, this product was susceptible to photo-aquation and as such , it is unsuitable as a photoreaction center in aqueous systems. 4-Methyl-4'-vinyl-2,2'-bipyridine was prepared and copolymerized with various monomers giving polymers containing pendant bipyridine groups that are subsequently reacted withH cis-Ru(bpy)Cl₂ giving polymers with pendant RuTBPy++. These polymer complexes showed slightly higher absorption maxima, about 460 nm, than the monomeric RuTBPy++. Their emission maxima were also at higher wavelength (about 610 nm) than the monomeric complexes and their emission intensities lower than monomeric RuTBPy++.

A comparison between the properties of monomeric RuTBPy++ and the requirements needed for luminophores and sensitizes shows the main drawbacks to be the
*relatively fast radiation-less decay of the 3 CT excited state to the ground state (resulting in a relatively short excited state lifetime and a small luminescence efficiency),
*occurrence of ligand photo-substitution reactions, and
*poor reproducibility.

Research continues on the synthesis and characterization of Ru-containing materials as photo reaction centers. Our work involves the synthesis and characterization of polymers containing ruthenium(II) bis-(2,2'-bipyridine) in their backbones. The current products are extensions of the reaction of ruthenium(II) bis-(2,2'-bipyridine) dichloride, DBR, with monothiols (16).

$$R\text{-}SH + Ru(bpy)_2Cl_2 \quad \rightarrow \quad (R\text{-}S)_2Ru(bpy)_2$$

The structure of the polymeric materials is believed to be analogous to that of the monomeric materials.

$$n \ HS\text{-}R\text{-}SH + n \ Ru(bpy)_2CL_2 \quad \rightarrow$$

$$\text{-}(\text{-}S\text{-}R\text{-}S\text{-}Ru(bpy)_2)_n\text{-} + nHCL$$

EXPERIMENTAL

The chemicals were used as purchased. Cis-Dichloro-bis(2,2'-bipyridine)ruthenium(II) dihydrate (CA# 15746-57-3) was purchased from Strem Chemical Company, Newburyport, MA; 1,8-octanedithiol was purchased from Alpha Johnson Mathey, Ward Hill, MA; and the remaining thiols were obtained from Aldrich Chemical Company, Milwaukee, WI.

Synthesis is straight forward and is based on the reaction between cis-dichloro-bis(2,2'-bypyridine)ruthenium(II), DBR, and monothiols (16). Thus, 2-mercaptoethyl ether, MEE (0.10 gram; 0.75 mmole) and an equivalent molar amount of sodium hydroxide were dissolved in 20 cc of a 80 % methanol/water mixture. This mixture was allowed to remain at room temperature for 10 minutes after which DBR (0.386 gram; 0.74 mmole) was added. The combined mixtures were heated to 90 C for one hour, after which the precipitate was repeatedly washed with toluene and chloroform to remove unreacted starting material. A dark brown precipitate (0.234 grams; 48 % yield) was obtained.

Infrared spectra were obtained using Mattson Instruments Galaxy 4020 FTIR spectrophotometers employing potassium bromide pellets. Scans were recorded using an instrument resolution of 4 l/cm employing 32 scans. UV-VIS spectra were obtained using a Cary/Varian UV-Visible spectrophotometer.

High resolution electron impact positive ion mass spectral analysis was carried out at the Midwest Center for Mass Spectrometry, Lincoln, Neb. employing a Dupont 21-496-B double focusing mass spectrometer. The samples were inserted as solids in a glass ampule using a direct insertion probe (DIP-HREI-MS). Spectra were obtained from rapid heating to about 450 C.

Thermogravimetric analysis was conducted using a Model 951 Dupont Thermogravimetric Analyzer with a heating rate of 20 C/min to 750 C. Differential scanning calorimetry was conducted using a Dupont DSC Cell Base II connected to a Dupont 990 Thermal Analyzer console employing a heating rate of 20 C/min to 450 C. Samples were analyzed using a gas flow rate of about 50 ml/min.

Solubility tests were carried out using about one mg of sample for 3 cc of liquid. Those materials that gave colorless liquid-sample combinations after one week were deemed as being "insoluble". The materials were cited as being soluble if solubilities were of the order of 10 grams per liter and more. Intermediate solubilities are cited as being "slightly soluble".

Weight average molecular weights were obtained using a Brice-Phoenix BP3000 Universal Light Scattering Photometer. The refractive index increment was determined using a Bausch & Lomb Abbe Model 3-L refractometer. Proton and carbon NMR spectra were obtained using a General Electric-QE 300 instrument at 23 C.

RESULTS AND DISCUSSION

The reaction is general occurring for both aromatic and aliphatic dithiols (Table 1).The yields are modest and the molecular weights correspond to degrees of polymerization in the range of 30 to 40. The product from dimercaptophenyl ether is insoluble in all tested liquids probably due in part to the rigidity of the mercaptophenyl portion. The remaining products are soluble in dipolar aprotic liquids (Table 2).

Table 1.   Yield and molecular weight values for products obtained from the condensation of dithiols and DBR.

Dithiol	Mol. Wt.	Yield(%)
Dimercaptophenyl Ether	-------	35
1,8-Octanedithiol	18,000	53
2,5-Dimercapto-1,2,3-thiadiazole	16,000	44
2-Mercaptoethyl Ether	19,000	48

Table 2. Solubility of Synthesized Polymers

Solvent	I	II	III
Methylenechloride	SL	SL	IS
Tetrahydrofuran	IS	IS	IS
N, N-Dimethylformide	S	S	IS
DMSO	S	S	IS
Toluene	S	S	IS
Nitrobenzene	SL	— SL	IS
Acetonitrile	S	S	IS

Note: IS... insoluble

SL...soluble slightly(partial)

S.....soluble

where: i   =   product from 1, 8-octanedithiol
ii =   product from 2-mercaptoethyl ether
iii =   product from dimercaptophenyl ether

For brevity and directness of data presentation, the polymer derived from reaction of 2-mercaptoethyl ether, MEE, will be emphasized in the following discussion.

For the UV-VIS spectra, there is a general shift in the bands with two of the bands coming together. DBR shows bands at 520 nm, 359 nm and 290 nm associated with an eg to t2g d-electron transition. For the product of DBR/MEE the 520 nm and 349 nm bands merge forming a band about 409 nm. The 290 nm band remains relatively unchanged. DBR and the products also exhibit combination bands at 220 and 250 nm. For DBR these bands are more distinctive while for the products they are more merged.

For the FTIR analysis, DBR exhibits bands in the region 1600-1350 (IR bands are given in 1/cm) characteristic of aromatic ring vibrations. The strong peak at 3435 results from the presence of water in the starting ruthenium compound. The C-H stretching vibration associated with the bipyridine ring occurs at 3049. The ring breathing and stretching bands associated with the bipyridine appear between 1400 and 1500. Absorptions from 600 to 800 are characteristic of C-H out-of-plane deformation. These bands and other assignments appear in Table 3.

Table 3.   **Infrared assignments for dichloro-bis(2 ,2'-bipyridine)ruthenium**

cm^{-1}	Intensity	Assignment
3400	wide & strong	hydrated water in the molecule
3000-3080	strong bands	C-H stretching of the bipyridine
1481-1390	strong bands	C-C stretching of the bipyridine ring
1020	strong bands	C-H deformation on the bipyridine ring
850	strong bands	C-H out of plane deformation
790	strong bands	

The assignment of important bands for the DBR/MEE product is given in Table 4. Bands around 3100, 2928 and 2850 are characteristic of the presence of C-H stretching for aromatic and aliphatic moieties. Pikes at 1599, 1477, 1444 and 1305 are characteristic of the C=C stretch in the bipyridine ring; bands at 1008 and 920 are characteristic of the "ring-breathing" of the bipyridines; a band at 1105 is assigned to the C-O-C stretching vibration; the band at 1425 is assigned to C-H scissoring of the methylene moiety; bands at 771 and 727 are assigned to the C-H deformation in the bipyridine ring (with possible overlap with the band(s) derived from C-S).

Table 4.   Infrared spectral assignment of bands found for the polymer DBR/MEE

Wavelength cm^{-1}	Intensity	Assignment
3402	wide & strong	hydrated water of ruthenium
3115	med to weak	C-H stretching vibration
3068		
2922	medium	C-H stretching of methylene
2864		
1599		
1477		
1444	strong	C=C stretching vibration of the
		bipyridine ring
1305		
1425 group	medium	C-H scissoring of the methylene
1105	strong	C-O-C stretching vibration
1008	strong	ring breathing of bipyridine
920		
771	strong	C-H deformation of bipyridine ring
727		(C-S may be overlapped with)
592	med to weak	new peak assigned to Ru-S

The Ru-S stretching vibration has been assigned to a weak band around 500 (41-43). New bands appear at 586 and 513. These bands are assigned to the presence of the Ru-S stretching consistent with the formation of the proposed structure.

Mass spectral assignments for DBR are given in Table 5 for the most abundant ion fragments. The most intense ion fragments correspond to the bipyridine moiety (m/e=156; all ion fragment intensities are given in terms of m/e=1). The peak at m/e=78 corresponds to pyridine. The presence of Cl is indicated by the presence of ion fragments at m/e=35-38. The natural isotopic abundance for Cl(35) to Cl(37) is 3:1. The ratio of m/e=36(HCl-35) to 38(HCl-37) is 3.0

Table 5. The most abundant ion fragments for DBR.

m/e	Intensity	Possible Assignments
157	10.8	bipyridine
156	100	
155	43	
78	12.5	bipyridine
51	14	
36	38	HCl (35)
38	12	HCl (37)
96	4.2	
97	5.3	
98	3.9	
99	0.5	
101	1.83	
102	0.6	
103	1.6	
104	0.85	

While there are ion fragments whose mass correspond to the presence of ruthenium-containing ion fragments the isotopic abundance ratios do not agree with this assignment. Further, for these ion fragments, there are reasonable alternative assignments. For instance, there are ion fragments at m/e=96-104 that could be assigned to ruthenium. The most abundant ion fragment from isotopic considerations should be m/e=102 yet the abundance of m/e=102 is less than that of 103,101,98 and 96. Further, the ion fragments about m/e=131-141 could be assigned to RuCl. The most abundant ion fragment should be 137 (combination of the most abundant isotope of Cl and Ru), yet this ion fragment is missing. The fragments at m/e=96-104 are derived from the decomposition of bipyridine. Ion fragments about 50 to 52 are derived from the decomposition of one of the pyridine rings. The ion fragments at m/e=131 to 141 are decomposition products of one or both of the pyridine units contained within the bipyridine.

Mass spectral results for the DBR/MEE product are given in Table 6. The most abundant ion fragments are at 156 and 155 and correspond to the mass of bipyridine (m and m-1). Ion fragments derived from the dithiol-containing moiety are also among the most abundant found and include the ion fragments at 76, 77, 91, 102 and 103. There are also present a number of less intense higher mass ion fragments including the following-332 from Ru(bpy)-S-HCH-HCH-O; about 398 from Ru(bpy)-S-HCH-HCH-O-HCH-HCH-S; about 490 from (bpy)Ru(bpy)-S-HCH-HCH-O; and about 590 from HCH-S-(bpy)Ru(bpy)-S-HCH-HCH-O-HCH-HCH-S (The term "about" signifies that there are several ion fragments where the cited mass corresponds the most abundant ion fragment mass). Moderate isotopic matches are found for the ruthenium-containing ion fragments of higher abundance.

Table 6.   **Most abundant ion fragments for the polymer DBR/MEE**

m/e	Relative Intensity	Possible Assignment
156	100	Bipyridine parts
155	42.6	
78	15.9	pyridine
76	1.8	$S-CH_2CH_2O$
77	2.7	
91	1.5	$S-CH_2CH_2O-CH_2$
102	2.9	$S-CH_2CH_2O-CH_2-CH_2$
332	0.2	$-Ru(bpy)-SCH_2CH_2O$
398	0.5	$-Ru(bpy)SCH_2CH_2OCH_2CH_2S$
490	0.4	$-Ru(bpy)_2-SCH_2CH_2O$
206	0.7	$-Ru-S-CH_2CH_2)-CH_2-CH_2$
594	0.4	
593	0.3	$CH_2S-Ru(bpy)_2-SCH_2CH_2OCH_2CH_2S$
591	1.1	
590	1.8	
589	0.3	
588	0.6	
587	0.2	
586	0.4	

The NMR results are inconclusive because of the low solubility of the product.

TGA thermograms for the DBR/MEE product vary according to the employed atmosphere. Initial weight losses in the 100 to 200 C range are similar for both air and nitrogen atmospheres. About 90 % weight retention occurs in nitrogen to about 450 C. From about 450 C to 500 C there is relatively rapid weight loss corresponding to about a 30 % weight loss. In air, more gradual weight loss occurs from 300 to 500 C corresponding to about 35 % loss. DSC results show similar thermogram traces in air and nitrogen to about 275 C after which there is a highly exothermic change occurring in air characteristic of oxidative-associated reactions. By comparison, in nitrogen there exists only (relatively) mild endothermic changes occurring over the 275 to 450 C range.

In summary, structural characterization is consistent with the proposed structure. Current efforts are underway to further evaluate these materials for use in solar energy conversion schemes.

We thank the ACS-PR for partial support of this work-Grant # 19222-B).

REFERENCES

1. M. Kaneko and A. Yamada, Adv. Polm. Sci. $\underline{55}$ (1983)
2. M. Gratel, Acc. Chem. Res., 14,376(1981)
3. T. K. Foreman, W. M. Sobol and D. G. Whitten, J. Am. Chem. Soc. $\underline{103,533(1981)}$
4. M. Cavin, Can. J. Chem. $\underline{601}$, 873(1983)
5. J. S. Connoly, (Editor), "Photochemical Conversion and Storage of Solar Energy", Academic Press, NY,1981
6. J. H. Fendler, J. Photochem., $\underline{17}$, 303(1981).
7. T. G. Meyers, Acc. Chem. Res.11,94,(1978).
8. R. R. Miller; W. W. Brandt; Marina J. Am. Chem. Soc., $\underline{77}$, 3178, 1955.
9. E. Dodsworth, S. Lever, Chem. Physic. Lett, $\underline{124}$, 152, 1986.
10. M. J. Root; B. P. Sullivan; T. J. Meyer; E. Deutsch, Inorg. Chem. $\underline{24}$, 2731, 1985.
11. E. C. Johnson, B. P. Sullivan, D. J. Salmon, Inorg. Chem. $\underline{17}$, 2211, 1978.
12. M. Kaneko, S. Nemoto, A. Yamada, and Y. Kurimura, Inorg. Chim. Acta, $\underline{44}$, L289(1980).
13. M. Kaneko, A. Yamada and Y. Kurimura, Inorg. Chim. Acta, $\underline{45}$, L73(1980).
14. J. M. Klear, J. M. Kelly, D. C. Peper and J. G. Vos; Inorg. Chim. Acta $\underline{33}$, L139(1979).
15. T. Shimidzu, K. Izaki, Y. Akai and T. Iyada, Polym. J., $\underline{13}$, 889(1981).
16. M. A. Geaney, L. Coyle, M. Harmer, A. Jordan and E. Stiefel, Inorg. Chem., $\underline{28}$(5), 912(1989).

SYNTHESIS AND CHARACTERIZATION OF ALIPHATIC

AROMATIC POLYAMIDE - METAL COMPLEXES

Udai D.N. Bajpai[1], Sandeep Rai[1], and Anjali Bajpai[2]

[1]Polymer Research Laboratory,
Department of Post Graduate Studies and Research in Chemistry
R.D.University Jabalpur 482 001, India.

[2]Department of Chemistry
Government Autonomous Science College
Jabalpur 482 001, India.

INTRODUCTION

Synthesis, characterization and physico-chemical studies of metal-chelate polymers are of great importance. Renewed interest in this field has been evinced in the context of obtaining materials with high temperature resistance. The potential applications of metal containing polymers are numerous and well documented. Schiff's base complexes involving low molecular weight diamines and aldehydes were extensively studied. Polyamides are versatile high performance polymers. A variety of aliphatic, aromatic and aliphatic-aromatic polyamides have been synthesised. In order to improve the thermal properties of existing polyamides, we have attempted to synthesise polyamide- Schiff's base complexes[1,2]. In the present paper, we are reporting the syntheses of polyamide schiff's base coordination polymers using amino terminated oligomers of hexamethylene terephthalamide (PHTA) and Co(II), Ni(II) and Cu(II) complexes with salicylaldehyde and 2-hydroxy-1-napthaldehyde.

EXPERIMENTAL

Materials.

Terephthalic acid (E.Merck, A.R. Grade) and hexamethylene diamine (BDH, A.R. Grade) were used without further purification. DMF and pyridine were GR grade products of Qualigen, Glaxo, India and were purified by distillation before use. Methanol (E.Merck, A.R. Grade) and triphenyl phosphite (TPP) (Wilson Chemicals India, A.R. Grade) were used as such.

Salicylaldehyde and 2-hydroxy-1-naphthaldehyde (Fluka, AG) were used as such. Hydrated acetates of Co(II), Ni(II) and Cu(II) of AR Grade were obtained from E.Merck, India and were used without further purification.

Preparation of Poly(Hexamethylene terephthalamide) (PHTA)

The PHTA oligomer was prepared by heating a mixture of terephthalic acid (0.1 mole), hexamethylene diamine (0.12 mole) in a DMF (150 mL) and pyridine (30 mL) solvent mixture at $110 \pm 1^{\circ}c$ with constant stirring for 90 min. The oligomer was precipitated in methanol, filtered and washed several times with methanol and dried in vacuum.

$$n \ HOOC\text{-}C_6H_4\text{-}COOH + (n+1) \ H_2N\text{-}(CH_2)_6\text{-}NH_2 \longrightarrow$$

$$H_2N\text{-}(CH_2)_6N\left[\ \underset{\underset{H}{|}\ \underset{O}{\|}}{C}\text{-}C_6H_4\text{-}\underset{\underset{H}{|}\ \underset{O}{\|}}{C}\text{-}N\text{-}(CH_2)_6\underset{\underset{H}{|}}{N}\right]_n H + 2H_2O$$

Preparation of Co(II), Ni(II) and Cu(II) Complexes of Salicylaldehyde and 2-Hydroxy 1-Naphthaldehyde

The complexes were prepared by the method reported earlier[1]. The aqueous ethanolic solution of $Co(CH_3COO)_2.4 \ H_2O$ (2.49g, 0.01 mole), $Ni(CH_3COO)_2.4H_2O$ (2.49g, 0.01 mole) or $Cu(CH_3COO)_2.H_2O$ (2.00g, 0.01 mole) and salicylaldehyde (2.1 mL, 0.02 mole) or 2- hydroxy 1-napthaldehyde (3.44 ml, 0.02 mole) were used in respective preparations. The solvent was a mixture of water and ethanol in 40:10 ratio. This solution was stirred thoroughly and refluxed for 1-2 hrs. The precipitated complexes were filtered, washed several times with ethanol and ether successively and dried. The reaction is shown below :

$$2\,Ar\text{-}OH\underset{CHO}{\overset{|}{\big|}} \quad + \quad M(CH_3COO)_2 \cdot x\,H_2O \quad \xrightarrow[\text{reflux}]{\text{ethanol}/H_2O} \quad \begin{array}{c} Ar\text{-}O \\ | \\ HC=O \end{array} \,M\, \begin{array}{c} O-Ar \\ | \\ O=CH \end{array}$$

Where   M   =   Co(II), Ni(II) or Cu(II)

and   Ar =   or

Synthesis of Poly(hexamethylene terephthalamide)- Metal Aldehyde Complexes

Solution of PHTA (1.25g, 0.001M) and respective metal complexes of aldehydes (0.001M) in 50 ml hot DMF were refluxed for 6-8 hrs. to obtain the resulting coordination polymers in solid forms.

$$H_2N\text{-}(CH_2)_6\text{-}NH_2 \quad + \quad \begin{array}{c} Ar\text{-}O \\ | \\ HC=O \end{array} \,M\, \begin{array}{c} O\text{-}Ar \\ | \\ O-CH \end{array} \quad \longrightarrow \quad \begin{array}{c} Ar\text{-}O \\ | \\ HC=N \end{array} \,M\, \begin{array}{c} O\text{-}Ar \\ | \\ N=CH \end{array}$$

$$\begin{array}{cc} | & | \\ & (CH_2)_6 \text{----} \end{array}$$

The coloured precipitates were filtered, washed thoroughly with hot DMF and alcohol and then dried under vacuum. The polymide- metal complexes were air stable, powdery, insoluble in water and common organic solvents.

Measurements

C, H, N, analyses were done with Perkin-Elmer elemental analyser 2400 model. Metal contents in the coordination polymers were determined by complexometric titrations with EDTA after decomposing them with fuming nitric acid. IR spectra were recorded on a Perkin-Elmer infrared spectrophotometer No. 1430 using CsI pellets in the range 4000-200 cm^{-1}. Magnetic susceptibility measurements were performed at room temperature with Guoy's balance using $Co(NCS)_4$ as a calibrant. Thermogravimetric analyses was done on a Perkin-Elmer Thermal analyser TGA-7 witth a professional computer 7700 in an inert atmosphere at a heating rate of $20^\circ C/min$.

RESULTS AND DISCUSSION

The analytical data (Table 1) for polyamide-metal complexes are in close agreement with the general formula $M(SAL)_2L$ or $M(NAPHTHAL)_2L$ where M = Co(II), Ni(II) or Cu(II), SAL = Salicyladehyde, NAPHTHAL = 2-hydroxy-1-naphthaldehyde and L = -[NH-(CH$_2$)$_6$-NH-CO- C$_6$H$_4$-CO]$_n$. The discrepancy observed between theoretical and calculated values can be attributed to the polymeric structure, which may have many structural defects and variable compositions at the end units. Further, the strong tendency of aromatic polyamides for moisture pick up[1] may also give rise to the discrepancy between theoretically calculated and experimentally determined values.

Table 1. Analytical data of PHTA and their metal complexes.

S.No.	Compounds	Elemental Analyses			
		% C	% H	% N	% M
1.	PHTA	66.14	07.40	13.20	–
		64.22	07.07	11.26	–
2.	PHTA- Co (II) Sal	65.80	06.80	10.90	03.70
		62.42	06.64	10.07	04.20
3.	PHTA - Ni (II). Sal	65.80	06.80	10.90	03.70
		61.02	06.24	09.20	04.50
4.	PHTA - Cu (II)- Sal	65.70	06.80	10.90	04.50
		63.50	04.98	07.90	04.60
5.	PHTA- Co (II) - Naphthal	66.90	06.70	10.50	03.46
		62.47	05.90	03.75	05.02
6.	PHTA - NI (II) - Naphthal	66.90	06.70	10.50	03.45
		64.22	06.02	09.80	04.24
7.	PHTA - Cu (II). Naphthal	66.80	06.70	10.50	03.70
		63.37	05.67	09.72	05.04

a.   The upper values represent the theoretical values and the lower values are the observed results.

The IR spectra of the ligand, PHTA and its complexes (Table 2) were compared for the elucidation of the nature of coordination. The assignment of the important IR bands were based on the literature data[3,4] for structurally similar compounds. The strong and sharp band due to NH at 3320 cm^{-1} in the spectrum of PHTA shifts towards lower frequency on complexation indicating involvement of amino group in bonding. The splitting of the amide II band of PHTA ligand at 1580 cm^{-1} upon complexation further supports nitrogen coordination. A new band appears at 1540 - 1565 cm^{-1} in the complexes which can be assigned to ν C = N, a characteristic of Schiff's base[1]. The Schiff's base formation should be accompanied by the disappearance of the carbonyl frequency of the bis(aldehyde) chelates, but it could not be identified due to the presence in the same region of the amide I or carbonyl absorptions due to the polyamide chain. Generally the ν C = N absorption appears in the range 1690 - 1590 cm^{-1}. The considerable lowering in frequency of this absorption due to this vibration may be attributed to the coordination through azomethine nitrogen and also a great extent of bonding between the metal atom and the ligand molecule. The appearance of M-O bands in the region 495 - 530 cm^{-1} and M - N bands in the region 495 - 530 cm^{-1} further confirms participation of nitrogen and oxygen atoms in complex formation.

The absorption due to the acetate group was not observed in the IR spectra of the polyamide-metal complexes.

Table 2. Important IR spectral assignments of PHTA and its metal complexes (wave number expresed in cm^{-1})

ASSIGN MENTS	PHTA	COMPOUND					
		PHTA Complexes					
		Co(II)- Sal	Ni(II)- Sal	Cu(II)- Sal	Co(II)- Naphthal	Ni(II)- Naphthal	Cu(II)- Naphthal
νNH	3320(s)	3310(s)	3310(s)	3300(s)	3310(s)	3300(s)	3300(s)
νCH(arom)	3060(w)	3080(w)	3080(w)	3080(w)	3080(w)	3080(w)	3080(w)
νCH(asym)	2920(s)	2940(s)	2940(s)	2940(s)	2940(s)	2920(s)	2920(s)
νCH (sym)	2860(s)	2860(s)	2860(s)	2860(s)	2860(s)	2860(s)	2860(s)
amide I or C=S	1630(s)	1630(s)	1630(s)	1620(s)	1625(s)	1630(s)	1625(s)
amide II	1530(s)	1530(s)	1530(s)	1535(s)	1530(s)	1530(s)	1530(s)
νC=N	–	1550(s)	15640(s)	1565(s)	1550(s)	1565(s)	1550(s)
δ CH$_2$	1420(s)	1410(s)	1415(s)	1420(s)	1420(s)	1410(s)	1420(s)
amide III	1280(s)	1280(s)	1285(s)	1285(s)	1285(s)	1285(s)	1280(s)
	1180(s)	1170(s)	1180(s)	1170(s)	1200(s)	1200(s)	1990(s)
amide V & (CH$_2$)$_n$ rocking	725(s)	730(s)	735(s)	750(br)	750(br)	730(s)	730(s)
νM-O	–	530(s)	525(s)	530(s)	500(s)	495(s)	540(w)
νM-N	–	420(s)	425(s)	430(s)	425(s)	430(s)	425(s)

The values of magnetic moments for different PHTA-metal complexes are given in Table 3 alongwith the proposed geometry[1].

Table 3 Magnetic moment values of PHTA-metal complexes

No.	Compounds	μeff^{BM}	Geometry Proposed
1.	PHTA- Co(II) - SAL	4.62	Tetrahedral
2.	PHTA- Co(II) - NAPHTHAL	4.62	Tetrahedral
3.	PHTA- Ni(II) - SAL	Diamagnetic	Square planar
4.	PHTA- Ni(II) - NAPHTHAL	Diamagnetic	Square planar
5.	PHTA- Cu(II) - SAL	1.78	Square planar
6.	PHTA- Cu(II) - NAPHTHAL	1.78	Square planar

Thermogravimetric analysis

The thermogravimetric analysis data is depicted in Table 4 and Fig1.

The thermal degradation mechanism of polyamides at temperatures above $300^{\circ}C$ has been described by several authors[5-7]. The main fraction of decomposition products generated by polyamides is non-valatile and the volatile fraction consists mainly of CO_2, H_2O, benzene, cyclopentanone and ammonia. It was assumed that the overall decomposition process for nylon is typical of a random decomposition of linear chains similar to that for linear polyethylene. A pure free-radical mechanism for the degradation can be ruled out. Both free radical and heterolytic decompositions (caused by water lightly bound to the peptide group) occur simultaneously. Evidently, both the N-C and C-CH_2 bonds break during degradation.

The degradation pattern of PHTA and its metal complexes is very similar to the mechanism described above. The ligand PHTA shows a two step degradation. The first step starts above $250^{\circ}C$ and second step decomposition temperature is above $400^{\circ}C$. A perusal of Table 4 indicates that the metal complexes have superior thermal stability than the ligand PHTA. This can be attributed to the chelate ring formation through coordination with metal ion and introduction of aromatic nuclei in the polymeric chain besides the increment in molecular weight.

Table 4. Thermogravimetric data of PHTA-metal complexes

Temp ($^{\circ}C$)	Percentage weight Loss			
	PHTA	PHTA-Co(II) NAPHTHAL	PHTA-Ni(II) NAPHTHAL	PHTA-Cu(II) NAPHTHAL
0	0.0	0.0	0.0	0.0
50	0.0	0.0	0.0	0.0
100	3.6	2.0	4.0	0.0
150	5.3	2.0	6.0	4.0
200	7.0	3.0	7.0	7.0
250	12.5	5.0	9.0	9.8
300	33.4	7.0	10.0	13.2
350	43.4	9.0	11.0	17.5
400	45.8	25.0	22.0	37.5
450	55.0	74.0	33.0	76.0
500	96.5	91.0	69.0	78.0
550		92.0	76.0	79.0
600		92.0	76.0	79.0
650		92.0	78.0	94.0
700		92.0	79.0	

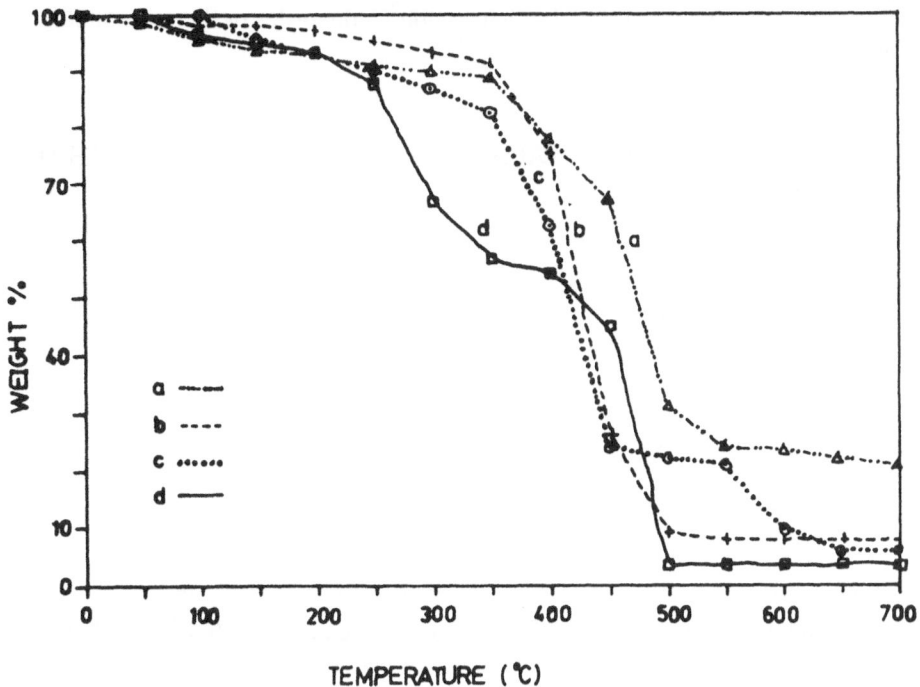

Fig. 1. Thermogravimetric curves of

        (a) PHTA - Ni (II) - NAPHTHAL

        (b) PHTA - Co (II) - NAPHTHAL

        (c) PHTA - Cu (II) - NAPHTHAL

        (d) PHTA

On the basis of all the above studies, structure proposed for the PHTA-metal complexes is shown in Figure 2.

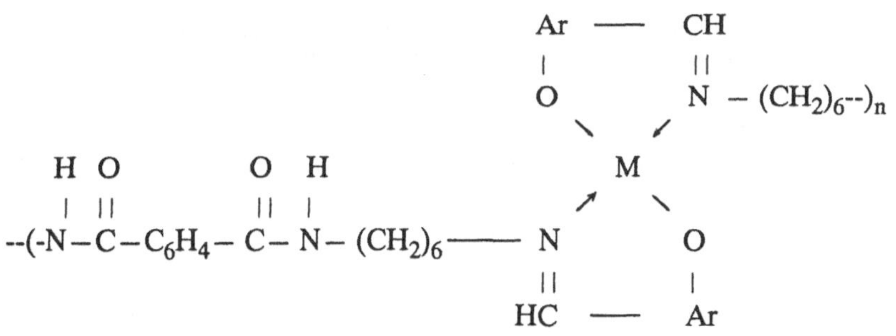

$$
\begin{array}{ccc}
Ar & - & CH \\
| & & \| \\
O & & N - (CH_2)_6\text{--})_n \\
& \searrow \quad \swarrow & \\
& M & \\
& \nearrow \quad \searrow & \\
--(-N-C-C_6H_4- \ C- N- (CH_2)_6 \text{------} \ N & & O \\
& \| & | \\
& HC & - Ar \\
\end{array}
$$

--(-N−C−C$_6$H$_4$− C− N− (CH$_2$)$_6$ ——— N

Fig. 2

125

On comparing the residual weight at 400°C, the order of thermal stabilities can be written as

PHTA < PHTA - Cu(II) - NAPHTHAL < PHTA - Co(II) - NAPHTHAL

< PHTA - Ni(II) - NAPHTHAL

However, if the residual weights at 500°C are considered, where the ligand decomposes almost completely, the order becomes.

PHTA < PHTA - Co(II) - NAPHTHAL < PHTA - Cu(II) - NAPHTHAL

< PHTA - Ni(II) - NAPHTHAL

Some of the kinetic parameters of thermal degradation viz., energy of activation (E) and frequency factor (z) were also calculated using Fuoss method[10] (Table 5).

Table 5. Activation Parameters of PHTA and its Metal Complexes

S. No.	Compounds	Ti °K	Wi mg	$(dw/dT)_i$	E	Z
					K cal/mol	
1.	PHTA	619	0.476	-0.0017	2.73	0.14
2.	PHTA-Co(II)-NAPHTHAL	783	0.486	-0.07	176.65	6.45
3.	PHTA-Cu(II)-NAPHTHAL	773	1.138	-0.043	45.16	1.56
4.	PHTA-Ni(II)-NAPHTHAL	773	3.058	-0.052	20.48	0.69

where Ti is inflection temperature, w_i weight at T_i, $(dw/dT)_i$, is rate change in weight (heating rate 20°C/8 min), E-energy of activation, Z-frequency factor (E and Z calculated by Fuoss equation).

The order of thermal stability on the basis of energy of activation was found to be as :

PHTA < PHTA - Ni(II) - NAPHTHAL < PHTA - Cu(II) - NAPHTHAL

< PHTA - Co(II) - NAPHTHAL

The order of thermal stabilities of PHTA-metal complexes is different under different criteria and none matches with Irving- Williams order of stability[9]. Exact reasons for such different results are not known at this stage but it may be due to the different degradation mechanisms operating at different stages. The participation of metal ion in catalyzing the thermal degradation can also not be ruled out.

REFERENCES

1. U.D.N.Bajpai, S.Rai, and A.Bajpai, Synthesis and characterization of metal containing coordination polymers of poly(methylene diphenylene terephthalamide) *J. Appl. Polym. Sci.,* 48: 124(1993).

2. U.D.N.Bajpai, S.Rai, and A.Bajpai, Heat resistant coordination polymers based on amino group terminated oligomer of hexamethyleneadipamide, *Synth. React. Inorg. Met. Org. Chem* 24:1719 (1994)

3. J. Brandrup and E.H. Immergut, "Polymer Hand Book", Inter-science publishers, John Wiley & Sons, New York, p. IV- 81, 1967.

4. L.J. Bellamy, Amides, proteins and polypeptides, *in*: "The Infra-red Spectra of Complex Molecules", John Wiley & Sons, New York, pp. 204-206 (1964).

5. B.G. Achhamimer, F.W. Reinhart, and G.M. Kline, Mechanism of degradation of some polyamides, *J. Res. Nat.Bur.Stand.,* 46: 381(1951).

6. B. Kamerbeek, G.H. Kroes, W. Grolle, Thermal degradation of some polyamides, *Khim. Tekhnol. Polim.,* 4: 53(1961).

7. J. Dipietro, H. Barda, and H. Stepniczka, Burning character-istics of cotton, polyester and nylon fabrics, *Text. Chem. Color.,* 3: 45(1971).

8. R.M. Fuoss, I.O.Salyer and H.S.Wilson, Evaluation of rate constants from thermogravimetric data, *J. Polym. Sci.,* Part A-2: 3147 (1971).

9. H. Irving and R.J.P. Williams, The stability of transition- metal compleres, *J. Chem. Soc.,* 3192: (1951).

SYNTHESIS AND CHARACTERIZATION OF POLY(THIOOXAMIDE) METAL COMPLEXES

Anjali Bajpai[1], Milind Khandwe[2], and Udai D.N.Bajpai[2]

[1]Department of Chemistry,
Government Autonomous Science College,
Jabalpur - 482 001, M.P., India.

[2]Polymer Research Laboratory,
Department of Post Graduate Studies and Research in Chemistry,
R.D.University, Jabalpur - 482 001, India.

INTRODUCTION

Zinc complexes of sulphur-nitrogen containing ligands are very important from the biological point of view. Zinc is one of the widely coordinated metals in biologically important macromolecules. Synthesis of novel macromolecular complexes of Zn, coordinated via S and/or N should be beneficial for the future research in this field. Among S and N containing ligands, dithiooxamide and its N, N'- substituted products are particularly interesting, since they may act as chelating agents and may form polymeric complexes. In our previous communication[1] we had reported Zn(II) complexes of N, N'- bis(carboxymethyl) dithiooxamide. In the present communication we are reporting the synthesis and characterization of Zn(II) complexes of poly(thiooxamides) viz., poly(hexane thiooxamide) [PHTO], poly(butane thiooxamide [PBTO], and Poly(ethane thiooxamide) [PETO].

EXPERIMENTAL

Materials.

Synthetic grade dithiooxamide (DTO) obtained from Baker Chemical Co. (N.J.) and diamines viz., 1,6 - diamino hexane, 1,4 - diaminobutane and 1,2 - diaminoethane (Merck) were used as such. Zinc acetate and chloride of A.R. grade were used. Solvents were distilled before use and double distilled water was used for all preparations.

Preparations.

The oligomeric ligands, poly (alkanethiooxamides), were prepared and purified by a known method[2]. Equimolar amounts of dithiooxamide and diamine in ethanolic solutions were constantly stirred for 3 hours at 50°C. Upon cooling polymers precipitated as pale coloured solids

$$n \, H_2N.CS.CS.NH_2 + n \, H_2N.R.NH_2 \rightarrow H - [-NH.CS.CS.NH.R-]_n-NH_2 + NH_3$$

For purification of oligomers the precipitates were dissolved in pyridine and reprecipitated by the addition of ethanol. The process was repeated twice. The calculated amount of ethanol was added to recover the high molecular weight fraction of the polymer. The molecular weights of polythiooxamides, prepared in this manner[2] were reported to be approximately 10^3.

The zinc complexes of these ligands were prepared by using one mole of metal ion per repeat unit weight of the ligand. The ligand solution in pyridine and a saturated aqueous solution of zinc acetate or zinc chloride were mixed with constant stirring. The complexes thus precipitated were purified by repeated washing with pyridine, water and acetone. The complexes were dried under vacuum. The Zn(II) complex of dithiooxamide (DTO) was also prepared under similar conditions and used for comparison.

$$H-[-NH.CS.CS.NHR-]_n-NH_2 + Zn(II) \xrightarrow{\text{Pyridine}} H-[-(Zn)N.CS.CS.NR]_nNH_2$$

Measurements.

C, H, N analyses were carried out with a Carlo - Erba elemental analyzer. The metal content was determined by decomposing the complex with conc. HNO_3 and titrating against EDTA. The IR spectra were recorded over the 4000 - 200 cm^{-1} range on a Perkin - Elmer model 1410 IR spectrophotometer. The IR spectra of some of the residues obtained by thermal decomposition of polymeric complexes were recorded over Perkin Elmer 1720 FTIR spectrophotometer. Thermogravimetric analyses were carried out on a Perkin Elmer TGA-4 thermal analyser in air at the heating rate of 15°C/min.

RESULTS AND DISCUSSION

The complexes were yellow coloured solid materials which decompose before melting and are insoluble in common organic solvents. Due to intractability molecular weights of the polymeric ligands could not be determined.

All the Zn(II) complexes of poly(thiooxamides) are non- electrolytic in nature and do not give qualitative tests for acetate and chloride ions. Therefore, thiooxamide group must also be satisfying the electrovalency of zinc by deprotonation of amino group. Experimental values of the elemental analyses (Table 1) fairly agree with the theoretical values calculated on the basis of formula $[Zn(N.CS.CS.NR)]_n$ for PHTO, PBTO and PETO complex and $[Zn(NH.CS.CS.NH).2H_2O]_n$ for DTO complexes. A slight deviation of the experimental values from the theoretical values may be due to the polymeric nature of the complexes. Further the exact amount of coordinated/lattice water molecule could not be determined. The posibility of some unreacted repeat unit also exists.

Table 1. Analytical Data of Poly(thiooxamides) and their Metal Complexes.

No.	Compound	Elemental Analysis[a]			
		% C	% H	% N	% M
1-	[DTO]	20.00	3.33	23.33	–
	$(C_2H_4.N_2S_2)$	20.14	3.57	23.49	–
2.	$[DTO-Zn(II).2H_2O]$	10.08	2.71	12.65	29.50
	$(Zn.C_2H_2N_2S_2.2H_2O)_n$	11.93	2.74	12.31	30.21
3.	[PHTO]	47.52	6.93	13.86	–
	$(C_8H_{14}N_2S_2)_n$	46.79	7.39	14.03	–
4.	[PHTO - Zn(II)]	36.17	4.52	10.55	24.63
	$(ZnC_8H_{12}N_2S_2)_n$	35.39	4.79	9.45	24.81
5.	[PBTO]	41.27	5.74	16.09	–
	$(C_6H_{10}N_2S_2)_n$	41.24	5.98	16.99	–
6.	[PBTO - Zn(II)]	30.33	3.37	11.79	27.54
	$(Zn\ C_6H_8N_2S_2)_n$	30.47	3.76	10.64	27.54
7.	[PETO]	32.87	4.10	19.17	–
	$(C_4H_8N_2S_2)_n$	32.52	3.88	20.59	–
8.	[PETO -Zn(II)]	22.92	1.91	13.37	31.23
	$(ZnC_4H_4N_2S_2)_n$	22.59	2.15	13.55	30.39
a.	For each compound, the upper values represent theoretical analysis and the lower values show the results obtained experimentally.				

Infrared Spectral Studies.

IR spectra of complexes were compared with those of ligands to elucidate the nature of coordination. The assignments of the important IR bands based on available literature data for dithiooxamide and similar compounds[3, 4] are depicted in

Table 2. The IR spectra of the ligands PHTO, PBTO and PETO and one of their complex viz. PHTO-Zn are illustrated in Figure 1.

The non-appearance of characteristic peaks of acetate ions, νM-O and νM-Cl in complexes prepared with zinc acetate and zinc chloride respectively further support the absence of these anions in the complexes. All the measurements show that complexes prepared from zinc acetate or zinc chloride with the same ligand are identical in all respects. It seems that the anions of the metal salts affect only the yield of complex. Under identical conditions zinc acetate gives higher yield of the complex than zinc chloride.

In the spectrum of dithiooxamide the bands observed between 3295 cm^{-1} correspond to symmetrical and assymetrical stretching of NH group. On complexation, a broad band is observed at 3400 cm^{-1} which may be due to OH stretching and a sharp band due to NH stretching is observed at 3240 cm^{-1} of the coordinated water molecules. In the poly(thiooxamides) a single band is observed at 3160 cm^{-1} due to NH stretching which on complexation appears as a broad band with greatly reduced intensity. Therefore, bonding of nitrogen with metal through deprotonation can be inferred.

Table 2. Important IR spectral assignment of poly(thiooxamides) and their metal complexes (wave numbers in cm^{-1}).

S. No.	Compounds	νNH	Mixed Vibrations of NCS group				νCS	νMN/ νMS
1.	[DTO]	3295s 3210s 3140s	1690s	1585s 1540s	1450m 1435m	1330m	835s	–
2.	[DTO.Zn(II)2H$_2$O]	3500m	1620 mb –	1520vs 1490s 1434m	–	1300s	830s	370m 270s
3.	[PHTO]	3108s	–	1530vs	1455s	1380sh 1360s	870s	–
4.	[PHTO-Zn(II)]	3190b	–	1520sb	1440mb	1380mb 1340mb	860m	390 w 280 w
5.	[PBTO]	3180s	–	1530vs	1455m	1385sh	870m	–
6.	[PBTO-Zn(II)]	3260b		1510s 1500s	1440m	1340m	840m	440 48,0
7.	[PETO]	3060s	1660s 1620m	1590m 1520s	1440m	1360m 1330m	770m	–
8.	[PETO-Zn(II)]	3190s	1650m 1600m	1540m	1440s 1420m	1320m	840m	340m 270m

v = very, s = strong, sh = shoulder, b = broad, m = medium, w= weak.

In the 1650-1300 cm^{-1} region the band observed in the spectra of ligands are due to the mixed vibrations of NCS group which have main contribution from NH bending and CS stretching. These vibrations are greatly influenced on complexation with Zn suggesting coordination through sulphur and nitrogen atoms of the thioamide group.

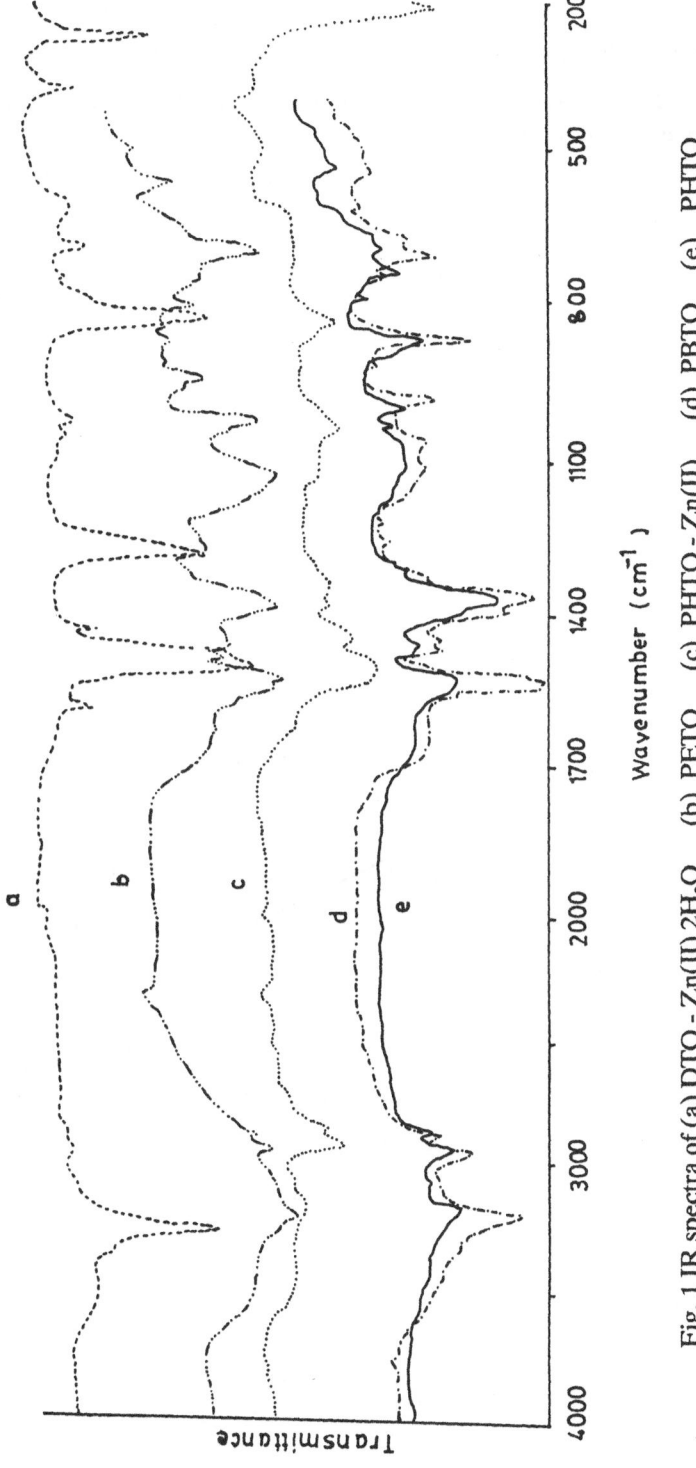

Fig. 1 IR spectra of (a) DTO - Zn(II) 2H$_2$O　(b) PETO　(c) PHTO - Zn(II)　(d) PBTO　(e) PHTO

The presence of bands in 400 cm^{-1} region can be assigned for vibration due to Zn - S and Zn - N containing chelate.

Thermogravimetric analyses

Thermogravimetric data for the oligomeric ligands and their Zn(II) complexes are depicted in Table 3 and Figure 2 and 3.

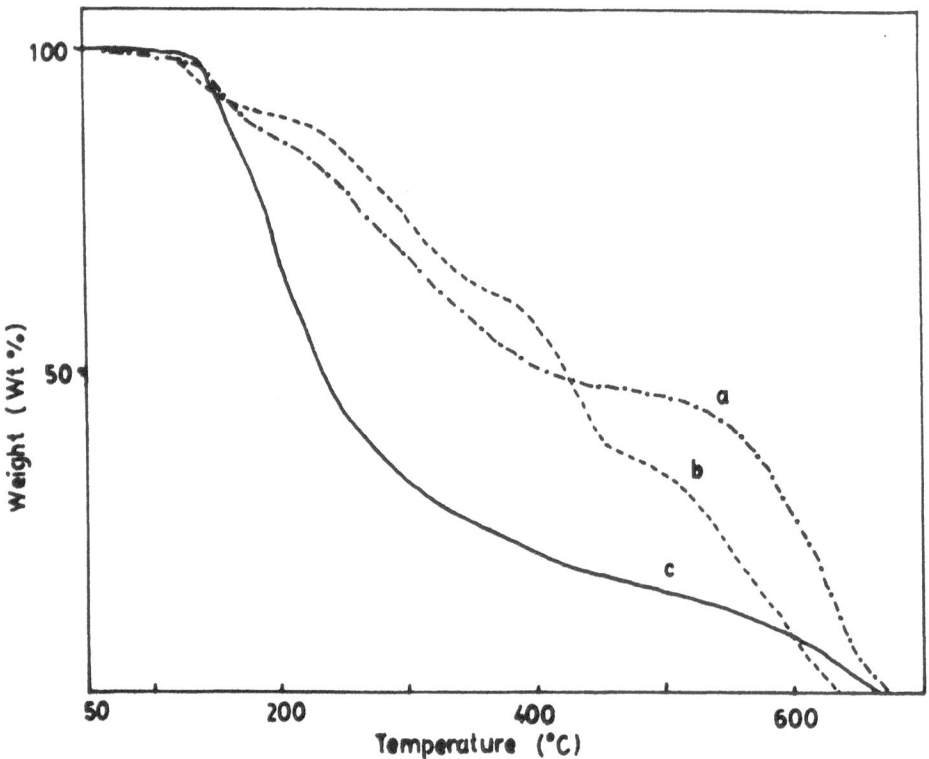

Fig. 2 . Thermogravimetric curves of poly(thiooxamides)

     (a) PBTO

     (b) PHTO

     (c) PETO

Table 3. TGA data of poly(thiooxamides) and their complexes prepared with zinc acetate

Temp °C	Percentage Weight loss					
	[PHTO]	[PHTO-Zn(II)]	[PBTO]	[PBTO-Zn(II)]	[PETO]	[PETO-Zn(II)]
50	0	0	0	0	0	0
100	1	0	1	2	0	2
150	7	0	5	2	5	3
200	10	0	14	5	31	6
250	15	1	22	8	56	16
300	26	33	33	14	66	33
350	37	36	42	34	74	40
400	44	42	50	45	78	44
450	60	46	52	48	82	46
500	67	48	53	51	84	47
550	78	51	57	56	86	48
600	93	56	73	61	93	50
650	100	70	95	62	98	50
700	100	70	100	62	100	51

TG analyses shows that thermal stabilities of ligands are enhanced on metal coordination. The TGA of the ligands show continuous decomposition above 140°C. At about 620-670°C the ligands completely decompose into volatile products. However, all the coordination polymers show high residual weight even at 700°C. The initial weight loss in the case of dithiooxamide complex is equivalent to the loss of two water molecules and the decomposition of polymeric structure sets in above 360°C. In the metal complexes of poly(thiooxamides), the sharp inflection points in TG curves can be attributed to the release of the gaseous products, resulting from the structural changes in the polymers. The PHTO complex shows about 3% weight loss upto 290°C and a steep weight loss of about 33% is observed between 300-310°C, which corresponds to the loss of hydrocarbon part, i.e., the hexamethylene, $(CH_2)_6$, unit of the polymeric ligand.

$$[(Zn)N.CS.CS.N(CH_2)_6]_n \xrightarrow[300°C]{\Delta} [(Zn)N.CS.CS.N]_n$$

The release of hydrocarbon part may be in the form of gaseous hydrocarbon or CO_2, the samples of PHTO and PBTO complexes were heated for 15 minutes at 300°C in air oven and the FTIR spectra of resulting samples were studied. The IR spectra of unheated complexes show bands due to CH stretching at 2920 and 2890 cm^{-1} but in the spectra of heated samples these bands were almost absent. This observation further supports our conclusion regarding the loss of hydrocarbon part.

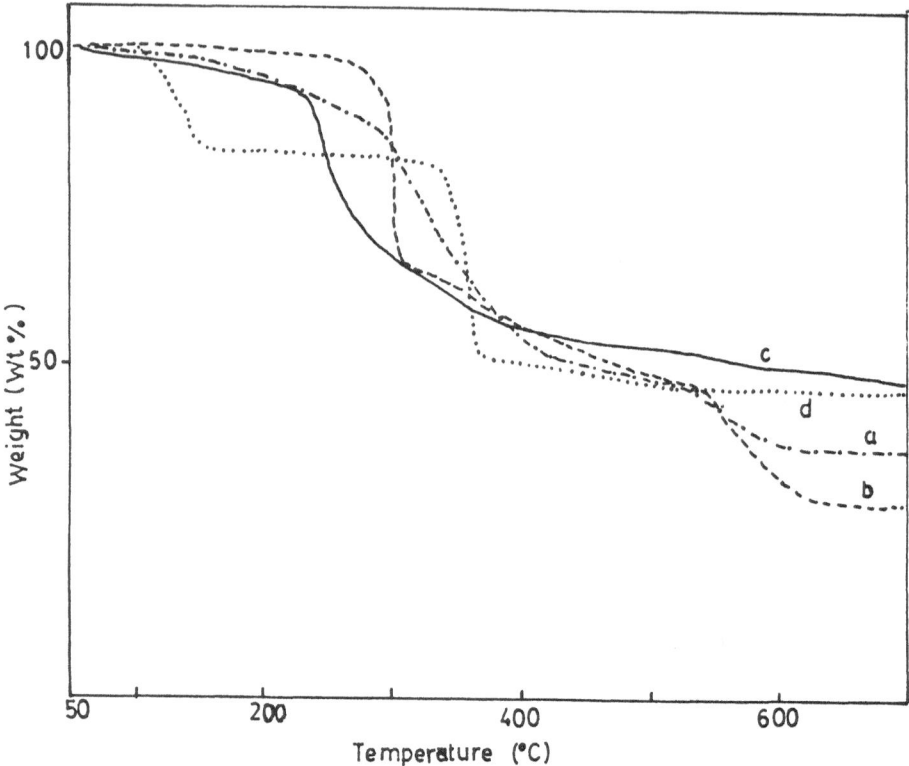

Fig. 3. Thermogravimetric curves of polythiooxamide - Zn complexes

(a) PBTO - Zn(II) (b) PHTO - Zn(II) , (c) PETO - Zn(II) , (d) DTO - Zn(II)

The thermal decomposition of hydrocarbon part (methylene units) further suggests that coordination with S and N atoms of thiooxamide group might have resulted in a stable chelate formation with zinc involving considerable electron

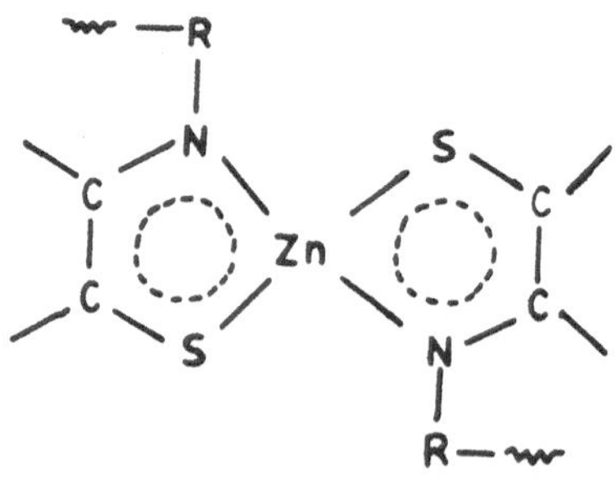

Fig. 4.

delocalization within the five-membered chelate ring. On the basis of all the above studies, a general structure may be proposed for the complexes (figure 4).

If the percentage weight of the methylene units in the respective complexes is subtracted from the residual weight at 700°C, i.e., 11.99% for $(CH_2)_2$ from PETO complex, 21.42% for $(CH_2)_4$ from PBTO complex and 29.02% for $(CH_2)_6$ from PHTO complex, the the total weight loss upto 700°C appears to be almost same ($\sim$ 40%) in all the complexes. This residual weight is greater than the metal content, therefore, the proposed structure finds further support. It is possible that the clelate rings which are part of the polymeric chain resist the thermal degradation.

The activation energy parameters calculated by Fuoss method[8] are given in Table 4.

Table 4. Activation energy parameters of poly(thiooxamides) and their Zinc(II) complexes.

S.No.	Compounds	Ti °K	Wi mg	(dw/dt)i	E Kcal/mole
1.	[DTO.Zn(II)2H₂O]	620	2.55	-0.0227	6.87
2-	[PHTO]	752	2.60	-0.016	7.13
3.	[PHTO-Zn(II)]	573	2.95	-0.049	11.12
4.	[PBTO]	658	2.02	-0.006	2.65
5.	[PBTO-Zn(II)]	688	1.50	-0.008	4.98
6.	[PETO]	543	1.20	-0.013	6.48
7.	[PETO-Zn(II)]	538	5.47	-0.050	28.94

Ti inflection temperature, i.e. DTG maximum temperature, w_i weight at T_i, $(dw/dT)_i$ rate of change in weight (heating rate 15°C/min),E energy of activation.

On the basis of activation energy the order of thermal stabilities for these ligands and their complexes can be given as

[PETO - Zn(II)] > [PHTO - Zn(II)] >[PHTO] >[DTO . Zn(II). 2H₂O] > [PETO] >[PHTO - Zn(II)] > [PBTO]

Coordination polymers of PBTO with Co(II), Ni (II), and Cu(II) have shown little electronic and protonic conductivities[9] but these zinc complexes were not found to be conductive.

REFERENCES

1. M.Khandwe, A.Bajpai, and U.D.N.Bajpai, Coordination polymers of Zn(II) and N,N'-bis (carboxymethyl)dithiooxamide,*Polymer Bulletin* 23: 51(1990).
2. M.Khandwe, A.Bajpai, and U.D.N.Bajpai, Synthesis and characterization of coordination polymers of Hg(II) with poly(thiooxamides), *J. Inorg. Organomet. Poly.*, 1: 417(1991).
3. A.Ray and D.N.Sathyanarayana, Reinvestigation of the normal vibrations of thioacetamide, *Bull. Chem. Soc. Jap.* 47(3): 729(1974).
4. A.C.Fabretti, G.C.Pellacani and G.Peyronel, Palladium(II) complexes with

N,N' -hydroxyethyl and N, N' -dibenzyl- dithiooxamide, *Gazz. Chem. Ital.,* 103(3): 397(1973).

5.   R.M.Fuoss, I.O.Salyer and H.S.Wilson, Evaluation of rate constants from thermogravemetric data, *J. Polym. Sci.* Part A-2, 3147(1964)

6.   M.Khandwe, A.Bajpai and U.D.N.Bajpai, Synthesis and characteriziation of coordination polymers of poly(butane thiooxamide) with Co(II), Ni(II), and Cu(II) salts, *Macromolecules,* 24:5203(1991).

MULTI-ELECTRON TRANSFER PROCESS OF A VANADIUM DINUCLEAR COMPLEX FOR MOLECULAR CONVERSIONS

Eishun Tsuchida*, Kimihisa Yamamoto†, and Kenichi Oyaizu

Department of Polymer Chemistry
Waseda University, Tokyo 169, Japan

INTRODUCTION

The chemistry of vanadium(IV) is dominated by the oxovanadium cation ($[V=O]^{2+}$) which remains unchanged during many chemical reactions.[1]   The reactivity is usually enhanced by the deoxygenation of oxo complexes of vanadium(IV) to produce   six-coordinate vanadium(IV) complexes.[2]   Since vanadium(V) is a strong oxidant and vanadium(III) and, sometimes, vanadium(IV) can be oxidized by O_2, redox processes involving the V(IV)/V(III) and V(V)/V(IV) couples are interesting in relation to the autoxidation of organic molecules.   In practice, oxovanadium(IV) derivatives have been used as catalysts in the epoxidation of olefins and in the oxidation of sulfides with peroxides.[3]

Facile oxidative polymerization utilizing the abundant and cheap oxidant O_2 to make aromatic polymers such as high-performance engineering plastics and electroconductive polymers provides a desirable clean process for upgrading the value of a material. The typical example is the commercial production of poly(2,6-dimethyl-1,4-phenylene oxide) through oxygen oxidative polymerization of 2,6-dimethylphenol with a copper amine catalyst.[4,5]

Vanadium complexes are of considerable interest as redox catalysts. Vanadium (III, IV, and V) complexes bridge a large potential gap between oxygen and aromatic compounds with high oxidation potential such as diphenyl disulfide, pyrrole, thiophene and benzene.   Vanadium complexes are excellent electron mediators because the redox potentials of V(III) and V(V) are located in the potential gap and they can act as powerful reducing and oxidizing agents, respectively. This article describes that a novel catalysis by vanadyl complexes is applicable to the synthesis of oligo(p-phenylene sulfide) (OPS) containing an S-S bond by an oxygen-oxidative polymerization of diphenyl disulfide.

Multi-electron transfer process[6] plays an important role in the oxidative polymerization. The first example of a vanadium binuclear complex which displays the reversible multi-electron transfer is also reported. The net reaction of this redox process is the transfer of two electrons, which can be regarded as being responsible for bridging the electron transfer from diphenyl disulfide to O_2 in the polymerization of the disulfide.

Metal-Containing Polymeric Materials
Edited by C.U. Pittman, Jr., *et al.*, Plenum Press, New York, 1996

EXPERIMENTAL

Preparation of Vanadium Binuclear Complex. Synthesis of μ-oxo-bis[(N,N'-ethylenebis(salicylideneaminato))vanadium(IV)] tetrafluoroborate ([{V(salen)}₂(μ-O)](BF₄)₂) was carried out under a dry argon atmosphere. VO(salen) (0.631 mmol, 0.21g) was dissolved in 20 mL of degassed dry dichloromethane. An equimolar amount of triphenylmethyl tetrafluoroborate (φ₃CBF₄) (0.21 g) in dichloromethane (5 mL) was added slowly to the solution at 0 ℃. The solution immediately turned dark blue, and black powder precipitated after a few minutes which was isolated under anaerobic conditions. The yield was >90 %. Anal. Calcd for [{V(salen)}₂(μ-O)](BF₄)₂: C, 46.64; H, 3.42; N, 6.80%. Found: C, 46.58; H, 3.66; N, 6.84%. IR (Kbr, cm⁻¹): 1124, 1183, 1033 (v_{B-F}). FABMS: *m/e* 650, 333, 316.

Polymerization Procedure. A typical procedure was as follows. Vanadyl acetylacetonate (0.066 g, 5 mmol/L), trifluoromethanesulfonic acid (0.075 g, 10 mmol/L), and trifluoroacetic anhydride (2.1 g, 0.2 mol/L) were dissolved in 1,1,2,2-tetrachloroethane (40 mL) in a closed vessel maintained at 20 ℃. The vessel was kept airtight. The mixture was stirred for 1h under a dry argon atmosphere before the polymerization. The atmosphere in the vessel was then replaced by oxygen. A solution of diphenyl disulfide (1.09 g, 0.1 mol/L) in 1,1,2,2-tetrachloroethane (10 mL) was added.

The reaction was carried out for 40 hr under oxygen atmosphere with constant stirring at 20 ℃. During the polymerization, a white powder precipitated in the solution. After the reaction, the mixture was poured into 200 mL of methanol containing 5 % hydrochloric acid. The precipitate was collected and washed repeatedly with methanol and water.

O₂ Uptake Masurements. The amount of O₂ consumed was determined by using modified Warburg's apparatus as follows. Vanadyl acetylacetonate, trifluoromethanesulfonic acid, and trifluoroacetic anhydride were dissolved in 1,1,2,2-tetrachloroethane in a closed vessel maintained at 20 ℃. The mixture was stirred under a dry argon atmosphere for ca. 30 min until a steady state was reached. A similar vessel equipped with a manometer and a buret was connected to the vessel with a pressure tube. The atmosphere in the vessel was replaced by O₂ and the whole instrument was kept airtight. A solution of diphenyl disulfide in 1,1,2,2-tetrachloroethane was added to the acidic solution of VO(acac)₂. 1,1,2,2-Tetrachloroethane was carefully added from the buret to the second vessel to keep the pressure at 1 atm. Oxygen consumption was measured from the net amount of 1,1,2,2-tetrachloroethane thus added during the reaction.

RESULTS AND DISCUSSION

Electron Transfer Process of Vanadium Dinuclear Complex. Electron transfer process of vanadium complexes were investigated by the use of stable (N,N'-ethylenebis(salicylideneaminato))oxovanadium(IV) (VO(salen)) as a model compound. VO(salen) forms a dinuclear complex [{V(salen)}₂(μ-O)](BF₄)₂ in the presence of trifluoromethanesulfonic acid or triphenylmethyl tetrafluoroborate as Eq. 1 (see Experimental Section).

$$2VO(salen) + 2\phi_3C^+ = [\{V(salen)\}_2(\mu-O)]^{2+} + \phi_3COC\phi_3 \qquad (1)$$

The cyclic voltammograms in Figure 1(a) shows the redox of [{V(salen)}₂(μ-O)](BF₄)₂ (0.5 mmol/L) in dichloromethane containing 0.5 mol/L of tetrabutylammonium tetrafluoroborate using the platinum disk electrode. Reversible redox wave is observed at 0.58 V(vs. Ag/AgCl)[7] with a potential separation (ΔE_p) of 75 mV between cathodic and anodic peaks which is smaller than that for VO(salen) (85 mV) at the same conditions.[8] The redox potential of the μ-oxo dimer shifts to more positive potential than the one-electron oxidation potential of VO(salen) (0.56 V vs. Ag/AgCl).[9]

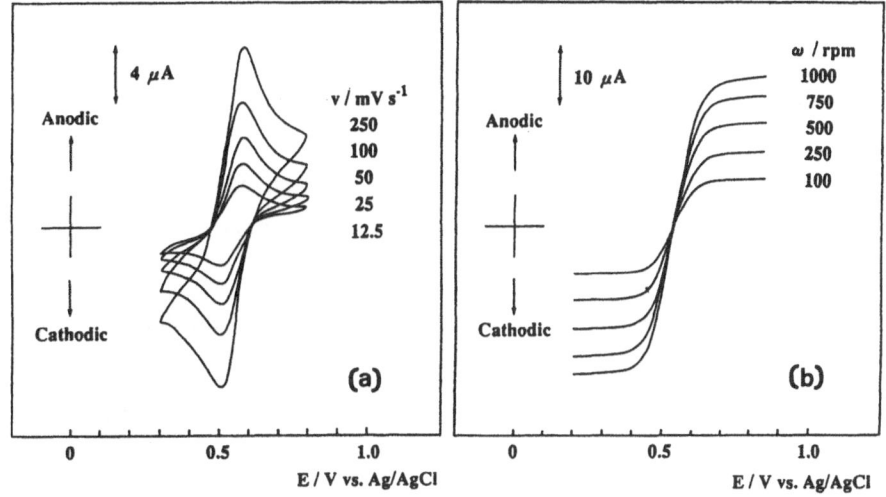

Figure 1. Cyclic voltammograms (a) and rotating disk voltammograms (b) of [{V(salen)}₂(μ-O)](BF₄)₂ (0.5 mmol/L) under anaerobic conditions in dichloromethane containing tetrabutylammonium tetrafluoroborate (0.5 mol/L) on Pt disk electrode (4 mmφ).

As an additional check on the system, the rotating disk voltammetry of VO(salen) and [{V(salen)}₂(μ-O)](BF₄)₂ were measured at the same conditions. The rotating disk voltammetry (Figure 1(b)) shows symmetric waves at $E_{1/2} = 0.58$ V vs. Ag/AgCl, i.e. the cathodic current is equal to the anodic one regardless of the rotation rate. The limiting currents, i_{la}, i_{lc}, are identified to be the oxidation and the reduction of the vanadium dinuclear complex, respectively. A plot of the plateau current ($i_l = i_{la} + i_{lc}$) vs. the square root of the rotation rate (Levich plot)[10] yields a straight line. The electrode potential at $i = 0$ was equal to the half wave potential. The plateau current for [{V(salen)}₂(μ-O)]²⁺ was about two times larger than the one-electron oxidation (V^{IV}O/V^{V}O) current of VO(salen), which confirmed the two-electron transfer process of the dinuclear complex [{V(salen)}₂(μ-O)]²⁺. The coulometric titration (exhaustive electrolysis) also supported that this dinuclear complex exhibits reversible two-electron transfer in a single voltammetric step ($n = 1.98$). The redox process of the oxovanadium involves the two-electron transfer of the VO complex based on vanadium(III) and vanadium(V) species in the acidic enviroments of this experiments. In the oxidative polymerization of diphenyl disulfide, the vanadium(V) species reacts with diphenyl disulfide to form the active species of the polymerization as will be described later. The vanadium(III) species is reoxidized with molecular oxygen. VO(acac)₂ acts as an electron mediator through two-electron transfer from disulfide to oxygen. The resulting redox system of VO complexes bridge the large potential gap between the reduction of oxygen and the oxidation of the disulfide.

Electrochemical confirmation that VO(salen) reacts with trifluoromethanesulfonic acid (CF₃SO₃H) or triphenylmethyl tetrafluoroborate (φ₃CBF₄) to form a deoxygenated complex, V^{IV}(salen)²⁺, and a μ-oxodinuclear complex, [(salen)VOV(salen)]X₂, (X = CF₃SO₃⁻ or BF₄⁻) has been presented.[11] Cyclic voltammograms of VO(salen) in the presence of CF₃SO₃H or φ₃CBF₄ exhibit reversible waves with formal potentials near 0.5 V and 0.8 V (vs. Ag/AgCl). The cathodic wave at 0.5 V arise from the combined reduction of V(salen)²⁺ and the μ-oxo dimeric complex and the wave at 0.8 V from the oxidation of the V(salen)²⁺ complex.   The dimerization of VO(salen) is initiated by deoxygenation of the V=O center by H⁺ or φ₃C⁺ to produce V(salen)²⁺ which enters into an equilibrium with a second VO(salen) complex to produce the μ-oxo dimer. The equilibrium contant for the formation of the μ-oxo dimer in acetonitrile was evaluated as 0.7 mM⁻¹.

Oxidative Polymerization.   In the presence of a catalytic amount of VO(acac)₂, the O₂-oxidative polymerization of diphenyl disulfide proceeded   accompanied by a quantitative oxygen uptake.   Oxygen is essential for the polymerization of diphenyl disulfides. For each mole of oxygen consumed, 2.0 mol of disulfide is consumed.

Table 1. O₂ Oxidative Polymerization of Diphenyl Disulfide with Catalyst.

catalyst	[Cat.] / [Monomer]	E$_{1/2}$ (V vs Ag/AgCl)	OPS yield (wt %)
DDQ	0.1	1.2[b] (Q/QH)	0
Pb(CH₃CO₂)₄	0.1	1.5[b] (III/IV)	0
Cu(acac)₂	0.1	-0.3 (I/II)	0
Fe(acac)₃	0.1	-0.65 (II/III)	0
VO(salen)	0.1	0.5 (IV/V)	0
VO(bzac)₂	0.1	1.1 (IV/V)	95
VO(acac)₂	0.1	1.1 (IV/V)	92
VO(acac)₂	0.01	1.1 (IV/V)	68
VOTPP^{c}	0.01	1.5 (IV/V)	95
V(acac)₃	0.1	0.8 (III/IV)	85
V₂O₅^{d}	2.0	e	97

[a] 2,3-Dichloro-5,6-dicyano-p-benzoquinone. [b] Oxidation peak potential. [c] Reaction time 120 h. [d] vanadium(V), under N₂ atmosphere. [e] Insoluble.

A variety of other metal complexes that are known to act as redox catalysts in many chemical reactions were utilized under the same conditions. However, all of them are not effective as a catalyst for the oxygen oxidative polymerization of diphenyl disulfide (Table 1).   Although an equimolar of 2,3-dichloro-5,6-dicyano-p-benzoquinone(DDQ) and lead tetraacetate can oxidize disulfide to yield OPS due to the high oxidizing ability, these oxidants do not act as a catalyst for the efficient polymerization of disulfides because the reduced species of the oxidant is not reoxidized with oxygen. The other complexes such as Cu(acac)₂, Fe(acac)₂, and VO(salen) can not oxidize disulfide due to their lower redox potentials.

In contrast, bis(benzoylacetonato)oxovanadium(IV) (VO(bzac)₂) with the same redox potential as VO(acac)₂ is effective as a catalyst in the polymerization. In addition, even in the presence of 0.2 % VOTPP complex as a catalyst, the polymerization also proceeds quantitatively for 120 h accompanied by oxygen uptake, though the polymerization rate is slow. These results may suggest that only vanadyl complexes are available as catalysts because the redox potentials match the system and they can transfer multiple electrons.

The OPS yield also depend on the acidity of the mixture. The oxidative polymerization does not proceed in the absence of acids. Strong acids such as trifluoromethanesulfonic acid or trifluoroacetic acid are effective in the VO-catalyzed polymerization (Table 2). The polymerization is facilitated by the high oxidizing ability of the acid-activated VO(acac)₂. Diphenyl disulfides are not oxidized with only oxygen or with only an equimolar amount of VO(acac)₂ in the absence of acid due to the high oxidation peak potential (1.7 V vs. Ag/AgCl) of dipenyl disulfide.[12] The VO catalyst is an excellent electron mediator through activation by acid to promote electron transfer between the disulfide and oxygen.

Table 2. VO-catalyzed oxidative polymerization of diphenyl disulfide at 20 °C under acidic environments

Monomer [a)	E [b) (V vs Ag/AgCl)	Solvent	Acid	OPS yield (wt %)	T_g (°C)	$\bar{M}_w$
(1)	1.65	(CHCl₂)₂	CF₃SO₃H	92	58	5450
(1)	1.65	CH₂Cl₂	CF₃COOH	76	-	1890
(2)	1.4	CH₂Cl₂	CF₃COOH	93	198	9200
(3)	1.5	CH₂Cl₂	CF₃COOH	98	298	6790

[a] Monomer: (1), diphenyl disulfide; (2), bis(3,5-dimethylphenyl) disulfide; (3), bis(2,5-dimethylphenyl) disulfide. [b] Oxidation peak potential.

The intermediate species produced from diphenyl disulfide have been studied using nonpolymerizable dimethyl disulfide as a model compound. Methylbis(methylthio) sulfonium cation salt has been isolated quantitatively by the oxidation of dimethyl disulfide. The structure has been confirmed by ^{1}H-NMR and elemental analysis.[13-15] The cation is the active species in the polymerization. The polymerization mechanism (Scheme I) has been determined.[16,17] Diphenyl disulfide is oxidized by the activated vanadium species to yield the phenylbis(phenylthio)sulfonium cation. The electrophilic attack of the sulfonium cation proceeds on the benzene ring of the monomer or the oligomer as a propagation process. The oxidation and the electrophilic reaction are continuously repeated to form OPS.

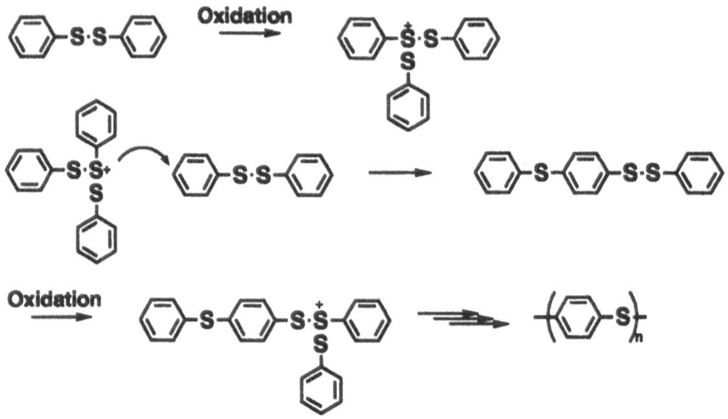

Scheme 1

Polymerization Kinetics. The initial rate of the oxygen uptake (V_0) was measured at various partial pressures of O_2 in the polymerization of diphenyl disulfide (0.05 mol/L). The dependence of V_0 on the O_2 pressure over the range of 0-0.5 atm is shown in Figure 2. At O_2 pressures above 0.6 atm, first-order disulfide substrate kinetics were observed : *i.e.* V_0 was proportional to the concentration of diphenyl disulfide and independent to the O_2 partial pressure. Below 0.5 atm of O_2 partial pressure, zero-order kinetics was found: *i.e.* V_0 was proportional to the partial pressure of O_2 and independent to the concentration of diphenyl disulfide. That is, at fixed concentration of diphenyl disulfide, V_0 became constant above 0.5 atm and was proportional to the O_2 partial pressure below 0.5 atm. These results suggests that the rate-determining step in the polymerization under atmospheric pressure is the oxidation process in which diphenyl disulfide is oxidized by the activated vanadyl complex. The observed zero-order dependence of the O_2 partial pressure is consistent with this mechanism since the activated vanadyl species is a better oxidant than O_2. This is also consistent with the two-electron transfer process in the catalytic system as indicated by the electrochemical measurements.

One intriguing aspect of these VO-catalyzed oxidations was the effect of substituents in a series of methyl-substituted diphenyl disulfides on the observed rate of the reaction. The VO-catalyzed oxidation of bis(3,5-dimethylphenyl) disulfide showed a zero-order substrate kinetics: *i.e.* V_0 was proportional to the partial pressure of O_2 and independent to the concentration of diphenyl disulfide (Figure 2). A linear plot of V_0 versus O_2 partial pressure was obtained in the range of 0.33-1 atm. At a low partial pressure (0.33 atm), V_0 agreed with that of diphenyl disulfide. These results suggests that, in this case, the rate-determining step in the polymerization is the reoxidation of the VO species.

Bis(3,5-dimethylphenyl) disulfide is oxidized in this reaction three times rapidly than the unsubstituted diphenyl disulfide. While this may be due in part to the fact that bis(3,5-dimethylphenyl) disulfide is easier to be oxidized than diphenyl disulfide by *ca.* 0.2 V, it is not clear at this point if other factors (*e.g.* steric) affect the interaction with the VO species.

The initial rates of polymerization determined by the O_2 uptake were independent to the concentration of CF_3SO_3H. Therefore, the dis-proportionation of the oxovanadium (IV) complex is not a rate-determining step in the oxidative polymerization of diphenyl disulfide.

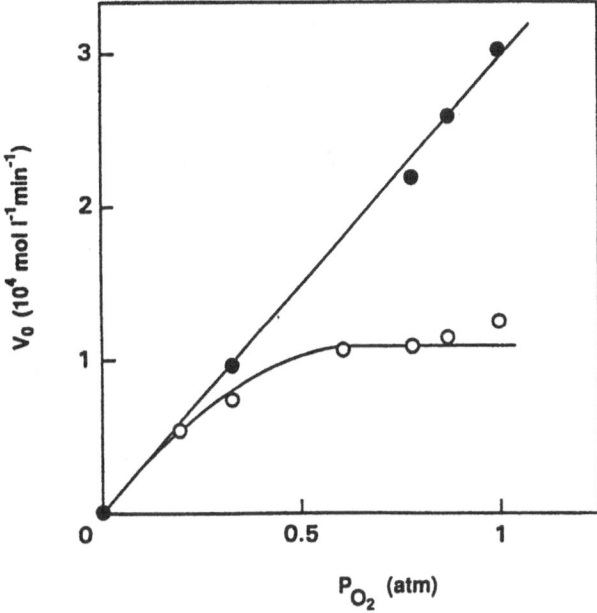

Figure 2. Effect of oxygen partial pressure (P_{O_2}) on oxygen uptake rate (V_0) in the VO-catalyzed polymerization of 0.1 mol/L of diphenyl disulfide ( ) and bis(3,5-dimethylphenyl) disulfide ( ) at 20 °C in 1,1,2,2-tetrachloroethane in the presence of CF_3SO_3H ($C_2H_2Cl_4$ (50 mL) with [disulfide] = 0.1 mol/L, [acid] = 0.01 mol/L, and sub/cat. = 40).

The initial polymerization rate (V_0) can be measured by O_2 uptake during the polymerization of a large excess of diphenyl disulfide with 2.5 mmol/L VO(acac)$_2$. Since V_0 tends to be a limiting value when the diphenyl disulfide concentration increases (Figure 3(b)), the polymerization can be classified to a Michaelis-Menten type reaction, the steady state being maintained.

Lineweaver-Burk plots (L-B plots) of the reaction were shown in Figure 3(a). The reciprocal of the initial rate of O_2 uptake vs. [diphenyl disulfide]$^{-1}$ is linear under 1 atm of O_2. The same L-B plots were observed at O_2 pressure above 0.6 atm. At lower partial pressure of O_2, polymerization rate depended on the O_2 pressure. V_0 is independent to the concentration of diphenyl disulfide at 0.33 atm of O_2 pressure.

V_{max} and K_n were calculated from the L-B plots (Table 3). The value of V_{max} for the polymerization of diphenyl disulfide was smaller than that for the polymerization of 2,6-dimethylphenol with copper-amine complex. While the K_n values of the two monomers are comparable, k_2 for the polymerization of diphenyl disulfide is much smaller than that for 2,6-dimethylphenol, which may arise from the fact that diphenyl disulfide shows higher oxidation potential (1.7V vs. Ag/AgCl) than 2,6-

dimethylphenol (0.45 V vs Ag/AgCl in methanol).   The increase in the rate of the oxidative polymerization of bis(3,5-dimethylphenyl) disulfide was observed whose oxidation potential (1.4 V vs Ag/AgCl) is lower than that of unsubstituted diphenyl disulfide, as   described in the following section.

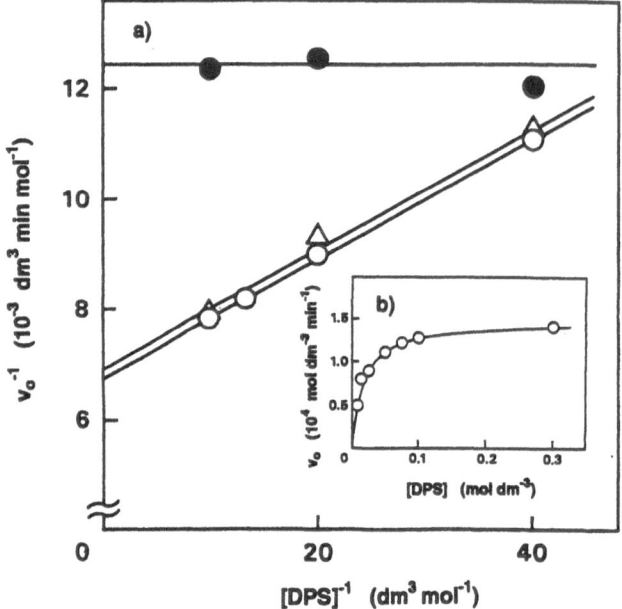

Figure 3. (a) Lineweaver-Burk plots for the polymerization of diphenyl disulfide with ( )
1.0 atm; ( ) 0.78 atm; ( ) 0.33 atm of partial pressure of oxygen. VO(acac)2: 2.5 mmol
dm-3, CF3SO3H: 0.01 mol dm-3, (CF3CO)2O: 0.5 mol dm-3. Solvent: (CHCl2)2 (50
cm3), 20 oC.   (b) Effect of the monomer concentration on   the initial rate of the
polymerization.

Figure 3 involves the effect of O₂ pressure on the polymerization.   The rate-determining step for the   polymerization of diphenyl disulfide depends on the   partial pressure of O₂. The L-B plots for 0.33 atm of O₂ pressure is parallel to X-axis.   That is, the polymerization rate is constant in the region of low partial pressure of O₂ regardless of the monomer concentration because the reoxidation of the catalyst is suppressed under the lower O₂ pressure.

V_0 was measured at various partial pressure of O₂ in the polymerization of diphenyl disulfide. The dependence of the   O₂ uptake rate on the partial pressure of O₂ was observed in   the range of 0-0.5 atm. V_0 became constant above 0.5 atm and   was proportional to the partial pressure of O₂ below 0.5 atm. These results also indicate that the rate-determining step in 1 atm O₂ atmosphere is the oxidation of diphenyl disulfide by the activated vanadyl complex.

Effect of Substituents on the Polymerization Behaviors.   The oxovanadium-catalyzed oxidative polymerization was applied to the synthesis of poly(arylene sulfide)s. The   oxidative polymerization of alkyl-substituted diphenyl disulfides, such as   bis(2-

methylphenyl) disulfide, bis(3-methylphenyl) disulfide, bis(2,5-dimethylphenyl) disulfide, bis(2,6-dimethylphenyl) disulfide, and bis(3,5-dimethylphenyl) disulfide, were carried out under the same conditions (Table 4). The polymerization of the alkyl-substituted monomers proceeded accompanied with the O_2 uptake, and the corresponding polymers were formed in high yield. V_{max} and K_m were calculated by using the Lineweaver-Burk (L-B) plots based on the measurements of the initial rate of O_2-uptake rates during the polymerization. The L-B plots of the monomers showed linear relationships. The values of V_{max} and K_m for the polymerization of bis(2,6-dimethylphenyl) disulfide were determined to be 1.3×10^{-4} mol $L^{-1}min^{-1}$ and 7.4×10^{-2} mol L^{-1}, respectively, from the L-B plots. K_m for the polymerization of bis(2,6-dimethylphenyl) disulfide is larger than that for unsubstituted diphenyl disulfide (Table 3). It suggests that the steric hindrance in case of the polymerization of bis(2,6-dimethylphenyl) disulfide may labilize the intermediate complex (Michaelis complex) which results in the larger K_m.

Table 3. Kinetic constants for the O_2-oxidative polymerization of diphenyl disulfide and 2,6-dimethylphenol

Catalyst	$10^4 V_{max}$ mol $L^{-1}min^{-1}$	$10^2 K_m$ mol L^{-1}	$10^{-1}K_I$ [a] L mol^{-1}	$10 k_2$ [b] min^{-1}
VO(acac)$_2$ [c]	1.5	1.6	6.3	0.6
Cu-Pyridine [d]	73	2.5	4.0	15

[a] $K_I = 1/K_m$. [b] $k_2 = V_{max}/[Catalyst]$, $[Catalyst] = 2.5 \times 10^{-3}$ (mol L^{-1}). [c] Monomer: diphenyl disulfide. [d] Monomer: 2,6-dimethylphenol.

K_m values for the polymerization of bis(3-methylphenyl) disulfide and bis(2,5-dimethylphenyl) disulfide were not determined because their oxovanadium-catalyzed oxidation showed zero-order substrate kinetics: *i.e.* the initial rate of O_2 uptake was independent to the concentration of the monomers. These results imply two possibilities that K_m is close to zero, or that the polymerization rate is independent to the monomer concentration.

Table 4. Oxidative polymerization of diaryl disulfides[a]

Substituent	Product		Kinetic constants		
	Yield (wt %)	$10^{-3}M_w$ [b]	$10^2 K_m$ [c] (mol L^{-1})	$10^4 V_{max}$ [c] (mol $L^{-1}min^{-1}$)	E [d] (V)
Unsubstituted	76	-	1.6	1.5	1.7
2-Methyl	85	2.6	-	-	1.6
3-Methyl	97	4.6	-	-	1.6
2,5-Dimethyl	100	-	-	-	1.5
3,5-Dimethyl	93	9.2	-	-	1.5
2,6-Dimethyl	73	1.6	7.4	1.3	1.7

[a] [Disulfide] = 0.1, [VO(acac)$_2$] = 0.005, [CF$_3$COOH] = 1.0, [(CF$_3$CO)$_2$O] = 0.2 (mol L^{-1}). [b] Determined by GPC. [c] Calculated from Lineweaver-Burk plots. [d] Oxidation peak potential (vs. Ag/AgCl).

147

Bis(3-methylphenyl) disulfide, bis(2,5-dimethylphenyl) disulfide, and bis(3,5-dimethylphenyl) disulfide exhibits lower oxidation potentials than that of unsubstituted diphenyl disulfide or bis(2,6-dimethylphenyl) disulfide (Table 3), which results in *ca.* 10 times larger rate of the oxidation of disulfide. The polymerization rate of bis(3,5-dimethylphenyl) disulfide increases in proportion to the partial pressure of O_2. Therefore, in the oxidative polymerization of monomers which exhibit lower oxidation potentials, the rate-determining step is estimated to be the reoxidation of the catalyst by O_2, as is the polymerization of unsubstituted diphenyl disulfide under low partial pressure of O_2.

ACKNOWLEDGEMENTS

This work was supported by a Grant-in-Aid for Scientific Research (Nos. 06226279, 05650865, 040850) and Developmental Scientific Research (No. 04555223) from the Ministry of Education, Science and Culture, Japan.

REFERENCES

† PRESTO investigator (1992- 1994)
1. (a) Cotton, F. A.; Wilkinson, G. "Advanced Inorganic Chemistry", 4th ed., Interscience, New York, 1987. (b) Boas, L. V.; Pessoa, J. C. in "Comprehensive Coordination Chemistry, The Synthesis, Reactions, Properties & Applications of Coordination Compounds", Wilkinson, G.; Gillard, R. D.; McCleverty, J. A., eds. Pergamon Press: Oxford, 1987, vol. 3, Ch. 33.
2. (a) Swinehart, J. H., J. Chem. Soc., Chem. Commun., 1971, 1443. (b) Pasquali, M.; Filho, A. T.; Floriani, C., J. Chem. Soc., Chem. Commun., 1975, 534. (c) Behzadi, K.; Thompson, A., J. Less-Common Met., 1987, 128, 281.
3. (a) Curci, R.; Furia, F. D.; Testi, R.; Modena, G., J. Chem. Soc., Perkin Trans., 1974, 2, 753. (b) Howard, J. A.; Tait, J. C.; Yamada, T.; Chenier, J. H. B., Can. J. Chem., 1981, 59, 2184.
4. A. S. Hay, H. S. Blanchard, G. F. Endres, J. W. Eustance, *J. Am. Chem. Soc.,* **81**, 6335 (1959).
5. A. S. Hay, *J. Polymer Sci.,* **58**, 581 (1967).
6. D. E. Richardson and H. Taube, *Inorg. Chem.,* **20**, 1278 (1981).
7. The electrode potential of the reference electrode (Ag/AgCl) was determined to be 0.21 V vs. NHE by the adjustment with a ferrocene/ferrocenium redox couple.
8. D. S. Polcyn and I. Shain, *Anal. Chem.,* **38**, 370 (1966). One of the reason that DEp for VO(salen) (VIV/VV) is larger than the calculated value for one-electron transfer is the uncompensated resistance of the electrolyte solution.
9. J. A. Bonadies and C. J. Carrano, *J. Am. Chem. Soc.,* **108**, 4088 (1986).
10. $i_\iota = 0.62nFAD^{2/3}w^{1/2}v^{1/6}C^*$(298 K). The diffusion coefficient of the binuclear complex is smaller than that of VIVO(salen) due to the increase in the charge and the molecular size. The two-electron transfer was concluded from the slope of the binuclear complex in the Levich plot which is larger than that of VO(salen).
11. E. Tsuchida, K. Yamamoto, K. Oyaizu, N. Iwasaki, F. C. Anson, *Inorg. Chem.,* **33**, 1056 (1994).
12. K. Yamamoto, S. Yoshida, H. Nishide, E. Tsuchida, *J. Elecrochem. Soc.,* **369**, 2401 (1992).
13. E. Tsuchida, K. Yamamoto, H. Nishide, S. Yoshida, M. Jikei, *Macromolecules,*

23, 2101 (1990).

14. G. Capozzi, V. Lucchini, G. Modena, F. Rivetti, *J. Chem. Soc., Perkin Trans.,* **2,** 900 (1974).

15. A. S. Gibin, W. A. Smitl, V. S. Bogdanov, *Tetrahedron,* **21,** 383 (1980).

16. K. Yamamoto, M. Jikei, K. Oi, H. Nishide, E. Tsuchida, *J Polymer Sci. Chem.* , **29,** 1359 (1991).

17. K. Yamamoto, M. Jikei, J. Katoh, H. Nishide, E. Tsuchida, *Macromolecules* , **25,** 2698 (1992).

SYNTHESIS OF HIGH MOLECULAR WEIGHT POLY(PHENYLENE SULFIDE) THROUGH OXIDATIVE POLYMERIZATION WITH OXYGEN

Kimihisa Yamamoto[+] and Eishun Tsuchida

Department of Polymer Chemistry

Waseda University, Tokyo 169 Japan

INTRODUCTION

The polymerization via selective oxidation utilizing oxygen as the oxidant presents potentially a desirable low-cost method for upgrading the value of raw materials. The most successful example is the oxidative polymerization of 2,6-dimethylphenol to yield poly (2,6-dimethyl-1,4-phenylene oxide) using a copper-amine catalyst in an oxygen atmosphere at room temperature(Eq. 1).[1] Our studies on the preparation of poly(p-phenylene sulfide) (PPS), which deserves much attention as a high-performance engineering plastic owing to its excellent chemical, thermal, and mechanical properties,[2] have revealed that thiophenol and diphenyl disulfide are oxidatively polymerized to PPS via a cationic mechanism by electrochemical or chemical oxidations(Eq. 2).[3-6] The oxovanadium catalyst acts as an efficient catalyst for the O_2 polymerization of diphenyl disulfide.[7]

The O_2-oxidative polymerization proceeds at room temperature and provides pure PPS but it results in a low molecular weight due to its poor solubility of PPS in dichloromethane at room temperature. In order to resolve the problem we employed a soluble precursor method as a new synthetic route of high molecular weight PPS. The excellent properties of PPS make it difficult to synthesize a high molecular weight PPS. Therefore, the preparation of PPS employs high-temperature and high-pressure process in order to improve poor solvent solubility. These synthetic difficulties are overcome by the use of a soluble precursor for the target polymer.

Metal-Containing Polymeric Materials
Edited by C.U. Pittman, Jr., *et al.*, Plenum Press, New York, 1996

151

$$n \ \text{(2,6-dimethylphenol)}\text{—OH} + n/2\ O_2 \longrightarrow \left[\text{(phenylene)—O}\right]_n + n\,H_2O \qquad (1)$$

$$n \ \text{(phenyl)—SS—(phenyl)} + n/2\ O_2 \longrightarrow \left[\text{(phenylene)—S}\right]_{2n} + n\,H_2O \qquad (2)$$

$$n \ \text{(phenyl)—S—(phenyl)—SCH}_3 + n/2\ O_2 + n\,HX \longrightarrow \left[\text{(phenylene)—S}\right]_{2n} + n\,H_2O + n\,CH_3X \qquad (3)$$

On the basis of the electronic structure of the sulfur atom, the poly(phenylsulfonium cation) should be employed as a precursor. We proposed a new synthetic route for the synthesis of high molecular weight PPS via a soluble precursor under mild conditions.[8] This paper describes the polymerization of methyl 4-phenylthiophenyl sulfoxide to make poly[methyl-(4-phenylthiophenyl)sulfonium trifluoromethanesulfonate] as a poly (sulfonium cation) in the presence of trifluoromethanesulfonic acid. Recently, we have found that sulfoxide was oxidatively polymerized directly to poly(sulfonium cation) catalyzed by cerium ammonium nitrate (CAN) with atmospheric oxygen in the presence of methanesulfonic acid. The present paper also describes the Ce-catalyzed O_2 polymerization of a sulfide as a novel and convenient synthetic route to high molecular weight PPS (Mw > 10^5) via a soluble precursor(Eq. 3).

EXPERIMENTAL

Measurements. The IR spectra were obtained using a Jasco FT-IR 5300 with a potassium bromide pellet. The ^{1}H-NMR spectra and ^{13}C-NMR spectra of the poly(sulfonium cations) were obtained using a JEOL FT-NMR GSX400. The CP/MAS spectrum of PPS was obtained using the same NMR spectrometer. The gas chromatograms were measured with a Shimadzu GC-12 gas chromatograph instrument. The molecular weight of the obtained PPS was obtained with a high-temperature gel permeation chromatograph (Senshu Scientific Co. Ltd., SSC VHT-7000) with α-chloronaphthalene as eluent at 210 $_o$C and equipped with Shodes GPC columns (AT-80M/s) and a UV-vis spectrophotometer (Senshu Scientific Co. Ltd., S-3750) as a detector. Polystyrene standard samples were used to corrective calibration curve.

Materials. Thioanisole, diphosphorus pentaoxide (P_2O_6), and cerium ammonium nitrate (CAN) (extra reagent grade) were purchased from Tokyo Kasei Co. and used without further purification. Methyl phenyl sulfoxide was prepared by the oxidation of thioanisole with nitric acid and purified by distillation under reduced pressure. Methyl 4-

phenylthiophenyl sulfide (MPS) was prepared by the demethylation of methyl-(4-methylthiophenyl)phenylsulfonium perchlorate (MSP) with pyridine, which was synthesized by the reaction of thioanisole and methyl phenyl sulfoxide in methanesulfonic acid. Anhydrous magnesium sulfate was obtained from Tokyo Kasei Co. and used without further purification. Methanesulfonic acid was also purchased from Tokyo Kasei Co. and used after purification by vacuum distillation.

Oxidation of Thioanisole in an O₂ Atmosphere.   CAN (109 mg, 0.2 mmol) and methanesulfonic acid (10 mL) were added to a 100-mL flask and stirred with a magnetic stirrer to give an orange solution or suspension. Thioanisole (1.24 g, 10 mmol) was added to the flask and stirred for 20 h at 25 ₀C under an oxygen atmosphere. Pyridine (30 mL) was added to the flask and the reaction mixture was refused for 30 min to demethylate the sulfonium cations. The mixture was cooled to room temperature (25 ₀C), neutralized with hydrochloric acid, extracted with chloroform, washed with water, dried over anhydrous magnesium sulfate, and concentrated by evaporation. The product was confirmed by means of gas chromatography (GC) and GC-mass spectrometry.

Synthesis of Methyl 4-(Phenylthio)phenyl Sulfoxide (MPS). A 200-mL, three-neck, round-bottom, flask which had a reflux condenser and N₂ gas inlet and was equipped with a Tefloncovered magnetic stirring bar, was charged with MPS (10 g, 28.8 mmol). Pyridine (50 mL) was added at room temperature and stirred for 30 min at room temperature. The temperature was increased to 100 °C, and reaction was continued for 10 min. The reaction mixture was poured into 10% HCl solution (300 mL) and was extracted with dichloromethane. The product was purified by flash column chromatography on silica gel using hexane-chloroform (3:1) as the eluent. After evaporation of the solvents, a colorless liquid (6.4 g, 95% yield) was obtained. The liquid was dried under vacuum at room temperature for 20 h

Polymerization of Methyl 4-(phenylthio)phenyl Sulfoxide. A   100-mL,   round-bottom flask with a Teflon-covered magnetic stirring bar was charged with methyl 4-(phenylthio) phenyl sulfoxide (1 g, 4 mmol). The flask was cooled to 0 °C. Trifluoromethane sulfonic acid (5 mL) was added at 0 °C and stirring. The temperature was increased slowly to room temperature over a period of 0.5 - 1 h. The reaction solution turned from colorless to pale blue. The reaction are continued for another 20 h at room temperature. The reaction was then quenched by pouring it into ice water. The precipitated polymer was then chopped in a blender, washed with water, and dried in vacuum at room temperature for 20 h. Yield: 1.53 g (100%).

Oxygen Oxidative Polymerization with CAN. *(i) Polymerization.* CAN (109 mg, 0.2 mmol) and P_2O_6 (1.4 g, 10 mmol) were dissolved in methanesulfonic acid (5 mL) in a closed vessel maintained at 25 $_o$C (or 70 $_o$C), and the resulting mixture was stirred for 1 h until a steady state was reached under a dry argon atmosphere. A similar vessel equipped with a manometer and a buret was connected to it with a tube. The atmosphere in the vessel was replaced by oxygen. The instrument was kept airtight. A solution of MPS (1.16 g, 5 mmol) in methanesulfonic acid (5 mL) was added. 1,1,2,2-Tetrachloromethane was carefully added from a buret to the second vessel to keep the pressure at 1 atm. The oxygen consumption was measured from the net amount of tetrachloromethane added during the reaction. After the reaction, the mixture was added to 100 mL of water. The reaction mixture was precipitated in perchloric acid, and the product was isolated as a pale yellow solid. The solid was filtered, thoroughly washed with water, and dried in vacuo over P2O6 for 10 h at 25 $_o$C to give poly[(methyl-(4-phenylthio)phenyl)sulfonium perchlorate] (PPSP; 1.59 g. Yield: 96.

(ii) Demethylation. The obtained poly(sulfonium cation) (PPSP; 0.7 g) was dissolved in 10 mL of pyridine at 25 $_o$C and stirred for 1 h. The clear reaction mixture became a white suspension after 10 min. The reaction mixture was refluxed for 12 h. The product was isolated and then poured into 200 mL containing 10% HCl. The precipitate was filtered washed with methanol, and refluxed in ethanol for 2 h, and the product was dried in vacuo for 12 h at 50 $_o$C to give 0.45 g of PPS. Yield: 98% .

Spectroscopic Data. *(i) Methyl 4-(Phenylthio)phenyl Sulfoxide(MPS).* IR(KBr,cm-l): 3057,2919 (v_{C-H}), 1580,1476,1437 ($v_{C=C}$),810 (δ_{C-H}) 739, 689. ^{1}H NMR (400 MHz, ppm, CDCl$_3$): 7.29-7.01 (phenyl 9 H, m), 2.26 (methyl, 3 H). ^{13}C NMR (400 MHz, ppm, CDCl~) 126.6,127.1,129.0,130.1,131.4,132.2,136.4, 138.2 (phenyl C) 15.6 (methyl C). Anal. Calcd for $C_{13}H_{12}S_2$: C, 67.2; H, 5.21; S 27.6. Found: C, 67.0; H, 5.24; S, 27.4.

(ii) Poly[methyl-(4-phenylthiophenyl)sulfonium perchlorate] (PPSP). IR(KBr, cm^{-1}): 3084, 3021, 2932 (v_{C-H}), 1568, 1476, 1395, 1092, 1009, 982, 816, 745, 698, 623 , 554, 494. lH NMR (DCOOD, TMS): 7.77.9 (d, 8H), 3.8 (8, 3H). ^{13}C NMR (DCOOD, TMS): 141.7, 132.3,130.2,124.0 (phenyl C), 27.5 (SCH$_3$, C). [η]: 1.7 [in CH$_3$CN/H$_2$O = 65/35 (v/v) with CF$_9$SO$_9$Na (50 mM)].

(iii) Poly(p phenylene sulfide) (PPS). IR (KBr, cm^{-1}): 3065,1572,1472,1389,1092,1074, 1009, 812, 743, 706, 556, 482. CP/MAS ^{13}C-NMR: 134.4,131.9 (phenyl C). Anal. Calcd for C_6H_4S: C, 66.63; H, 3.73; S, 29.64. Found: C, 66.70; H, 3.69; S, 30.32.

RESULTS AND DISCUSSION

Coupling Reaction of Thioanisole with Methyl Phenyl Sulfoxide.   The control experiment for the formation of sulfonium compounds were carried out in acidic media Quantitative formation of the methylphenyl [4-(phenylthio)phenyl]sulfonium cation was confirmed by the model reaction between thioanisole and methyl phenyl sulfoxide in trifluoromethanesulfonic acid (Eq. 4)). The methylphenyl[(phenylthio)phenyl]sulfonium cation is isolated as a stable perchlorate salt having the empirical formula, $C_{14}H_{15}S_2ClO_4$ after the exchange of the counter anion from $CH_3SO_3^-$ by precipitation into perchloric acid (60%). ^1H NMR shows two peaks at 3.7 and 2.4 ppm attributed to methyl groups. The former peak is assigned to methyl protons binding the sulfonium cation because the peak was observed at afield lower than that of the neutral methyl one due to the positive charge. A total of 10 peaks attributed to two methyl carbons and eight phenyl carbons in the ^{13}C NMR spectrum also indicates the formation of the MSP structure. Furthermore, the IR spectrum shows peaks at 1088 and 625 cm^{-1} which are ascribed to the perchlorate anion. The absorption band at 816 cm-1 indicates the formation of atypical 1,4-phenyl linkage. Combination of the COSY ^1H-^{13}C NMR and IR spectra reveal the formation of MSP.

$$\text{(structures)} \quad (4)$$

The electrophilic substitution which proceeds through the protonation of the sulfoxide was accompanied by the elimination of water as a byproduct. Therefore, the reaction was influenced by the acidity of the mixture. Trifluoromethanesulfonic acid, which is the strongest protic acid, is the most effective for the formation of the sulfonium cation. The conversion to MSP in the reaction between thioanisole and methyl phenyl sulfoxide depends on the acidity function (H°). In weak acids such as CH_3COOH and CF_3COOH, the electrophilic reaction scarcely proceeds.

Polymerization of Aryl Sulfoxide Derivatives.   Methyl(p-thiophenoxy)phenyl sulfoxide was polymerized in trifluoromethanesulfonic acid at room temperature for 24 h to produce the polymer quantitatively. The mixture was poured into water to precipitate the polymer as a trifluoromethanesulfonate salt. The resulting polymer was isolated as a white powder having the empirical formula $C_{13}H_{11}S_2CF_3SO_3$ and was soluble in sulfolane, nitrobenzene, dimethyl sulfoxide, pyridine, and formic acid. The methyl group in the resulting polymer was confirmed by the IR spectrum. The IR spectrum of the polymer also shows a strong absorption at 1258, 638 (ν_{CF}) and 1161, 1067 cm($\delta_{S=O}$), which means that the polymer contains trifluoromethanesulfonate as a counter anion. In the 500 MHz COY ^{1}H-^{13}C NMR spectrum, methyl groups are observed at 3.78 and 28.93 ppm, respectively, whose shifts are located in a lower field than that of a neutral thiomethyl group. These results support the formation of poly(methyl(4-phenylthio)phenylsulfonium cation)(Eq. 5).

$$(5)$$

The activated sulfoxide is well-known for use as an electrophile (Swern method). The polymerization of the sulfoxide compound is initiated via the sulfonium cation as an active species by the protonation on the oxygen of the sulfoxide bond . The active species electrophilically substitutes on the benzene ring to eliminate water.

The demethylation of poly(methyl(4-phenylthio)phenyl sulfonium cation) was carried out in pyridine by refluxing(Eq. 5). The precursor polymer is completely soluble in the mixture at room temperature before the reaction. The white insoluble powder precipitated during time passage. N-Methylpyridinium trifluoromethane sulfonate salt was isolated quantitatively in the reaction mixture(Table 1). The resulting polymer has the empirical formula of C_6H_4S and shows the same IR spectrum and CP/MAS ^{13}C NMR spectra of commercially available PPS. No absorption bands of methyl, sulfone, and sulfoxide bonds are detected in the IR spectrum.

Table 1 Synthesis of Poly(phenylene sulfide)
through Oxidative Polymerization of Thioanisole Derivatives

Monomer	Reaction Temp.(°C)	Solvent	Mw (x 10^4)	$Td_{10\%}$ (°C)
	25	CF_3SO_3H/P_2O_5	24.9	530
	25	CF_3SO_3H	12.0	
a	20	CH_2Cl_2	2.0	515
b	20	CH_3SO_3H/P_2O_5	20.4	526
	70	CH_3SO_3H/P_2O_5	10.4	525

a Ref. 9   b Ce-catalyzed Oxidative Polymerization

Oxidation of Thioanisole in Acid in the presence P₂O₅.　　Thioanisole is oxidized quantitatively to yield methyl phenyl sulfoxide with oxygen in the presence of a catalytic amount of Ce(NH₄)₂(NO₃)₆ (CAN)[10] under high pressure.　　In the presence of methanesulfonic acid, the Ce-catalyzed oxidation of thioanisole in an oxygen atmosphere at ambient pressure for 20 h at 25 °C yields 22% of MPS and 78% of methyl 4-phenylthiophenyl sulfoxide after refluxing with pyridine as the nucleophile (Scheme 1).

Control experiments reveal that, under acidic conditions, methyl phenyl sulfoxide couples with thioanisole to yield the methyl-(4-methylthiophenyl)phenylsulfonium cation quantitatively via formation of methyl phenyl sulfoxide in the Ce-catalyzed oxidation. MPS is produced by demethylation of the methyl-(4-methylthiophenyl)phenylsulfonium cation and methyl 4-phenylthiophenyl sulfoxide is produced by demethylation of the methyl [4-(methylsulfinyl)phenyl]phenylsulfonium cation which is formed by the O₂ oxidation of the methyl(4-methylthiophenyl)phenylsulfonium cation in the reaction mixture. The O₂ oxidation is influenced by the acid strength. In a weak acid such as acetic acid, the coupling reaction does not proceed because the formed sulfoxide is not activated through the protonation on the oxygen atom. The predominant formation of the dimer supports the fact that Ce(IV) acts as a catalyst for oxidation of the sulfide in acid and the resulting sulfoxide is allowed to electrophilicity react with thioanisole to form the sulfonium cation in high yield.

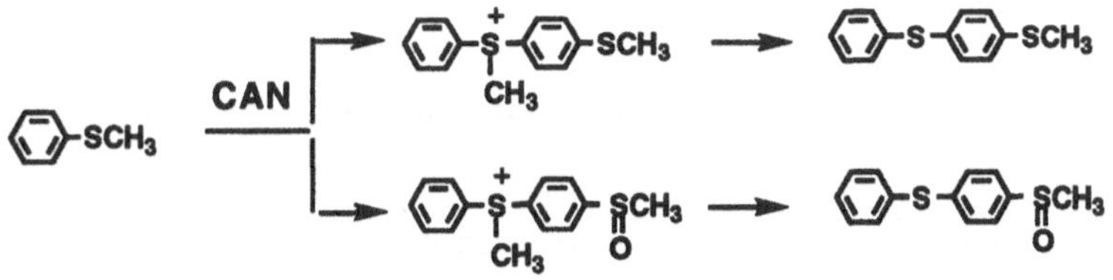

Scheme 1

Oxidative Polymerization MPS.　　On the basis of the model reactions, the polymerization of MPS was carried out in methanesulfonic acid in an oxygen atmosphere in the presence of a catalytic amount of CAN. In the reaction, water was formed in an amount equivalent to MPS and suppressed the polymerization because of the decreasing acidity of the mixture. P₂O₅ was added to the reaction mixture to remove the water. The polymerization proceeded rapidly with a quantitative O₂ uptake of oxygen to give the poly (sulfonium cation) (PPSP) using methanesulfonic acid (10 mL). The polymerization mechanism of methyl 4-phenylthiophenyl sulfide is believed to follow Scheme 2. Methyl 4-phenylthiophenyl sulfoxide which was formed by oxidation of methyl 4-phenylthiophenyl sulfide is protonated by methanesulfonic acid.　　The (hydroxymethyl)

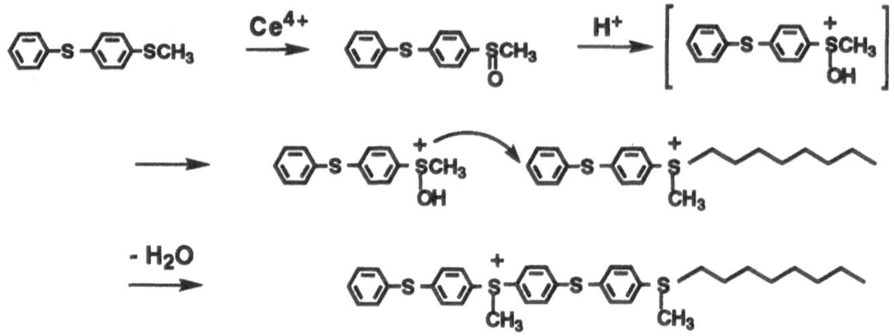

Ce^{4+} H$^+$

- H$_2$O

Scheme 2

phenylsulfonium cation electrophilically attacks the phenyl ring of the monomer. The coupling reaction provides PPSP. The oxidation of 1 mol of sulfide consumed 1/2 mol of O$_2$ to give 1 mol of the sulfoxide. oxygen is essential for the polymerization. The polymer was isolated by precipitation in perchloric acid (60%) as a white resin having the empirical formula C$_{13}$H$_{11}$S$_2$ClO$_4$.

The absorption peak of the resulting polymer at 2925 cm^{-1} is attributed to the C-H of the methyl group, and the absorption peaks at 1100 and 625 cm-^{1} indicate the counter anion ClO$_4$ in the IR spectrum. The peaks at 3.8 ppm in the ^{1}H-NMR spectrum and 27.5 ppm in the ^{13}CNMR spectrum reveal the existence of the S$^+$-CH$_3$ group. An absorption band attributed to a C-H out-of-plane vibration of the benzene ring was observed at 810 cm^{-1} in the IR spectrum. ^{1}H NMR of 5 shows AB-quartet peaks attributed to phenyl protons at 7.7-7.9 ppm. These results clearly indicate the formation of a linear chain of a poly (sulfonium cation).

After the demethylation of PPSP by refluxing in pyridine as a nucleophile, PPS is isolated as a white powder having the empirical formula C$_6$H$_4$S. In the IR spectrum, the demethylated polymer has an absorption peak at 810 cm^{-1} attributed to a C-H out-of-plane vibration of the benzene ring. The IR spectrum of the resulting PPS was consistent with that of commercially available PPS (Ryton). The IR absorption peaks of the methyl group and peaks of ClO$_4$ disappear in the IR spectrum. The demethylation proceeds through a trans-methylation mechanism from the sulfonium cation to pyridine, which and the N-methylpyridinium cation can be isolated. Two peaks at 131.9 and 134.4 ppm were observed in the CP/ MAS ^{13}C-NMR spectrum for the resulting PPS. The molecular weight of the resulting PPS was determined as 20.4 x 10^4 (Table 1). DSC measurement shows Tm = 260 °C, Tg = 93 °C, and Tc= 158 °C. Tm did not change upon annealing at 200 °C, which indicates the absence of acid contamination. Even though Tm was lower than that of the

Ryton V-1 grade (280 °C), detectable differences in structure between the resulting polymer and Ryton were not indicated by the IR and NMR spectra. It cannot be denied that the lower melting point may be caused by the presence of a small amount of the sulfonium cation unit or ortho substitution of the sulfide bond in the main chain of the PPS.

The molecular weight of the resulting PPS is influenced by,the reaction temperature. A higher temperature reaction temperature of 70 °C results in a lower molecular weight product. We propose that the polymerization proceeds via a cationic mechanism. MPS is oxidized to methyl 4-phenylthiophenyl sulfoxide with oxygen. Methyl 4-phenylthiophenyl sulfoxide is protonated to form the (hydroxymethyl)(4-phenylthiophenyl)sulfonium cation which is the active species in the polymerization. The cation electrophilically attacks the benzene ring of MPS at the para position, yielding the dimer sulfonium cation accompanied by water elimination. The oxidation and the electrophilic reactions are continually repeated to yield the poly(sulfonium cation).

ACKNOWLEDGMENT

This work was partially supported by a Grant-in-Aid for Development Scientific Research (No. 04655223) and Scientific Research No. 05650865 from the Ministry of Education, Science and Culture, Japan.

REFERENCES

⁺ PRESTO JRDC Investigator

1. A. S. Hay, H. S. Blanchard, G. F. Endres, and J. W. Eustance, *J. Am. Chem. Soc.* **81**, 6335(1969,).

2. J. T. Edmonds Jr., H. W. - Hill Jr, U.S. Patent 3354129,1967 -Chem. Abstr. 1968, 68,13598. Campbell, R. W., Edmonds, J. T. U.S. Patent 4038259, Chem. Ahtr. 1977, 87, 854v. Hill, H. W., Jr. *Int. Eng. Chem., Prot. ReJ. Dev.* **18**, 262(1979)

3. K. Yamamoto, E. Tsuchida, H. Nishide, S. Yoshide, Y. S. Park, J. *Electrochem. Soc.* **139**, 2401(1992).

4. K. Yamamoto, M. Jikei, K. Oi, H. Nishide, and E. Tsuchida , *E. J. Polym. Sci.*, Part A **29**, 1359(1991).

5. E. Tsuchida, K. Yamamoto, M. Jikei, E. Shouji, H. Nishide, *J. Macromol. Sci., Chem.* , **11**, 1285(1991).

6. K. Yamamoto, S. Yoshida, H. Nishide, AND E. Tsuchid, *Bull. Chem. Soc. Jpn.* **62**, 3655(1989)

7. E. Tsuchida, K. Yamamoto, M. Jikei, H. Nishide, *Macromolecules* **22**, 4138 (1989).

8. K. Yamamoto, E. Shouji, H. Nishide, E. Tsuchida, *J. Am. Chem. Soc.*, **115**, 5819 (1993).

9. E. Shouji, K. Yamamoto, and E. Tsuchida, *Chem. Lett.* **1993**, 1927.

10. D. P. Rilcy, AND P. E. Correh, *J. Chem. Soc., Chem. Commun.* 1986, 1057.

CHEMICAL MODIFICATIONS OF HALOMETHYLATED POLY(METHYLPHENYLSILANE): A NEW AND FACILE ROUTE TO FUNCTIONALIZED POLYSILANES

A.C.Swain, S.J.Holder, R.G.Jones[*], A.J.Wiseman, M.J.Went and
R.E.Benfield

Centre for Materials Research, Chemical Laboratory,
University of Kent, Canterbury,
Kent CT2 7NH, Great Britain

INTRODUCTION

Soluble polysilanes are most commonly prepared using the *Wurtz* type reaction between dichloroorganosilanes and molten sodium metal in refluxing toluene[1] or by the action of sodium dispersions in refluxing diethylether.[2]

$$n \quad Cl-\underset{R'}{\overset{R}{Si}}-Cl \;+\; 2n\;Na \;\xrightarrow{\text{Reflux}}\; \left(\!\underset{R'}{\overset{R}{Si}}\!\right)_{\!n} \;+\; 2n\;NaCl$$

The properties, molecular weights, and yields of polysilanes are strongly dependent upon the substituents R and R'. They show strong UV-absorption's, σ_{Si-Si} HOMO to σ_{Si-Si} LUMO, and readily undergo Si-Si bond cleavage when exposed to UV-irradiation[3-5] or rearrange to carbosilanes when heated.[6] These interesting properties are utilised in their applications as UV-photoresists,[7,8] as radical photo-initiators,[9,10] and as pre-ceramic materials.[11,12] Recent interest in polysilane syntheses has concentrated on the preparation of high molecular weight polysilanes bearing functional groups.[13-15] However, the functional groups that would survive the harsh conditions (110°C, molten sodium metal) commonly employed to prepare polysilanes are severely restricted to those that are chemically robust. Nonetheless, number of functionalized polysilanes prepared in this way have been reported[6,16-19] including those bearing arylether,[1,20,21] amino,[20] and silyl[22] groups.

Functional group protection has recently been employed to introduce the phenolic moiety as a side-group substituent of a polysilane homopolymer,[23] albeit in low yield; in this

---

[*] Author to whom correspondence should be addressed.

synthesis the trimethylsilyl group was used to protect the phenoxy substituent of dichlorophenyl-2-(3-hydroxyphenyl)propylsilane prior to polymerization.

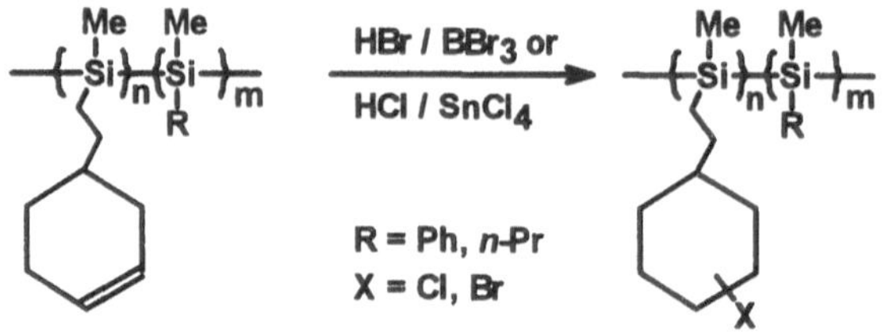

Scheme 1. Friedel-Crafts halogenation of a polysilane with olefinic side-groups.

However, perhaps the most useful chemical modification thus far reported is the *Friedel-Crafts* chloromethylation of poly(methylphenylsilane)[24,25] (PMPS) and poly(β-phenethylmethylsilane)[26] using the mild Lewis-acid, stannic chloride, in chloromethyl-methylether. The chloromethylation of the phenyl groups was reported to be in excess of 95%. The attraction of this preparative route is somewhat reduced since chloromethyl-methylether is a highly potent carcinogen.[27] However, the use of *in situ* preparations of chloromethylmethylether have been documented for the chloromethylation of polystyrene[28] and it is this approach that we have adopted to effect the chloromethylation of poly(methyl-phenylsilane) with a high level of control.[25] Halogen atoms have been introduced into polysilanes with pendent olefinic residues by the addition of HCl or HBr, using the mild Lewis acid stannic chloride as exemplified in Scheme 1.

In this paper we outline the important aspects of chloro- and bromomethylation reactions of PMPS together with some important reactions involving the subsequent replacement of the halogen atoms with other functionalities, including metallic substituents. These preparations are the prelude to investigations into the modifying effect such substituents have on the reaction chemistry of polysilanes. These studies will be reported in a future publication.

RESULTS AND DISCUSSION.

The Halomethylation of Poly(methylphenylsilane).

Chloromethylmethylether can be generated *in situ* from a solution of thionylchloride and dimethoxymethane in chloroform and this can subsequently be reacted with PMPS in the presence of stannic chloride.[25] Chloromethylated PMPS (CPMPS) prepared by this method is usually obtained as a light brown/cream powder with yields in the range 40-70% (expressed as the recovered product weight, following reprecipitation, as a fraction of the initial the PMPS weight). The variation of the extent of chloromethylation with time is illustrated in Figure 1; these results are reproducible and thus the percentage of chloromethylated phenyl groups on CPMPS can be controlled. However, chloromethylation of PMPS by the above procedure has been observed to lead to a significant reduction in the molecular weight parameters of the polymer as illustrated in Figure 2. This degradation occurs within the first hour of the reaction timed from the addition of the stannic chloride. However, it has been observed that degradation of the PMPS also occurs before the addition of the stannic chloride the effect is presumably a result of the action of chloromethylmethylether or thionyl chloride instead of, or in addition to, the stannic chloride. It is clear that the degradation affects the high molecular weight fraction of the polymer preferentially. It has tentatively been ascribed to the cleavage of the siloxane linkages in the PMPS backbone.[25] These linkages are thought to form during the product isolation following *Wurtz* synthesis of the PMPS [31,32] but there is strong evidence that at least part of the degradation is caused by Si-C bond cleavage (see below).

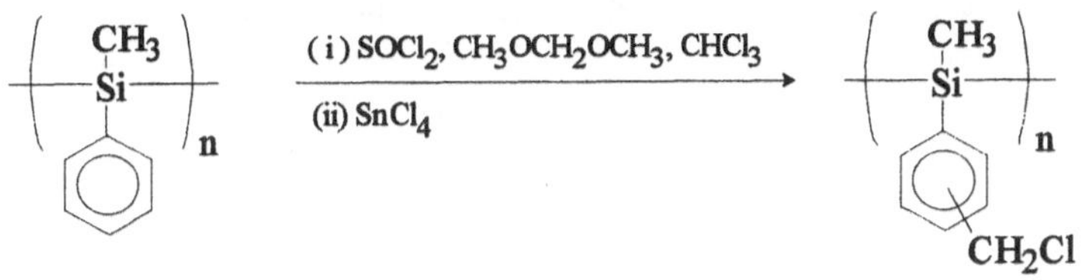

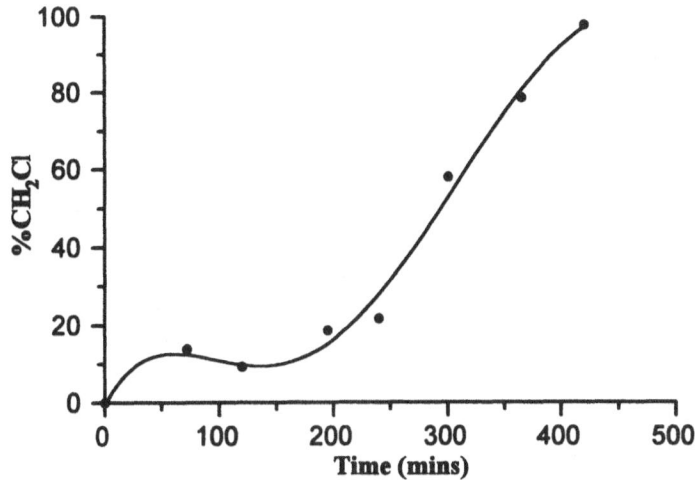

Scheme 2. Chloromethylation of PMPS

Figure 1. The variation of the extent of the chloromethylation of PMPS with time.

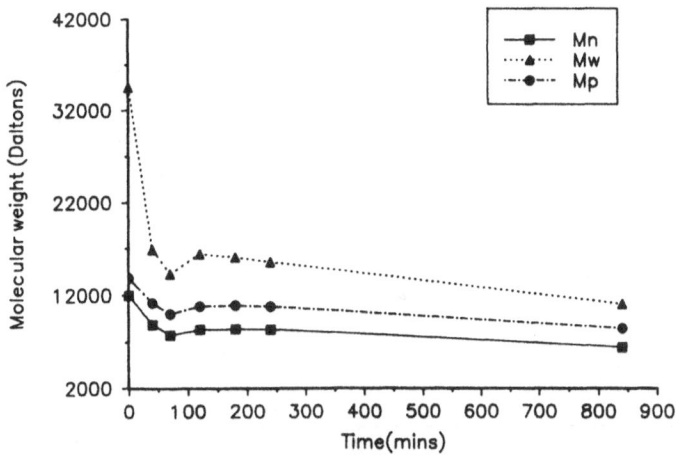

Figure 2. Variation of the molecular weight parameters with percentage chloromethylation of PMPS in chloroform solution at 0°C.

To complement the chloromethylation of PMPS as a first step functionalisation of phenyl bearing polysilanes, and to introduce a potentially more reactive group onto the polymer, a method analogous to that employed for chloromethylation can be utilised. Bromomethylmethylether, like chloromethylmethylether, can be generated *in situ* from the reaction of thionylbromide and dimethoxymethane in dichloromethane solution and this can subsequently be reacted with PMPS. At 0°C however, the reaction leads to low yields (less than 10%) and extensive degradation the molecular weight drops to below 3,000 from 126,000 in the resultant bromomethylated PMPS (BPMPS). However, a reduction in the reaction temperature from 0°C to -18°C leads to consistently high yields (>80%) and even less degradation than that observed in the chloromethylation reaction.

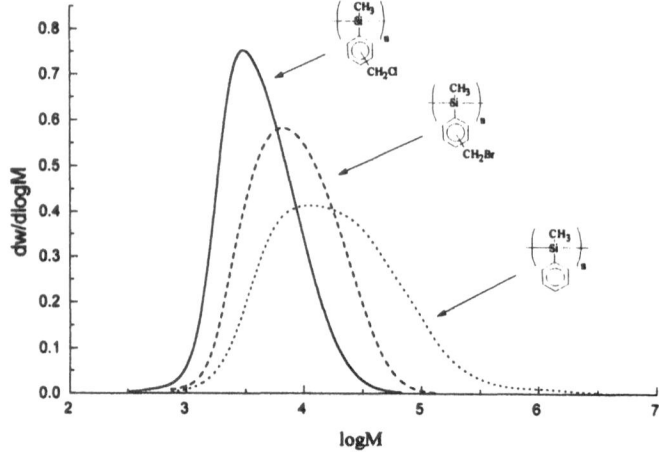

Figure 3.   Differential molecular weight distributions comparing the initial PMPS sample with its halomethylated derivatives.

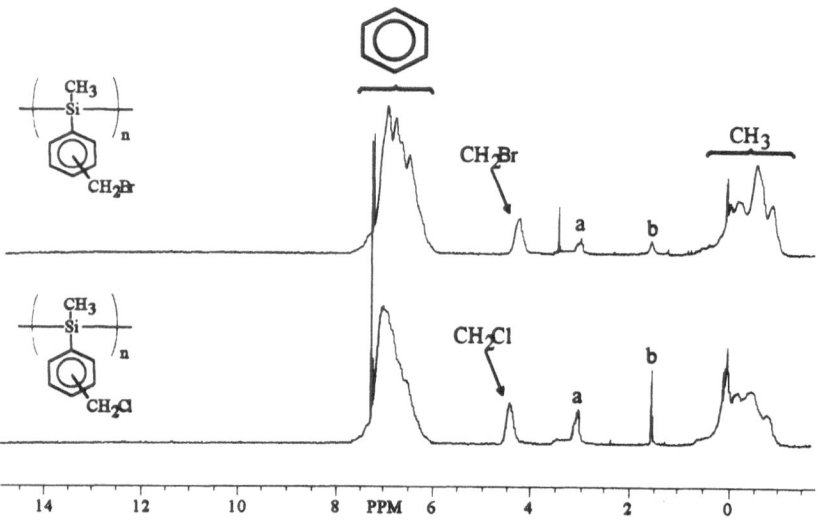

Figure 4. ^{1}H NMR spectra for chloro- and bromomethylated PMPS. **a** is thought to represent methoxy-groups; **b** represents unknown impurities (the sharp peaks are probably due to occluded solvent).

Overlaid differential molecular weight distributions facilitate the comparison of the initial PMPS with its halomethylated derivatives as shown in Figure 3. Preliminary results indicate that the degree of bromomethylation, like chloromethylation, increases gradually with time and that the reaction can be similarly controlled. The ^{1}H and ^{13}C$\{^1$H$\}$ spectra of CPMPS and BPMPS are shown in Figures 4 and 5 respectively and the peaks for the chloromethyl (^{1}H δ=4.6) and bromomethyl (^{1}H δ=4.5) groups occur at nearly identical chemical shifts. Noticeable features in the ^{1}H NMR spectra are the broad jagged peaks between ~3.0 - 3.5 ppm (a in Figure 4). These are thought to arise between the substitution of methoxy- groups onto the PMPS backbone via Si-C cleavage induced by the halomethylmethylether reagents, a possible mechanism for which is shown in Scheme 3. Further investigations on the degradative processes occurring in the halomethylation reactions are underway.

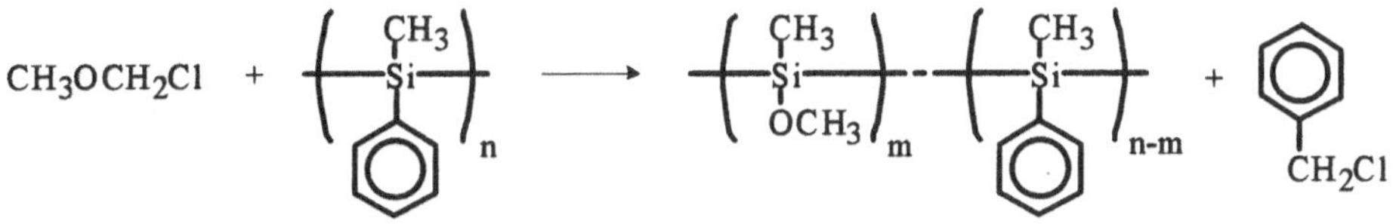

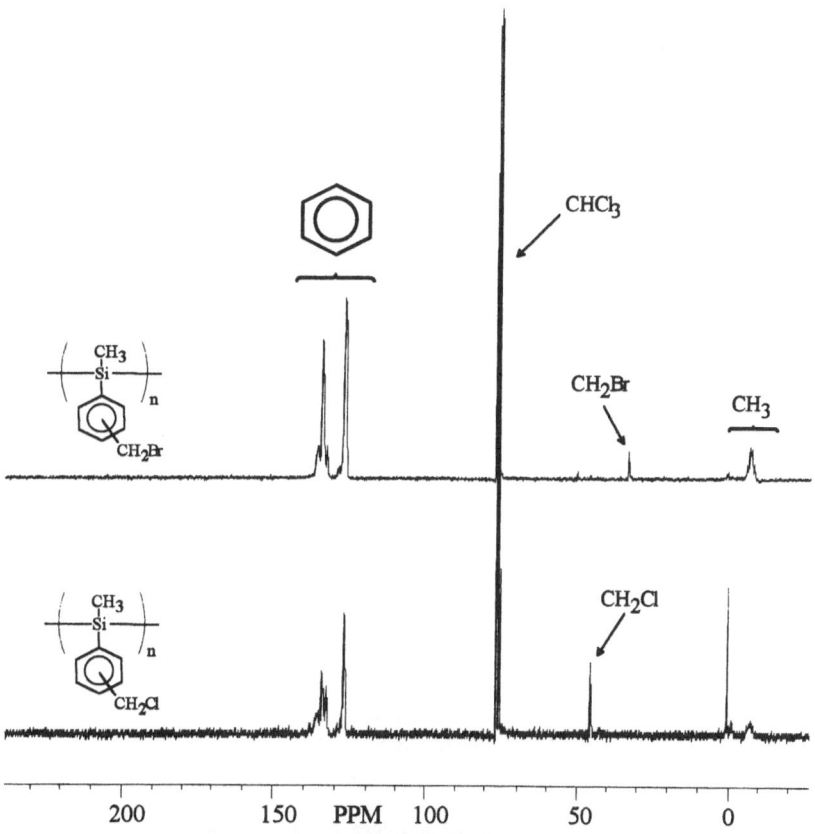

Scheme 3. Postulated reaction for the formation of methoxy groups; evidence for Si-C activation.

Figure 5. ^{13}C $\{^1$H$\}$ spectra for chloro- and bromomethylated PMPS.

Reactions of Halomethylated Poly(methylphenylsilane).

The chloromethyl- and bromomethyl groups of halomethylated PMPS undergo a number of simple organic transformations in high yield, providing access to a range of new polysilane structures, see Scheme 4. Described below are typical chemical transformations performed on 15-20% halomethylated PMPS.

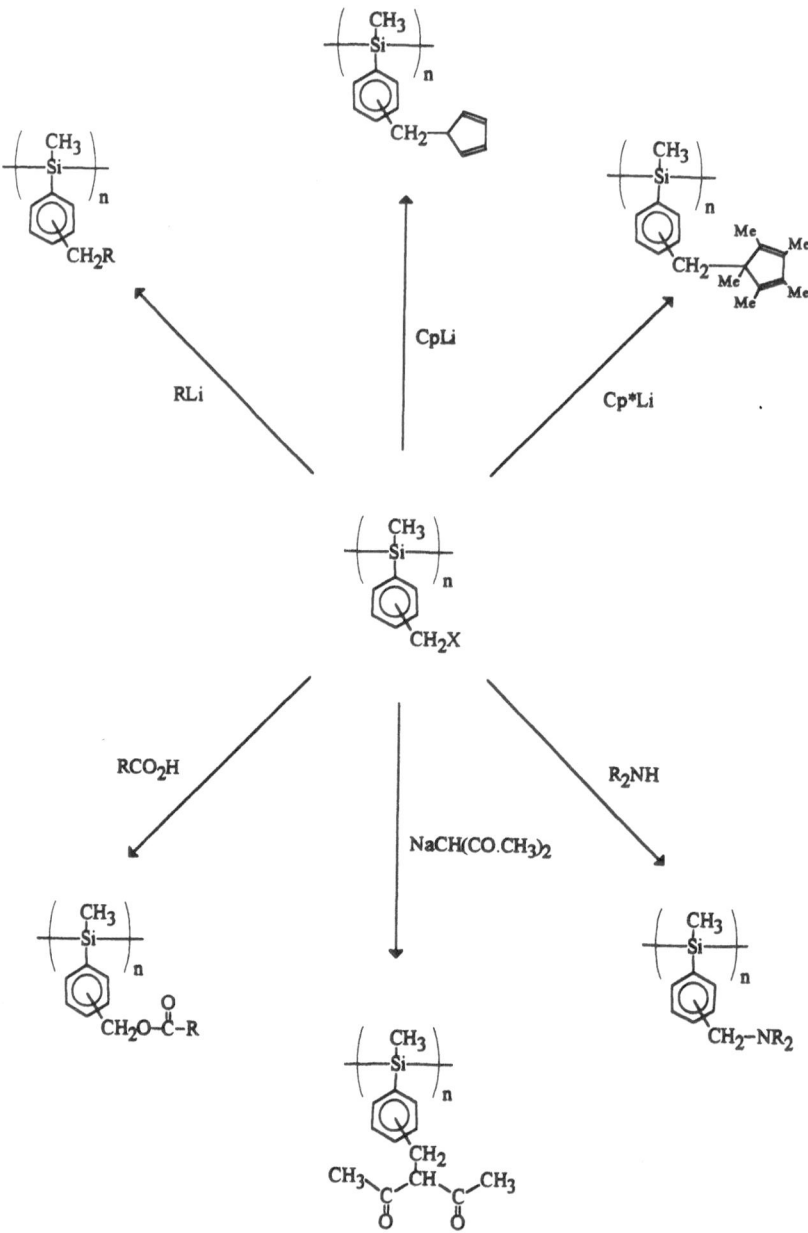

Scheme 4. Some chemical transformations of halomethylated PMPS are shown. Cp* denotes 1,2,3,4,5-pentamethylcyclopentadienide.

Reaction of CPMPS with either lithium butyl or lithium phenyl in stoichiometric ratio results in the expected nucleophilic displacement product copolymers i.e., butylated- and phenylated-poly(methylphenylsilane)'s respectively. The product polymers are isolated in high yield as solid white tractable materials. A typical $^{13}C\{^1H\}$ NMR spectrum of poly(methyl(p-butylphenyl)silane) is shown in Figure 6.

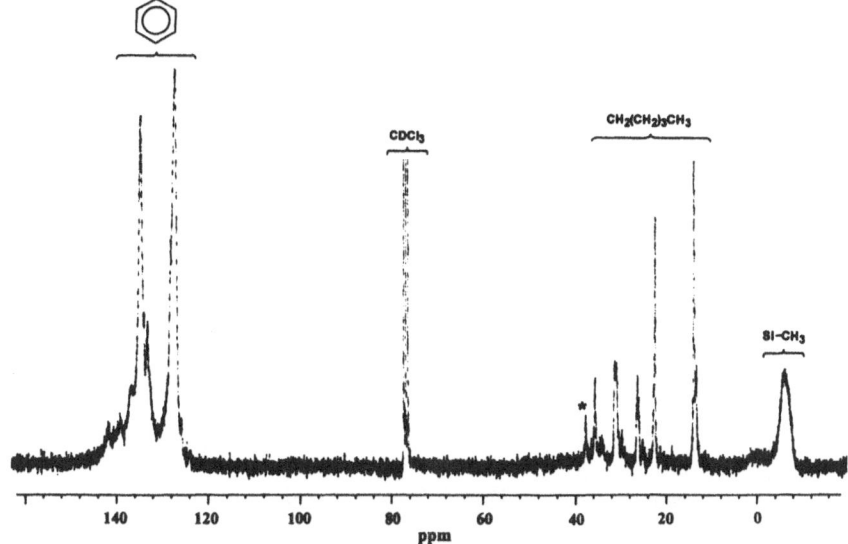

Figure 6. A typical $^{13}C\{^1H\}$ NMR spectrum of the product polymer isolated from the reaction of lithium butyl and CPMPS. * indicates impurity.

The GPC molecular weight distributions of the butylated (Figure 7) and the phenylated (Figure 8) product polymers shows that the reaction is accompanied by an increase in both the polydispersity and molecular weight parameters of the polymers. The increase in the molecular weight parameters might be expected since the addition of a pendant group to the phenyl moiety would conceivably increase the hydrodynamic volume of the polymer. However, the large increase in polydispersity and the evident polymodal

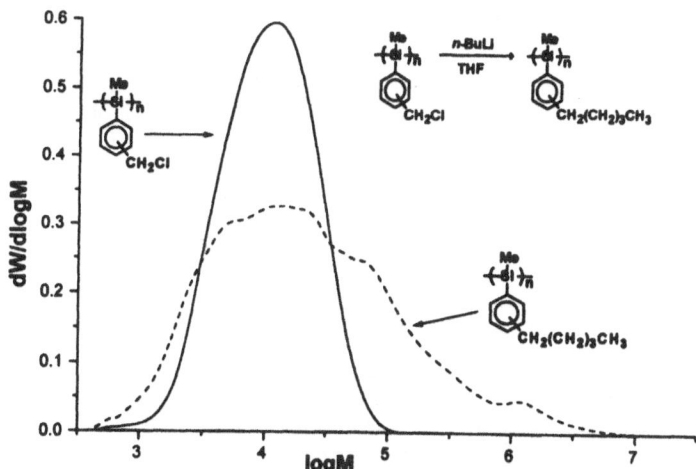

Figure 7. Differential molecular weight distributions of the product polymer obtained from the reaction between CPMPS and lithium butyl, showing an increase in molecular weight accompanied by cross-linking.

nature of the molecular weight distributions of the product polymers, we interpret as resulting from a minor cross-linking reaction operating in parallel with the substitution reaction. The nature of the cross-linking reaction is currently under investigation but it may arise from the coupling of the chloromethyl group of one polymer chain with an adventitiously formed lithiomethyl group (*vide infra*) of a neighbouring polymer chain, as shown in Scheme 5.

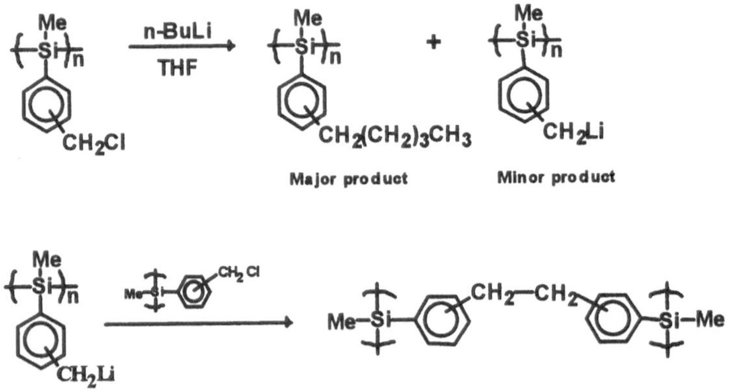

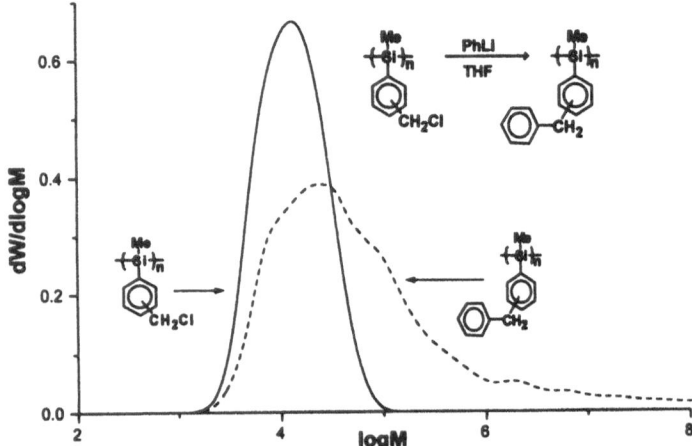

Scheme 5. The reaction between CPMPS and lithium butyl together with a postulated cross-linking mechanism are shown.

It is known that metal halogen exchange can occur between lithium alkyls and aryl halides and is favoured at low temperatures.[33] It should also be noted that the Si-Si bond within the polysilane chain can be cleaved by the action of strong nucleophiles.[29] When lithium phenyl is employed in excess, Figure 8, the resulting diarylmethyl group can be deprotonated *in situ* to form a lithiated polymer which can then also undergo a cross-linking reaction, see Scheme 6.

In complete contrast to the above reactions, if a Grignard reagent is employed e.g., *cy*-hexylMgBr only substitution of the chloromethyl group is observed, see Figure 9.

Figure 8. Differential molecular weight distribution of the product polymer obtained from the reaction between CPMPS and lithium phenyl, showing an increase in molecular weight accompanied by cross-linking.

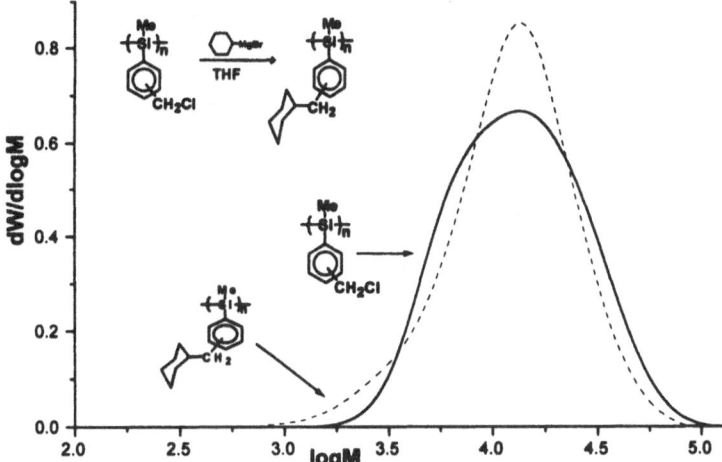

Figure 9. Differential molecular weight distributions of the product polymer obtained from the reaction between CPMPS and cyclohexyl Grignard.

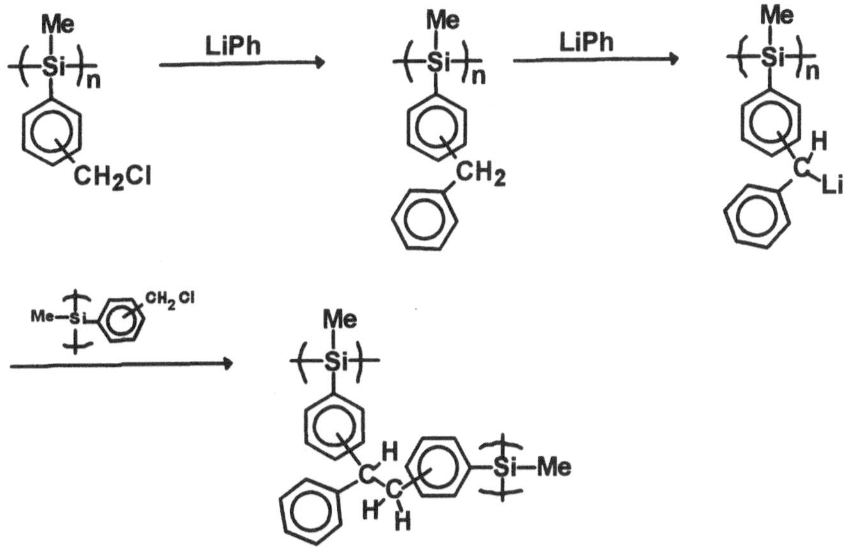

Scheme 6. Proposed cross-linking mechanism for the reaction between CPMPS and excess lithium phenyl.

The reaction between CPMPS and either sodium cyclopentadienide or lithium pentamethylcyclopentadienide results in the expected nucleophilic substitution of the chloro group. The resulting polymers were isolated as white solids in high yield with the pentamethylcyclopentadienyl substituted polymer stable for many weeks in air. The GPC data of the product polymers also showed a movement toward higher molecular weight after reaction, a small quantity of cross-linking was observed for the pentamethylcyclopentadienyl substituted polymer, Figure 10. However, the cyclopentadienyl substituted polymer remained tractable only for a short period of time after isolation, thereafter it became intract-

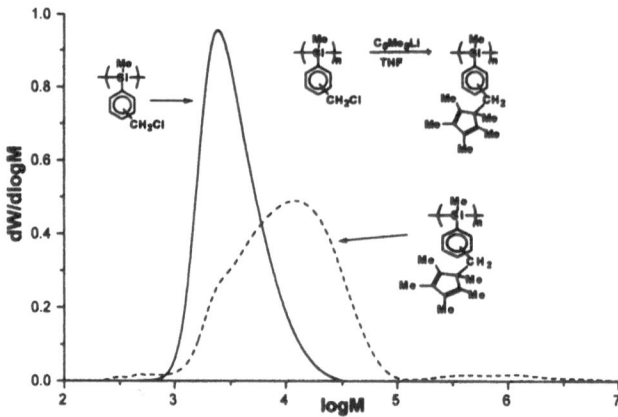

Figure 10. Differential molecular weight distribution of the product polymer obtained from the reaction between CPMPS and lithium pentamethylcyclopentadienide. Showing an increase in molecular weight parameters and a small quantity of cross-linked material.

able. Analysis of the ^{13}C{CPMAS} NMR spectrum, Figure 11, shows an incidence of methylene groups which would be consistent with an inter-molecular Diels-Alder reaction between cyclopentadienyl groups on neighbouring chains causing cross-linking.

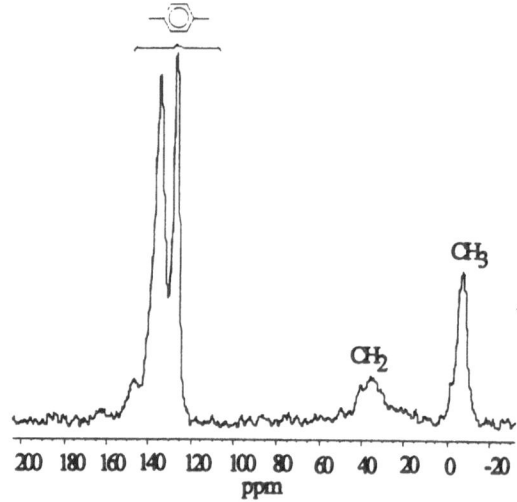

Figure 11. The ^{13}C{CPMAS} NMR spectrum of the intractable product polymer obtained from the reaction between CPMPS and sodium cyclopentadienide.

The reaction of BPMPS with a threefold excess of sodium acetylacetonoate, prepared *in situ* from sodium ethoxide and 2,4-pentanedione, resulted in the nucleophilic displacement of the bromide by the acetylacetonoate group as illustrated in Scheme 7, After reprecipitation from toluene solution with methanol, the product polymer was obtained as a creamy-white powder. Recovered yields were typically in the range 70-95% and the reaction is not quite quantitative, even at prolonged reaction times, as evidenced by a slight peak attributable to the -CH$_2$Br group in the ^{1}H NMR spectrum and an extraneous peak in the ^{13}C{^{1}H} NMR spectra. The ^{13}C{^{1}H} spectra for a sample of acacPMPS in deuterated

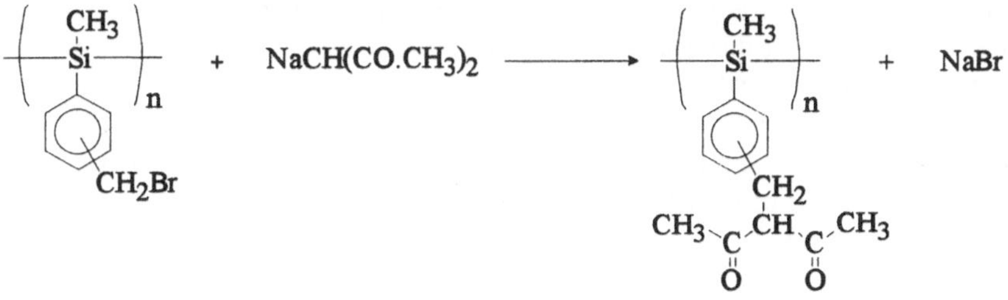

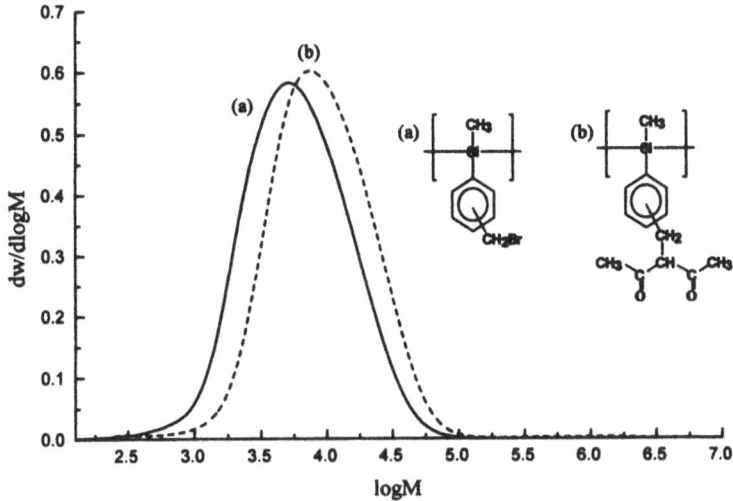

Scheme 7. The reaction of sodium acetylacetonoate with BPMPS.

Figure 12 . Differential molecular weight distribution of the product polymer obtained from the reaction between BPMPS and sodium acetylacetonoate.

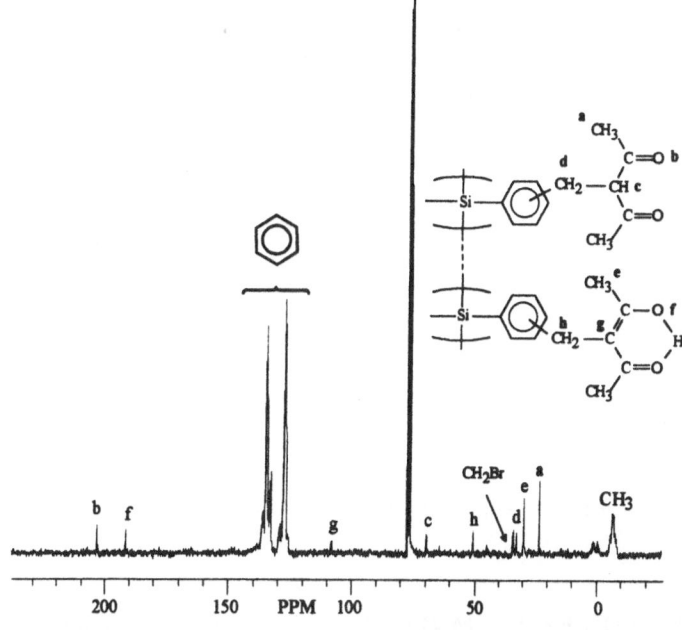

Figure 13. A typical $^{13}C\{^1H\}$ NMR spectrum of the product polymer obtained from the reaction between sodium acetylacetonoate and BPMPS showing the peak assignments.

chloroform is shown in Figure 13; in this peaks attributable to both keto- and enol-tautomers are clearly visible. The gel permeation chromatogram for the 2,4-pentanedione substituted PMPS (acacPMPS) shows the predictable increase in the molecular weight parameters when compared with that of the precursor BPMPS (Figure 12). Consistent with the proposed structure the infra-red spectrum shows a broad peak at 1700 cm^{-1} attributable to the carbonyl-stretch of the acetylacetonoate group as seen for a similar acetylacetonoate substituted polystyrene[34].

Metallation of Poly(methylphenylsilane).

Reaction of CPMPS with freshly prepared NaMn(CO)$_5$ resulted in the complete displacement of chloride and formation of a manganese-carbon bond. This was evidenced by the complete disappearance of the chloromethyl resonance (^{1}H, δ=4.6) and the appearance of a CH$_2$Mn(CO)$_5$ resonance (^{1}H, δ=2.25) in the same integral ratio. An infra-red spectrum of the polymer showed carbonyl resonances at 2103 and 2013 cm^{-1} consistent with those reported by Pittman[30] for an analogous manganese pentacarbonyl bound polystyrene. The GPC data of this polymer shows the expected increase in molecular weight with no evidence of cross-linking or degradation (Figure 14), this we attribute to the weak nucleophilicity of the Mn(CO)$_5^-$ anion. The stabilisation of this metal-carbon bond is facilitated by the absence of β-hydrogen atoms and represents an example of a rare class of metal containing polymer.

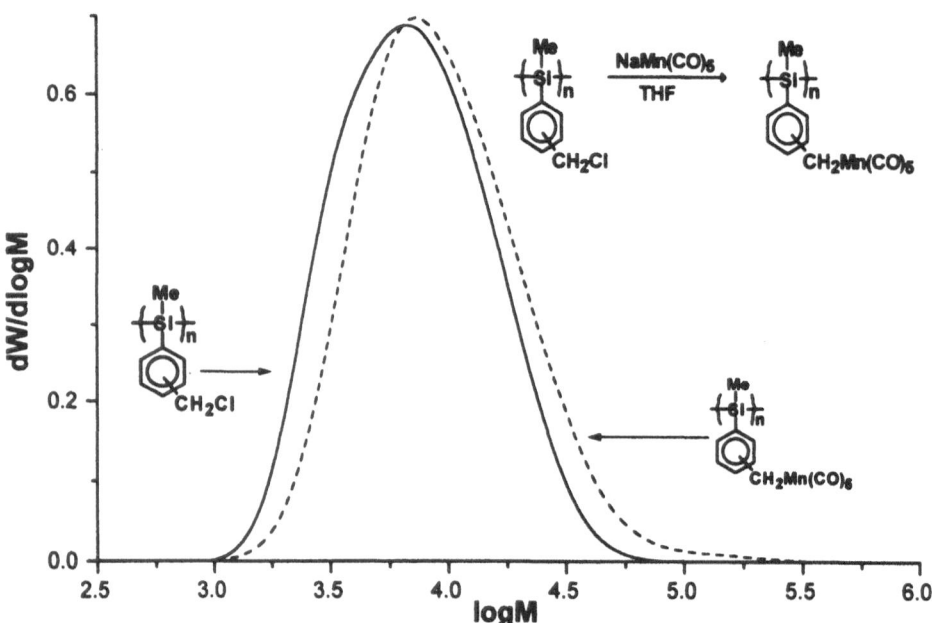

Figure 14. Differential molecular weight distributions of CPMPS and of the manganese pentacarbonyl polysilane.

Reaction of PMPS with Mo(CO)$_3$(py)$_3$ in the presence of three equivalents of BF$_3$•OEt$_2$, affords the copolymer poly(methylphenylsilane-*co*-(η^6-phenyl-tricarbonylmoly-

172

denum)-methylsilane) in moderate yield as a yellow air-sensitive solid. The presence of molybdenum was confirmed by atomic absorption and the molybdenum-tricarbonyl group by comparison of the ultra-violet and infra-red spectra with those obtained from molybdenum(η^6-trimethyl-silylphenylsilane)tricarbonyl. Reaction of the polymer with trimethyl-N-oxide regenerated PMPS without degradation. A comparison of the gel permeation chromatograms, Figure 15, shows the predictable increase in molecular weight and/or hydrodynamic volume with little change in polydispersity.

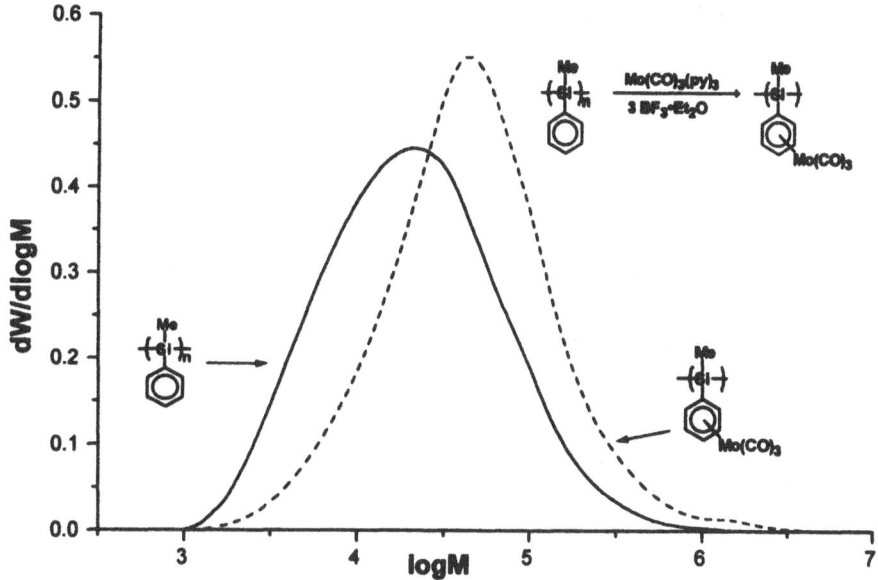

Figure 15. Differential molecular weight distributions for the reaction between Mo(CO)$_3$(py)$_3$ and PMPS showing the expected increase in molecular weight.

Preliminary investigations into the complexation of transition metals with acacPMPS (ca. 10% loading of 2,4-pentanedione) have been made with iron(III) chloride. A sample of acacPMPS was added to a solution of FeCl$_3$ (100-fold excess) in tetrahydrofuran and the solution was stirred for 15 min whereupon addition of an excess of methanol gave a precipitate which was filtered, washed with methanol and dried. The resultant powder was a deep pink colour and soluble in chloroform and tetrahydrofuran. The infrared spectrum of the polymer showed a large reduction in the intensity of the carbonyl stretching band at 1700 cm^{-1}, a reduction in the intensity of the broad ~1600 cm^{-1} band and the appearance of a band at 1566 cm^{-1} attributable to the co-ordinated diketone groups. The IR data is therefore consistent with, at the very least, limited co-ordination of the Fe^{3+} ions and when considered with the tractability of the product polymer, suggests that not all of the diketone groups are complexed and therefore exist as ketone, enol and enolate.

The gel permeation chromatogram shows a dramatic increase in the molecular weight parameters and the polydispersity on forming the product polymer as illustrated in Figure 16 attributable to the formation of cross-links *via* di- and tri-coordination of the Fe^{3+}. After 24 h the solution of the polymer in THF was reanalysed by GPC and the molecular weight parameters of the polymer were identical to those of the 'parent' acacPMPS. This suggests that the THF effectively leached the Fe^{3+} ions from the polymer demonstrating the reversibility of the complexation. Studies into the complexation of iron and other metals by this polymer are ongoing.

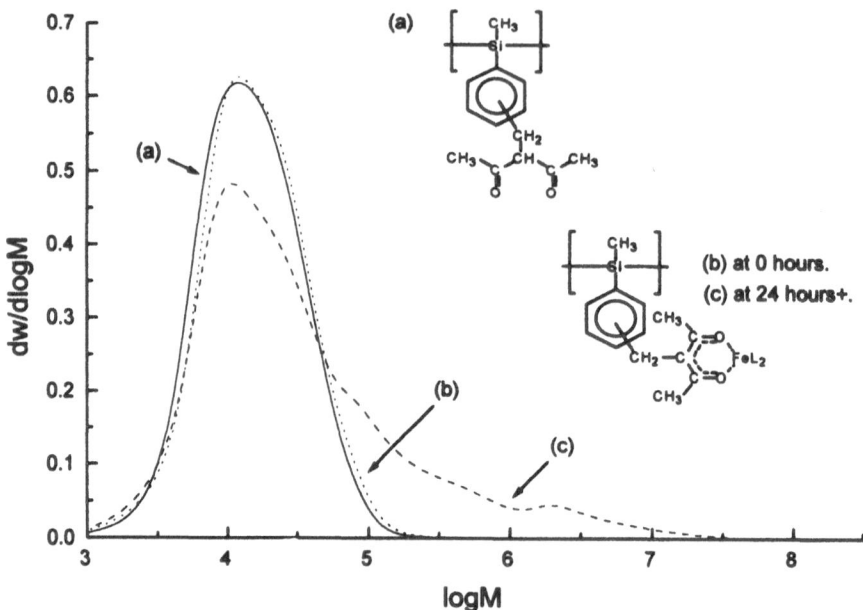

Figure 16. Differential molecular weight distributions of acacPMPS and of the product polysilane from the reaction with FeCl$_3$ after O hr and 24 hr.

CONCLUSIONS

The esterification and amination reactions of Scheme 4 have not been detailed in this paper. This is not because they cannot be accomplished but rather because it was wished to lend emphasis to the more novel aspects of our synthetic studies. Accordingly, it has been demonstrated that CPMPS and BPMPS can be used as precursors to a range of otherwise inaccessible polysilane structures. Although, in some of these reactions concomitant crosslinking occurs, the reasons for this are evident and this understanding allows its avoidance in other reactions. In none of the reactions have the conditions used been such as would lead to chain scission, which is an important consideration when dealing with such sensitive materials as polysilanes. Three examples of the incorporation of metals in the substituent groups of polysilanes are demonstrated; of these, one is sigma-bonded, one is through π-complexation, and one is through reversible complexation. To the best of our knowledge, the only other example of a metal being bound to the substituent of a polysilane being iron within poly(methylphenyl-co-methylferrocenylsilane) which, by virtue of the stability of the ferrocenyl group, can be formed in the conventional fashion from the corresponding dichlorosilane precursor. In contrast our method presents three entirely new and novel routes to the binding of metals in a polysilane.

ACKNOWLEDGEMENTS

We gratefully acknowledge the Royal Society for a University Research Fellowship for R.E.B., and the United Kingdom Engineering and Physical Science Research Council for a studentship for A.J.W. and Research Fellowships for S.J.H. and A.C.S. We also thank Miss C.S.Simnett for her invaluable assistance.

REFERENCES

1. R.D.Miller and J.Michl, *Chem.Rev.*, (1989), **89**, 1361 and references therein.
2. R.H.Cragg, R.G.Jones, A.C.Swain and S.J.Webb, *J.Chem.Soc.,Chem.Commun.*, (1990), 1147.
3. P.Trefonas, R.West, R.D.Miller and D.Hofer, *J.Polym.Sci.,Polym.Lett.Ed.*, (1983), **21**, 823.
4. P.Trefonas, R.West and R.D.Miller, *J.Am.Chem.Soc.*, (1985), **107**, 2737.
5. R.West, L.D.David, P.L.Djurovich, K.L.Stearly, K.S.V.Srinivasan and H.Yu, *J.Am.Chem.Soc.*, (1981), **103**, 7352.
6. R.West, L.D.David, P.I.Djurovich, H.Yu and R.Sinclair, *Ceram.Bull.*, (1983), **62**, 899.
7. S.Gauthier and D.Worsfold, *Macromolecules*, (1989), **22**, 2213.
8. R.D.Miller, J.Rabolt, R.Sooriyakumaran, W.Fleming, G.N.Fickes, B.L.Farmer and H.Kuzmany, Inorganic and Organometallic Polymers; M.Zeldin, K.H.Wynne, H.R.Allcock, Eds, ACS Symposium Series; Am.Chem.Soc.:Washington, DC, (1988); Vol. **360**, pp 43-60.
9. R.West, A.R.Wolff and D.J.Peterson, *J.Radiat.Curing*, (1986), **13**, 35.
10. A.R.Wolff and R.West, *Applied Organomet.Chem.*, (1987), **1**, 7.
11. S.Yajima, Y.Hasegawa, J.Hayashi and M.Iiomora, *J.Mater.Sci.*, (1980), **15**, 720.
12. S.Yajima, J.Hayashi and M.Omori, *Chem.Lett*, (1975), 931.
13. J.P.Wesson and T.C.Williams, *J.Polym.Sci.,Polym.Chem.Ed.*, (1980), **15**, 720.
14. J.P.Wesson and T.C.Williams, ibid., (1981), **19**, 65.
15. T.Uhsirogauchi, T.Tada and H.Watanabe, Proceedings of Conference on RadiationCuring Ashia, (1986), 101.
16. P.Trefonas, P.I.Djurovich, X.-H.Zhang, R.West, R.D.Miller and D.Hofer, *J.Polym.Sci., Polym.Lett.Ed.*, (1983), **21**, 819.
17. X.-H.Zhang and R.West, *J.Polym.Sci.,Polym.Chem.Ed.*, (1984), **22**, 159.
18. X.-H.Zhang and R.West, *ibid.*, (1984), **22**, 225.
19. H.Stuger and R.West, *Macromolecules*, (1985), **18**, 2349.
20. S.Ho.Yi, N.Maeda, T.Suzuki and H.Sato, *Polymer J*, (1992), **24**, 865.
21. R.D.Miller, D.Thompson, R.Sooriyakumaran and G.N.Fickes, *J.Polym.Sci.,Polym.Chem.Ed.*, (1991), **29**, 813.
22. L.A.Harrah and J.M.Zeigler, *Macromolecules*, (1987), **20**, 2037.
23. R.Horiguchi, Y.Onishi and S.Hayase, *ibid*, (1988), **21**, 304.
24. H.Ban, K.Sukegawa and S.Tagawa, *ibid*, (1987), **21**, 304.
25. R.E.Benfield, R.H.Cragg, R.G.Jones, A.C.Swain, S.J.Webb and M.J.Went, *Polymer*, in press.
26. T.Seki, T.Tamaki and K.Ueno, *Macromolecules*, (1992), **25**, 3825.
27. B.L.Van Duisen, A.Sivak, B.M.Goldschmidt, C.Katz and S.Melchionne,*J.Natl.Cancer Inst.*, (1969), **43**, 481.
28. M.E.Wright, E.G.Toplikar and S.A.Svejda, *Macromolecules*, (1991), **24**, 5879.
29. H.Sakurai and F.Kondo, *J.Organomet.Chem.*, (1975), **92**, C46.
30. C.U.Pittman and R.F.Felis, J.Organometallic.Chem., (1974), **72**, 389.
31. H.Sakurai, persional communication.
32. H.K.Kim and K.Matyjaszewski, *J.Polym.Sci.*, *Part A: Polym.Chem*, (1993), **31**, 299.
33. Ch.Elschenbroich and A.Salzer, Organometallics,VCH Verlagsgesellschaft/VCH Publishers, Weinheim/New York (1989)
34. S.Bhaduri, H.Khwaja and V.Khanwalkar, J.Chem.Soc.,Dalton Trans., (1981), 445

SILOXANE POLYMERS WITH PENDANT
METAL CARBONYL GROUPS

Fred B. McCormick[*], Bradford B. Wright, and Jerry W. Williams

Corporate Research Laboratories
3M Company
201-2N-21, 3M Center
St. Paul, MN 55144-1000

INTRODUCTION

The first transition metal organometallic polymer, polyvinylferrocene, was reported in 1955.[1] Since then, organometallic polymers have been extensively studied due to their potential utility in applications such as catalysts, UV stabilizers, ceramic precursors, and electrode modifiers.[2] While investigating the use of organometallic polymers as adhesion promoters, we found that metal carbonyl complexes with vinylcyclopentadienyl ligands could be hydrosilated with polysiloxanes containing Si-H bonds. During the course of our work, a similar hydrosilation of vinylferrocene and electrode modification with the resulting polymer was reported.[3] We have found that pendant metal carbonyl groups may be used to photo-adhere polymers to substrates having basic surface groups, such as many metal oxides. This chapter details the adhesion chemistry of metal carbonyl containing polymers, the synthesis and characterization of $CpMn(CO)_3$ and $(arene)Cr(CO)_3$ containing polysiloxanes, the conversion of α-hydroxyalkylcyclopentadienyl complexes to ether byproducts during the synthesis of vinylcyclopentadienyl complexes, and an unexpected product from the hydrosilation of vinylferrocene with triphenylsilane.

EXPERIMENTAL RESULTS

Preparation of vinylcyclopentadienylmanganese tricarbonyl

A 0.50g (2.02 mmol) sample of α-hydroxyethylcyclopentadienylmanganese tricarbonyl[4] was dissolved in 20 mL of toluene in a 50 mL round bottom flask fitted with a reflux condenser. The system was flushed with nitrogen and 0.15g of $KHSO_4$ was added to the toluene solution. The reaction was magnetically stirred and refluxed for 1.5 hrs. The reaction mixture was then cooled to room temperature and dried with MgSO$_4$. After filtration and removal of solvent on a rotary evaporator, 0.44g (96%) of crude vinylcyclopentadienylmanganese tricarbonyl was obtained as a yellow oil. This material was dissolved in a minimum amount of dichloromethane and the solution was filtered through a small (0.5g) plug of silica gel. Evaporation of the solvent from the filtrate gave 0.42g (92%) of the purified product as a bright yellow oil. Vinyl(methyl)cyclopentadienylmanganese tricarbonyl was also prepared by this method.

Dehydrative coupling of α-hydroxyethylcyclopentadienylmanganese tricarbonyl

A 0.50g (2.02 mmol) sample of α-hydroxyethylcyclopentadienylmanganese tricarbonyl and 0.21g (1.02 mmol) of 1,3-dicyclohexylcarbodiimide were placed in a 1 dram glass vial fitted with a screw cap septum lid. The vial was evacuated and back-filled with N$_2$ three times using syringe needle techniques and was placed in a 100°C oven for 10 hrs. The oily solid obtained upon cooling was extracted with chloroform to give a yellow solution and colorless, sparingly-soluble crystals of 1,3-dicyclohexylurea. After filtration, the chloroform solution was evaporated to a viscous yellow oil containing a small amount of colorless crystals. The product was separated from the crystals by extraction with hexanes. After concentrating the hexane extracts and cooling to -10°C, fine yellow crystals of 1,1'-bis-(cyclopentadienylmanganese tricarbonyl)ethyl ether were isolated. Spectral characterization: IR (cm^{-1}, hexanes) 2025 (s), 1941 (s); ^{1}H-NMR (δ, CDCl$_3$) 4.59, 4.67, 4.84, and 4.89 (m, C$_5$H$_4$R), 4.35 (q, J = 6.2 Hz, OCHCH$_3$), 1.38 (d, J = 6.2 Hz, OCHCH$_3$); ^{13}C-NMR (ppm, CDCl$_3$) 107.4 (RCC$_4$H$_4$), 80.0, 81.7, 82.0, and 83.4 (s, RCC$_4$H$_4$), 68.5 (s, OCHCH$_3$), 21.7 (s, OCHCH$_3$); MS (daltons) 478 (12%, M$^+$), 394 (20%, M$^+$ - 3 CO), 231 (80%, (CO)$_3$Mn-C$_5$H$_4$CHCH$_3$$^+$), 55 (100%, Mn$^+$).

Hydrosilation reactions with model siloxanes and silanes

Vinylcyclopentadienylmanganese tricarbonyl (1.0 g, 4.3 mmol), pentamethyldisiloxane (0.9 g, 6.1 mmol), and chloroplatinic acid (100 µL, 0.10 molar isopropanol solution) were refluxed in 25 mL of heptane for 5 hours. The yellow solution was cooled and rotary evaporated to an oil. The oil was vacuum distilled to give 0.7 g (57%) of (α- and β-pentamethyldisiloxyethyl)cyclopentadienylmanganese tricarbonyl. Integration of the ^{1}H-NMR spectrum showed the α/β ratio to be 1/2.1. Hydrosilation reactions of vinylferrocene with pentamethyldisiloxane and triphenylsilane were performed under similar conditions.

Synthesis of polysiloxanes containing metal carbonyl groups

Methyl(vinyl)cyclopentadienylmanganese tricarbonyl (0.48 g, 2.0 mmol), a copolymer of 3-4% methylhydro and 96-97% dimethylsiloxane (2.0 g, Hüls PS124.5 13,300 MW), and chloroplatinic acid (0.15 mL, 0.10 molar isopropanol solution) were refluxed in heptane (25 mL) for 18 hours. After cooling, the heptane was removed by rotary evaporation. The crude polymer residue was dissolved in a minimum amount of dichloromethane and filtered through silica gel to remove residual platinum catalyst. The dichloromethane was vacuum evaporated from the filtrate to give the manganese derivatized polysiloxane as a yellow fluid (IR: 2015 and 1925 cm^{-1}). The polymer could be further purified by dissolution in

dichloromethane and reprecipitation with methanol. Several other Si-H containing polysiloxanes were similarly derivatized.

Chromium hexacarbonyl (0.5 g, 2.3 mmol) and 2.0 g of polyphenylmethylsiloxane
(Hüls PS160, 2,600 MW) were refluxed in 25 mL of 1,2-dimethoxyethane under a N_2 atmosphere for 16 hours. After filtration to remove a small amount of greenish precipitate, the
solution was rotary evaporated to a viscous yellow fluid. The fluid was dried under high
vacuum to remove residual solvent and unreacted $Cr(CO)_6$. The IR spectrum of the resulting $Cr(CO)_3$ derivatized polysiloxane showed CO stretches at 1976 and 1909 cm^{-1}.
Chromium tricarbonyl was also incorporated into a methylhydro-phenylmethylsiloxane copolymer (Hüls PS129.5) by this method.

Peel tests on coatings of polysiloxanes containing metal carbonyl groups

Manganese functionalized siloxanes were coated from hexanes (1% solids) onto aluminum Q-panels using a #10 wire wound bar. The coated panels were air dried, irradiated
from a distance of 1/2 inch for 2 minutes with two 15 watt fluorescent blacklights (366nm),
and rinsed with copious quantities of hexanes. Adhesion of the coatings to the aluminum
Q-panels was assessed by $180°$ peel tests (6"/min) using Scotch™ brand Magic™ tape and
Macdermid silicone PSA tape. Each panel was first tested with Magic™ tape and then with
Macdermid tape (Table 1, initial peel columns). Twenty peels with Macdermid tape were
then done and the panels were retested with Magic™ tape (Table 1, retest peel column).
All peels were performed on the same area of the test panel.

DISCUSSION

Synthesis of vinylcyclopentadienyl monomers

The traditional synthesis[4] of vinylcyclopentadienylmanganese tricarbonyl and methyl-
(vinyl)cyclopentadienylmanganese tricarbonyl from the corresponding α-hydroxyethyl precursors utilizes a one-pot, solvent-free dehydration/vacuum distillation technique. The neat
α-hydroxyethylcyclopentadienylmanganese tricarbonyl is placed in the boiling flask of a
short-path vacuum distillation apparatus along with an acid catalyst, usually $KHSO_4$, and
a free-radical inhibitor such as t-butylcatechol. This mixture is heated under vacuum to
remove the water generated in the dehydration of the hydroxyethyl group and the resulting
vinyl compound is vacuum distilled. Although the literature reports 90% yields of the vinyl
compound,[4a] we have experienced variable results with this technique. Vigorous bumping
during the dehydration sometimes occurred thereby contaminating the desired product.
Traces of the free radical inhibitor often seemed to distill with the product which complicated later reactions with the material. The yields were quite variable and, on numerous
occasions, little or no product could be distilled before the reaction mixture became a high
boiling, viscous material. Some of these difficulties have been pointed out in the literature
and the reaction residue was thought to be a homopolymer of the desired vinylcyclopentadienylmanganese tricarbonyl.[4b] Because of these difficulties, we sought to develop alternative, more reliable syntheses of vinylcyclopentadienylmanganese tricarbonyl derivatives.

The $KHSO_4$ catalyzed dehydration of α-hydroxyethylcyclopentadienylmanganese tricarbonyl complexes in refluxing toluene, with the optional removal of water by a Dean-

Stark trap, appeared to be the desired improved synthesis of vinylcyclopentadienylmanganese tricarbonyl derivatives. Yields greater than 90% of vinylcyclopentadienylmanganese
tricarbonyl were thus obtained from α-hydroxyethylcyclopentadienylmanganese tricarbonyl
when this reaction was carried out on a small scale. However, attempts to scale up this reaction lead to poor conversion to the desired vinyl compound and produced unidentified Mn
compounds which were not readily separated from the reaction mixture. Fortunately, yellow crystals eventually precipitated from the reaction mixture upon storage at -10°C. These
were isolated and shown to be the symmetrical ether, 1,1'-bis(cyclopentadienylmanganese
tricarbonyl)ethyl ether, resulting from the dehydrative coupling of α-hydroxyethylcyclopentadienylmanganese tricarbonyl.

Low yields of the ether, characterized by its IR spectrum and elemental analysis, were
reported by Nesmeyanov[5] when α-hydroxyethylcyclopentadienylmanganese tricarbonyl was
reacted with sulfuric acid. We independently prepared the ether in high yield by the dehydration of neat α-hydroxyethylcyclopentadienylmanganese tricarbonyl with 1/2 equivalent
of dicyclohexylcarbodiimide (DCC) in a sealed vial held at 100°C for several hours (eq 1).

$$(1)$$

The methyne carbons of this ether are chiral centers and thus the product is a mixture
of diastereomers, a d,l pair and a meso form, which complicates the spectral characterization of the product. The DCC dehydrative coupling reaction appeared to be a general reaction and α-hydroxyethylcyclopentadienylrhenium tricarbonyl, α-hydroxyethylferrocene, and
hydroxymethylferrocene were all converted to the corresponding ethers. Reaction of α-hy-
droxyethyl(methyl)cyclopentadienylmanganese tricarbonyl with DCC also appeared to produce the corresponding ether as a viscous oil. Since the starting alcohol was a mixture of
1,2 and 1,3 ring substituted isomers, the resulting ether should be a mixture of 10 isomers
(diastereomers and regioisomers). The complex nature of the mixture hampered characterization of the products, but the expected 506 dalton molecular ion was observed in the mass
spectrum of the product mixture.

The ether formation observed in the scale up of the solvent based dehydration was
likely due to concentration effects. High dilution techniques such as dropwise addition of
dilute toluene solutions of the alcohol to dilute refluxing toluene solution of the acid catalyst generally eliminated ether byproduct formation.

Since the ether byproduct is readily produced in the acid catalyzed dehydration of the alcohol under higher concentration conditions, it was of interest to revisit the viscous materials often generated in our attempts to reproduce the literature preparations of vinylcyclopentadienylmanganese tricarbonyl and vinyl(methyl)cyclopentadienylmanganese tricarbonyl.[4] One such residue from the boiling flask of an attempted $KHSO_4$ catalyzed dehydration/vacuum distillation of neat α-hydroxyethyl(methyl)cyclopentadienylmanganese tricarbonyl did not yield any meaningful NMR or IR spectra due to its complex nature. However, the mass spectrum of the residue showed the same molecular ion (506 dalton) and fragmentation pattern as the ether obtained from the dehydrative coupling of α-hydroxyethyl(methyl)cyclopentadienylmanganese tricarbonyl using DCC. It thus seems clear that dehydrative coupling of the alcohol can occur in the previously reported synthesis of vinylcyclopentadienylmanganese tricarbonyl complexes. Based on our experience, any reduced yields obtained in the literature synthesis of vinylcyclopentadienylmanganese tricarbonyl derivatives is more likely due to the formation of the ether byproducts than homopolymerization of the desired vinyl monomers.

Hydrosilation reactions with model siloxanes and silanes

Hydrosilation of vinylcyclopentadienylmanganese tricarbonyl, methyl(vinyl)cyclopentadienyl tricarbonyl, and vinylferrocene with model silanes and siloxanes was undertaken in order to determine optimum conditions for the hydrosilation reactions of polysiloxanes. It was also of interest to determine the regiochemistry of the Si-H bond addition to the C=C bond of a vinylcyclopentadienyl ligand. This had not been addressed in the previous reports on the hydrosilation of vinylferrocene.[3] Although an exhaustive screening of hydrosilation catalysts was not undertaken, chloroplatinic acid, H_2PtCl_6 (Speier's catalyst), and bis(divinyltetramethyldisiloxane)platinum (Karstedt's catalyst) appeared to be the most effective while $(Ph_3P)_3RhCl$ (Wilkinson's catalyst) showed little or no activity. Reaction of vinylcyclopentadienylmanganese tricarbonyl with pentamethyldisiloxane and Speier's catalyst gave both the α, $(Si(CH_3)_3OSi(CH_3)_2CH(CH_3)C_5H_4)Mn(CO)_3$, and β, $(Si(CH_3)_3O-Si(CH_3)_2CH_2CH_2C_5H_4)Mn(CO)_3$, isomers of the hydrosilation product. Integration of the distinctive ethylidene and ethylene multiplets[6,7] in the ^{1}H-NMR indicated the product mixture was 32% α and 68% β (eq 2). The methylene groups of the α product appear as two

(2)

well separated AA'XX' multiplets[7] at about 2.25 and 0.75 ppm. The methyl and methyne groups of the β product appear as the expected quartet (~1.65 ppm) and doublet (~1.15 ppm). Similar α/β ratios were seen in styrene hydrosilations using Karstedt's catalyst.[6]

Vinylferrocene appeared to give considerably less α isomer, approximately 12%, upon hydrosilation with pentamethyldisiloxane. In an attempt to obtain crystalline hydrosilation products, vinylferrocene was reacted with triphenylsilane in the presence of Speier's catalyst. To our surprise, none of the expected hydrosilation product, $Ph_3SiCH_2CH_2C_5H_4Fe$-C_5H_5, was observed in this reaction. The vinylferrocene was consumed and a good yield of an apparent ferrocene derivative was obtained and it was purified by column chromatography. Analysis by ^1H and ^{13}C-NMR and mass spectroscopy revealed the product to be $CH_3CH_2C_5H_4FeC_5H_4SiPh_3$, an isomer of the expected product (eq 3). The mechanism for

$$(3)$$

the formation of this isomer is not currently understood. The efficiency of its formation leads us to believe the expected hydrosilation product, $Ph_3SiCH_2CH_2C_5H_4FeC_5H_5$, forms initially and undergoes an intramolecular rearrangement where the ethylenic Ph_3Si group is exchanged with a hydrogen of the unsubstituted cyclopentadienyl ligand. Other mechanisms, such as radical or catalyzed redistribution reactions, cannot be ruled out, but these would be expected to provide a more complex mixture of products.

Preparation of siloxane polymers containing metal carbonyl groups

Virtually any Si-H bond containing polysiloxane can be derivatized with vinylcyclopentadienyl-containing organometallic compounds by the hydrosilation chemistry described herein. We have been primarily concerned with methyl(vinyl)cyclopentadienylmanganese tricarbonyl due to its simple preparation, low cost, and photolability. The hydrosilation is quite efficient and the nature of the resulting polymer is readily controlled by adjusting the reaction stoichiometry. The Si-H band of the siloxane (~2160 cm^{-1}) and the two CO bands of the $CpMn(CO)_3$ groups (~2015 and 1930 cm^{-1}) are quite distinct in IR spectra and provide a ready method to monitor the progress of the hydrosilation reactions. Numerous Si-H containing polysiloxanes are commercially available, such as polymethylhydrosiloxanes, methylhydro/dimethylsiloxane copolymers, and polydimethylsiloxanes with $HSi(CH_3)_2$- end-caps, and all of these have been successfully hydrosilated with methyl(vinyl)cyclopentadienylmanganese tricarbonyl (eq 4). The resulting polymers are light to medium yellow in color. The physical properties of the polysiloxanes, such as viscosity, are not significantly changed upon derivatization with $CpMn(CO)_3$ groups.

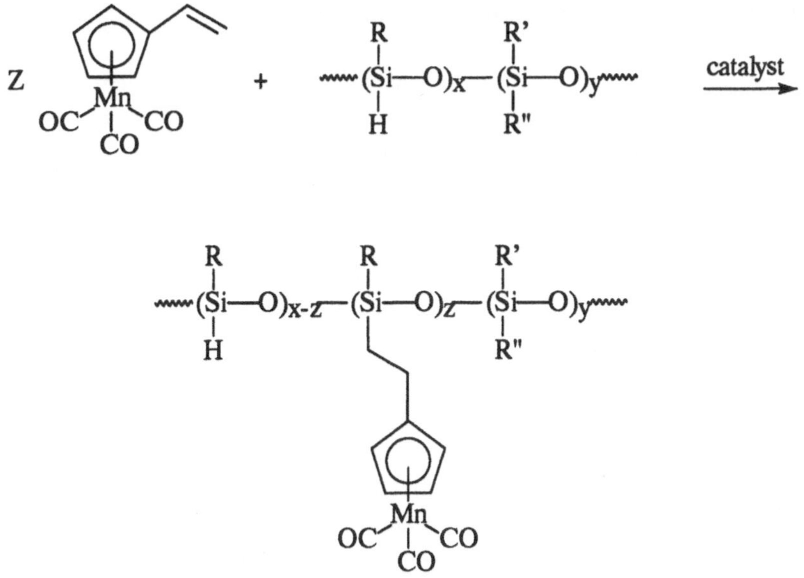

$$Z \overset{\text{(cyclopentadienyl-vinyl)}}{\underset{OC-\underset{CO}{\overset{Mn}{|}}-CO}{}} \quad + \quad \text{\textasciitilde\textasciitilde}(\underset{H}{\overset{R}{Si}}-O)_x-(\underset{R''}{\overset{R'}{Si}}-O)_y\text{\textasciitilde\textasciitilde} \quad \xrightarrow{\text{catalyst}}$$

$$\text{\textasciitilde\textasciitilde}(\underset{H}{\overset{R}{Si}}-O)_{x-z}-(\overset{R}{Si}-O)_z-(\underset{R''}{\overset{R'}{Si}}-O)_y\text{\textasciitilde\textasciitilde}$$

with the $-(Si-O)_z-$ unit bearing a $-CH_2CH_2-$ tether to a cyclopentadienyl ring coordinated to $OC-\underset{CO}{\overset{Mn}{|}}-CO$.

(4)

A modification of a literature preparation of (styrene)Cr(CO)$_3$/styrene copolymers,[8] utilizing the well known thermal reaction of Cr(CO)$_6$ with an arenes to give (arene)Cr(CO)$_3$ complexes, is useful in derivatizing polysiloxanes containing phenyl groups. Thus, refluxing preformed polyphenylmethylsiloxane or methylhydro/phenylmethylsiloxane copolymers with Cr(CO)$_6$ in 1,2-dimethoxyethane (glyme) provided the corresponding Cr(CO)$_3$ containing co- and terpolymers (eq 5). The Cr(CO)$_6$ has a tendency to sublime and collect in the

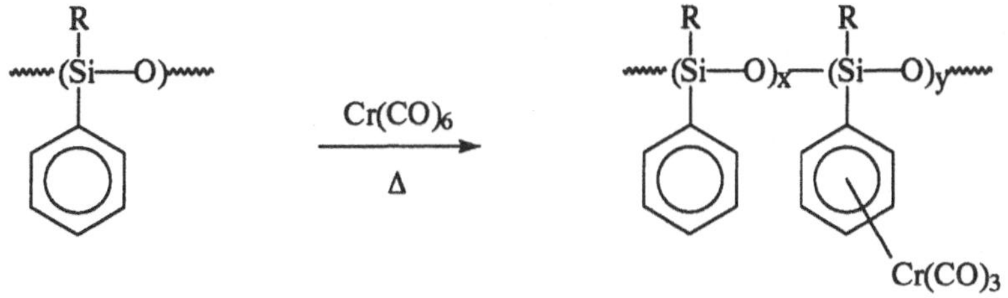

$$\text{\textasciitilde\textasciitilde}(\underset{Ph}{\overset{R}{Si}}-O)\text{\textasciitilde\textasciitilde} \quad \xrightarrow[\Delta]{Cr(CO)_6} \quad \text{\textasciitilde\textasciitilde}(\underset{Ph}{\overset{R}{Si}}-O)_x-(\underset{Ph\cdot Cr(CO)_3}{\overset{R}{Si}}-O)_y\text{\textasciitilde\textasciitilde}$$

(5)

reflux condenser which makes it difficult to control the stoichiometry of the reaction. Glyme seems to minimize the accumulation of Cr(CO)$_6$ in the condenser. The polymers show the characteristic two CO stretches in their IR spectra at about 1980 and 1900 cm^{-1}. Compared to the CpMn(CO)$_3$ containing polysiloxanes, these (arene)Cr(CO)$_3$ containing polysiloxanes were less stable and slowly precipitated chromium oxides on storage at room temperature.

Adhesion of metal carbonyl-containing polysiloxane coatings

Coatings of siloxane polymers typically provide low energy surfaces which have found great industrial utility as release materials in, for example, adhesive tapes or labels and polymer molds. However, this very release property can make it difficult to keep the siloxane coating on the desired substrate and migration of the siloxane can create problems in various manufacturing processes. When the metal carbonyl containing polysiloxanes described above are coated onto aluminum panels and the coatings are irradiated with UV light, the polysiloxane coatings adhere remarkably well to the panels. These polysiloxanes may be useful as protective coatings and durable release coatings. Irradiation of the coatings through a photomask followed by solvent development give adherent coatings in the negative image of the mask and high resolutions have been achieved. Good polysiloxane images have also been obtained by focusing the image of a photographic negative onto a Mn/polysiloxane coating using a standard 2" x 2" slide projector.

Adhesion of the irradiated polysiloxane coatings has not been directly measured but the peel test data in Table 1 provides some insights. Acrylic pressure-sensitive adhesive (PSA) tape (Magic[TM]) and silicone PSA tape (Macdermid) adhere equally well to bare Al. After 20 peels with Macdermid, the adhesion of Magic[TM] tape, to the same portion of the

Table 1. Peel tests using PSA tapes on aluminum panels treated with polysiloxanes having pendant -CpMn(CO)$_3$ groups.

Siloxane [b]	Initial peel [a]		Retest peel [a]
	Magic[TM]	Macdermid	Magic[TM]
none	35	35	18
PS120 (1%)	1.4	22	7.1
PS120 (10%)	3.6	26	6.5
PS123 (5%)	2.0	28	7.1
PS123 (50%)	4.9	26	5.1
PS125.5 (5%)	2.0	30	6.8

a) peels in oz/in.
b) Hüls catalog #s and the percentage of Si-H bonds hydrosilated.

test panel, was lowered, probably due to silicone transfer from the Macdermid tape. On Al panels coated with several different Mn-derivatized polysiloxanes and irradiated, Magic[TM] tape showed very low adhesion while Macdermid tape showed only slightly lower adhesion compared to bare Al. After 20 repeated peels with Macdermid tape, the adhesion of Magic[TM] tape, to the same portion of the test panel, was somewhat increased but remained significantly less than the values to bare Al. This indicates that the Macdermid tape, which would be expected to adhere well to a siloxane surface, does not readily remove the irradiated Mn-derivatized polysiloxane coatings. Thus a lower limit for the adhesion of these polysiloxane coatings can be placed at about 26 oz/in.

Adhesion mechanism for metal carbonyl-containing polymers

The photoinduced polymer adhesion is not limited to the polysiloxanes containing metal carbonyl groups or to aluminum substrates. A wide variety of polymers with pendant metal carbonyl groups were prepared by several routes including 1) copolymerization of vinylcyclopentadienylmanganese tricarbonyl with styrene and various acrylate monomers, 2) reaction of chloroformyl(methyl)cyclopentadienylmanganese tricarbonyl[9] with preformed polymers containing hydroxyl groups, and 3) reaction of $Cr(CO)_6$ with polystyrene[8]. Coatings of these polymers on aluminum plates were exposed to UV light through a photographic mask as rapid screening test for adhesion to the plate. After exposure, the coated plates were rinsed with solvent and inspected for image formation. All of the above polymers containing pendant metal carbonyl groups gave abrasion resistant polymer coatings in the negative image of the photographic mask which was indicative of photoinduced adhesion. Coatings of a styrene/vinyl(methyl)cyclopentadienylmanganese tricarbonyl copolymer were made on a variety of substrates. Photoinduced adhesion to the substrates was screened as described above for the various polymers on aluminum. Metallic substrates known to have basic oxide surfaces, such as Al, Fe, Ni, Cu, and Ti, all gave adherent coatings in the negative image of the photographic mask. Metallic substrates such as Au, Pt, and Ta, all generally lacking oxide surfaces, did not form any image which indicated that there was no photoinduced adhesion of the polymer to the substrate.

Although the adhesion mechanism for the metal carbonyl containing polysiloxanes has not been definitively established, we believe covalent bonding of the photogenerated, coordinatively unsaturated organometallic groups to basic sites on the substrate is of key importance as shown schematically in Figure 1. Support for such a mechanism is found in that

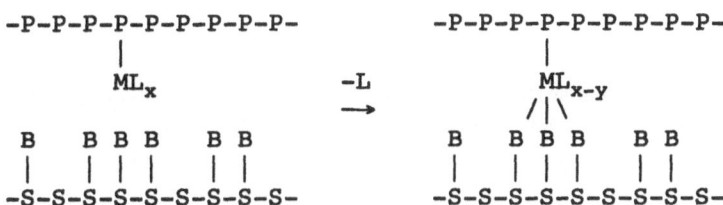

Figure 1. Schematic diagram of the proposed bonding mechanism where S = substrate, B = basic site on substrate, P = polymer backbone, M = transition metal, L = CO ligand.

adhesion is seen only on metals known to have basic oxide surfaces. Specular reflectance IR studies on polished aluminum substrates also support this adhesion mechanism. A thin coating of a styrene/vinyl(methyl)cyclopentadienylmanganese tricarbonyl copolymer on the polished Al showed CO stretches at 2013 and 1930 cm^{-1}. Irradiation of the coating with 366nm light for a short period in a N_2 atmosphere produced an IR spectrum with bands at 2151, 2014, 1960, 1930, 1905, and 1865 cm^{-1}. Considering the relative intensities of the bands, three Mn species were postulated. The 2014 and part of the 1930 cm^{-1} band were due to the starting copolymer. The 2151, 1960, and 1905 cm^{-1} bands correspond to the known polymer bound $CpMn(CO)_2N_2$ complex.[10] The 1865 and part of the 1930 cm^{-1} bands are consistent with a $CpMn(CO)_2L$ species.[11] We assign this species to a polymer-$CpMn(CO)_2$-OAl-surface species which would correspond to the "P-ML_{x-y}-B-S" linkage in

Figure 1. Similar changes in the IR spectra were seen when a CpMn(CO)₃ derivative was photo-immobilized on a layered α-Zr(HPO₄)₂•H₂O substrate; -CpMn(CO)₂-O(H)-P≡ species were proposed for the surface-supported intercalated organometallic complex.[12] The photosubstitution chemistry of CpMn(CO)₃ and (arene)Cr(CO)₃ complexes to give (polyene)M(CO)₂L species is a well established and often used synthetic method. Decomposition of organometallic compounds on inorganic supports such as alumina, silica, and magnesia has long been used as a means of generating heterogeneous catalysts. Although this often leads to metal or metal oxide particles, a number of surface bound organometallic species have been identified or postulated as intermediates. Figure 2 shows such an example where a Re(CO)₃ fragment bonds to a MgO substrate by Re-O covalent bonds.[13] This

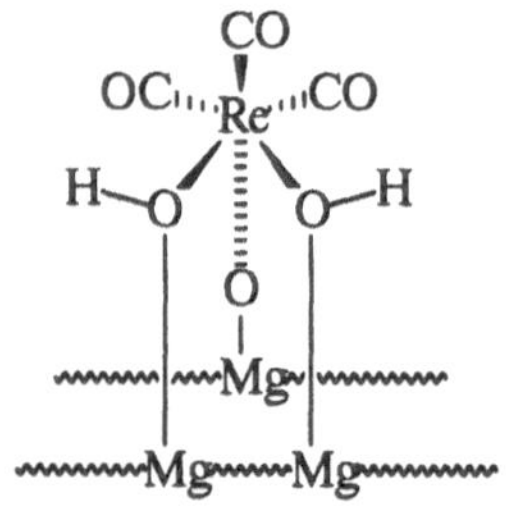

Figure 2. Proposed structure of Re(CO)₃ immobilized on magnesia (after reference 13).

is the same type of bonding that we propose as the source of our observed photoinduced polymer adhesion.

Applications of metal carbonyl-containing polysiloxanes

Polysiloxanes with pendant metal carbonyl groups have been applied as thin coatings to metallic printing plates such as 3M's Viking™ brand (silicated aluminum) plate. The plate was then exposed through a mask and developed with solvent to give a printing plate with a negative silicone image. Depending on the type of ink used, positive or negative printed images may be obtained. A hydrophilic ink would wet the metallic portions of the printing plate made with the metal carbonyl containing polysiloxanes. These are the portions of the printing plate which were not exposed to light and thus would print a positive image of the exposure mask. Hydrophobic inks would tend to wet the portions of the plate coated with polysiloxane. The inking of the polysiloxane image could be enhanced by first coating the plate with an aqueous fountain solution. The hydrophobic ink would thus print a negative image of the exposure mask.

Particulates may also be printed with these polysiloxane plates. Small particles like glass beads, glass bubbles, and toner powder are attracted to the polysiloxane image and are held in place fairly strongly. They may be brushed off, but are generally not dislodged by solvent rinses or strong streams of compressed air. The particles do transfer to higher energy surfaces, such as PSA tape, softened thermoplastic films, or rubber offset rollers, when the plate is pressed against these surfaces. After particle transfer, especially to offset rollers, the plate will again accept particles in the silicone image areas which makes continuous printing with these plates possible. This type of particle printing has been used to produce coated abrasive sheets where the abrasive particles occur in printed patterns.[14]

ACKNOWLEDGMENTS

The technical assistance of Michelle M. Heil, Donna A. Frisvold, and Paul Humpal is gratefully acknowledged.

REFERENCES

1. Arimoto, F. S.; Haven, A. C. *J. Am. Chem. Soc.* **1955**, *77*, 6295.

2. Sheats, J. E.; Carraher, C. E.; Pittman, C. U.; Zeldin, M.; Currell, B., Eds. *Inorganic and Metal-Containing Polymeric Materials*; Plenum Press, NY, 1990.

3. (a) Hale, P. D.; Inagaki, T.; Karan, H. I.; Okamoto, Y.; Skotheim, T. A. *J. Am. Chem. Soc.* **1989**, *111*, 3482. (b) Inagaki, T.; Lee, H. S.; Skotheim, T. A.; Okamoto, Y. *J. Chem. Soc., Chem. Commun.* **1989**, 1181.

4. (a) Kurimura, Y; Ohta, J.; Nishide, H.; Tsuchida, E. *Makromol. Chem.* **1982**, *183*, 2889. (b) Pittman, C. U.; Marlin, G. V.; Rounsefell, T. D. *Macromolecules* **1973**, *6*, 1. (c) Kozikowski, J.; Cais, M. *U. S. Patent* 3,290,337 **1966**.

5. Nesmeyanov, A. N.; Anisimov, K. N.; Kolobova, N. E.; Zlotina, I. B. *Dokl. Akad. Nauk SSSR* **1964**, *154*, 391.

6. Lewis, L. N. *J. Am. Chem. Soc.* **1990**, *112*, 5998.

7. Colborn, *J. Chem. Ed.* **1990**, *67*, 438.

8. Pittman, C. U.; Grube, P. L.; Ayers, O. E.; McManus, S. P.; Rausch, M. D.; Moser, G. A. *J. Poly. Sci., A-1* **1972**, *10*, 379.

9. Gowal, H.; Schlögl, K. *Monatsh. Chem.* **1968**, *99*, 972.

10. Nishide, H.; Kawakami, H.; Kurimura, Y.; Tsuchida, E. *J. Am. Chem. Soc.* **1989**, *111*, 7175.

11. Connelly, N. G.; Kitchen, M. D. *J. Chem. Soc., Dalton Trans.* **1977**, 931.

12. Thompson, M. E.; Lee, C. F. *Inorg. Chem.* **1991**, *30*, 4.

13. Lamb, H. H.; Gates, B. C.; Knözinger, H. *Angew. Chem. Int. Ed. Engl.* **1988**, *27*, 1127.

14. Calhoun, C. D.; Foss, G. D.; Fleming, M. J.; Bruxvoort, W. J. *U. S. Patent* 4,930,266, **1990**.

REACTION OF TRIMETHYLSILYLMETHYL SUBSTITUTED SILANES WITH ALKALI METALS: SYNTHESES OF POLYCARBOSILANES AND POLYSILANES*

Gan Ouyang, Richard Simons, and Claire Tessier

Department of Chemistry
The Unversity of Akron
Akron, OH 44325-3601

INTRODUCTION

During the last decade, silicon–containing polymers such as polysilanes and polycarbosilanes have attracted much attention because of their potential applications as ceramics precursors,[1] binders for ceramics[2] and for powder metallurgy,[3] photoresistors[4] and photoinitiators,[5] and non–linear optical materials.[6] The first practical use of a polysilane was in the production of β-silicon carbide by the Yajima process (Eq. 1).[7]

$$(Me_2Si)_n \xrightarrow{\sim 400\ ^\circ} \left[\begin{matrix} Me \\ | \\ -SiCH_2- \\ | \\ H \end{matrix} \right]_n \xrightarrow{\sim 1000\ ^\circ} \beta\text{-SiC} \qquad (\text{ Eq. 1 })$$

The two-step process involves the conversion of a linear polysilane to a polycarbosilane which, in turn, is converted to β-SiC. Many studies have been conducted to improve the technology.[8] Most have focused on elimination of one of the steps shown in Eq.1; either direct pyrolysis of polysilanes or direct syntheses of polycarbosilanes.

Direct syntheses of polycarbosilanes can be realized by many routes. This paper will focus on polycarbosilanes which have alternating carbon and silicon atoms in the backbone. The major route to such polycarbosilanes involves the use of an elemental metal to couple Si-Cl and C-Cl functionalities in a heterogeneous process. The coupling produces Si-C bonds and a metal halide by-product. Examples of Wurtz-type coupling synthesis of polycarbosilanes are shown in Eqs. 2[9] and 3.[10]

---

*Parts of this work were published in *J. Inorg. Organomet. Polym.*, 1995, 5, 87–94.

$$\underset{\underset{R'}{|}}{\overset{\overset{R}{|}}{Cl-Si}}-CH_2Cl \ + \ 2\,Na \ \longrightarrow \ \underset{\underset{R'}{|}}{\overset{\overset{R}{|}}{(Si}}-CH_2)_n \ + \ 2\,NaCl \qquad (\,Eq.\,2\,)$$

$$R_2SiCl_2 \ + \ CH_2Cl_2 \ + \ 4\,Na \ \longrightarrow \ (R_2SiCH_2)_n \ + \ 4\,NaCl \qquad (\,Eq.\,3\,)$$

Highly branched polycarbosilanes are prepared from chloromethyltrichlorosilane by magnesium coupling (Eq. 4).[11]

$$Cl_3SiCH_2Cl \ + \ Mg \ \xrightarrow{\ -MgCl_2\ } \ "(SiCl_2CH_2)_n" \xrightarrow{\ LiAlH_4\ } \ "(SiH_2CH_2)_n" \qquad (\,Eq.\,4\,)$$

Another elemental-metal induced coupling has used copper metal.[12] An electrochemical variant of these coupling methods uses a sacrificial metal electrode to effect Si-C bond formation.[13] A different approach to carbosilanes makes use of soluble transition metal catalysts with a variety of silane monomers. Some noble metal (Pt, Pd, Ru, Ir) compounds can catalyze ring–opening polymerization of cyclic carbosilanes[14] (Eq. 5):

$$\underset{(CH_3)_2Si-\!\!-CH_2}{\overset{CH_2-Si(CH_3)_2}{|\qquad\quad|}} \ \xrightarrow{\ catalyst\ } \ \underset{\underset{CH_3}{|}}{\overset{\overset{CH_3}{|}}{(Si}}-CH_2)_n \qquad (\,Eq.\,5\,)$$

Activation of C-H bonds of trimethylsilane by ruthenium phosphine complexes gives a catalytic synthesis of polycarbosilanes (Eq. 6).[15]

$$HSiMe_3 \ \xrightarrow{\ [Ru]\ } \ H_2 \ + \ HSiMe_2CH_2SiMe_3 \ + \ HSiMe(CH_2SiMe_3)_2$$
$$+ \ HSiMe_2CH_2SiMe_2CH_2SiMe_3 \ + \ higher\ oligomers \qquad (\,Eq.\,6\,)$$

Self hydrosilation of vinylsilane, induced by titanocene complexes, has also yielded polycarbosilanes (Eq. 7).[16]

$$H_3SiCH=CH_2 \ \xrightarrow{\ [Ti]\ } \ H_3Si(\underset{\underset{CH_3}{}}{\overset{\overset{CH_3}{|}}{C}}H-SiH_2)_n \qquad (\,Eq.\,7\,)$$

The most common methods used to make linear/cyclic (Eq. 8) and network (Eq. 9) polysilanes involve the reduction of chlorosilanes with alkali metals.[17,18]

$$R_2SiCl_2 \ + \ 2\,M \ \longrightarrow \ (R_2Si)_n \ + \ 2\,MCl \qquad (\,Eq.\,8\,)$$

$$RSiCl_3 \ + \ 3\,M \ \longrightarrow \ (RSi)_n \ + \ 3\,MCl \qquad (\,Eq.\,9\,)$$

As with Eqs. 2 and 3, Eqs. 8 and 9 are examples of Wurtz-type couplings. There appears to be only one reported example in which a Wurtz coupling of a chlorosilane resulted in a *major* product that was not a polysilane.[31b] It should be noted that small amounts of polysiloxane by-products have often been reported from Wurtz couplings. Catalytic dehydrogenative coupling of silanes by early transition complexes (Eq. 10) has also been used to prepare polysilanes.[19, 8a]

$$\text{RR'SiH}_2 \xrightarrow[- \text{H}_2]{\text{catalyst}} \text{H}-(\text{RSiR'})_n\text{H} \quad + \quad (\text{RSiR'})_m \quad\quad (\text{Eq. 10})$$

For the purposes of this paper it is important to understand the spectral differences between polysilanes, polycarbosilanes, and polysiloxanes. *Unlike other silicon polymers*, polysilanes give characteristic electronic spectra which reflect the delocalized σ_{Si-Si} bonds in the backbone.[20] Linear and cyclic polysilanes show maximum absorbance in UV spectrum about 300 nm to 400 nm, with the longer chains absorbing the lower energy.[17,21] Network polysilanes (polysilynes) show characteristic UV-visible spectrum, in which the absorption tails into the visible region with no real maximum.[18,21c,22] One can expect this feature, because the σ_{Si-Si} bonds are delocalized in two dimension in polysilynes. In polycarbosilanes, the silicon atom is insulated from other silicon atoms by carbon atoms and no absorption is expected at lower energy. A peak around 220 nm to 250 nm is observed for polycarbosilanes, which may be attributed to the Si–C–Si fuctionality,[23] although there has not been much discussion of this absorbance in the literature.

The ^{29}Si NMR resonances for organosilicon polymers are observed in the range of -120 ppm to 120 ppm and are informative.[24] In a typical carbosilane structure, by definition, the silicon atoms are coordinated by four carbon atoms. The chemical shifts for ^{1}H NMR, ^{13}C NMR, and[29]Si NMR interestingly all fall in a narrow range around 0 ppm (compare the chemical shifts of Me$_4$Si, used as standards for these three nuclei). Linear or cyclic polysilanes, (R$_2$Si)$_n$, have chemical shifts at -10 to -40 ppm.[17,24] Chemical shifts for polysilynes, (RSi)$_n$, are usually between -50 to -70 ppm.[18,24] In general, lower field shifts are observed for silicons bound to two other silicon atoms and higher field shifts for silicons which are bound to three other silicons.[17,18,24] Polysiloxanes have chemical shifts extending from -120 ppm to 40 ppm. The more oxygen atoms the silicon atom is bound to, the further upfield the chemical shift is observed, with the silicates at -80 to -120 ppm.[24a,b] It is difficult to distinguish polysiloxanes from polycarbosilanes and polysilanes solely by NMR spectra. Almost all silicon hydrides are shielded relative to TMS. Generally, silicon trihydrides (SiH$_3$) appear at -30 to -70 ppm, dihydrides at -20 to -40 ppm, and monohydrides at 0 to -30 ppm.[24b] But exceptions are not rare. The hydrogen coupled ^{29}Si NMR will make a distinction among different silicon hydrides for simple molecules. In the case of polymers, the spectrum responsible for SiH$_n$ often becomes broad and featureless. Application of DEPT technique has been useful in providing structural information for hydride containing polymers.[19b]

FTIR also provide valuable structural information for organosilicon polymers.[25] Polysiloxanes show Si–O bands in the range 1000 cm^{-1} to 1130 cm^{-1}. A single sharp and strong band around 1020~1060 cm^{-1} is an indication of a disiloxane or a small ring cyclic siloxane structure. For the more complicated siloxane structures, the Si–O band becomes very broad and featureless.[25] Polycarbosilanes show a band at about 1040 cm^{-1} assigned to the Si–C–Si functionality. The sharpness of this band distinguishes it from Si–O, a band at 1350 cm^{-1} provides further proof of Si–C–Si.[23,25]

Generally speaking, Si–Cl bonds reduced by alkali metal will form Si–Si bonds, as shown by Eqs. 8 and 9. In this chapter we discuss a unique system associated with trimethylsilylmethyl, (CH$_3$)$_3$SiCH$_2$–, substituted chlorosilanes. Our work demonstrates that there is an unusual reaction of trimethylsilylmethyltrichlorosilane, (CH$_3$)$_3$SiCH$_2$SiCl$_3$,[26] with alkali metals, leading to the formation of a polycarbosilane instead of the expected polysilyne,

In contrast the reduction products of trimethylsilylmethyldichlorosilane, $(CH_3)_3SiCH_2SiHCl_2$,[27] are the expected cyclic and linear polysilanes (Eqs 11 and 12).

$$(CH_3)_3SiCH_2SiCl_3 \quad + \quad 3\ M \quad \longrightarrow \quad \text{Polycarbosilanes}$$
$$(M = Li,\ Na,\ Na/K)$$
$$(\text{Eq. 11})$$

$$(CH_3)_3SiCH_2SiHCl_2 + 2\ M \longrightarrow H[(CH_3)_3SiCH_2SiH]_mH + [(CH_3)_3SiCH_2SiH]_n$$
$$(M = Li,\ Na,\ Na/K, \quad m = 3\text{-}9, \quad n = 5\text{-}9)$$
$$(\text{Eq. 12})$$

The trimethylsilylmethyl substituent has been used extensively in organometallic chemistry. The stabilization of low coordination number, the enhancement of solubility, and the inducement of novel electronic properties have all been attributed to the introduction of the trimethylsilylmethyl substituent into a compound.[28] A Wurtz-type coupling of a trimethylsilylmethyl substituted silane has been reported. The reduction of $(Me_3SiCH_2)_2SiCl_2$, with lithium resulted in the four-membered ring $[(Me_3SiCH_2)_2Si]_4$ in low yield (12%) (Eq.13) and an uncharacterized oily liquid phase.[29]

$$[(CH_3)_3SiCH_2]_2SiCl_2 \quad + \quad 2\ Li \quad \longrightarrow \quad [(CH_3)_3SiCH_2Si]_4 \qquad (\text{Eq. 13})$$

The formation of higher polymer was probably prevented due to the steric hindrance induced by two bulky trimethylsilylmethyl substituents. For this reason we chose to study the Wurtz couplings of silanes with only one trimethylsilylmethyl substituent.

RESULTS AND DISCUSSION

The compounds $(CH_3)_3SiCH_2SiCl_3$ or $(CH_3)_3SiCH_2SiHCl_2$, were reduced with three or two molar equivalents of alkali metals (lithium, sodium, and sodium/potassium alloy), respectively, in toluene or tetrahydrofuran under sonication or at reflux. The reaction was quenched with one of the following reagents: methyl lithium, lithium aluminum hydride, and methanol to neutralize unreacted Si–Cl bonds, followed by aqueous workup steps. The reducing agents, reaction condition, and solvents did not show much effect on the products resulted. The product from reduction of $(CH_3)_3SiCH_2SiCl_3$ was a colorless glassy solid with excellent film forming properties. The product from $(CH_3)_3SiCH_2SiHCl_2$ was pale yellow viscous liquid.

As expected, the oligomeric materials obtained from $(CH_3)_3SiCH_2SiHCl_2$ had distinctive features of polysilanes. The UV-Visible spectra showed a strong and broad absorbance band at 320~340 nm consistent with linear or cyclic Si–Si structures.[20] A stronger band at 240 nm can be attribute to $(CH_3)_3SiCH_2$ side group. In the FTIR spectrum a band near 2100 cm^{-1} was observed for the Si–H bonds and the Si–CH$_2$–Si moieties showed bands at 1040 cm^{-1} (sharp) and 1350 cm^{-1}.[23,25] FI–MS of the polymer showed both linear oligomers, $H[CH_3)_3SiCH_2Si]_mH$ (m = 3–9), and cyclic oligomers, $[CH_3)_3SiCH_2Si]_n$ (n = 5–9), present. Little fragmentation of the oligomers and no evidence of cross-linking was observed. ^{29}Si NMR confirmed a polysilane structure, mulplets at 2.4 ppm and 0 ppm for $(CH_3)_3SiCH_2$,[29a] a complex mulplet at –70 ppm for Si–H, and linear and cyclic Si–Si at –7 ppm and –35 ppm.[17]

Presumably, the Wurtz reduction of $(CH_3)_3SiCH_2SiCl_3$ would lead to the formation of $(RSi)_n$ (R = $(CH_3)_3SiCH_2$), or, if the reduction was incomplete, $(RSiCl)_n$ or $(RSiCl)_n(RSi)_m$.

In contrast to the results of $(CH_3)_3SiCH_2SiHCl_2$, the polymer obtained from $(CH_3)_3SiCH_2SiCl_3$ lacked any absorption in the UV-Vis spectra which would be expected for an oligomer or polymer with network or chain Si–Si structure.[20] The polymers from $(CH_3)_3SiCH_2SiCl_3$ showed a UV band 220-240 nm which may be consistent with short (1 - 3) Si-Si linkages but the $(CH_3)_3SiCH_2$ fragment (and polycarbosilanes in general) also shows absorbances at such wavelengths.[23] At first we speculated that the lack of the higher wavelength absorption may be due to a reaction induced by the quench or the aqueous work-up. However, spectra of the product before quenching and work-up also shows the lack of an the absorption expected for an Si-Si backbone. Figure 1 shows a comparison of UV-Vis spectra for the products from $(CH_3)_3SiCH_2SiCl_3$ and $(CH_3)_3SiCH_2SiHCl_2$.

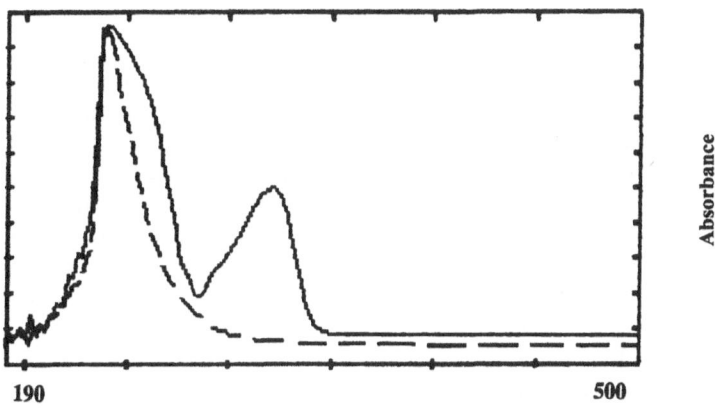

Figure 1. UV–Vis spectra: polysilane and polycarbosilane. —— polysilane from $(CH_3)_3SiCH_2SiHCl_2$ reduced with Na/K alloy in THF at 0°C under sonication, followed by quenching with CH_3Li; – – – polycarbosilane from $(CH_3)_3SiCH_2SiCl_3$, reduced with Na/K alloy in THF under sonication at 0°C, quenched with CH_3Li.

More significant differences are revealed in the mass spectra of the polymers derived from the Wurtz coupling of $(CH_3)_3SiCH_2SiHCl_2$ and $(CH_3)_3SiCH_2SiCl_3$. A complicated FD-MS spectrum (similar results were obtained by FI-MS) of the product obtained from $(CH_3)_3SiCH_2SiCl_3$ is shown in Figure 2. Most of the 418 peaks in the spectrum are distributed in the mass range of 900-2200, a range of only 1300 mass units. A hydrogen redistribution process is suggested by the presence of so many peaks. This spectrum indicates a very complex, probably cross-linked structure.[30] The similarity of the molecular weight obtained by FD-MS to the molecular weight obtained by GPC (e.g., M_n = 904 and M_w/M_n = 1.24 for a sample quenched with MeLi, and M_n = 975 and M_w/M_n = 1.18 for a sample quenched with methanol) suggests that the FD-MS in Fig. 2 shows actual oligomers and not just a complicated fragmentation pattern. The complexity of the mass spectrum makes a simple formula for the polymer hard to derive.

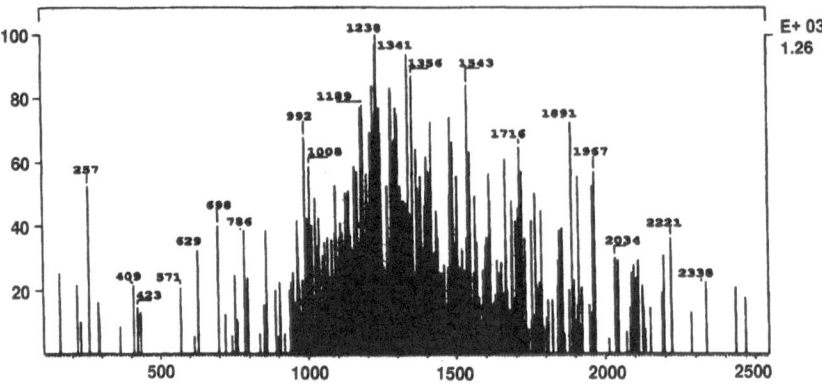

Figure 2. FD–MS pattern of the polycarbosilane from $(CH_3)_3SiCH_2SiCl_3$ reduced with Na/K alloy in THF under sonication at 0°C, quenched with CH_3Li.

Further evidence for a hydrogen redistribution process from the reduction of $(CH_3)_3SiCH_2SiCl_3$ was provided by the FTIR spectrum. In addition to the bands at 1040 cm^{-1} (sharp) and 1350 cm^{-1} for Si–CH$_2$–Si deformation,[23] a band at 2065 cm^{-1} to 2085 cm^{-1} for Si–H was observed.[25] FTIR spectra of products *before* quenching and work-up also showed this band. The formation of Si–H has been reported in the synthesis of $(RSi)_n$ polymers from $RSiCl_3$ under the conditions of over-reduction[18b]. But in the case of $(CH_3)_3SiCH_2SiCl_3$, the Si–H band was observed when reduction was carried out with only one or two molar equivalents of the alkali metal. The relative intensity of the Si–H band *decreased* with increasing molar equivalents of the reducing agent, suggesting that once Si–H was formed a second process lead to its partial eradication.

NMR spectra for the polymers obtained from $(CH_3)_3SiCH_2SiCl_3$ supported neither a Si–Si network nor a Si–Si chain structure. The polymer from $(CH_3)_3SiCH_2SiCl_3$ showed a very strong featureless broad band from –0.5 ppm to +4.5 ppm, plus a very weak and broad resonance about –55 ppm to –65 ppm. It should be noted that the later absorbance may be assigned to the Si–Si network or Si–H or both.[18,19,24] But Si–H is consistent with FTIR spectrum and with the coupling constant indirectly obtained by the DEPT experiment (J_{Si-H} ~185 Hz). The existence of Si–H was also confirmed in the ^1H NMR, which showed a weak and broad peak centered at 4 ppm in addition to a strong broad peak at –0.5 ppm to 0.4 ppm assigned to C(H)$_x$ adjacent to silicon. ^{13}C NMR showed only one broad peak from –1 ppm to +5 ppm. DEPT ^{13}C NMR indicated the presence of many CH$_3$, few CH$_2$, and no CH groups, suggesting a process in which C-H of $(CH_3)_3SiCH_2SiCl_3$ is activated.

In attempting to form a network polymer $[RSi]_n$ from $RSiCl_3$ with a bulky R group like $(CH_3)_3SiCH_2$– incomplete reduction may reasonably expected. The number of unreduced Si–Cl before the quench has been estimated for a polymer quenched with methyl alcohol. The integration of ^1H NMR spectrum indicated the ratio of $(CH_3)_3SiCH_2$ groups (assuming they remained unchanged) to CH$_3$O groups was 20 to 1. This method overestimates the number of unreacted Si–Cl because, as the above discussion indicates, the $(CH_3)_3SiCH_2$ groups have reacted and integration of the peaks near zero has been reduced. The ^{29}Si NMR spectra also showed no detectable Si–Cl resonance at lower field for the unquenched samples.

We can conclude that the products synthesized from $(CH_3)_3SiCH_2SiCl_3$ are polycarbosilanes with some Si–H and probably some Si–Si segments based on the characterization data. The existence of a significant quantity of siloxanes can be ruled out by the elemental analysis which showed that Si, C, and H added up to 98.24% of the mass.

The mechanism of the formation of polycarbosilanes from $(CH_3)_3SiCH_2SiCl_3$ by Wurtz reduction is not very clear and more detailed studies are under progress. Some comments are appropriate at present. Thermodynamically, the Si–C bond (bond energy ~314 kJ/mol) and Si–H bond (bond energy ~314 kJ/mol) are favored over the Si–Si bond (bond energy ~196 kJ/mol),[31] though the formation Si–Si bond from Si–Cl is kinetically dominant under most Wurtz condensation conditions. The bulky trimethylsilylmethyl substituent may prevent the formation of long chains and networks. In the case of $[(CH_3)_3SiCH_2]_2SiCl_2$ (Eq. 13) only the four-membered ring was isolated[29] and substituting one R group with hydrogen $[(CH_3)_3SiCH_2SiHCl_2]$ led to the formation of larger rings (m = 4~9) and longer chains (n = 3~9). For $(CH_3)_3SiCH_2SiCl_3$ the formation of a network of Si–Si bonds may be sterically impossible. The most important reason, however, may be due to the acidic character of C-H when the carbon is bound to silicon. The removal of acidic protons in $SiCH_2Si$ or even in $SiCH_3$ fragments has been shown to give carbanions.[32, 34c] The stabilization of α–silicon carbanions is attributed to the high polarizability of Si and the presence of low lying $\sigma^\bullet$ orbitals.[34c] The removal of acidic protons could be triggered by the silyl anion which is predictably formed in the course of Wurtz reduction,[20a,33] though the alkali metal could also serve such a function. The carbanion thus generated can in turn react with Si–Cl to form a Si–C bond and salt (Eqs. 14 and 15).

$$SiCH \quad + \quad Si^- M^+ \quad \longrightarrow \quad SiC^- M^+ \quad + \quad SiH \qquad (\text{Eq. 14})$$

$$SiC^- M^+ \quad + \quad SiCl \quad \longrightarrow \quad SiCSi \quad + \quad MCl \qquad (\text{Eq. 15})$$

The results are the formation of Si–H bonds and Si–C–Si bonds instead of Si–Si bonds. There may be substantial difference in the acidity of such protons in $SiCH_2SiCl_3$ and $SiCH_2SiHCl_2$ groups and this may explain the differences that are observed between $(CH_3)_3SiCH_2SiCl_3$ and $(CH_3)_3SiCH_2SiHCl_2$. Alternatively, highly reactive intermediate silenes $RHC=SiCl_2$ (or α–metallated silyl trichlorides) have been formed by treating RCH_2SiCl_3 with base.[34] Opening the double bond may engender further crosslinking between Si and C.

ACKNOWLEDGMENTS

We are very grateful to Wiley J. Youngs, Peter Rinaldi, Dale Ray and William Kroenke for suggestions and discussions, Richard Seeger for GPC measurements, and Robert Lattimer for mass spectra. Gan Ouyang and Richard Simons acknowledge The University of Akron for research fellowships. The research was funded by the Research (Faculty Projects) Committee of The University of Akron, a State of Ohio Board of Reagents Academic Challenge Grants, grants from the National Science Foundation (CHE-9309160 and Chemical Instrumentation Program CHE–8820644).

REFERENCES

1. (a) Laine, R. M.; Babonneau, F., *Chem. Mater.*, 1993, **5**, 260. (b) Soula, G., in *Inorganic and Organometallic Polymers with Special Properties*, Laine, R. M. Ed., Kluwer Academic Publishers: the Netherlands, 1992, p 31. (c) Baney, R.; Chandra, G. in *Encyclopedia of Polymer Science and Engineering*, Mark, H. F.; Bikales, N. M.; Overberger, C. G.; Menges, G., Eds.; John Wiley & Sons: New York, 1988; Vol. 13, p 312. (d) Wynne, K. J.; Rice, R. W., *Ann. Rev. Mater. Sci.*, 1984, **14**, 297.

(e) Schilling, Jr., C. L.; Wesson, J. P.; Williams, T. C., *Ceram. Bull.*, 1983, **62**(8), 912. (f) Seyferth, D.; Yu, Y., in *Design of New Materials*, Coke, D. L.; Clearfield, A. Eds., Plenum Press: New York, p 79.

2. Sartori, P.; Habel, W.; Aefferden, B. van, *et al*, *Eur. J. Solid State Inorg. Chem.*, 1992, **29** 127.

3. Seyferth, D.; Czubarow, P., *Chem. Mater.*, 1994, 6(1), 10.

4 (a) Weidman, T. W.; Joshi, A. M., *Appl. Phys. Lett.*, 1993, **62**(4), 372. (b) Miller, R. D., *Polym. Prep.*, 1990, **31**(2), 252. (c) Kunz, R. R.; Horn, M. W., Goodman R. B., *et al, J. Vac. Sci. Tech.*, 1990, **B8**(6), 820. (d) Hornak, C. A.; Weidman T. W.; Kwock, W. J., *Appl. Phys.*, 1990, **67**.(5) 2235. (e) Miller, R. D.; Hofer, D.; McKean, D. R.; Willson, C. G., in *Materials for Microlithography*, Thompson, L. F.; Willson, C. G.; Frechet, J. M. J., Eds., American ChemicalSociety:Washington D. C.,1984, p 293.

5. (a) West, R.; Wolff, A. R.; Peterson, D., *J. Radiat. Curing*, 1986, **13**, 35. (b) Wolff, A.; West, R., *Appl. Organomet. Chem.*, 1987, **1**, 7.

6. (a) Hasegawa, T.; Iwasa, Y.; Kishida, H., *et al, Mol. Cryst. Liq. Cryst.*, 1992, **217**, 25. (b) Schellenberg F. M.; Byer, R. L.; Miller, R. D. *et al*, in *Inorganic and Organometallic Oligomers and Polymers*, Harrod, J. F.; Laine, R. M., Eds., Kluwer Academics Publishers: the Netherlands, 1991, p 73. (c) Kajzar, K.; Messier, J.; Rosili, C., *J. Appl. Phys.*, 1986, **60**(9), 3040.

7. (a) Yajima, S.; Hasegawa, Y.; Okamura, K.; Matsuzawa, T., *Nature*, 1978, **273**, 525. (b) Yajima, S.; Okamura., K.; Hayashi, J.; Omori, M., *J. Amer. Ceram. Soc.*, 1976, **59**(7–8), 324. (c) Yajima, S.; Okamura., K.; Hayashi, J.; Omori, M.; Okamura., K., *Nature*, 1976, **261**, 683.

8. Examples: (a) Masnovi, J.; Bu, X. Y.; Conroy, P.; Andrist, H.A.; Hurwitz,F. I.; Miller, D., *Mater. Res. Soc. Symp. Proc.*, 1992, **271**, 771. (b) Seyferth, D.; Lang, H., *Organometallics.*, 1991, **10**, 551. (c) Larkin, D. J.; Interrante, L. V., *Chem. Mater.*, 1992, **4**, 22. (d) Bialecka-Flojañczyk, E.; Ganicz,T.; Pluta, M.; Stañczyk, W., *J. Organomet. Chem.*, 1993, **444**, C9. (e) Habel, W.; Judenau, P.; Sartori, P., *J. Prakt. Chem./Chem.-Ztg.*,1992, **334**, 391. (f) Liao, C. X.; Chen, M. W.; Weber, W. P., *Polym. Prepr.*,1993, **34**(1), 23. (g) Corriu, R. J. P.; Leclercq, D.; Mutin, P. H.; Planeix, J.–M.; Vioux, A., *Organometallics.*,1993, **12**, 454. (h) Bacqué, E.; Pillot, J.-P.; Birot, M.; Dunogués, J. *Macromolecules*, 1988, **21**, 30, 34. (i) Jung, I. N.; Yeon, S. H.; Han, J. S., *Organometallics.*,1993, **12**, 2360. (j) Lee, G.–H.; Song, C.–H.; Jung, I. N., *Bull. Korean Chem. Soc.*,1993, **14**, 5. (k) West, R.; David, L. D.; Djurovich, P. I.; Yu, H., *Ceram. Bull.*, 1983, **62**(8), 899. (l) Abu–eid, M. A.; King, R. B.; Kotliar, A. M., *Eur. Polym. J.*, 1992, **28**(3), 315. (m) Zhang, Z.; Babonneau, F.; Laine, R. W., *J. Amer. Ceram. Soc.* 1991, **74**(3), 670.

9. Goodwin, J. T.; Baldwin, W. E.; McGregor, R. R., *J. Amer. Chem. Soc.*, 1947, **69**, 2247.

10. Kotaka, H.; Hayashi, S.; Saito, H., *J. Ceram. Soc. Jpn.*, 1992, **100**(3), 332.

11. Whitmarsh, C. K.; Interrante, L. V., *Organometallics*, 1991, **10**, 1336.

12 Jung, I, N.; Yeon, S. H.; Han, J. S., *Organometallics*, 1993, **12**(6), 2360.

13 (a) Umezawa, M.; Kojima, M.; Ichikawa, H.; Ishikawa, T.; Nonaka, T., *Electrochim. Acta*, 1993, **38**(4), 529. (b) Bordeau, M.; Biran, P.; Pons, P.; Leger, M.–P.; Dunnoques, J., *J. Organomet. Chem.*, 1990, **382**, C21.

14. Banford, W. R.; Lovie, J. C.; Watt, J. A. C., *J. Chem. Soc. C*, 1966, 1137.

15. Procopio, L. J.; Berry, D. H., *J. Amer. Chem. Soc.*, 1991, **113**, 4039.

16 Masnovi, J; Bu, X. Y.; Beyene, K., *et al, Mat. Res. Soc. Sym. Pro.*, 1992, **182**,

17. (a) West, R.; Maxka, J., in *Inorganic and Organometallic Polymers*, Zeldin, M.; Wynne, K. J.; Allcock, H,. R., Eds. American Chemical Society: Washington, D. C.,1988, p 6. (b) Miller, R. D., *Chem. Rev.*, 1989, **89**, 1359. (c) West. R., *J. Organomet. Chem.*, 1986, 300, 327. (d) Matsumoto, M.; Watanabe, H.; Nagai, Y., in *New Aspects of Organic Chemistry I*, Yoshicha, Z. Ed., Kodansha: Tokyo, 1989, p 303. (e) Trefonas, P., in *Encyclopedia of Polymer Science and Engineering*, Mark, H. F.; Bikales, N. M.; Overberger, C. G.; Menges, G., Eds.; John Wiley & Sons: New York, 1988; Vol. 13, p 162. (f) West, R., in *Comprehensive Organometallic Chemistry*, Stone, F. G. A.; Abel, E. W., Eds., Pergamon Press: Oxford, 1983, Vol. 9, p 365.

18. (a) Bianconi, P. A.; Weidman, T. W., *J. Amer. Chem. Soc.*, 1988, **110**, 2342. (b) Bianconi, P. A.; Schilling, F, C.; Weidman, T. W., *Macromolecules*, 1989, **22**(4), 1697. (c) Smith, D. A.; Freed, C. A.; Bianconi, P. A., *Chem. Mater.*, 1993, **5**, 245. (d) Matsumoto, H.; Higushi, K.; Kyushin, S.; Goto, M., *Angew. Chem. Int. Ed. Engl.*, 1992, **31**(10), 1354. (e) Sekiguchi, A.; Yatabe, T.; Kamatani, H., *et al, J. Amer. Chem. Soc.*, 1992, **114**(15), 6260. (f) Weidman, T. W.; Bianconi, P. A.; Kwock, E. W., *Ultrasonics*, 1990, **28**, 310.

19. (a) Aitken, C.; Harrod, J. F., *J. Organomet. Chem.*, 1985, **279**, C11–C13. (b) Harrod, J. F., in *Inorganic and Organometallic Polymers with Special Properties*, Laine, R. M., Ed., Kluwer Academic Publishers: the Netherlands, 1992, p 87. (c) Campbell, W. H.; Hilty, T. K.; Yurga, L., *Organometallics*, 1989, **8**, 2615. (d) Harrod, J. F., in *Transformation of Organometallics into Common and Exotic Materials*,

Laine, R. M., Ed., Martinus Nijhoff Publishers, 1988, p 103. (e) Woo, H.; Walzer, J. F.; Tilley, T. D., *J. Amer. Chem. Soc.*, 1992, **114**, 7047. (f) Tilley, T. D., *Acc. Chem. Res.*, 1993, **26**, 22. (g) Aitken, C.; Harrod, J. F.; Gill, U. S., *Can. J. Chem.*, 1987, **65**, 1804.

20. (a) Mark, J. E.; Allcock, H. R.; West, R., *Inorganic Polymers*, Prentice Hall: New Jersey, 1992, Chapter 5. (b) Bock, H.; Ensslin, W., *Angew. Chem. Int. Ed. Engl.*, 1971, **10**, 404.

21. (a) Yuan, C.; West, R., *Macromolecules*, 1994, **27**, 629. (b) Harrah, L. A.; Zeigler, J. M., *Macromolecules*, 1987, **20**, 601. (c) Watanabe, A.; Miike, H.; Tsutsumi, Y.; Matsudu, M., *Macromolecules*, 1993, **26**, 2111.

22. Furukawa, K.; Fujino, M.; Matsumoto, N., *Macromolecules*, 1990, **23**, 3423.

23. (a) Hasegawa, Y.; Okamura, K., *J. Mater. Sci.*, 1986, **21**, 321. (b) Yajima, S.; Hasegawa, Y.; Hayashi, Y.; Iimura, M., *J. Mater. Sci.*, 1978, **13**, 2569.

24. (a) Marsmann, H., in *NMR Basic Principles and Progress*, Diehl, P.; Fluck, E.; Kosfeld, R., Eds., Springer-Verlag: New York, 1981, Vol. 17, p 66. (b) Williams, E. A., in *The Chemistry of Organic Silicon Compounds*, Patai, S.; Zappoport, Z., Eds., John Wiley & Sons: New York, 1989, Chapter 8. (c) Blinka, T. A.; Helmer, B. J.; West, R., in *Advances in Organometallic Chemistry*, Vol. 23, p 193. (d) Kupce, E.; Lukevics, E., in *Isotopes in the Physical and Biomedical Science*, Buncel, E.; Jones, J. R., Eds, Elsevier: Amsterdam, 1991, Chapter 5.

25. (a) Launar, P. J., in *Silicon Compouds Register and Reviews*, Petrarch Systems Silanes and Silicones, 1987, p 69. (b) Baqué, E.; Pillot, J.-P.; Birot, M.; Dunogués, J., in *Transformation of Organometallics into Common and Exotic Materials: Design and Activation*, Laine, R. M., Ed., Martinus Nijhoff: Boston, 1988, p 116. (c) Belamy, L. J., *The Infrared Spectra of Complex Molecules*, 3rd ed., Chapman and Hall: London, 1973, Chapter 20. (d). Crompton, T. R., in *The Chemistry of Organic Silicon Compounds*, Patai, S.; Rappoport, Z., Eds., John Wiley & Sons: New York, 1989, Chapter 6.

26. Sommer, L. H.; Murch, R. M.; Mitch, F. A., *J. Amer. Chem. Soc.*, 1954, **76**, 1619.

27. Daniels, B. F. ; Post, H. W., *J. Org. Chem.*, 1957, **22**, 748.

28. (a) Cundy, C. S.; Kingston, B. M.; Lappert, M. F., *Adv. Organomet. Chem.*, 1973, **11**, 253. (b) Eller, P. G.; Bradley, D. C.; Hursthouse, M. B. ; Meek, D. W., *Coord. Chem. Rev.*, 1977, **24**, 1. (c) Beachley Jr., O. T.; Banks. M. A.; Churchill, M. R.; *et al*, *Organometallics*, 1991, **10**, 3036. (d) Wells, R. L., *Cord. Chem. Rev.*, 1992, **112**, 273. (e) Traven, V. F.; Shpakin, S. Y., *Adv. Organomet. Chem.*, 1992, **34**, 149. (f) Bock, H.; Meuret, J., *J. Organomet. Chem.*, 1993, **459**, 43,

29. (a) Watanabe, H.; Kato, M.; Okawa, T., *et al*, *Appl. Organomet. Chem.*, 1987, **1**, 157. (b) Watanabe, H.; Kougo, Y.; Kato, M., *et al*, *Bull. Chem. Soc. Jpn.*, 1984, **57**, 3019.

30. Schulten, H.-R.; Lattimer, R. P., *Mass Spec. Rev.*, 1984, **3**, 231.

31. (a) Corey, J. Y., in *The Chemistry of Organic Silicon Compounds*, Patai, S.; Rappoport, Z., Eds., John Wiley & Sons: New York, 1989, Chapter 1. (b) Fritz, G.; Matern, E., *Carbosilanes*, Springer-Verlag: New York, 1986, p 1,

32. (a) Bassindale, A. R.; Taylor, P. G., in *The Chemistry of Organic Silicon Compounds*, Patai, S.; Rappoport, Z., Eds., John Wiley & Sons: New York, 1989, Chapter 12. (b) Fritz, G.; Matern, E., *Carbosilanes*, Springer-Verlag: New York, 1986, pp 169–174. (c) Seyferth, D.; Rodison, J. L.; Mercer, J., *Organometallics*, 1990, **9**, 2677, (c) Seyferth, D.; Lang, H., *Organometallics*, 1991, **10**, 551. (d) Seyferth, D.; Lang, H, in *Ultrastructure Processing of Advanced Materials*, Uhlmann, D. R.; Ulrich D. R., Eds., John Wiley & Sons: New York, 1992, p 667. (e) Robison J. L.; Davis W. M.; Seyferth D., *Organometallics*, 1991, **10**, 3385. (f) Lappert M. F.; Engelhardt, L. M.; Raston, C. L.; White, A. H., *J. Chem. Soc., Chem. Commun.* 1982, 1323. (g) Al-Juaid, S. S.; Eaborn, C.; Hitchcock, P. B., *et al, Angew. Chem. Int. Ed. Engl.*, 1994, **33**(12), 1268.

33. (a) Jones, R. G.; Benfield, R. E.; Cragg, R. H.; Swain, A. C.; Webb, S. J., *Macromolecules*, 1993, **26**, 4878. (b) Matyjaszewski, K.; Cypryk, M.; Frey, H., *et al*, *J. Macromol. Sci. –Chem.*, 1991, **A28**(11 &12), 1151. (c) Kim, H. K.; Matyjaszewski, K., *J. Amer. Chem. Soc.*, 1988, **110**(10), 323. (d) Jones, R. G.; Benfield, R. E.; Cragg, R. H.; Swain, A. C., *J. Chem. Soc., Chem. Commun.*, 1992, 112.

34. (a) Sewald, S.; Ziche, W.; Wolff, A.; Auner, N., *Organometallics*, 1993, **12**, 4123, and references therein. (b) Auner, N.; Heikenwälder, C.-R.; Wagner, C., *Organometallics*, 1993, **12**, 4135. (c) Barton, T. J.; Boudjouk, P., in *Silicon –Based Polymer Science: a Comprehensive Resource*, Zeigler, J. M.; Fearon, F. W. G., Eds., American Chemical Society: Washington, DC, 1990, p 3.

PREPARATION AND PROPERTIES OF NOVEL TITANOSILOXANE AND ZIRCONOSILOXANE

Takahiro GUNJI and Yoshimoto ABE

Department of Industrial Chemistry,
Faculty of Science and Technology,
Science University of Tokyo
2641 Yamazaki, Noda, Chiba 278 Japan

INTRODUCTION

A Preparation of ceramic materials from organometallic or metalorganic compound is called as "a precursor method": Ceramic materials are prepared by sintering the ceramic precursor of desired shape such as fiber, film or bulk body. "Sol-gel process" is one of the precursor method and normally a mixture of metal alkoxides is subjected to the hydrolysis under a moderate condition followed by sintering. The formation of ceramic materials having excellent properties is, therefore, achieved by controlling the molecular structure of precursor polymer.

Metalloxane compound is composed of metal-oxygen-metal bond and it is a good precursor for oxide ceramics. Scheme 1 shows the schematic figure of SiO_2-TiO_2 oxide fiber: SiO_2-TiO_2 oxide fiber is prepared by pyrolysis of SiO_2-TiO_2 precursor fiber formed by spinning of polytitanosiloxane, which consists of silicon-oxygen-titanium backbone.

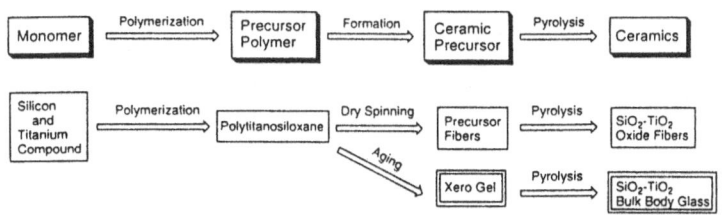

Scheme 1 Schematic figure of the preparation of SiO_2-TiO_2 ceramics by precursor method.

We have investigated the preparation of oxide materials from polymetalloxane compounds in order to find out a reasonable way to control the physical, electrical, and mechanical properties of oxide ceramics. Our project shows a possibility that we can control the characteristics of oxide ceramics by designing the molecular structure of metalloxane compound: The microstructure of metalloxane compound may greatly affect on the properties of ceramic materials.

SiO_2-TiO_2 and SiO_2-ZrO_2 ceramics are our first target, because a phase separation behavior observed in this system is suitable examples for our purpose. We have reported the preparation and properties of SiO_2-TiO_2 fibers from polytitanosiloxane, which was synthesized by the reaction of silicic acid with bis(2, 4-pentanedionato)titanium diisopropoxide in methanol according to the equation (1). In this reaction, polytitanosiloxane was a random block copolymer, which was formed by the reaction of silicic acid oligomers with bis(2, 4-pentanedionato)titanium diisopropoxide.[1-11] On the other hand, polytitanosiloxane was prepared by another reaction: Di-t-butoxysilanediol with bis(2, 4-pentanedionato)titanium diisopropoxide according to the equation (2).[12] As a result, cyclic titanosiloxane compound was formed quantitatively, and polytitanosiloxane was not formed.

$$x \; HO{-}Si(OH)_2{-}O{-}H + x \; Pr^iO{-}M{-}OPr^i \xrightarrow[\cdot \; 2x\text{-}1 \; Pr^iOH]{MeOH} HO{-}[Si(OH)_2{-}O{-}M{-}O]_x{-}Pr^i \quad (1)$$

M=Ti, Zr

$$2 \; (Bu^tO)_2Si(OH)_2 + 2 \; Pr^iO{-}M{-}OPr^i \xrightarrow[\cdot \; 4 \; Pr^iOH]{Benzene} [{-}Si(OBu^t)_2{-}O{-}M{-}O{-}]_2 \quad (2)$$

M=Ti, Zr

In this work, therefore, preparation of titanosiloxane or zirconosiloxane compounds were investigated according to the equations (3) and (4) to offer new monomers for the preparation of polytitanosiloxane or polyzirconosiloxane with well defined molecular sequences.

$$2 \; MeO{-}Si(OBu^t)_2{-}OH + Pr^iO{-}M{-}OPr^i \xrightarrow[\cdot \; 2 \; Pr^iOH]{Hexane} MeO{-}Si(OBu^t)_2{-}O{-}M{-}O{-}Si(OBu^t)_2{-}OMe \quad (3)$$

M=Ti, Zr

TS12 (M=Ti)
ZS12 (M=Zr)

$$4 \; MeO{-}Si(OBu^t)_2{-}OH + M(OPr^i)_4 \xrightarrow[\cdot \; 4 \; Pr^iOH]{Hexane} M[{-}O{-}Si(OBu^t)_2{-}OMe]_4 \quad (4)$$

M=Ti, Zr

TS14 (M=Ti)
ZS14 (M=Zr)

EXPERIMENTAL

Preparation of Di-t-butoxymethoxysilanol (DBMS)

All reactions were carried out under nitrogen atmosphere.

To a mixture of tetrachlorosilane and benzene was added dropwise a benzene solution of t-butanol and pyridine in the molar ratio of tetrachlorosilane/t-butanol/pyridine=1/2/2 at 0 °C, then refluxed 1 h. After filtration of salt and the evaporation of solvents, di-*t*-butoxydichlorosilane (DDS) was distilled under reduced pressure. (Yield: (86 %); bp: 74.5-75.5 °C/ 17 mmHg)

To a mixture of DDS and benzene was added dropwise a benzene solution of methanol and pyridine in the molar ratio of DDS/methanol/pyridine=1/1/1 at 0 °C, then refluxed 1 h. After filtration of salt and the evaporation of solvents, di-*t*-butoxymethoxychlorosilane (DMCS) was distilled under reduced pressure. (Yield: 88 %; bp: 75.5-77.4 °C/ 21 mmHg). Finally, to a mixture of DMCS and ether was added dropwise a ether solution of water and aniline in the molar ratio of DMCS/water/aniline=1/1/1 at -5 °C, and refluxed 30 min. After filtration of salt and the evaporation of solvents, di-*t*-butoxymethoxysilanol (DBMS) was distilled under reduced pressure. (Yield: 76 %; bp 48.2 °C/ 0.5 mmHg).

Synthesis of TS12 and ZS12

All reactions were carried out under nitrogen atmosphere. To a hexane solution of bis(2, 4-pentanedionato)titanium diisopropoxide (PTP) or bis(2, 4-pentanedionato)zirconium diisopropoxide (PZP) was added dropwise a hexane solution of DBMS in the equivalent molar ratio to PTP or PZP at room temperatures. Solvents were removed under reduced pressure to give a brown viscous solution of TS12, $(acac)_2Ti[OSi(OBu^t)_2(OMe)]_2$, and ZS12, $(acac)_2Zr[OSi(OBu^t)_2(OMe)]_2$. TS12 was purified by distillation under reduced pressure.

Synthesis of TS14 and ZS14

All reactions were carried out under nitrogen atmosphere. To a hexane solution of titanium tetraisopropoxide (TPT) or zirconium tetraisopropoxide (TPZ) was added dropwise a hexane solution of DBMS in the equivalent molar ratio to TPT at 0 °C (for TS14) or TPZ at room temperatures (for ZS14). Solvents were removed under reduced pressure to give a transparent viscous solution of TS14 and ZS14. TS14 was purified by sublimation.

Measurements and Analysis

NMR spectra were measured by using JEOL PMX-60 Si and JEOL FX-90Q type spectrometers in carbon tetrachloride or chloroform-*d*. IR spectra were measured by using HITACHI IR spectrophotometer 260-50 type. Mass spectroscopy was measured by using Shimadzu GCMS-QP2000A type.

The silicon, titanium, and zirconium contents were determined by wet method described in our previous paper.[4]

RESULTS AND DISCUSSIONS

Synthesis of di-*t*-butoxymethoxysilanol (DBMS)

It is well known that alkoxysilanols, $(RO)_nSi(OH)_{4-n}$, are produced in the sol-gel process by the hydrolysis of tetraethoxysilane or tetramethoxysilane.[13] In this process, however, these alkoxysilanols are difficult to isolate, because they lead to condensation during isolation process such as distillation, recrystallization, or concentration. t-Butoxyl group is considered to be an effective protecting groups to prevent condensation of silanol group to form polysiloxane. If we want to isolate these alkoxysilanols or alkoxysiloanols, it is important to decrease the electron density of the silicon atom and protect the hydroxyl group by bulky substituents. Therefore, trimethylsilanol (Me_3SiOH), diphenylsilanediol (($Phenyl)_2Si(OH)_2$), Phenylsilanetriol ($PhenylSi(OH)_3$), tri-t-butoxysilanol(($Bu^tO)_3SiOH$), and di-t-butoxysilanediol (($Bu^tO)_2Si(OH)_2$) were used. In this work, we tried to prepared a silanol, which have methoxyl and t-butoxyl groups on a silicon atom according to the equations (5) - (7), in order to investigate the difference in hydrolysis rates between these two alkoxyl groups.

$$SiCl_4 \ + \ 2\ Bu^tOH \ + \ 2\ \underset{}{\bigcirc\!\!N} \ \xrightarrow{\text{Benzene}} \ (Bu^tO)_2SiCl_2 \ + \ 2\ \underset{}{\bigcirc\!\!NHCl} \quad (5)$$

$$(Bu^tO)_2SiCl_2 \ + \ MeOH \ + \ \underset{}{\bigcirc\!\!N} \ \xrightarrow{\text{Benzene}} \ (Bu^tO)_2Si(OMe)Cl \ + \ \underset{}{\bigcirc\!\!NHCl} \quad (6)$$

$$(Bu^tO)_2Si(OMe)Cl \ + \ H_2O \ + \ \underset{}{\bigcirc\!\!N} \ \xrightarrow{\text{Benzene}} \ (Bu^tO)_2Si(OMe)OH \ + \ \underset{}{\bigcirc\!\!NHCl} \quad (7)$$

Physical and analytical data of di-t-butoxymethoxysilanol (DBMS) are summarized in Table 1. DBMS was prepared from tetrachlorosilane via three steps: Di-t-butoxydichlorosilane and di-t-butoxymethoxychlorosilane were intermediate compounds. In this procedure, di-t-butoxydimethoxysilane, $(Bu^tO)_2Si(OMe)_2$, was produced as a minor product. As a result, DBMS was isolated in reasonable yield. A white crystal for this compound, which is same as white crystals of $(Bu^tO)_3SiOH$ and $(Bu^tO)_2Si(OH)_2$. DBMS was expected. However, a transparent liquid was isolated by distillation.

Table 1. Physical and analytical data of DBMS.

Bp (^oC)	48.2/ 0.5
^{1}H-NMR [δ/ppm] (Solv. CCl_4)	1.3 (18H,s,-**OBu**^{t}), 3.3 (3H,s,-**OMe**), 4.3 (1H,s,-**OH**)
^{29}Si-NMR [δ/ppm] (Solv. $CDCl_3$)	-86.0
IR [cm^{-1}] (CCl_4 soln. method)	3400 (v_{O-H}), 1200 (v_{Si-O-C}), 1070 (v_{Si-O-C})
MS (m/z) (EI method)	221 (M$^+$-H)
Elemental analysis (%)[a]	Si 12.6 (12.6)

a) Wet method.

Synthesis of Titanosiloxane Compounds: TS12 and TS14

Physical, analytical, and spectral data of 12 and TS14 are summarized in table 2. Titanosiloxane compounds, TS12 and TS14, were isolated by distillation or sublimation in a reasonable yield. TS14 was liquefied by decomposition after its isolation.

Reaction progress was investigated by NMR and IR spectra of products. In ^{29}Si-

NMR spectrum, a signal at -86.0 ppm due to DBMS disappeared and an another signal at higher magnetic field appeared. ^{1}H-NMR spectrum indicated the disappearance of the signal due to the silanol group of DBMS. The IR spectrum showed the disappearance of the absorption peak at 3400 cm^{-1} ascribed to ν_{SiOH} and an appearance of a new peak due to the $\nu_{Si-O-Ti}$ bond near 1000 cm^{-1}. These spectral data clearly indicates that the reaction takes place.

Table 2. Physical and analytical data of TS12 and TS14.

Compound	TS12[b)]	TS14
State	Orange liquid	White paste[c)]
Bp (°C)	151.0-152.0/ 0.075	135-160/0.05[d)]
^{1}H-NMR [δ/ppm] (Solv. CCl$_4$)	1.3 (9H,s,-OBu^t) 1.9 (6H,s,C-CH$_3$) 3.3 (1H,s,-OMe) 5.0 (1H,s,-CH=)	1.3 (6H,s,-OBu^t) 3.3 (1H,s,-OMe)
^{13}C-NMR [δ/ppm] (Solv. CDCl$_3$)	24.9 (CH$_3$CO-) 26.6 (CH$_3$CO-) 31.2 ((CH$_3$)$_3$CO-) 50.5 (CH$_3$O-) 72.0 ((CH$_3$)$_3$CO-) 103.9 (-CH=) 185.7 (CH$_3$CO-) 191.7 (CH$_3$CO-)	31.6 ((CH$_3$)$_3$CO-) 50.8 (CH$_3$O-) 72.8 ((CH$_3$)$_3$CO-)
^{29}Si-NMR [δ/ppm] (Solv. CDCl$_3$)	-97.0	-97.5
IR [cm^{-1}] (CCl$_4$ soln. method)	1600 ($\nu_{C=O}$) 1520 ($\nu_{C=C}$) 1380 (ν_{C-C}) 1070 (ν_{Si-O-C}) 920 ($\nu_{Si-O-Ti}$)	1380 (ν_{C-C}) 1070 (ν_{Si-O-C}) 950 ($\nu_{Si-O-Ti}$)
Elemental analysis (%)[a)]	Si 8.04 (8.15) Ti 7.17 (6.95) C: 48.8 (48.7) H: 8.0 (8.2)	Si 12.06 (12.13) Ti 5.21 (5.13)

a) Silicon and titanium contents were analyzed by wet method.

b) Mass spectrum data: 859 (M$^+$-Bu^{t}O), and 710 (M$^+$-DBMS).

c) Deliquescent on aging. d) Sublimation point.

The reactions were also monitored by GC analysis of the reaction mixture. In every reaction, isopropyl alcohol was formed in quantitative yield, while t-butyl alcohol was not detected. The GC analysis shows that the silanol group of DBMS attacks the metal atom of the alkoxide or chelate compound to form titanosiloxane bond. On the other hand, DBMS was stable to self-condensation or degradation. As a result, DBMS reacted selectively with titanium or zirconium alkoxides.

TS12 was isolated as an orange viscous liquid in 96 % yield by distillation under reduced pressure. The IR spectrum of TS12 indicates the disappearance of the absorption peak at *ca.* 3400 cm^{-1} due to silanol and the appearance of the absorption peak at 920 cm^{-1}. The ^{1}H-NMR spectrum shows signals due to *t*-BuO, MeO and acac groups with the expected proton ratio. In addition, eight signals were observed by the ^{13}C-NMR spectrum of TS12 and were ascribed to *t*-BuO, MeO, and acac groups. The ^{29}Si-NMR spectrum indicated a signal at -97.0 ppm. The metal analysis data were also in good agreement with the calculated value for TS12 by the reaction in equation (3).

Fig. 1 shows the ^{13}C-NMR spectrum of TS12. The signals at 24.9 ppm and 26.6 ppm and those at 185.7 ppm and 191.7 ppm indicate that two kinds of methyl and carbonyl groups exist in different spatial positions. Since our recent research revealed that the acac groups in a bis(2, 4-pentanedionato)titanium diisopropoxide exist mainly in *cis* form and are perpendicular to one another,[14] those in TS12 are considered to be perpendicular to one another (fig. 2) so that two kinds of NMR signals appear.

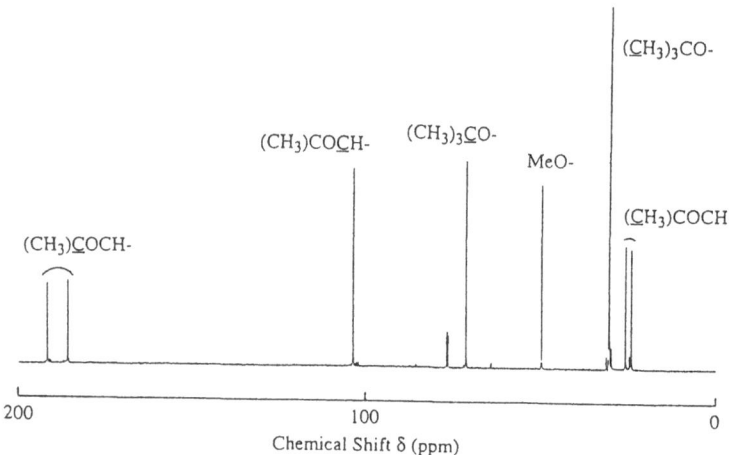

Fig. 1 ^{13}C-NMR spectrum of TS12 in CDCl$_3$ with complete decoupling.

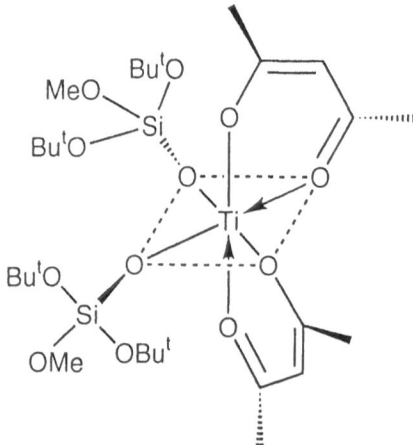

Fig. 2 Estimated structure for TS12.

TS14 was isolated as a colorless solid by sublimation at 135-160 °C/0.05 mmHg in 85 % yield. TS14 was soluble in many common organic solvents and it was stable in such a solution. TS14, however, turns to a transparent paste when isolated and stored in a sealed glass tube even under the reduced pressure. TS14, therefore, was confirmed by ^{1}H-, ^{13}C-, and ^{29}Si-NMR spectra, IR spectrum, mass spectrum and metal analysis for the crude material. In the ^{1}H-NMR spectrum, two signals were observed at 1.3 ppm and 3.3 ppm and were ascribed to t-BuO group and MeO group, respectively. The disappearance of the signal at 4.3 ppm is due to silanol group.

Fig. 3 shows the GPC traces of TS14 before (a) and after (b) aging. TS14 was stored at room temperature in a sealed bottle under reduced pressures for one week after isolation. Fig. 3 (a) was measured just after isolation and indicates a large peak at an elution volume of 20 ml with a small peak at 25.5 ml. Fig. 3(b), however, shows a broad peak at 17 ml with a peak at 25.5 ml. Fig. 3 (b) indicates decomposition of TS14. Fig. 4 shows the IR spectra of TS14 before (a) and after (b) aging. The IR spectrum supports the structure with a large absorption peak at 920 cm^{-1} due to Si-O-Ti bond. On the other hand, an increase of the absorption peak due to Si-O-Si bond with the decrease of absorption peak due to Si-O-Ti bond indicates the cleavage of Si-O-Ti bond in TS14 and the formation of Si-O-Si bond. The spectral change shown in fig. 4 supports the decomposition of TS14 and formation of polysiloxane.

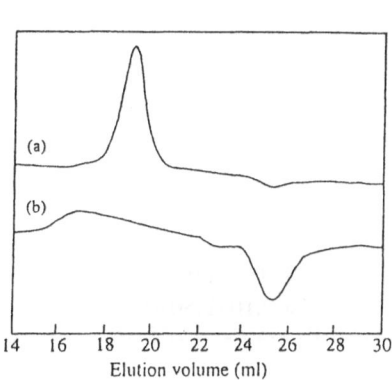

Fig. 3 GPC traces of TS14 before (a) and after (b) storage a room temperature for one week.

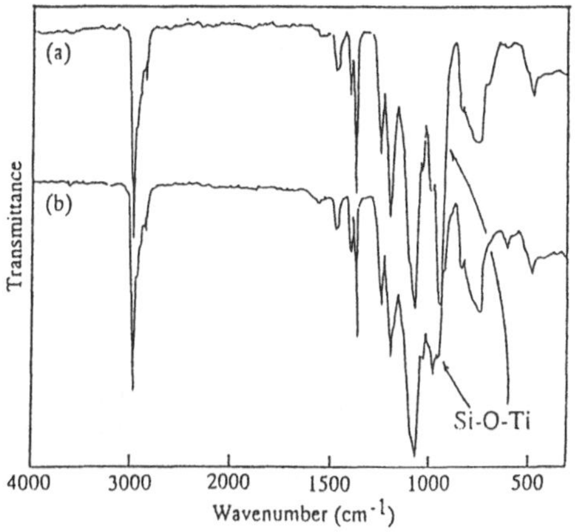

Fig. 4 IR spectra of TS14 before (a) and after (b) storage a room temperature for one week.

Synthesis of Zirconoxane Compounds: ZS12 and ZS14

Physical, analytical, and spectral data of ZS12 and ZS14 are summarized In table 3. Zirconosiloxane compound, ZS12 and ZS14, were not isolated by conventional methods.

Reaction progress was investigated by NMR and IR spectra of products. In ^{29}Si-NMR spectrum, a signal at -86.0 ppm due to DBMS disappeared and an another signal at higher magnetic field appeared. ^{1}H-NMR spectrum indicated the disappearance of the signal due to the silanol group of DBMS. The IR spectrum showed the disappearance of the absorption peak at 3400 cm^{-1} ascribed to ν_{SiOH} and an appearance of a new peak due to the $\nu_{Si-O-Zr}$ bond near 1000 cm^{-1}. These spectral data clearly indicates that the reaction takes place.

Table 3 Physical and analytical data of ZS12 and ZS14.

Compound	ZS12	ZS14
State	Brown liquid[b)]	White paste[b)]
Bp ($^\circ$C)	117 (Decompd.)	90 (Decompd.)
^{1}H-NMR[δ/ppm] (Solv. CCl_4)	1.3 (9H, s, -OBu^t) 1.9 (6H, s, C-CH_3) 3.3 (1H, s, -OMe) 5.4 (1H, s, -CH=)	1.3 (6H,s,-OBu^t) 1.3 (OCH $(CH_3)_2$) 3.3 (1H,s,-OMe) 4.5 (OCH $(CH_3)_2$) 6.0 (-OH)
^{13}C-NMR[δ/ppm] (Solv. $CDCl_3$)	26.4 (CH_3CO-) 31.4 ($(CH_3)_3CO$-) 50.5 (CH_3O-) 71.7 ($(CH_3)_3CO$-) 104.4 (-CH=) 191.1 (CH_3CO-)	24.8 ($(CH_3)_2CHO$-) 31.6 ($(CH_3)_3CO$-) 50.8 (CH_3O-) 67.8 ($(CH_3)_2CHO$-) 72.8 ($(CH_3)_3CO$-)
^{29}Si-NMR[δ/ppm] (Solv. $CDCl_3$)	-94.4	-94.4
IR [cm^{-1}] (CCl_4 soln. method)	1580 ($\nu_{C=O}$) 1520 ($\nu_{C=C}$) 1370 (ν_{C-C}) 1060 (ν_{Si-O-C}) 960 ($\nu_{Si-O-Zr}$)	3300 (ν_{OH}) 1380 (ν_{C-C}) 1070 (ν_{Si-O-C}) 950 ($\nu_{Si-O-Zr}$)
Elemental analysis (%)[a)]	Si 7.82 (7.67) Zr 12.82 (12.46)	Si 10.81 (10.83)[c)] Zr 9.10 (8.80)[c)]

a) Wet method. b) Deliquescent on aging.
c) Calculated value based on the formula [(Bu^{t}O)$_2$(MeO)SiO]$_4$Zr(Pr^{i}OH).

ZS12 was obtained as an orange viscous liquid which is difficult to purify by usual methods such as reprecipitation method, recrystallization, or distillation. The DTA-TG analysis under nitrogen atmosphere showed two endothermic peaks at 117 $^\circ$C and 400 $^\circ$C. Total weight loss was in fairly good agreement with the theoretical zirconium content of ZS12. In addition, no sublimation was observed at the temperatures of below 117 $^\circ$C under the reduced pressures even below 0.05 mmHg. As a result ZS12 was characterized as a crude product without further purification. The results of ^{1}H-, ^{13}C-, ^{29}Si-NMR, and IR spectra and metal analysis showed the formation of ZS12.

The ^{13}C-NMR spectrum of ZS12 showed six signals due to t-butoxyl and acac groups. In contrast to TS12, methyl and carbonyl groups in acac groups gave single signals, which indicate that acac groups are in the same spatial positions.

ZS14 was obtained as colorless solid. ZS14 decomposed at 90 $^\circ$C under nitrogen atmosphere. No sublimation was observed below the decomposition temperatures. The IR spectrum of ZS14 showed the absorption peaks due to hydroxyl group and Si-O-Zr linkage.

The ^{1}H-NMR spectrum (shown in fig. 5) showed the signals due to t-BuO and MeO groups at 1.3 ppm and 3.3 ppm, respectively, together with the signals due to i-PrOH at 4.5 and 6.0 ppm. In addition, the metal analysis was in good agreement with the structure of ZS14 with one added isopropyl alcohol. Thus, the formation of an isopropyl alcohol adduct of ZS14 is proposed: $Zr[OSi(OBu^t)_2(OMe)]_4(Pr^iOH)$.

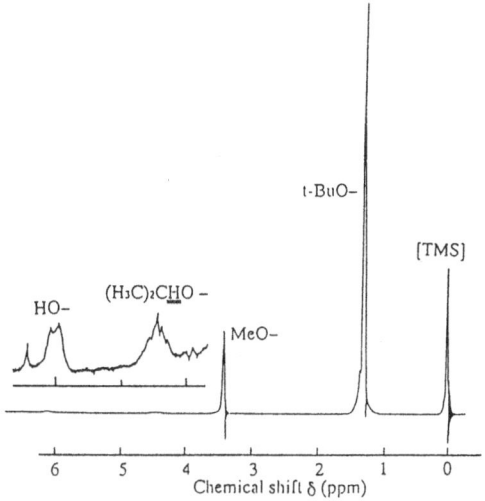

Fig. 5 ^{1}H-NMR spectrum of ZS14 in CCl$_4$.

Acknowledgment

This work was supported by Iketani Science and Technology Foundation and Research Foundation For Materials Science.

References

1. Y. Abe, T. Gunji, M. Hikita, Y. Nagao, and T. Misono, *Yogyo-Kyokai-Shi*, **94**, 1243-1245 (1986)
2. T. Gunji, Y. Nagao, T. Misono, and Y. Abe, *Journal of Non-Crystalline Solids*, **107**, 149-154 (1989)
3. T. Gunji, Y. Nagao, T. Misono, and Y. Abe, *Seramikkusu Ronbunshi*, **99**, 178-179 (1991)
4. T. Gunji, Y. Nagao, T. Misono, and Y. Abe, *J. of Polym. Sci.: Part A: Polym. Chem.*, **29**, 941-947 (1991).
5. T. Gunji, Y. Nagao, T. Misono, and Y. Abe, *J. of Polym. Sci.: Part A: Polym. Chem.*, **30**, 371-377 (1992)
6. T. Gunji, I. Sopyan, and, and Y. Abe, *J. of Polym. Sci.: Part A: Polym. Chem.*, **32**, 3133-3140 (1994)
7. Y. Abe, T. Gunji, Y. Kimata, Y. Nagao, and T. Misono, *Seramikkusu Ronbunshi*, **96**, 221-224 (1988)
8. Y. Abe, Y. Kimata, T. Gunji, Y. Nagao, and T. Misono, *Seramikkusu Ronbunshi*, **97**, 596-597 (1989)
9. T. Gunji, H. Goto, Y. Kimata Y. Nagao, T. Misono, and Y. Abe, *J. of Polym. Sci.: Part A: Polym. Chem.*, **30**, 2295-2301 (1992)

10. H. Goto, H. Tomioka, T. Gunji, Y. Nagao, T. Misono, and Y. Abe, *Seramikkusu Ronbunshi*, **101**, 336-341 (1993)

11. Y. Abe, H. Tomioka, T. Gunji, Y. Nagao, and T. Misono, *J. Mat. Sci. Lett.*, **13**, 960-962, (1994)

12. Y. Abe, T. Gunji, Y. Kimata, M. Kuramata, A. Kasgöz, and T. Misono, *Journal of Non-Crystalline Solids*, **121**, 21-25 (1990)

13. A. Yamamoto and S. Kambara, *J. Amer. Chem. Soc.*, **79**, 4344 (1957); D. C. Bradley and C. E. Holloway, *J. Chem. Soc. (A)*, 282 (1969)

14. T. Gunji and Y. Abe, *Unpublished data*

SYNTHESIS OF THERMOSETTING PRECERAMIC COPOLYMER

O.Funayama, T.Aoki, and T.Isoda

Tonen Corporation
Corporate Research & Development Laboratory
1-3-1, Nishi-Tsurugaoka Ohi-machi, Iruma-gun,
Saitama 356, Japan

INTRODUCTION

Silicon carbide is a well-known ceramic material with high thermal and chemical stability, high mechanical strength and hardness, and high thermal conductivity.
There has been much interest in the use of organosilicon polymers as precursors to silicon carbide based ceramics. Some of the advantages in the preparation of ceramic materials from these precursors include, compositional homogeneity in the final products, high purity ceramic products with uniform microstructure, amorphous to microcrystalline, and the preparation of refractory ceramics at relatively low temperatures.

Polycarbosilane is a precursor for silicon carbide. Yajima and his co-workers developed polycarbosilane by thermal rearrangement of polydimethylsilane [1]. This polycarbosilane was melt-spun into fibers, curing in air, and converted to predominantly β-SiC, mixed with excess carbon and oxygen, on heating in N_2 [2].

Continuous SiC fibers with small diameter had high tensile strength and high tensile modulus. However, applications for this fiber were sometimes limited by degradation at elevated temperatures. Weight loss of the fiber occured at temperatures as low as 1200° C and decrease in tensile strength occured at temperatures below 1000°C. The degradation of the fiber was attributed to the presence of excess carbon and oxygen which resulted in the formation of gaseous products, such as CO and SiO, at relatively low temperatures. Improvement of the high temperature mechanical properties can be expected by the reduction of oxygen content. Okamura et al. reported the low oxygen SiC fibers derived from polycarbosilane [3, 4]. Melt-spun polycarbosilane fibers were cured by electron beam irradiation in a stream of He and were pyrolyzed in an inert atmosphere. Weight loss of the fiber couldn't be observed at 1500°C. The fiber had a high tensile strength and a tensile modulus at 1500°C. Despite the excellent properties achieved, an electron beam irradiation equipment is necessary to prepare this low oxygen SiC fiber.

Thermosetting precursor allows the curing process to be eliminated. Furthermore, boron is a well-known sintering additive for SiC. The present paper provides information on the synthesis of thermosetting preceramic polymer by copolymerization of polycarbosilane with thermosetting polyborosilazane and properties of pyrolysis products.

EXPERIMENTAL PROCEDURES

Copolymer preparation

Si-Cl bonds were introduced to polycarbosilane (Shin-Etsu Chemical Co.,Ltd.) (PCS) as functional groups by refluxing PCS with CCl_4 for 24h [5]. Cl content of the chlorinated polycarbosilane (Cl-PCS) was 2.0 wt%.

Table 1. Composition of polyborosilazanes (wt%).

	Si	B	N	O	C	H
BSZ-1	44.8	4.2	28.5	3.6	12.9	6.0
BSZ-2	47.9	2.3	27.6	2.9	13.0	6.3
BSZ-3	49.4	1.8	24.1	2.6	13.2	8.9

Polyborosilazane (BSZ) was derived from perhydropolysilazane and trimethyl borate [6]. Boron content was controlled by the amount of trimethyl borate. Perhydropolysilazane was synthesized by ammonolysis of a dichlorosilane-pyridine adduct followed by heat-treated at 120°C in pyridine with ammonia. Composition of BSZs used in this study are shown in Table 1.

40 g of Cl-PCS and 20 g of BSZ were mixed with 400 mL of pyridine and 400 mL of m-xylene to form a homogeneous solution. A 1000 mL autoclave was charged with the solution and 15 g of anhydrous NH_3. The solution was heated with ammonia at 120°C for 1h. After being cooled to room temperature, the reaction mixture was filtered in N_2 to remove NH_4Cl. The solvent was removed by vacuum distillation over a rotary evaporator to give white powder (copolymer (PCS/BSZ)).

Pyrolysis

Copolymer samples were pyrolyzed by heating from room temperature to 1700°C in 150 min, holding at 1700°C for an additional 60 min, and cooling to room temperature in 150 min. All heating was completed under flowing N_2.

Characterization

Infrared (IR) spectra were recorded on polymer-coated KBr plates. ^{29}Si nuclear magnetic resonance (NMR) spectra were recorded on polymers in solution at room temperature. Elemental analyses for Si, N, B, C, O, and H were obtained for the polymers and pyrolyzed samples. Thermal behavior in N_2 up to 500°C was investigated by using differential scanning calorimeter (DSC).

X-ray diffraction (XRD) measurements were performed on the pyrolyzed powder samples with CuKα radiation using an automated powder diffractometer equipped with a monochrometer.

RESULTS

The IR spectrum of PCS (Fig. 1(A)) shows absorptions at 2900, 2950 cm^{-1} (C-H), 2100 cm^{-1} (Si-H), 1350-1450 cm^{-1} (C-H), 1250 cm^{-1} (Si-CH$_3$), 1020 cm^{-1} (Si-CH$_2$-Si), and 840 cm^{-1} (Si-C) [1]. Cl-PCS (Fig. 1(B)) has a new absorption at 500cm^{-1} assigned to Si-Cl bonds [5]. This band disappears in PCS/BSZ-1. PCS/BSZ-1 (Fig. 1(C)) exhibits new absorptions at 3400 cm^{-1} (N-H), 2150 cm^{-1} (Si-H), 1350-1550 cm^{-1} (B-O/B-N), and 1200 cm^{-1} (N-H). These absorptions were derived from BSZ-1 [6]. PCS/BSZ-1 presents bands due to PCS and BSZ-1.

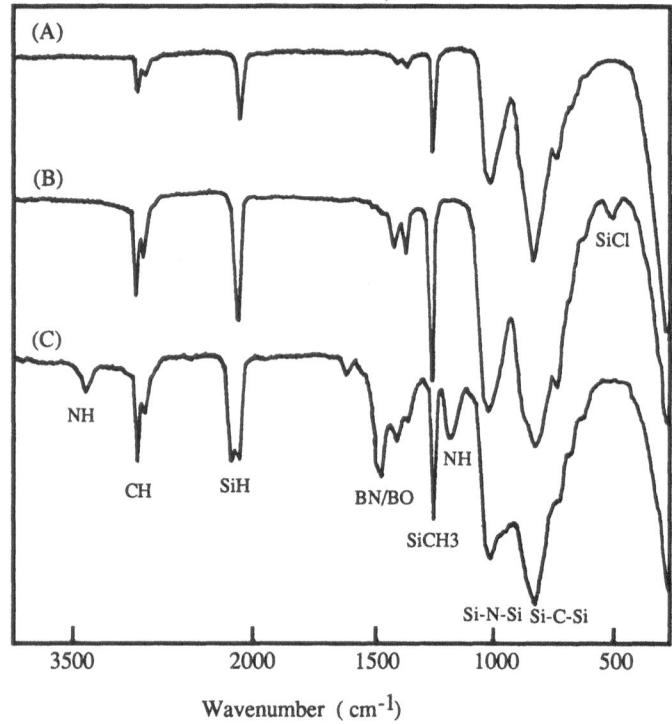

Figure 1. IR spectra of (A) PCS, (B) Cl-PCS, and (C) PCS/BSZ-1.

The ^{29}Si NMR spectra of PCS, Cl-PCS, and PCS/BSZ-1 are shown in Fig. 2. PCS shows two peaks at -0.8 and -18 ppm assigned to SiC_4 and $HSiC_3$, respectively. In Cl-PCS, a new peak appears at +26 ppm which can be assigned to some Cl containing units. Peak intensity of $HSiC_3$ units in Cl-PCS decreases compared with that of PCS. Cl containing units should be $ClSiC_3$. PCS/BSZ-1 presents two extra peaks at +3 and -37 ppm assigned to SiC_3N and $SiH_2N_2/SiHN_3$, respectively. $SiH_2N_2/SiHN_3$ units are derived from BSZ-1. SiC_3N should be bridging units in PCS blocks. $ClSiC_3$ units cannot be observed in PCS/BSZ-1.

DSC measurement of powder PCS shows an endothermic peak at 280°C (Fig. 3 (A)). After this measurement, powder PCS turned into a glassy solid. The DSC curves for the copolymers do not exhibit an endothermic peak at 280°C (Fig. 3(B)). The copolymers do not melt or soften in N_2 up to 500°C.

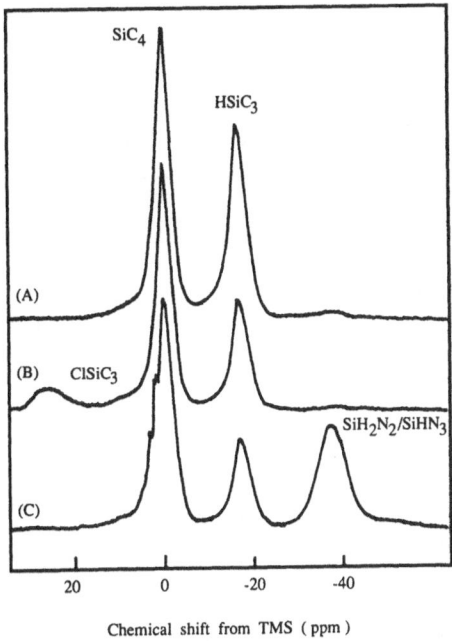

Chemical shift from TMS (ppm)

Figure 2. ^{29}Si NMR spectra of (A) PCS, (B) Cl-PCS, and (C) PCS/BSZ-1.

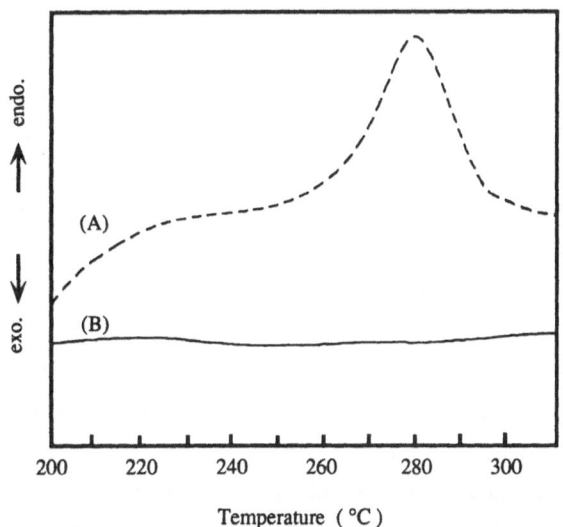

Temperature (°C)

Figure 3. DSC curves of (A) PCS and (B) PCS/BSZ-1 (N_2, 50°C/min).

Composition of the copolymers are summarized in Table 2. The Si/B ratio for PCS/BSZ-1, PCS/BSZ-2, and PCS/BSZ-3 was expected to be 14, 26, and 34 and found to be 15.2, 24.4, and 34.9, respectively.

Table 2. Composition of the copolymers (wt%).

	Si	B	N	O	C	H
PCS/BSZ-1	46.4	1.2	9.4	1.4	30.4	11.2
PCS/BSZ-2	49.6	0.8	8.8	1.5	30.6	8.7
PCS/BSZ-3	53.3	0.6	8.3	1.2	31.3	5.3

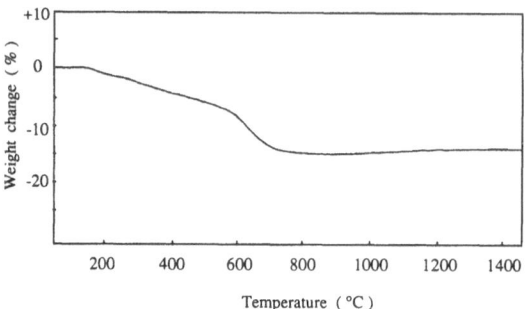

Figure 4. TGA curve of PCS/BSZ-3 (N_2, 5°C/min).

The TGA curves for the three copolymers have the same shape (Fig.4). Three region are apparent in the TGA curve. The first region, below 300°C, shows a weight loss of 2%, which results from the evaporation of residual solvent and hydrogen. The second region of 5% weight loss, from 300°C to 550°C, and the third region of 10% weight loss, from 550°C to 750°C, are thought to correspond to the breaking of organic bonds with the formation of gaseous species. Another 10% weight loss are observed in the temperature range between 1500°C and 1700°C. Pyrolysis of PCS/BSZ-1, PCS/BSZ-2, and PCS/BSZ-3 at 1700°C gave dark gray powder with 76 wt % ceramic yeild, greenish gray powder with 74 wt % ceramic yeild, and green powder with 71 wt % ceramic yeild, respectively. XRD patterns are shown in Fig. 5. β- and α-SiC are observed in the pyrolysis products. Ceramics from PCS/BSZ-1 exhibits poor crystallinity and high α/β ratio. However, pyrolysis product of PCS/BSZ-3 has good crystallinity and low α/β ratio. Composition of the pyrolysis products are summarized in Table 3. The pyrolysis product of PCS/BSZ-1 contains larger amount of boron and nitrogen than that of PCS/BSZ-2 and PCS/BSZ-3.

Table 3. Composition of the pyrolyzed copolymers at 1700°C in N_2 (wt%).

	Si	B	N	O	C
pyrolyzed PCS/BSZ-1	59.7	1.6	8.4	1.7	28.3
pyrolyzed PCS/BSZ-2	62.5	1.0	6.2	0.9	29.1
pyrolyzed PCS/BSZ-3	62.9	0.7	5.5	0.8	30.0

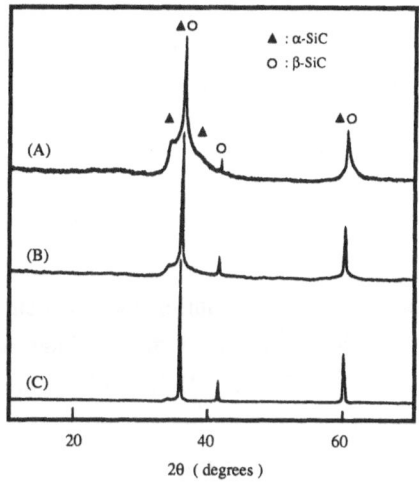

Figure 5. XRD patterns of copolymers pyrolyzed at 1700°C in N_2 [(A) PCS/BSZ-1, (B) PCS/BSZ-2, and (C) PCS/BSZ-3].

DISCUSSION

Due to the presence of the functional groups (Si-H and B-OCH$_3$) in BSZ, the copolymer may have bondings between PCS block and BSZ block. The formation of the bondings may be explained by two reactions :

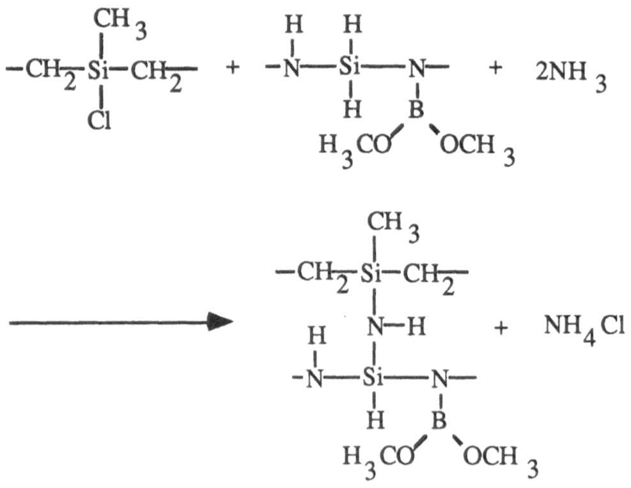

$$
\begin{array}{c}
\underset{\underset{Cl}{\overset{CH_3}{|}}}{-CH_2\text{-Si-}CH_2-} \quad + \quad \underset{\underset{\underset{H_3CO}{}}{\overset{H}{|}}\underset{B}{\overset{}{|}}OCH_3}{-N\text{-}\overset{H}{\underset{|}{Si}}\text{-}N-} \quad + \quad 2NH_3
\end{array}
$$

$$
\xrightarrow{\hspace{2cm}}
\begin{array}{c}
\underset{}{\overset{CH_3}{|}} \\
-CH_2\text{-Si-}CH_2- \\
\underset{\underset{H_3CO}{}}{-N\text{-}\overset{N-H}{\underset{|}{Si}}\text{-}N-} \\
H \quad \underset{B}{\overset{}{}} \\
H_3CO \quad OCH_3
\end{array}
\quad + \quad NH_4Cl \qquad (1)
$$

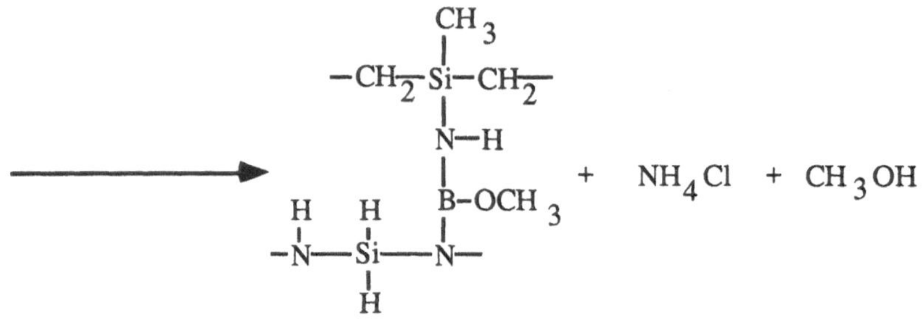

$$
\xrightarrow{\hspace{2cm}}
\begin{array}{c}
\overset{CH_3}{|} \\
-CH_2\text{-Si-}CH_2- \\
N-H \\
\underset{H}{\overset{H\ H}{}}\ B\text{-}OCH_3 \\
-N\text{-}Si\text{-}N-
\end{array}
\quad + \quad NH_4Cl \quad + \quad CH_3OH \qquad (2)
$$

However, CH3OH cannot be observed during the reaction. It seems that the first mechanism (eq.(1)) leads to the formation of the bondings in the copolymer.

Functional groups (Si-H, N-H and B-OCH3) in the copolymer are easily cross-linked to tridimensional network during the pyrolysis process. The release of gaseous species may result from the reaction of these functional groups :

$$
\equiv\!\text{Si-H} \quad + \quad \text{H-N}\!\!\begin{array}{c}\nearrow\text{Si}-\\ \searrow\text{Si}-\end{array} \quad\longrightarrow\quad \equiv\!\text{Si}-\text{N}\!\!\begin{array}{c}\nearrow\text{Si}-\\ \searrow\text{Si}-\end{array} \quad + \quad H_2 \qquad (3)
$$

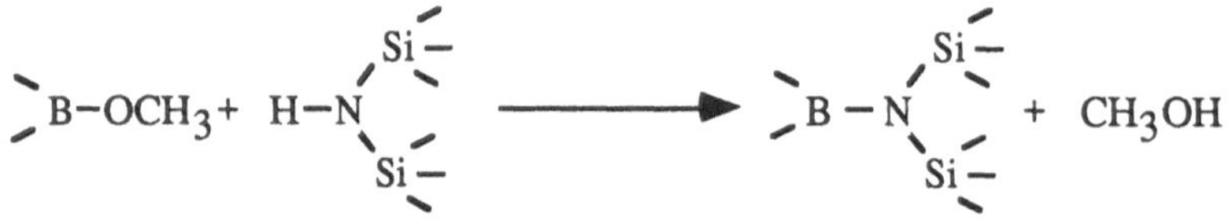

$$
>\!\!\text{B-OCH}_3 + \quad \text{H-N}\!\!\begin{array}{c}\nearrow\text{Si}-\\ \searrow\text{Si}-\end{array} \quad\longrightarrow\quad >\!\!\text{B}-\text{N}\!\!\begin{array}{c}\nearrow\text{Si}-\\ \searrow\text{Si}-\end{array} \quad + \quad CH_3OH \qquad (4)
$$

These cross-linking reaction, especially the first reaction (eq.(3)), at the early stage of pyrolysis prevent the copolymer from melting and softening.

The decomposition of PHPS in Ar is complete around 600°C [7]. Hasegawa et al. reported that PCS shows weight loss due to evaporation of low molecular weight PCS below 550°C and from 550°C to 800°C with evolution of H_2 and CH_4, which are mainly due to the decomposition of the side chains such as Si-H and Si-CH_3 [8]. The second region in the TGA curve, from 300°C to 550°C, is due to the decomposition of PHPS and PCS blocks, while the third region, from 550°C to 750°C, is certainly due to the decomposition of PCS block.

The composition of the pyrolysis products at 1700°C are $SiC_{1.11}N_{0.28}B_{0.07}O_{0.05}$ (PCS/BSZ-1), $SiC_{1.09}N_{0.20}B_{0.04}O_{0.03}$ (PCS/BSZ-2) and $SiC_{1.11}N_{0.17}B_{0.03}O_{0.02}$ (PCS/BSZ-3). The Si/B ratio of the pyrolysis products for PCS/BSZ-1, PCS/BSZ-2, and PCS/BSZ-3 is 14.7, 24.6, and 35.3, respectively. These values are close to those of the copolymers. Decomposition of the pyrolysis products and evaporation of Si and B do not occur up to 1700°C in N_2. Ceramics derived from the copolymers are stable up to 1700°C in N_2. The poor crystallinity and high α/β ratio of pyrolysis product of PCS/BSZ-1 has to be related to the impurities (B, N and O) introduced into the cubic structure of SiC. Defect in the cubic SiC may prevent the crystal growth and promote the β to α transformation. The impurities remain in the pyrolysis product because of high bonding energy between boron and nitrogen.

CONCLUSION

Thermosetting precursor for silicon carbide was synthesized by copolymerization of PCS with BSZ. Si-Cl bonds were introduced to PCS as functional groups to form bondings between PCS block and BSZ block in the copolymer. β- and α-SiC were obtained by pyrolysis of the copolymer at 1700°C in N_2. Boron was effective for suppression of crystal growth as well as promotion of β to α transformation in polymer derived silicon carbide.

ACKNOWLEDGEMENT

This work is a part of the Automotive Ceramic Gas Turbine Development Program conducted by Petroleum Energy Center.

REFERENCES

1 S.Yajima, J.Hayashi, and M.Omori, *Chem. Lett.* 931 (1975).

2 S.Yajima, K.Okamura, J.Hayashi, and M.Omori, *J.Am.Ceram.Soc.* 59:324 (1976).

3 K.Okamura, M.Sato, T.Seguchi, and S.Kawanishi, *Proc. 1st Jap. Int. SAMPE Symp.* 929 (1989).

4 M.Takeda, Y.Imai, H.Ichikawa, T.Ishikawa, N.Kasai, T.Seguchi, and K.Okamura, *Ceram. Eng. Sci. Proc.* 13 (7-8):209 (1992).

5 Y.C.Song, Y.Hasegawa, S.J.Yang, and M.Sato, *J.Mater.Sci.* 23:1911 (1988).

6 O.Funayama, T.Kato, Y.Tashiro, and T.Isoda, *J.Am.Ceram.Soc.* 76:717 (1993).

7 N.Kawamura and T.Isoda, *JETI* 38:104 (1990).

8 Y.Hasegawa and K.Okamura, *J.Mater.Sci.* 18:3633 (1983).

CONVERSION OF MOLECULES AND CLUSTERS TO EXTENDED 3-D CAGE AND CHANNEL NETWORKS

Omar M. Yaghi

Department of Chemistry and Biochemistry
Goldwater Center for Science and Engineering
Arizona State University
Box 871604
Tempe, Arizona 85287-1604

INTRODUCTION

Inorganic solids with inner cavities such as zeolites have found widespread use as industrial sorbents, ion-exchangers and shape-selective catalysts due to their ability to reversibly bind small molecules and ions.[1] In an effort to explore the possibility of producing zeolite-like materials that are not based on oxides, we and others have prepared coordination solids composed of organic ligands linked to metal ions to give extended structures. Bifunctional rod-like ligands such as, cyanides[2] and 4,4'-bipyridyl[3] have been utilized as anchors in linking transition metal ions to support the formation of structures with cavities where organic molecules or charge balance ions are accommodated. This synthetic strategy is schematically shown in Figure 1 for the formation of a diamond lattice from a rod-

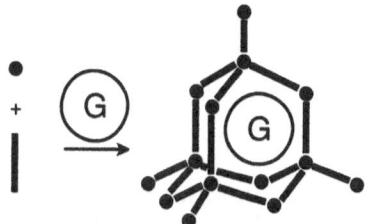

Figure 1. Schematic illustration of the assembly of molecular building blocks to form extended open-framework solids. A fragment of the solid is shown here with a cavity where a guest molecule G resides. In principle, the building blocks may assume other shapes and functionalities leading to diverse solid architectures.

like ligand and a metal ion capable of tetrahedral coordination. In principle, an extensive number of other frameworks with diverse topologies may be assembled by choosing the appropriate units of building blocks and guest species.

The construction of such solids from soluble molecular building blocks offers three important advantages towards the design of extended polymeric structures having the desired geometric and topological arrangement. First, the molecular chemistry of a large number of clusters and coordination complexes is extensive and well-studied, thus providing numerous starting materials. The second key advantage is that since the starting materials are soluble, the reactions may be performed at or near room temperature, thus preserving the structural integrity of the building block. Third, symmetry and functionality which are critically important for the design of rigid frameworks are both exploitable at the molecular level.

Using this synthetic approach, a number of 3-D frameworks with inner cavities have been discovered. Notable examples are the: (1) $Cu(4,4'-bpy)_2 \cdot PF_6$,[3] composed of four interpenetrating diamond-like copper bipyridyl frameworks with PF_6 acting as a non-coordinating counterion residing in the channels, vide infra, (2) $Cd(CN)_2$ family[4] of inclusion compounds which have channels, where organic molecules such as CCl_4 reside, and (3) $Cu[C(C_6H_4CN)_4] \cdot BF_4 \cdot xC_6H_5NO_2$,[5] having a diamond-like copper tetracyano-framework with cavities that are occupied by BF_4 anions and nitrobenzene solvent.

A number of open-framework solids which are based on the 4,4'-bpy, a, ligand have been discovered in our laboratory. However, we found that the formation of cavities is

complicated by the presence of interpenetrating lattices. Several new classes of open-framework structures, which utilize either multifunctional ligands such as, squaric acid, b, 1,3,5-benzenetricarboxylic acid, c, or inorganic tetrahedral clusters such as, $Ge_4S_{10}^{4-}$, d, have been synthesized showing rigid frameworks and no interpenetrating lattices. This chapter presents a summary of the synthetic procedures and the structural aspects of these solids; it is not intended as a comprehensive review of the work in the area. However, selected examples from the literature have been chosen to illustrate certain discussion points.

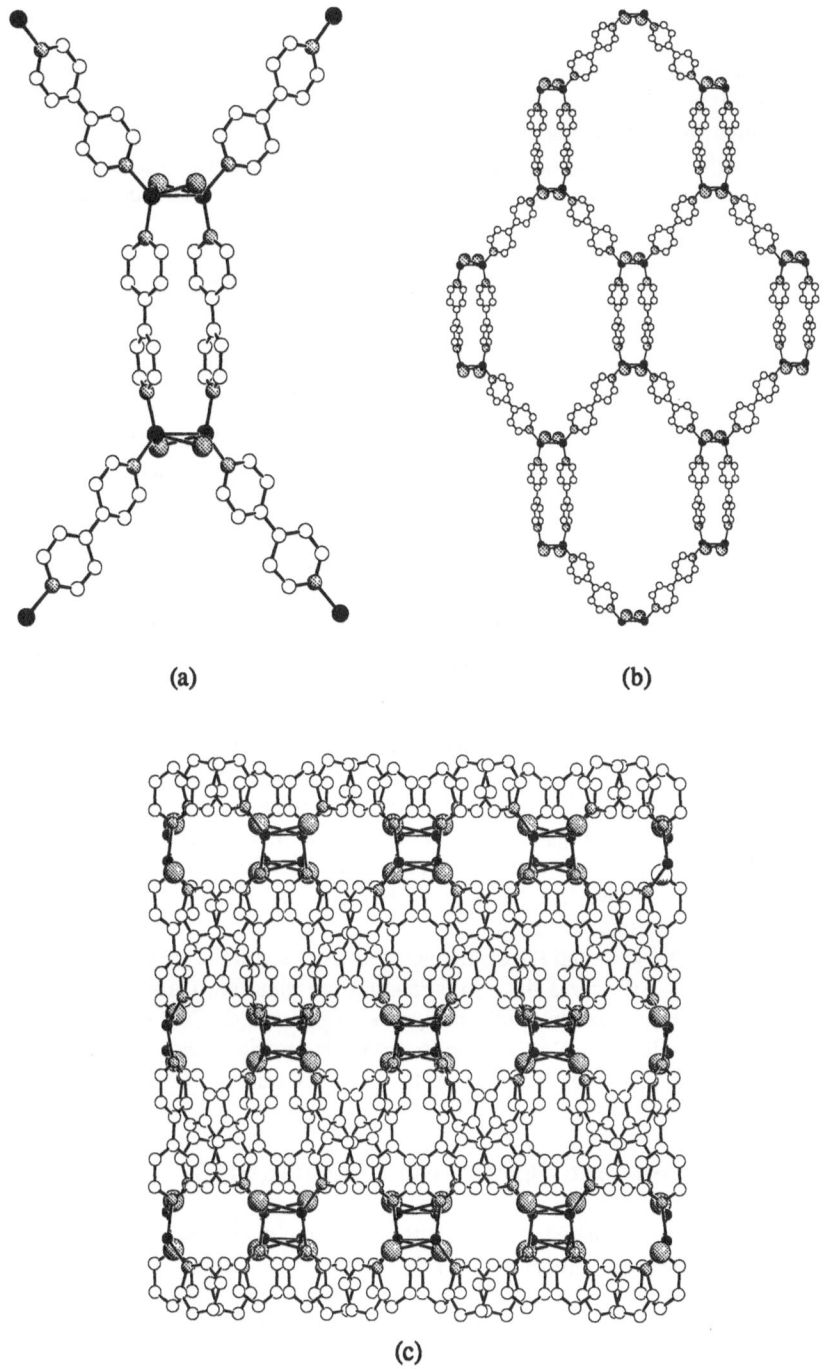

(a)

(b)

(c)

Figure 2. Drawings from single crystal data showing how the organization of the building block unit [Cu(4,4'-bpy)Cl]$_4$ (a) results in the construction of infinite porous sheets (b) that are a fragment of the Cu(4,4'-bpy)Cl structure, where these infinite sheets are interlocked to fill the pores leaving only small size cavities (c). The hydrogen atoms on the bipyridine units have been omitted for clarity. Dark spheres, Cu; shaded large spheres, Cl; shaded small spheres, N; and open spheres, C.

RESULTS

The reactions described here have been performed at room temperature and the solid products were isolated as crystalline homogeneous materials. Their elemental composition was determined from elemental microanalysis and their atomic structure was obtained by performing single crystal x-ray diffraction studies. This section briefly presents the synthetic procedures and identifies key structural aspects of some frameworks recently prepared and studied in our laboratory.

Open-Framework Solids with 4,4'-Bpy: Cu(4,4'-bpy)Cl and Cu(4,4'-bpy)$_2$· PF$_6$

The combination of equimolar amounts of CuCl and 4,4'-bpy in acetonitrile gives Cu(4,4'-bpy)Cl as a dark red microcrystalline material. Single crystals are obtained by diffusion of an acetonitrile solution of CuCl and a dmso solution of 4,4'-bpy through ethylene glycol.[6] The structure of this solid is made up of M-M bonded copper (I) centers, each linking four bpy ligands (Figure 2a) to form a sheet structure (Figure 2b) containing voids with 16 × 26 Å dimensions. However, these voids are almost completely filled with other mutually perpendicular interpenetrating sheets that are not chemically bonded but mechanically linked. Thus, the overall structure is a 3-D solid containing two interlocked sheets leaving only small channels with 2 × 4 Å dimensions as shown in Figure 2c. There is no evidence for any guest inclusions in the channels.

Larger channels are obtained with Cu(4,4'-bpy)$_2$· PF$_6$, which was prepared by reacting acetonitrile solutions of Cu(CH$_3$CN)$_4$PF$_6$, and 4,4'-bpy in 1:3 mole ratio. Orange crystals of the product are obtained by diffusion of acetonitrile solutions of the reacting reagents in poly(ethylene oxide) gel matrix.[3b] The structure of this compound is composed of tetrahedral copper (I) bpy centers (Figure 3a) which are linked in a diamond-like framework with 20 Å diameter voids (Figure 3b). However, most of this void space is occupied by three other interpenetrating diamond-like frameworks. Consequently, the structure is made up of four interlocked networks leading to a cationic framework with 6 Å channels that are occupied by the PF$_6$ anions (Figure 3c).

Addition of KX (X = Br and I) to either an aqueous or alcoholic suspension of the yellow solid, Cu(4,4'-bpy)$_2$· PF$_6$, gives the corresponding red material, Cu(4,4'-bpy)X, in 85-90 % yield. Conversely, the addition of NaPF$_6$ to a suspension of Cu(4,4'-bpy)X gives back the first solid in quantitative yield. The two compounds described above are stable in boiling water for up to 1-2 hr, and they are found to be thermally stable in He atmosphere to nearly 130 °C.

A 3-D Cage Network with Squaric Acid: ZnC$_4$O$_4$(OH$_2$)$_4$· 0.2 Py

This material is prepared either by suspending the chain-like solid, ZnC$_4$O$_4$(OH$_2$)$_2$ (DMSO)$_2$, in water at room temperature for 30 minutes or by boiling an aqueous mixture of the molecular solid, Zn(HC$_4$O$_4$)$_2$(OH$_2$)$_4$, for 10 minutes.[7a] The pyridine complex is prepared by refluxing[7b] an equimolar aqueous mixture of H$_2$C$_4$O$_4$ and Zn(NO$_3$)$_2$·6H$_2$O in the presence of pyridine. The overall structure of this material may be viewed as composed of large cubes (Figure 4) sharing faces with octahedral zinc ions occupying the edges and squarate ligands occupying the faces of the cubes to yield a 3-D cage network. Each squarate is linked to four zinc centers, and each zinc is linked to four different squarates and to two water molecules in *trans*-configuration. As shown in Figure 4 these water molecules are suspended into the cavities, which are occupied by pyridine as confirmed by solid state ^{13}C NMR data.

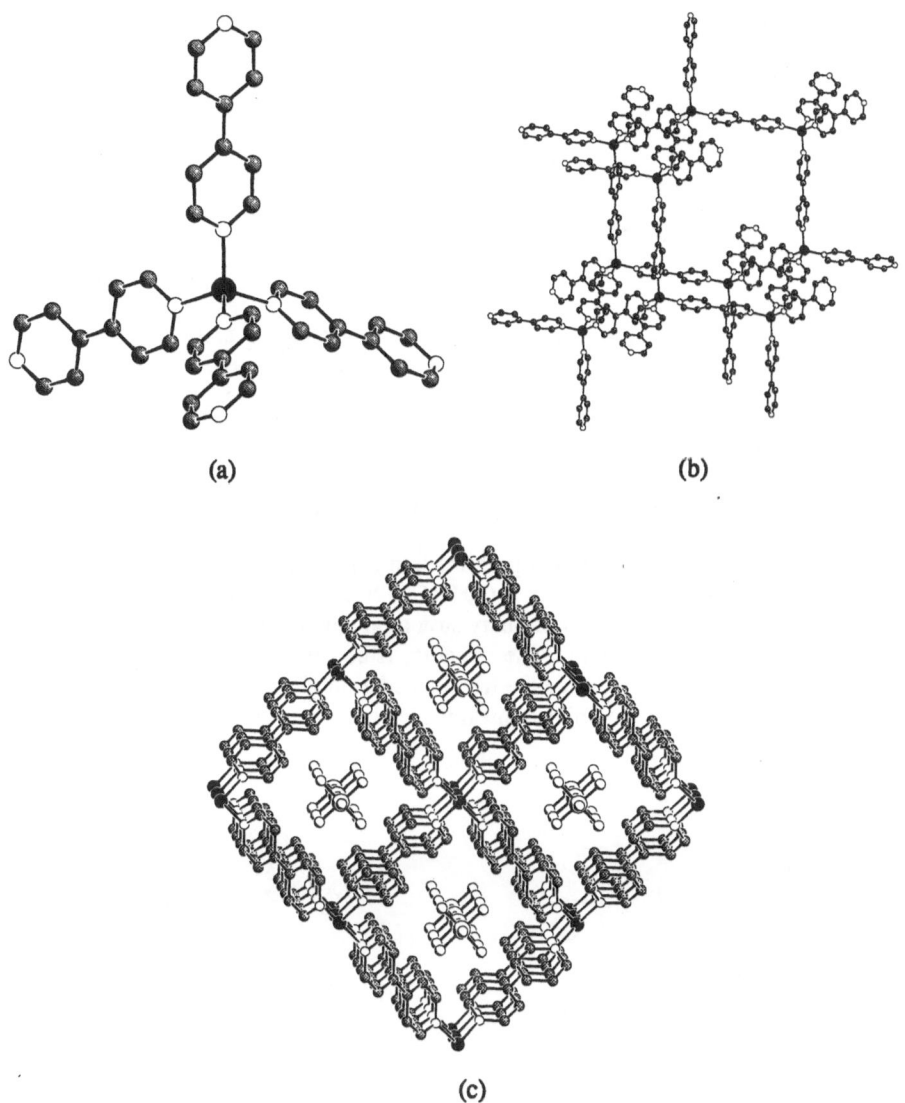

(a)　　　　　　　　　　　　　　(b)

(c)

Figure 3. The tetrahedral coordination geometry of the Cu(I) center (a) present as a building block in Crystalline $Cu(4,4'-bpy)_2 \cdot PF_6$, where these units are arranged in an adamatane-type motif (b). The copper atoms (dark spheres) and the 4,4'-bpy (C, shaded spheres; N, open spheres) may be visualized as replacing, respectively, the C and the C-C bonds in adamatane. In spite of the presence of four interlocking diamond-like networks, cavities of moderate size still form as channels along the crystallographic direction [001] (c). The channels are occupied with PF_6 anions (the P and F atoms shown respectively as larger and smaller partially shaded spheres). The hydrogen atoms on the bipyridine units have been omitted for clarity.

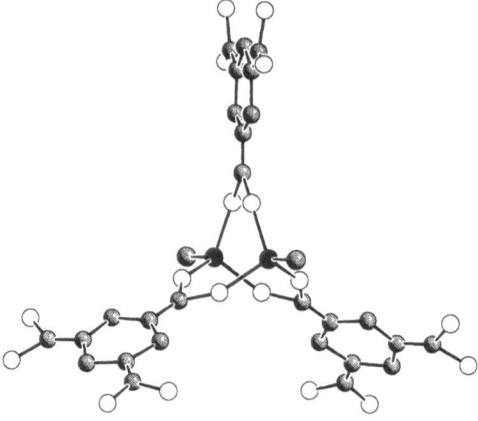

Figure 4. Perspective plot of the structural repeat unit existing in the crystal of $MC_4O_4(OH_2)_2$ phase (M = Mn, Zn). Dark spheres with octahedral coordination are Zn and those linked to it are O of H_2O molecules; shaded spheres, C; open spheres, O. The plot for $MC_4O_4(OH_2)_2$ was prepared using single crystal data obtained from Reference 7b.

A 3-D Channel Network: $Zn_2[1,3,5-C_6H_3(COO)_3]\cdot NO_3$

Diffusion of ethanol solution of pyridine into a nonaqueous PEO gel loaded with 1:2 mole ratio of $Zn(NO_3)_2\cdot 6H_2O$ to 1, 3, 5-$C_6H_3(COOH)_3$, gives large colorless crystals of $Zn_2[1, 3, 5-C_6H_3(COO)_3]\cdot NO_3$. Preliminary structural data indicate that the compound has a 3-D intersecting channel system. The structure is constructed from the units shown in Figure 5a, which form large rings that are fused together to give the channel network (Figure 5b) where the nitrate groups are pointed towards the voids.

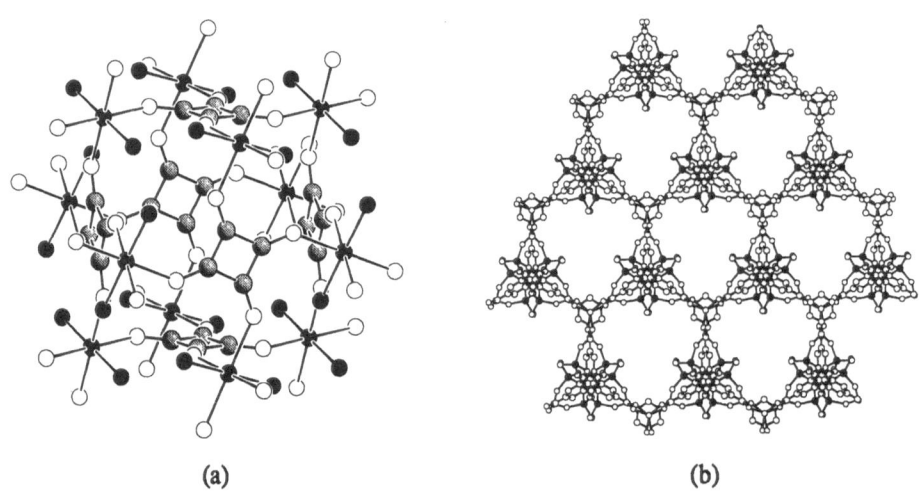

(a)   (b)

Figure 5. The structure of $Zn_2[1, 3, 5-C_6H_3(COO)_3]\cdot NO_3$ is composed of units (a) that agreggate to form large ring motifs, which are linked in 3-D to form large channels (b). The nitrate ions are pointing towards the centers of the channels. Dark spheres, Zn; open spheres, O; shaded spheres linked to O, C; shaded spheres linked to Zn, N. The hydrogens on the carbon atoms and the oxygens on the nitrogens are omitted for clarity.

A Diamond-Like Framework from Tetrahedral Inorganic Clusters: $MnGe_4S_{10}\cdot2(CH_3)_4N$

To prepare this solid an aqueous solution of $Mn(CH_3CO_2)_2\cdot4H_2O$ is diffused into another aqueous solution of $Ge_4S_{10}[(CH_3)_4N]_4$ in 1:2 mole ratio to give the product as yellow crystals. The structure of the solid shows that $Ge_4S_{10}{}^{4-}$ clusters (Figure 6a) have been linked with Mn(II) to form a diamond-like framework containing 3-D intersecting channel system, where the channels are occupied by the tetramethylammonium cations as shown in Figure 6b. It has been established that the cations may be removed without destructions of the metal sulfide framework. Thermal gravimetric and residual gas analysis on this material reveal that the cations decompose to at 350 °C to give nitrogen, methane, ammonia and some simple amines. No hydrogen sulfide was detected and the XRD data on the remaining sample show that the integrity of the framework is maintained up to 500 °C.[8]

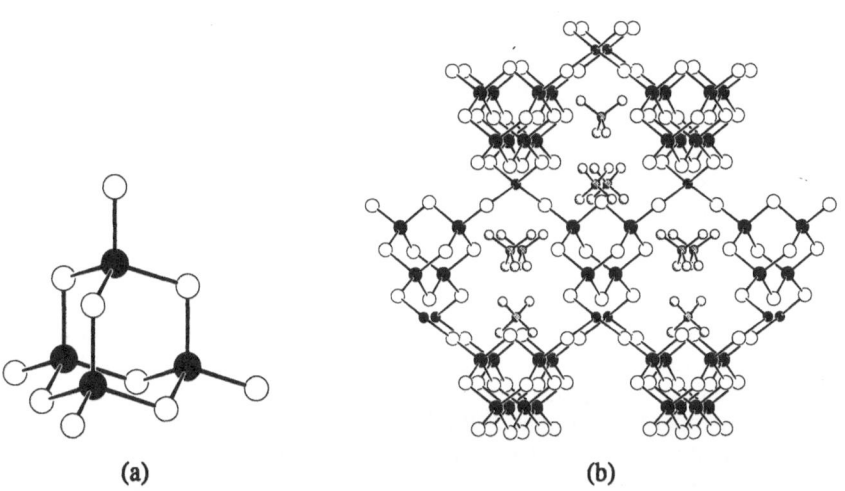

(a)   (b)

Figure 6. The terminal sulfides of the tetrahedral $Ge_4S_{10}{}^{4-}$ anion cluster (a) found in crystalline $Ge_4S_{10}[(CH_3)_4N]_4$ are linked with Mn(II) to form the diamond-like structure of $MnGe_4S_{10}\cdot2(CH_3)_4N$ (b) shown down the crystallographic direction [100], which contains $Ge_4S_{10}{}^{4-}$ anions as building blocks linked by manganese which is occupying a distorted tetrahedral coordination: dark spheres, Ge; open spheres, S; shaded spheres, Mn. The tunnels thus formed are occupied by $(CH_3)_4N^+$ cations represented here without the hydrogens and with shaded spheres for N and partially shaded spheres for C.

DISCUSSION

At the outset of this investigation two challenges to the synthesis of open-framework covalent solids were recognized. The first is the difficulty of obtaining the desired orientation of the building blocks in the solid state that would favor the formation of cavities. The second issue is that the attainment of the synthesized materials in crystalline form was found to be either difficult or at best fortuitous in nature. This section discusses the effectiveness of some of the strategies undertaken to address the first issue. The second issue was addressed by using polymeric matrices such as organic and inorganic gels as diffusion media to slow down the nucleation rate and to increase the possibilities for obtaining single crystals. These procedures are presented and discussed in an upcoming publication.[9]

The cubic diamond lattice is composed of tetrahedral carbon atoms bonded together to give a 3-D network containing no cavities. Cavities are generated from such structural arrangement by employing building units of larger size than that of carbon. The structure of the compound, $Cu(4,4'\text{-bpy})_2 \cdot PF_6$, is composed of four networks. Each network can be viewed in terms of the structure of diamond, where the C atoms are replaced by copper (I) and the C-C bonds are replaced by the 4,4'-bpy. In this way, the distance between the two tetrahedral Cu(I) centers is nearly 11 Å, thus creating a 3-D intersecting channel system with cross-sectional dimensions of 20 Å. In this structure, the large voids thus created are occupied by three other diamond-like frameworks identical to the one just described, and the PF_6 anions occupy the remaining channel space (6 Å) to balance the framework charge. Interpenetration of two sheet arrays results when Cu-Cu doubly bridged dimers are used as building units with 4, 4'-bpy to give $Cu(4,4'\text{-bpy})X$, which has very small cavities.

The formation of two-, three-, four-, five- and six-fold interpenetration have been observed. Their existence is likely to be due to a combination of two factors: (1) the availability of large enough space within the structure to allow their formation, and (2) the absence of an appropriate guest molecule to occupy this empty space. For example, $Cd(CN)_2$ is composed of two diamond-like frameworks. However, in the presence of CCl_4 it can be induced to crystallize with only one such framework. In general, as the cavities' size increases for rod-like ligands, controlling interpenetration becomes difficult due to the added flexibility in the framework and the problems associated with characterizing large noncoordinating guest molecules in these cavities.

Attempts to prepare $Cu(4,4'\text{-bpy})_2 \cdot PF_6$ and $Cu(4,4'\text{-bpy})X$ with fewer interpenetrating lattices by using larger guests were frustrated by the presence of a high degree of guest disorder in the cavities. In principle, the mobility of the guest molecules may be reduced significantly by hydrogen-bonding as was done in the diamond-like network of pyridone which enclathrates organic acids that are hydrogen-bonded to its large rectangular channels.[10] However, once the material is formed it is difficult to remove the guests due to the thermal instability of the framework, thus prohibiting access to the channels and rendering these materials nonmicroporous. In light of these observations it is interesting to note that the diamond-like structure of the solid, $Cu[C(C_6H_4CN)_4] \cdot BF_4 \cdot xC_6H_5NO_2$, is found not to have interpenetrating lattices in spite of the presence of large enough space to allow it. It is claimed that these channels are occupied by BF_4 anions and nitrobenzene solvent molecules.

We found that more rigid diamond-like frameworks containing no interpenetrating lattices may be prepared by the addition copolymerization of tetrahedral clusters with metal ions as in the compound $MnGe_4S_{10} \cdot 2(CH_3)_4N$. Here, each one of the four terminal sulfides on the tetrahedral cluster, $Ge_4S_{10}^{4-}$, are held tightly by a manganese ion that is bound tetrahedrally to these clusters. The channels thus created are approximately 9 Å wide and are occupied with the cations, which can be removed by their thermal decomposition or by ion-exchange without the destruction of the framework. The rigidity of this framework is seen as a direct consequence of the cluster ability to bind as a tetradentate ligand in contrast to the bidentate bpy or CN ligands. It is expected that larger cluster units will afford larger voids by inhibiting the formation of interpenetrating lattices.

Investigation of organic tetradentate ligands such as squarate is not an ideal choice for making open-structure due to its flat topology. However, it may offer opportunities in the construction of voided coordination solids. This is due to its rigidity and the number of diverse coordination modes it provides with transition metals. It has been observed that metal squarate complexes are converted to a 3-D cage network, $ZnC_4O_4(OH_2)_4 \cdot 0.2$ Py, that has a similar structure to that possessed by silicate clathrasils. These types of structures may not be useful as microporous solids. However, trapping organic molecules of highly polarizable nature may provide the basis for achieving nonlinear effects as observed in the silicate-based analogues.[11]

The synthesis of $Zn_2[1,3,5-C_6H_3(COO)_3] \cdot NO_3$ show that by using a hexadentate ligand such as a tricarboxylate frameworks with large cavities are possible. Although we have not fully characterized this solid, it appears that the framework is rigid and the nitrate ions pointing towards the center of the channels may be exchanged. We believe that this the first in an extensive class of materials that can be prepared using this ligand. Close examination of the structure shows the advantage of using this ligand in assembling a cluster of metal centers in the framework, thus enhancing the rigidity and preventing the formation of interpenetrating frameworks.

CONCLUSIONS

The results of this work show the feasibility of the deliberate design and construction of open-framework structure from soluble molecular building blocks. However, it also reveals certain problems associated with using rod-like ligands to make assemblies with large voids in that they are unable to support the formation of rigid frameworks. As a direct consequence of the availability of empty space and structure flexibility, the formation of large voids is suppressed by the formation of interpenetrating lattices. The large size of the cavities and the rigidity of the framework are both desirable features for the use of these solids as microporous materials. We found that the use of inorganic clusters or multifunctional organic ligands to assemble metal centers provides materials that have both large cavities and rigid frameworks.

ACKNOWLEDGMENTS

The financial support of the National Science Foundation, and the donors of the Petroleum Research Fund administered by the American Chemical Society and Arizona State University is acknowledged. I am grateful for the efforts of my coworkers C. E. Davis, T. L. Groy, G. Li, D. A. Richardson and Z. Sun, whose work is referenced herein. The assistance of Ms. E. Houseman in the preparation of the manuscript is greatly appreciated.

REFERENCES

1. A. Dyer. "An Introduction to Zeolite Molecular Sieves," Wiley, Chichester (1988).
2. For example: (a) J. Kim, D. Whang, J.I. Lee, and K. Kim, Guest-dependent $[Cd(CN)_2]_n$ host structures of cadmium cyanide-alcohol clathrates: two new $[Cd(CN)_2]_n$ frameworks formed with Pr^nOH and Pr^iOH guests, *J. Chem. Soc., Chem. Commun.* 1400 (1993). (b) K.-M. Park and T. Iwamoto, Urea- and thiourea-like host structures of *catena*-[(1,2-diaminopropane)cadmium(II) tetra-μ-cyanonickelate(II)] accommodating aliphatic guests, *J. Chem. Soc., Chem. Commun.* 72 (1992). (c) B.F. Abrahams, B.F. Hoskins, J. Liu, and R. Robson, The archetype for a new class of simple extended 3D honeycomb frameworks. The synthesis and x-ray crystal structures of $Cd(CN)_{5/3}(OH)_{1/3} \cdot 1/3(C_6H_{12}N_4)$, $Cd(CN)_2 \cdot 1/3(C_6H_{12}N_4)$, and $Cd(CN)_2 \cdot 2/3H_2O \cdot tBuOH$ ($C_6H_{12}N_4$ = hexamethylenetetramine) revealing two topologically equivalent but geometrically different frameworks, *J. Am. Chem. Soc.* 113:3045 (1991). (d) B.F. Hoskins and R. Robson, Design and construction of a new class of scaffolding-like materials comprising infinite polymeric frameworks of 3D-linked molecular rods. A reappraisal of the $Zn(CN)_2$ and $Cd(CN)_2$ structures and the synthesis

and structure of the diamond-related frameworks $[N(CH_3)_4][Cu^IZn^{II}(CN)_4]$ and $Cu^I[4,4',4'',4'''$-tetracyanotetraphenylmethane]$BF_4 \cdot xC_6H_5NO_2$, *J. Am. Chem. Soc.*, 112:1546 (1990).

3. For example: (a) L.R. MacGillivray, S. Subramanian, and M.J. Zaworotko, Interwoven two- and three-dimensional coordination polymers through self-assembly of Cu^I cations with linear bidentate ligands, *J. Chem. Soc., Chem. Commun.* 1325 (1994). (b) O.M. Yaghi, D.A. Richardson, G. Li, C.E. Davis, and T.L. Groy, Open-framework solids with diamond-like structures prepared from clusters and metal-organic building blocks, *Mater. Res. Soc. Symp. Proc.*, in press (1995).

4. T. Kitazawa, H. Sugisawa, M. Takeda, and T. Iwamoto, Structural characterisation of infinite three-dimensional $[Cd_4(CN)_9]^-$ ion, *J. Chem. Soc., Chem. Commun.* 1855 (1993), and references therein.

5. B.F. Hoskins and R. Robson, Infinite polymeric frameworks consisting of three dimensionally linked rod-like segments, *J. Am. Chem. Soc.* 111:5962 (1989).

6. O.M. Yaghi and G. Li, Presence of mutually interpenetrating sheets and channels in the extended structure of Cu(4, 4'-bipyridine)Cl, *Angew. Chem., Int. Ed. Engl.*, in press (1995).

7. (a) O.M. Yaghi, G. Li, and T.L. Groy, Conversion of hydrogen-bonded Mn(II) and Zn(II) squarate molecules, chains and sheets to 3-D cage networks, *J. Chem. Soc., Dalton Trans.*, in press (1995). (b) A. Weiss, E. Riegler, and C. Robl, Transition metal squarates, II: On the structure of cubic $(MC_4O_4 \cdot 2H_2O)_3 \cdot CH_3COOH \cdot H_2O$ (M = Zn^{2+}, Ni^{2+}), *Z. Naturforsch.* 41b:1329 (1986).

8. O.M. Yaghi, Z. Sun, D.A. Richardson, and T.L. Groy, Directed transformation of molecules to solids: synthesis of a microporous sulfide from molecular germanium sulfide cages, *J. Am. Chem. Soc.* 116:807 (1994).

9. O.M. Yaghi, G. Li, and T.L. Groy, Preparation of single crystals of coordination solids in silica gels: synthesis and structure of $Cu^{II}(1,4-C_4H_4N_2)(C_4O_4)(OH_2)_4$, *J. Solid State Chem.*, in press (1995).

10. X. Wang, M. Simard, and J.D. Wuest, Molecular tectonics. Three-dimensional organic networks with zeolitic properties, *J. Am. Chem. Soc.* 116:12119 (1994) and references therein.

11. H.K. Chae, W.G. Klemperer, D.A. Payne, C.T.A. Suchicital, D.R. Wake, and S.R. Wilson, Clathrasils: new materials for nonlinear optical applications, *in*: "Materials for Nonlinear Optics: Chemical Perspectives," S.R. Marder, J.E. Sohn, and G.D. Stucky, eds., Publisher, City (1991).

POLY(PHENYLMETALLOSILOXANE)S:
SYNTHESIS, STRUCTURE AND PROPERTIES

Olga I. Shchegolikhina, Inessa V. Blagodatskikh,
Yulia A. Pozdnyakova, Alexandre A. Zhdanov

Institute of Organo-Element Compounds,
Russian Academy of Sciences
Vavilova str. 28, Moscow, 117813

INTRODUCTION

The synthesis and the study of metallosiloxane polymers is of a considerable interest due to their unique properties, which include high thermal stability and electrical
conductivity. The chemistry of metallosiloxane polymers has already been studied for
25–30 years. The question about their structure was open until the beginning of 1990
when we have first obtained and structurally characterized poly(phenylmetallosiloxane)s
(PPMS), containing bivalent transition metal atoms: nickel, manganese, copper and
cobalt[1-4], which were named the cage-like poly(phenylmetallosiloxane)s[8].

In this paper we consider the problems of the synthesis and the structure of two
types of PPMS, produced from polyfunctional sodium phenylsiloxanolate (SPS). The
first type is PPMS based on the bivalent transition metals and the second one is
poly(phenylironsiloxane) (PPIS) which is an example of an amorphous PPMS with
a trivalent transition metal.

Structure investigation of the first type of PPMS is based on the combination
of X-ray diffraction analysis, decomposition trimethylsilylation and gel permeation
chromatography (GPC). A complex approach, including GPC, preparative fractionation, spectral techniques and elemental analysis in combination with decomposition
trimethylsilylation is employed to investigate poly(phenylironsiloxane) (PPIS) which is
not able to crystallize.

The last part of the paper considers the magnetic properties and thermo-conden-
sation of PPIS which is of interest from the practical point of view.

RESULTS AND DISCUSSION

The poly(phenylmetallosiloxane)s, containing transition metal atoms (Ni, Mn, Co,
Cu, Fe and Ln) are obtained by the exchange interaction between sodium phenylsiloxanolate with transiton metal chlorides. Their synthesis includes three stages:

(i) the hydrolysis of phenyltrichlorosilane in benzene at low temperature, giving poly(phenylsilsesquioxane):

$$nPhSiCl_3 \xrightarrow{H_2O} [PhSiO_{1.5}]_n;$$

(ii) the alkaline splitting of poly(phenylsilsesquioxane) by the mixture of metallic sodium and its hydroxide, which leads to the formation of sodium phenylsiloxanolate:

$$m/n[PhSiO_{1.5}]_n + m/2(NaOH + Na) \longrightarrow [PhSi(O)ONa]_m;$$

(iii) the exchange interaction between sodium phenylsiolxanolate with the corresponding metal chloride:

$$x/m[PhSi(O)ONa]_m + MCl_x \longrightarrow \{[PhSiO_{1.5}]_x(M_{2/x}O_{x/2})\}_y.$$

Structure of Cage-Like Poly(phenylmetallosiloxane)s

The crystalline products were isolated from poly(phenylmetallosiloxane)s containing the bivalent metal atoms (Ni, Mn, Cu) with yields of 30–70% . X-Ray diffraction data[1-3] show that these metallosiloxanes molecules have a sandwich-type structure (fig. 1), wherein two cis-hexaphenylhexasiloxanolate fragments are linked through oxygen atoms with metalloxide layer containing six transiton metal atoms. The Cl-anion is encapsulated in the centre of the cavity and coordinates with six metal atoms. The Na-counterion is located outside the siloxanolate molecule and coordinated with O atoms of siloxane cycles. In the case of copper-containing metallosiloxane the encapsulated Cl-ion is absent.

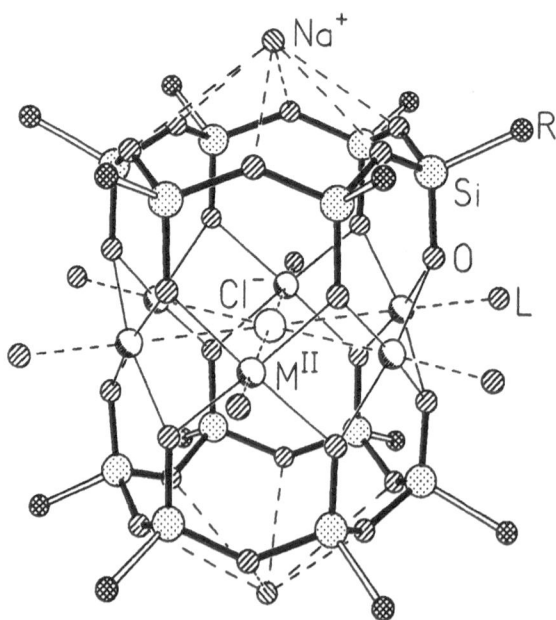

Figure 1. The X-ray structure of the poly(phenylmetallosiloxane)s framework: R=Ph, L=BuOH, $M^{II}=Ni^{II}$, Mn^{II}.

Another type of crystalline compound was isolated from the incomplete exchange reaction between SPS and $CuCl_2$:

$$12/m[PhSi(O)ONa]_m + 4CuCl_2 \longrightarrow \{[PhSiO_{1.5}](CuO)_4(NaO_{0.5})_4\}$$

Its structure (fig. 2)[4] is based on the dodecaphenyldodecasiloxanolate fragment with tris(*cis*)-tris(*trans*)-configuration. Its "horse saddle" conformation is fixed by four Cu atoms and four Na atoms.

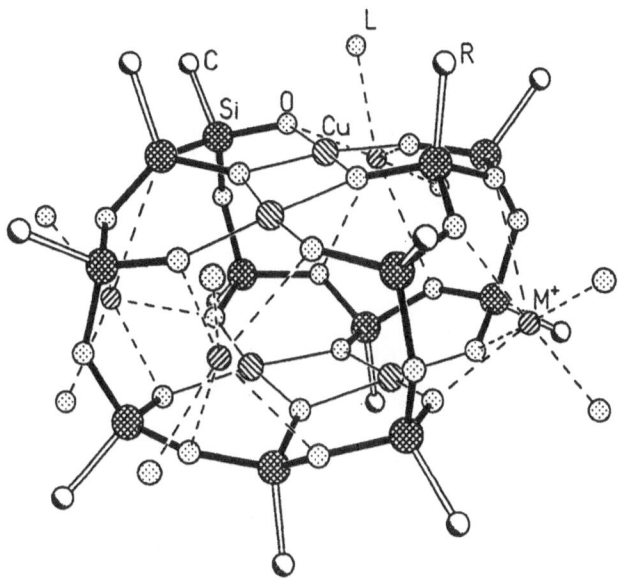

Figure 2. The X-ray structure of poly(phenylcoppersodiumsiloxane): R=Ph, L=HOBu, M^+=Na.

Organosiloxanolate fragments of cage-like PPMS were analysed additionally with the combination of decomposition trimethylsilylation and GPC. The composition of an amorphous PPMS as well as of crystalline compounds were investigated in this way. The decomposition trimethylsilylation is the treatment of the corresponding PPMS with trimethylchlorosilane and consists in the rupture of the Si-O-M bond and the replacement of the metal atom with the Me_3Si-group, according to schemes (1) or (2):

$$\{[(PhSiO_{1.5})_6]_2(NiO)_6(NaCl)\} + 12Me_3SiCl \longrightarrow 2[PhSi(O)(OSiMe_3)]_6 \qquad (1)$$

$$\{[PhSiO_{1.5}]_{12}(CuO)_4(NaO_{0.5})_4\} + 12Me_3SiCl \longrightarrow [PhSi(O)(OSiMe_3)]_{12} \qquad (2)$$

The mild trimethylsilylation technique with the use of pyridine[5] enables us to isolate with high yields crystalline stereoregular cyclosiloxanes which have been characterized by X-ray difraction and NMR. As we have shown by GPC[6], this technique of trimethylsilylation allows us to obtain siloxane rings practically without side reactions.

Pure cyclic polyphenyl(trimethylsiloxy)siloxanes (CPPTS) of general formula $[PhSi(O)(OSiMe_3)]_n$, with n=4,6 and 12, were used as calibration standards in GPC.

As a result we established the following regularities in the composition of trimethylsiloxy-derivatives (TMS derivatives) of cage-like PPMS: in the case of crystalline nickel- and manganesesiloxanes they are almost a pure six-membered siloxane ring with a small amount of its condensation products, probably bi- and tricycles (fig. 3a); products of trimethylsilylation of crude reaction masses of Ni- and Mn-containing siloxanes are also rather narrow mixtures of siloxane cycles (Mw/Mn≈1.2) which contain small amounts of condensation products (fig. 3b).

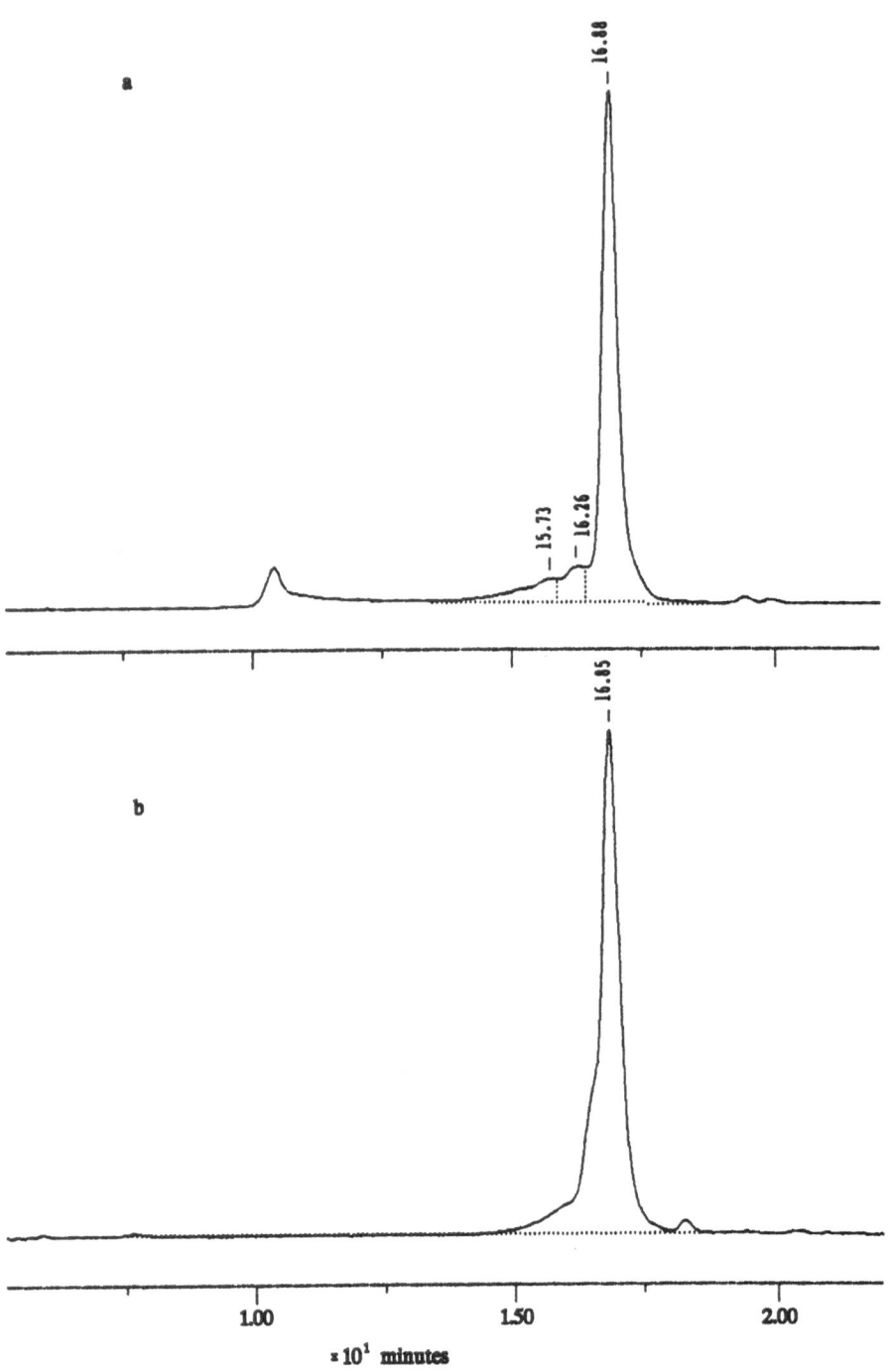

Figure 3. The chromatograms of TMS derivatives of crystalline poly(phenylnickelsiloxane) (a) and of the product of trimethylsilylation of crude reaction mass (b). 500 and 1000 Å U-styragel columns, chloroform, $\lambda=260$ nm.

The TMS derivative of crystalline poly(phenylcoppersodiumsiloxane) contains twelve-membered and six-membered siloxane rings in the ratio 84:16 (fig. 4) and the TMS derivatives of the mother liquor, after isolaton of the PPMS crystals (the yield is 67%), consists of two fractions with maxima near to those of twelve- and six-membered rings.

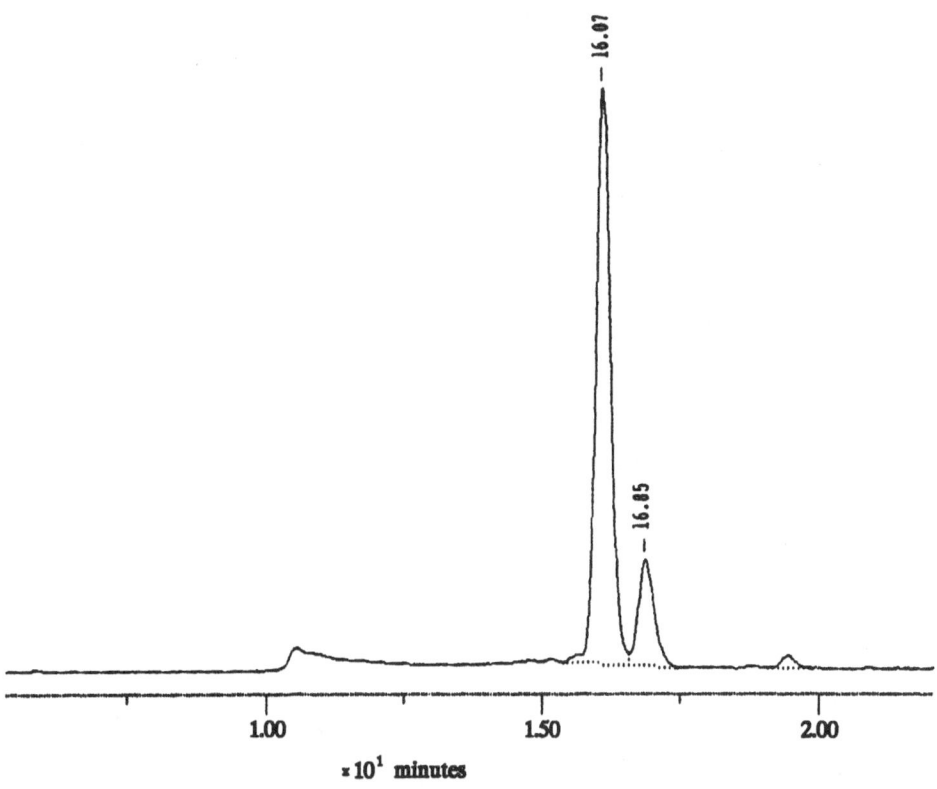

Figure 4. The chromatogram of TMS derivative of crystalline poly(phenylcoppersodiumsiloxane).

We conclude that during the exchange interaction of sodiumphenylsiloxanolate with the above mentioned bivalent transition metal chlorides six- and twelve-membered siloxane rings, depending on the metal origin, are generated with high selectivity. Such selectivity is evidently a result of an active coordinating role of the metal atom. On the contrary under polymerization or polycondensation processes, generating siloxane rings and chains, such selectivity is never observed and relative amounts of various cycles in the mixtures obey other regularities[7].

We attempted to formulate a possible mechanism for the formation of such poly-(phenylmetallosiloxane)s[8] assuming that the substitution of the sodium atom for the transition metal occurs via a transition complex rather than by free ions.

The structure of sodium phenylsiloxanolate in the reaction solution remains an open question until now. According to the data of different authors SPS can be identified as sodium monosilanolate $PhSi(OH)_2ONa$[9] or as a cyclotetrasiloxane with Ph and ONa groups bound to the silicone atom[10]. X-Ray data[11] show that crystalline SPS is a cyclotrisiloxane containing phenyl and ONa groups at the silicone atom and 8 molecules of crystallisation water.

In our opinion, the sodium siloxanolate molecules in the butanol solution exist in associate forms which are in dynamic equilibrium:

$$nPhSi(O)ONa \rightleftharpoons [(Ph)Si(ONa)O]_m \qquad n = 3, 4, 5 \text{ etc.}$$

These associates also contain butanol molecules, linked reversibly with hydrogen bonds. The introduction of metal halide into such a system should facilitate the exchange reaction in the transition complex:

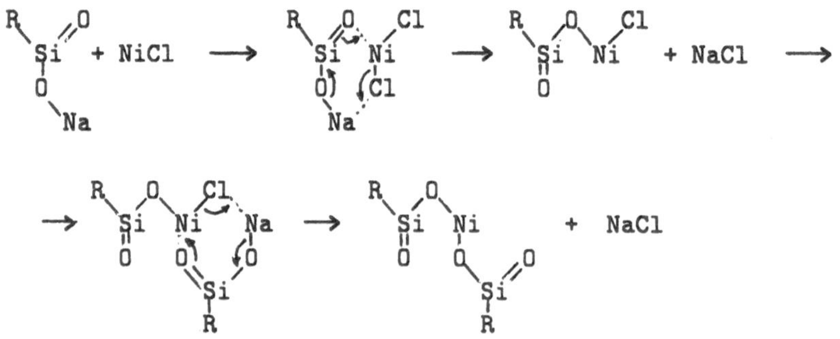

Two of the molecules formed in the reaction are capable of binding together to form a coordination pair:

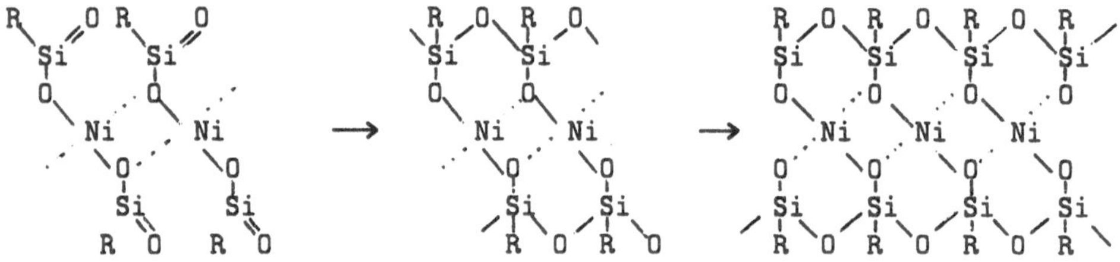

The possibility of the Ph-Si=O group free rotation about the Si-O axis in such coordinated pair may be ruled out because of the steric hindrance created by the phenyl group. We consider the behaviour of the coordinately bound pair in which the phenyl groups are located in the *cis*-position relative to the NiO$_4$ plane. In this case the stabilization of the system due to a rapid interaction of the Si=O group with the neighboring Si atom is feasible. This facilitates the subsequent addition of disubstituted Ni and hence, the chain growth. The distance between Si and O atoms is the most favorable for the reaction and this accounts for the predominant formation of six-membered rings which is the basis of the framework structure.

Structure of Poly(phenylironsiloxane)

At present a series of PPMS, containing trivalent transition metals, is obtained as amorphous materials, soluble in some organic solvents. The examples are PPIS, poly(phenylchromsiloxane) and some PPMS containing lanthanide metals (Dy, Nd, Gd). Structural study of such compounds is complicated since the direct X-Ray diffraction study is impossible.

The detailed investigation of PPIS, performed recently[12] using a set of chemical and physical-chemical methods, characterizes the main features of their structure.

The use of two-wave detection GPC, preparative fractionation and elemental analysis enabled us to estimate composition- and molecular-size heterogeneity. The analysis of decomposition trimethylsilylation products of crude polymer and fractions by means of GPC and H^1-NMR allows us to elucidate the main features of the PPIS structure.

Figure 5 presents the two-wave detection chromatogram of PPIS, synthesized by a standard method (the schemes on page 2). It can be seen from the figure 5 that this sample has composition heterogeneity and consists of two structural types of macromolecules with different absorbance spectra. The comparison of fractions chromatograms and elemental analysis data shows that the narrow peak (RT=16.65 at fig. 5), which does not manifest itself at 320 nm, is related to molecules, containing no iron atoms, i.e. poly(phenylsiloxane) molecules. This fraction was isolated by extraction with hexane from crude PPIS. The chromatogram at 320 nm reflects actually poly(phenylironsiloxane) and shows that this polymer has high molecular size heterogeneity. Its characterizaton according to PS-standards gives: $\bar{M}_n = 700$ and $\bar{M}_w = 6200$.

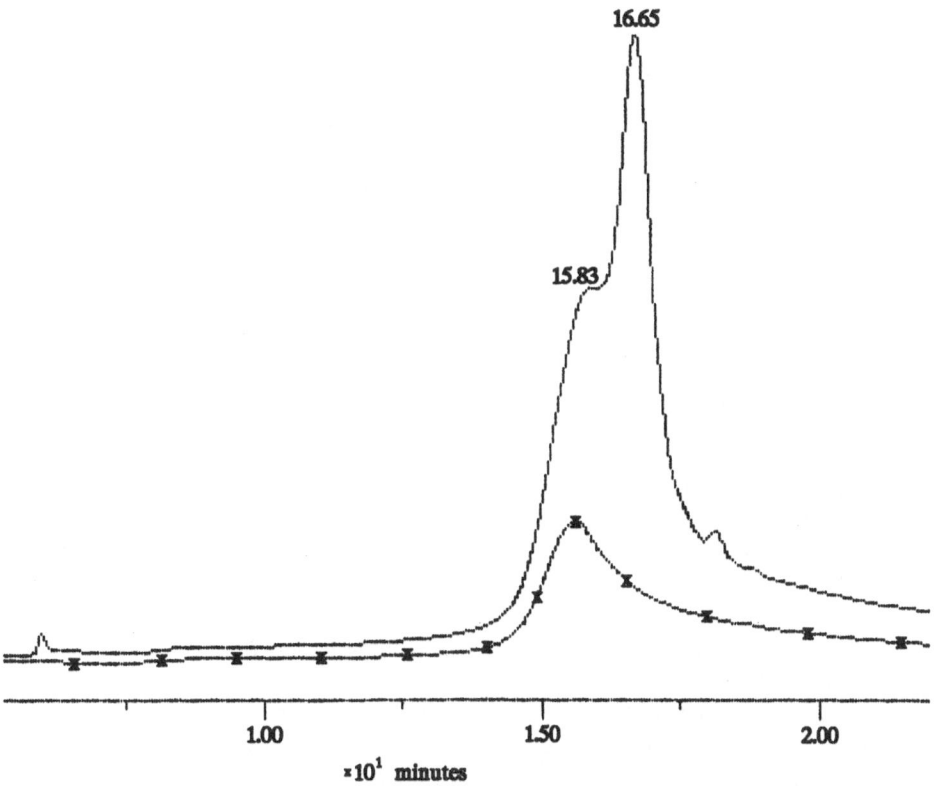

Figure 5. The two-wave detection chromatogram of poly(phenylironsiloxane): (——) 260 nm, (–x–) 320 nm, 500 and 1000 Å U-styragel columns, chloroform.

Almost pure PPIS are isolated by fractional precipitation in the first and second fractions. The comparison of Fe and Si contents in the initial polymer and the above mentioned fractions enabled us to estimate the ratio between poly(phenylironsiloxane) and oligo(phenylsiloxane) as about 60:40.

GPC traces of the trimethylsilylation products of initial sample and its fractions have shown that oligosiloxanes peak (RT = 16.65 at fig. 5) does not change under trimethylsilylation. One can conclude from this fact, that oligosiloxanes do not practically contain silanol groups and consist apparently of closed ladder fragments. Their molecular weights according to CPPTS calibration are: $\bar{M}_n = 1225$, $\bar{M}_w = 2100$. The peak maximum is near to that of six-membered CPPTS.

Poly(phenylironsiloxane) molecules on the decomposition break up into the fragments of a narrower molecular weight distribution without the high molecular weight tail. This means that the oligosiloxane fragments are linked in macromolecules through the ironoxide bridges. The comparison of Fe and Si contents in PPIS fractions, which are almost free of oligosiloxanes, and SiMe$_3$/SiPh ratios from H^1 NMR in their trimethylsilylation products (table 1) indicate, that about 1/3 of silicon atoms in PPIS is not bound up with iron atoms and, on the other hand, only approximately 1/3 of iron valencies is used for Fe-O-Si bonds, the rest forms apparently Fe-O-Fe bonds. So one can conclude that PPIS macromolecules are branched consequences of alternated oligosiloxane and ironoxide fragments. The structure of oligosiloxane fragments must be similar to that of free siloxane oligomers.

Table 1. Elemental analysis and ^{1}H-NMR data for poly-(phenylironsiloxane) and its trimethylsilylation products.

Sample	% Fe	% Si	Si/Fe	SiMe$_3$/SiPh
Crude polymer	11.11	15.50	2.78	0.36
Fract.1	16.88	13.55	1.60	0.56
Fract.3	16.64	14.27	1.94	0.56

We have also found, that carrying out the PPIS synthesis with FeCl$_3$ sublimated just before synthesis decreases the content of high polymer fractions and yields the narrower oligomeric PPIS, that has however the free phenylsiloxane fraction. At the same time materials, produced on a pilot plant in ethanol solution[13] are high molecular weight polymers and contain free siloxane oligomers as well. Their decomposition trimethylsilylation yields fragments similar to those described above.

These structural data enable us to make some assumptions about chemical processes which take place under interaction of SPS with iron chloride. If we consider this process as polycondensation, we can note the following: the average functionality of monomers is above 2 (FeCl$_3$ functionality is 3 and SPS is considered as trifunctional). In this case one can expect the formation of a polymeric network. However, we know that analogous polycondensation process of some bivalent metal chlorides with SPS, where the average functionality is also above 2, produces in reality closed sandwich-type molecules. This peculiarity is due to some trends: 1) high ability of siloxanes to form rings and 2) high coordinative ability of transition metal ions, which allows them to mount between siloxane rings.

In the case of SPS with iron chloride interaction there is the third tendency to Fe-O-Fe linking, which competes with the second one and remove part of the iron valencies from the Si-O-Fe linkage. The air and reaction medium humidity contributes to this process:

$$2 > \text{FeCl} + \text{H}_2\text{O} \longrightarrow > \text{Fe} - \text{O} - \text{Fe} < \ +2\text{HCl}.$$

Escaping HCl causes the hydrolysis and condensation of silanolate groups:

$$> \text{Si(Ph)(ONa)} + \text{HCl} \xrightarrow{-\text{NaCl}} > \text{Si(Ph)(OH)} \longrightarrow > \text{Si(Ph)} - \text{O} - \text{Si(Ph)} <$$

Due to this reason we observe incomplete linking of silicon atoms with iron via oxygen and incomplete employment of iron valencies for Fe-O-Si bonds.

Properties of Poly(phenylironsiloxane)

The magnetic properties of poly(phenylironsiloxane)s, produced in a pilot plant[13], and their evolutions during the pyrolysis have been investigated[14,15,16]. PPIS has been found to be paramagnetic[14]. It was established that the magnetization dependence of these PMOS at 77 and 296 K obeys the Curi law. The effective magnetic moments of a series of PPIS fractions with the ratio Si/Fe in the range from 1.2 to 2.5 are found to be much lower than the theoretical value 5.95 mB and were in the interval of 3.75 ÷ 4.73 mB. This could indicate an antiferromagnetic interaction of moments of neighboring $Fe(3+)$-ions, in the Fe-O-Fe structural fragments. Similar phenomenon was observed earlier in inorganic compounds[17].

Polymers containing paramagnetic metal atoms are interesting as precursors for materials with ferromagnetic properties. One of the ways to remove the organic fragments and to unite the paramagnetic atoms in clusters is to carry out the pyrolysis of these polymers.

The thermal condensation process was investigated[15,16] using some PPIS fractions in inert atmosphere under various temperatures from 400 to 800° C. This study has shown that the pyrolysis at 800° C yields ferromagnetic materials with magnetization values of 7.9 to 35 G/g (table 2). The thermocondensation is accompanied with the elimination of a large part (50–83%) of the phenyl groups[16] in the form of benzene and biphenyl and small amounts of H_2, CO and CO_2. Some of the phenyl groups remain in the composition of the pyrolyzed residue. As it follows from IR spectroscopic data polysubstituted aromatic nuclei are formed at the temperatures of 600° C and higher.

Table 2. Specific saturation magnetizations (σ) and molar magnetizations (σ') of pyrolyzed poly(phenylironsiloxane) samples.

Sample	Pyrol. temp. (° C)	Pyrol. time (min)	% Fe	Si/Fe	σ (G/g)	$\sigma' \times 10 - 4$ (G/mole)
1	800	10	34.60	1.3	7.9	0.13
2	800	10	30.26	1.8	21.0	0.39
3	800	10	25.79	2.0	27.0	0.59
4	800	10	23.61	2.4	35.0	0.83

X-Ray phase analysis of the solid ferromagnetic materials shows that they contain a mixed iron carbide-silicide with low crystallinity, iron carbide, metallic iron phase with cluster sizes about 350 Å. The whole amount of oxygen is concentrated in the amorphous SiO_2 phase.

Data on the quantitative distribution of iron in the variuos phases obtained by means of Mossbauer spectroscopy show that the process of pyrolysis transfers into reduced forms [Fe(2+), Fe(0)] almost 90% of the initial $Fe(3+)$ (table 3). The latter fact determines the ferromagnetic properties of the products of high-temperature condensation. The molar magnetization values (table 2), accounted using concentrations of metal in the materials and experimental values of the specific saturation magnetizaton, indicate the fraction of the reduced iron. It goes up with increasing Si/Fe ratio and approaches closely the corresponding value of pure iron (12000 G/mole). Following all these data we have proposed the radical process scheme which could explain the formation of reduced forms of iron, carbides and carbide-silicides of iron and silicon dioxide in the course of PPIS pyrolysis.

Table 3. Parameters of Mössbauer spectra of pyrolyzed poly(phenylironsiloxane) (sample 4).

Pyrolysis temp. (° C)	Pyrolysis time (min)	Spectral band	Parameters of spectral bands		Distribution of iron among forms (%)
			shift	splitting	
800	10	Doublet Fe^{3+}	0.41	1.00	9.6
800	10	Doublet Fe^{2+}	1.21	3.18	44.4
800	10	Sextet α-Fe	0.10	0.0	46.0

The primary stage is generation of radicals by thermal decomposition of Si-C bond:

$$ \geqslant Si - Ph \longrightarrow > Si \cdot + \cdot Ph $$

The action of phenyl radical on PPIS leads to the formation of benzene, then to biphenyl and to polysubstituted aromatic nuclei according to the following scheme:

$$ Ph \cdot + PhSi \leqslant \xrightarrow{Ph\cdot} PhH + PhC_6H_4Si \leqslant $$

$$ Ph \cdot + Ph \cdot \longrightarrow Ph - Ph $$

$$ Ph \cdot + PhSi \leqslant \longrightarrow PhH + \cdot PhSi \leqslant $$

$$ \geqslant Si \cdot + \cdot PhSi \leqslant \longrightarrow \geqslant SiC_6H_4Si \leqslant $$

$$ Ph \cdot + \cdot PhSi \leqslant \longrightarrow PhC_6H_4Si \leqslant $$

Reduction of Fe(3+) to Fe(2+) and Fe(0) could happen under the action of $\geqslant Si\cdot$ (silyl radical) on the Fe-O bond.

$$ > Fe - O - Si \leqslant + \geqslant Si \cdot \longrightarrow -Fe - O - + \geqslant Si - O - Si \leqslant $$

And the gradual realization of this scheme leads to the formation of structures -Fe-Fe-Fe-, which provide for exchange processes and for the formation of ferromagnetic material.

Acknowledgments

The authors thank the Russian Foundation for Basic Research (grant 93-03-19121) for partial support of this work.

REFERENCES

1. O.I.Shchegolikhina, A.A.Zhdanov, V.A.Igonin, Yu.E.Ovchinnikov, V.E.Shklover, Yu.T.Struchkov, Synthesis and structure of unusual skeletal cylindrical nickel cyclo-hexasiloxanolates, Organomet. Chem. USSR. 4:39(1991).
2. V.A.Igonin, S.V.Lindeman, K.A.Potekhin, V.E.Shklover, Yu.T.Struchkov, O.I.Shchegolikhina, A.A.Zhdanov, I.V.Razumovskaya, The structure of sandwich complexes of Ni with macrocyclic 12-membered cis-hexaphenylcyclohexasiloxanolate ligands, Organom. Chem. USSR. 4:383(1991).

3. V.A.Igonin, O.I.Shchegolikhina, S.V.Lindeman, M.M.Levitsky, Yu.T.Struchkov, A.A.Zhdanov, Novel class of transition metal coordination compouds with macrocyclic organosiloxanolate ligands, their synthesis and crystal structure, J.Organomet. Chem. 423:351(1992).

4. V.A.Igonin, S.V.Lindeman, Yu.T.Struchkov, O.I.Shchegolikhina, A.A.Zhdanov, Yu.A.Molodtsova, I.V.Razumovskaya, Structures of complexes of copper with macrocyclic organosiloxanolates ligands, Organomet. Chem. USSR. 4:672(1991).

5. O.I.Shchegolikhina, Yu.A.Molodtsova, Yu.A.Pozdnyakova, T.V.Strelkova, A.A.Zhdanov, Framework metallosiloxanes for synthesis stereoregular siloxanes cycles, Dokl. Akad. Nauk, 325:1186(1992), [Dokl. Chem. 325(1992)] (Engl. Transl.).

6. I.V.Blagodatskikh, O.I.Shchegolikhina, Yu.A.Pozdnyakova, Yu.A.Molodtsova, A.A.Zhdanov, Applicaton of size exclusion chromatography to the structural study of polyorganometallosiloxanes, Russ. Chem. Bull. 43:(1994), in press.

7. P.V.Wright and M.S.Beevers, Preparation of cyclic polysiloxanes, in: "Cyclic Polymers", J.A.Symlyen, ed., Elsevier Applied Science Publishers, N-Y (1986), p. 85-134.

8. A.A.Zhdanov, O.I.Shchegolikhina, Yu.A.Molodtsova, Peculiarities of the synthesis of cage-like metallosiloxanes, Russ. Chem. Bull. 42:917(1992).

9. K.A.Andrianov and A.A.Zhdanov, Synthesis of polyphenylalumosiloxanes. The reaction of the exchange interaction of sodium alkylsilantriols with aluminium chloride., Dokl. Akad. Nauk. SSSR 114:1005(1957) (in Russian).

10. M.Bartolin, A.Guyot, Preparaton de quelques polymetallophenylsiloxanes. Compt. Rend. C264:1694(1967).

11. I.L.Dubchak, V.E.Shklover, M.M.Levitskii, A.A.Zhdanov, Yu.T.Struchkov, Crystal structure of siloxanes and silazanes. XIX. Oktahydrate of sodium cis-1,3,5-triphenylcyclotrisiloxane-1,3,5-triolate, Zh. Strukt. khimii, 15:103(1980).

12. I.V.Blagodatskikh, Yu.A.Pozdnykova, O.I.Shchegolikhina, A.A.Zhdanov, Investigation of polyphenylironsiloxane structure, Russ. Chem. Bull. 1995, in press.

13. L.M.Khananashvilli and K.A.Andrianov. "Technology of Elementoorganic Monomers and Polymers". Khimia, Moscow (1983), 415 p. (in Russian).

14. A.Yu.D'yakonov, O.I.Shchegolikhina, A.D.Kolbanovskii, A.G.Knizhnik, A.L.Buchachenko, A.A.Zhdanov, M.M.Levitskii, Heteroorganic metal-containing paramagnetic and ferromagnetic polymers. 1. Polyferro- and polycobaltsiloxanes, structure and magnetic properties, Bull. Acad. Sci. USSR, Div. Chem. Sci. 39:2271(1990).

15. A.A.Zhdanov, A.L.Buchachenko, O.I.Shchegolikhina, R.A.Stukan, A.G.Knizhnik, M.M.Levitskii, Heteroorganic metal-containing paramagnetic and ferromagnetic polymers. 2. Investigation of thermal condensation of polymetalloorganosiloxanes, Bull. Acad. Sci. USSR, Div. Chem. Sci., 40:520(1991).

16. A.A.Zhdanov, A.L.Buchachenko, A.Yu.D'yakonov, O.I.Shchegolikhina, M.M.Levitskii. Heteroorganic metal-containing paramagnetic and ferromagnetic polymers. 3. Magnetic properties of the thermal condensation of polymetallic organosiloxanes, Bull. Acad. Sci. USSR, Div. Chem. Sci. 40:680(1991).

17. M.Tikadzumi, Physics of Ferromagnetism. Mir, Moscow, 1983, 245 p. (Russian translation).

POLYMER-SUPPORTED ZIEGLER-NATTA CATALYSTS
FOR THE POLYMERIZATION OF α-OLEFINS AND BUTADIENE

Ruicheng Ran* and Charles U. Pittman, Jr.

Department of Chemistry, University/Industry Chemical Research Center,
Mississippi State University, Mississippi State, MS 39762

INTRODUCTION

Over the past ten years, we have developed a series of polymer-supported Lewis acid catalysts for use in such synthetic reactions as esterification, ketone formation and etherification etc.[1-8] These catalysts exhibited very good catalytic activity. The catalysts were also used for cationic polymerization of styrene and α-methylstyrene[10-11]. A polymer-supported Ziegler-Natta catalyst, polystyrene-titanium chloride-diethylaluminum chloride (**PS-TiCl$_4$/Et$_2$AlCl**) for the polymerization of isoprene was further developed.[12] In this chapter two polymer-supported Ziegler-Natta catalysts, PS-TiCl$_4$/Et$_2$AlCl and poly(biphenylaminomethylstyrene)-titanium chloride diethylaluminum chloride (**PDPAS-TiCl$_4$/Et$_2$AlCl**) are reported. Their use in polymerization and copolymerization of ethylene, 1,3-butadiene, isoprene and isobutylene, is also described.

There have been many reports in the past few years about the highly active supported Ziegler-Natta catalysts for polymerization of olefins [13-29]. Most of these systems were held on inorganic supports. Examples include magnesium chloride or oxides of silica and aluminum. Magnesium chloride was shown to provide an increased number of active sites in these catalysts and its use enhanced propagation rate constants for olefin polymerization. On the other hand, electron donors (acting as Lewis bases) such as ethylbenzoate, p-cresol, 4-phenylphenol and methyl-p-toluate, have been used in classic Ziegler-Natta catalysts for olefin polymerization. Even though these inorganic supported catalysts have high activity, a major problem is that they introduce metal ions as impurities into the polyolefins they generate. These impurities can harm the physical properties and chemical stability of the polymer.

The polymer-supported Ziegler-Natta catalysts, PS-TiCl$_4$/Et$_2$AlCl and PDPAS-TiCl$_4$/Et$_2$AlCl, were prepared from polystyrene-titanium chloride (PS-TiCl$_4$) and poly(biphenylaminomethylstyrene)-titanium chloride (PDPAS-TiCl$_4$) based on our previous work[1,6] on polymer-supported Lewis acid catalysts. Then these catalysts were used for polymerization of gaseous olefins and exhibited reactivities as high as the classic catalysts. Furthermore, they have much better stability. Also, these polymer-supported catalysts avoid the introduction of inorganic impurities into the product polyolefins. This reduces the harmful chemical and physical properties associated with such impurities. In addition, it is interesting that these polymerizations are consistent with **living polymerizations**, similar to the polymerization of isoprene catalyzed by PS-TiCl$_4$/Et$_2$AlCl reported previously.[12]

EXPERIMENTAL

Materials

Polystyrene beads (2-4% divinylbenzene, DVB, 100-200 mesh, Bioproducts) were washed with 5% NaOH and water and then dried in vacuum. Chloromethylated styrene-divinylbenzene (2%) copolymer beads (100-200 mesh) were obtained from Bioproducts Company. Biphenylamine, TiCl$_4$, Et$_2$AlCl, ethylene, butadiene, isobutylene and isoprene were obtained from Aldrich Chemical Company and were used directly without further treatment. All solvents used in this research project were treated with CaH$_2$ and then distilled.

Preparation of Polystyrene-TiCl$_4$ Complex (PS-TiCl$_4$)

TiCl$_4$ (10 mL) was added to poly(styrene-divinylbenzene-2%) beads(20 g, 100-200 mesh) which had been pre-swollen with chloroform (180 mL) and stirred for 24 h at room temperature. The polymer beads were filtered and washed six times with chloroform (6x50 mL) and then twice with acetone (2x50 mL). The polymer beads were dried under vacuum for 72 h at 35 ^{0}C to give yellow catalyst beads. The beads of polystyrene-co-DVB-2% complexed with TiCl$_4$ were found to contain 1.34% Ti or 0.28 mmol Ti/g.

Preparation of Poly(4-diphenylaminomethylstyrene)-TiCl$_4$ Complex (PDPAS-TiCl$_4$)[6]

Chloromethylated styrene-divinylbenzene (2%) copolymer beads (30 g, 3.2 mmol Cl/g, 100-200 mesh) swollen in chloroform (150 mL) were added to a solution of diphenylamine (24.2 g, 0.144 mol) in chloroform (150 mL). After stirring for 6 h at 50 ^{0}C, pyridine (10 mL) was added to the reaction system. The reaction was stirred for an additional 4 h. The reaction mixture was cooled to room temperature, then filtered and washed with THF (6x50 mL) followed by water (300 mL) and acetone (200 mL). The polymer beads were dried in vacuum oven for 72 h at 40 ^{0}C. Elemental analysis of the polymer beads obtained gave 81.55% C and 4.02% N.

The polymer beads thus prepared (25 g) were swollen in chloroform (100 mL) and then stirred with TiCl$_4$ (10 mL) for 8 h at room temperature. The beads changed from yellow to emerald green. After cooling to 0-5 ^{0}C, the mixture was filtered and washed with chloroform (5x60 mL), and acetone (100 mL). The complexed beads were dried under vacuum for 72 h at 35 ^{0}C to give green complex catalyst beads. Elemental analysis of the complex found 24.9% Cl, and 8.81% Ti by weight, equivalent to 1.84 mmol TiCl$_4$/g catalyst.

Preparation of Polymer-Supported Zieger-Natta Catalyst PS-TiCl$_4$/Et$_2$AlCl[12]

Et$_2$AlCl (6 mL, 1.8 M in toluene) was added to the yellow beads of the PS-TiCl$_4$ complex (5 g) in benzene (50 mL) for 15 min to give the dark green complex:

polystyrene-TiCl$_4$/Et$_2$AlCl catalyst (PS-TiCl$_4$/Et$_2$AlCl). This catalyst was used directly without drying in the polymerization of an olefin. Elemental analysis of the catalyst PS-TiCl$_4$/Et$_2$AlCl found 3.92% by weight Al or 1.43 mmol Al/g complex catalyst. The mole ratio of Al/Ti in this polymer-supported catalyst was approximately 5:1.

Preparation of Polymer-Supported Zieger-Natta Catalyst PDPAS-TiCl$_4$/Et$_2$AlCl

Et$_2$AlCl (8 mL, 1.8 M in toluene) was added to the green PDPAS-TiCl$_4$ complex (5 g) in benzene (50 mL) for 15 min to give the active dark green complex, poly(4-diphenylaminomethylstyrene)-TiCl$_4$Et$_2$AlCl (PDPAS-TiCl$_4$/Et$_2$AlCl). This catalyst was used directly, without drying, as a catalyst in the polymerization of α-olefins. Elemental analysis of PDPAS-TiCl$_4$/Et$_2$AlCl found 9.2% by weight Al or 6.5 mmol Al/g complex catalyst. The mole ratio of Al/Ti in this catalyst was approximately 4.2:1.

Similarly, PDPAS-TiCl$_4$/Et$_2$AlCl catalysts with mole ratios of Al/Ti = 10, 15, 40 and 60, were obtained by using 10, 30, 50, 70 mL of Et$_2$AlCl, respectively.

Olefin Polymerizations Catalyzed by Polymer-supported Ziegler-Natta Catalysts at Atmospheric Pressure

The apparatus used for olefin polymerizations that were catalyzed by the polymer-supported Ziegler-Natta catalysts under atmospheric pressure is shown in Figure 1.

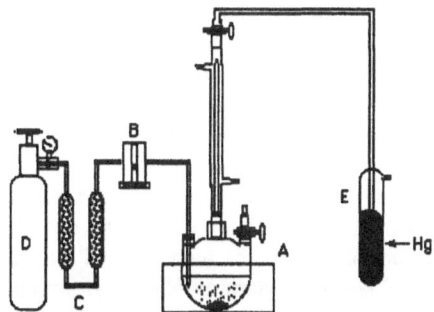

Figure 1. The apparatus for polymerization of gas olefin.   A. Reactor system with a magnetic stirrer; B. Gas flowmeter; C. Drying tubes; D. Cylinder of gaseous olefin; E. Airtight seal (bubbler) with mercury (7 cm high).

After adding the catalyst beads (0.5 g) and benzene (20 mL), the reactor system was degassed (10-20 mmHg). Et$_2$AlCl (1.5 mL, 1.8 M solution in toluene) was then added to the flask and stirred for 20 min. A gaseous olefin was passed into the system at a rate of 0.05 to 0.10 L/min for 60-70 min. The polymerization was carried out at 0 °C. Then the catalyst beads were separated from the polymerization system by filtration and the polymerization was quenched with methanol. The conversion and yield of the polymerization were determined by weighing the isolated polymer. Molecular weights were determined by the viscosity method in toluene at 30°C using an Ubbelohde viscometer. The microstructure of the polymer obtained was determined by proton NMR spectral analysis. These NMR spectra were obtained on a GE NMR, QE-300 spectrometer at 300 mHz.

RESULTS AND DISCUSSIONS

Preparation of the Polymer-Supported Ziegler-Natta Catalyst PS-TiCl$_4$/Et$_2$AlCl

It was reported[1] that the polystyrene-TiCl$_4$ complex (PS-TiCl$_4$) was a stable polymer-supported Lewis acid catalyst which showed good catalytic activity in ketal formation, esterification, acetylation and in the cationic polymerization of α-methylstyrene. The polymer-supported Ziegler-Natta catalyst PS-TiCl$_4$/Et$_2$AlCl was prepared by combining PS-TiCl$_4$ with Et$_2$AlCl[12] (see Scheme 1).

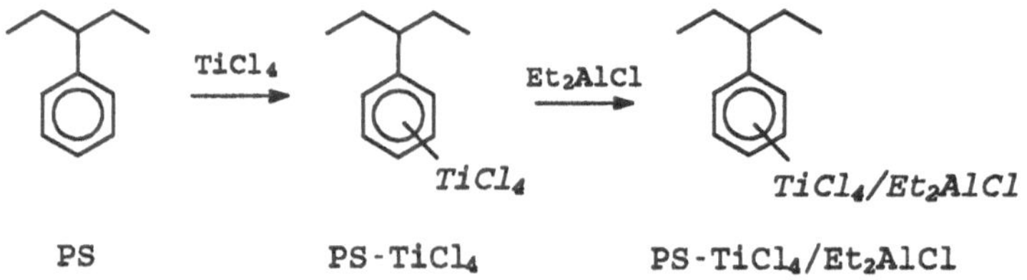

Scheme 1. Preparation of polymer-supported Ziegler-Natta catalyst PS-TiCl$_4$/Et$_2$AlCl

The mole ratio of Al/Ti in the polymer-supported catalysts was controlled by the amount of Et$_2$AlCl which was added to the PS-TiCl$_4$ before use. The PS-TiCl$_4$ is very stable. This catalyst did not lose its catalytic activety after a year of storage. Thus, the PS-TiCl$_4$/Et$_2$AlCl catalyst is easy to make and to store. However, the titanium in PS-TiCl$_4$ catalyst is easily washed out by solvents in these polymerization systems. After 60 hours of use, 85.7% of the titanium originally present was lost in these reactions.

Preparation of the Polymer-Supported Ziegler-Natta Catalyst PDPAS-TiCl$_4$/Et$_2$AlCl

To improve stability, a new polystyrene carrier containing diphenylamine groups was prepared and combined with TiCl$_4$ to form a very stable complex. Chloromethylated styrene-divinylbenzene (4%) copolymer beads were treated with diphenylamine to give a new polymer carrier, poly(4-N,N-diphenylamino-N-methyl)styrene (PDPAS)[6]. The new polymeric complex, PDPAS-TiCl$_4$, was then prepared by reaction with TiCl$_4$. This material was much more stable than PS-TiCl$_4$ due to the strong coordinate bond formed between the diphenylamine groups in PDPAS carrier and TiCl$_4$ (see Scheme 2). This stability has been demonstrated by UV-spectral analysis and by catalytic activity tests in organic reactions[6]. The titanium supported by PDPAS lossed only 8% of original amount after 60 h of use. A new polymer-supported Ziegler-Natta catalyst PDPAS-TiCl$_4$/Et$_2$AlCl was then readily prepared by treatment of PDPAS-TiCl$_4$ with Et$_2$AlCl (see Scheme 2).

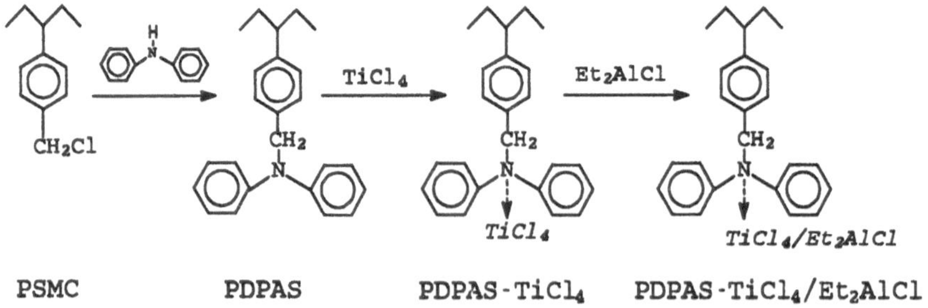

Scheme 2. Preparation of polymer-supported Ziegler-Natta catalyst PDPAS-TiCl$_4$Et$_2$AlCl

Elemental analysis of PDPAS-TiCl$_4$ found 1.84 mmol TiCl$_4$/g complex catalyst. The mole ratio of Al/Ti in the PDPAS-TiCl$_4$/Et$_2$AlCl catalysts was controlled by the amount of Et$_2$AlCl which was added to PDPAS-TiCl$_4$ before use. In order to study the dependence of catalytic activity and yield on mole ratio of Al/Ti, PDPAS-TiCl$_4$/Et$_2$AlCl catalysts with mole ratios of Al/Ti=4.2, 15, 40 and 60, respectively, were prepared by treating with the appropriate amount of Et$_2$AlCl.

Polymerization of 1,3-Butadiene Catalyzed by the Polymer-Supported Ziegler-Natta Catalysts

1. Catalytic Activity. The polymerization of 1,3-butadiene catalyzed by polymer-supported Ziegler-Natta catalysts, PS-TiCl$_4$/Et$_2$AlCl and PDPAS-TiCl$_4$/Et$_2$AlCl, and unsupported normal Ziegler-Natta catalyst, TiCl$_4$/Et$_2$AlCl, respectively, was carried out at 0 °C and atmospheric pressure. The polybutadiene (PBD) obtained in the polymerization consisted of 84-96% by weight of cis-1,4 units and <1% by weight 1,2-units according to NMR analysis. The relative activities of these catalysts for butadiene polymerization were evaluated by determining the percent conversion verses reaction time. The results are shown in Figure 2. It is evident that the butadiene polymerization rate induced by the polymer-supported Zieglar-Natta catalyst PS-TiCl$_4$/Et$_2$AlCl was much greater than that achieved using either PDPAS-TiCl$_4$/Et$_2$AlCl or unsupported TiCl$_4$/Et$_2$AlCl.

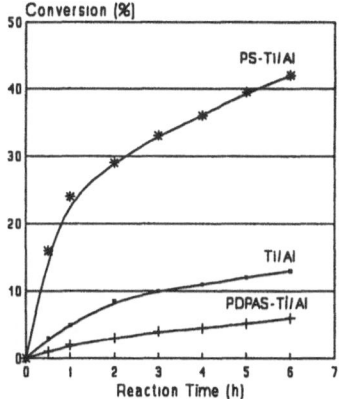

Figure 2. The relative catalytic activity of the polymer-supported and unsupported Ziegler-Natta Catalysts for polymerization of butadiene. Polymerization conditions: 0.5 g of catalyst beads, 20 mL of benzene, 1.2 mL of Et$_2$AlCl, butadiene (7 L), at 0°C.

2. Stability of the Catalysts. PS-TiCl$_4$/Et$_2$AlCl and PDPAS-TiCl$_4$/Et$_2$AlCl can be reused at least three times in the polymerization of butadiene, but significant activity loss occurs (see Figure 3). The spent catalyst can be regenerated by the addition of new Et$_2$AlCl. PDPAS-TiCl$_4$/Et$_2$AlCl is more stable to recycle and exhibits less loss of activity than PS-TiCl$_4$/Et$_2$AlCl in the polymerization of butadiene. The PS-TiCl$_4$/Et$_2$AlCl has a substantially higher catalytic activity during the first cycle.

Elemental analysis showed that there was a loss of titanium from the catalysts. After 60 h of catalyst use in these polymerization reactions, the amount of TiCl$_4$ is reduced to only 14.3% of that originally present for PS-TiCl$_4$/Et$_2$AlCl and to 92% for PDPAS-TiCl$_4$/Et$_2$AlCl.

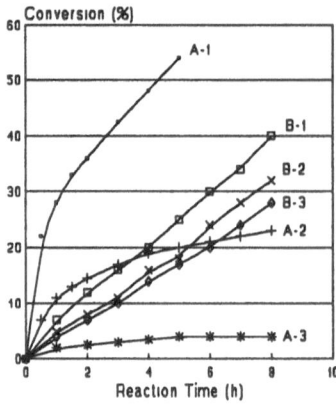

Figure 3. Dependence of conversion on reaction time for recycled catalysts in the polymerization of butadiene.   A-1, A-2, A-3, the first, second and third recycle of catalyst PS-TiCl$_4$/Et$_2$AlCl (Al/Ti=10). B-1, B-2, B-3, the first, second and third time use of catalyst PDPAS-TiCl$_4$/Et$_2$AlCl (Al/Ti=8.9).

The dependence of titanium content on reaction time for these supported catalysts in the polymerization of butadiene was shown in Figure 4. It is clear that the catalyst PDPAS-TiCl$_4$/Et$_2$AlCl is much more stable than PS-TiCl$_4$/Et$_2$AlCl. The TiCl$_4$ in PS-TiCl$_4$ was protected in the polystyrene beads by π-complex formation between TiCl$_4$ and the aromatic benzene-ring[1]. In PDPAS-TiCl$_4$, however, the TiCl$_4$ formed a coordinate bond with diphenylamino groups (ligand) in the PDPAS carrier beads (see Scheme 2)[6]. This stability differennce between the two polymer-supported catalysts has been comfirmed in catalytic organic synthesis reactions[1,6].

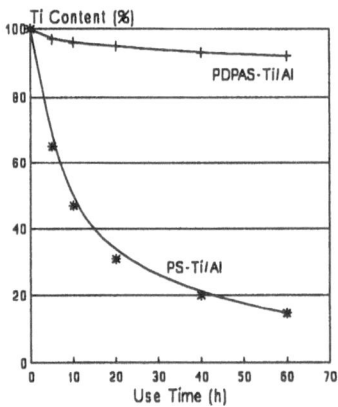

Figure 4. Loss of titanium content in the polymer-supported catalysts versus reaction time in the polymerization of butadiene

3. Yield and Molecular Weight in 1,3-Butadiene Polymerizations.   The polymerization of butadiene was carried out in the presence of different catalyst systems. The monomer conversion and polybutadiene (PBD) yield were obtained in varying polymerization conditions. Molecular weights of the polymers obtained were measured by viscosity method. The results are summarized in Table 1.

Table 1. The results of butadiene polymerization catalyzed by various catalyst systems*

Catalyst system	Time (h)	Conversion (%)	Yield (g PBD/g Ti)	$\bar{M}w$ x1000 of PBD
PS-TiCl$_4$	6	0		
PDPAS-TiCl$_4$	6	0		
Et$_2$AlCl	6	0		
TiCl$_4$/Et$_2$AlCl (Al/Ti=5)	6	13.2	158	2.86
	24	35.1	888	3.13
PS-TiCl$_4$/Et$_2$AlCl (Al/Ti=5)	6	41.2	1040	3.15
	24	52.4	1761	6.13
	70	60.7	2164	18.66
PDPAS-TiCl$_4$/Et$_2$AlCl	6	6.5	25	1.21
(Al/Ti=4.2)	24	18.2	69	4.42
	48	35.3	134	8.33
	70	50.1	192	12.36

* Polymerization conditions: in all reactions, butadiene (0.1 L/Min x 70 min=7 L); catalyst: 0.5 g of complex beads with Et$_2$AlCl (1.5 mL, 1.8 M in toluene); temperature 0 °C. Molecular weights of polybutadiene (PBD) were obtained by measuring the intrinsic viscosity with an Ubbelohde viscometer in toluene at 30 °C (k=33.9 x10^3 mL/g, α=0.688).

Table 1 illustrates that the rate and the yield in polymerizations catalyzed by PS-TiCl$_4$/Et$_2$AlCl are much higher than those catalyzed by PDPAS-TiCl$_4$/Et$_2$AlCl or by the unsupported catalyst TiCl$_4$/Et$_2$AlCl. The productivity of the polymer-supported catalyst PS-TiCl$_4$/Et$_2$AlCl (Al/Ti=5) was 1761 g PBD/g Ti for 24 h. Based on the loss of 85.7% of the titanium after 60 h of use, over 20 Kg of PBD would be obtained from each gram of supported Ti. When the Al/Ti mole ratio increased to 75, the catalytic yield reached 4.75 Kg PBD/g Ti for 6 h, and 51.6 Kg PBD would be produced from each gram supported Ti. The yield over PDPAS-TiCl$_4$/Et$_2$AlCl (Al/Ti=4.2) was 192 g PBD/g Ti for 70 h. Based on the loss of the original TiCl$_4$ of 8% in 60 h of use, over 2.4 Kg of PBD can be produced from each gram of supported Ti by this catalyst.

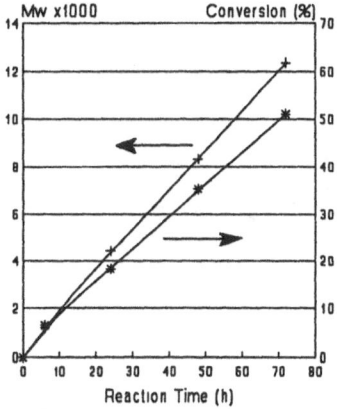

Figure 5. Dependence of monomer conversion and molecular weight of PBD in the polymerization of butadiene catalyzed by PDPAS-TiCl$_4$/Et$_2$AlCl

At Al/Ti=60, the yield of PDPAS-TiCl$_4$/Et$_2$AlCl catalyst could reach 1080 g PBD/g Ti (for 6 h, see Figure 6), and over 135 Kg PBD would be generated for each gram of polymer-supported titanium. Thus, the residual titanium, which remained in the polymer obtained, would be less than 7 ppm and it will not influence the properties of the polymer obtained.

The polymerization of butadiene doesn't occur in the presence of either $TiCl_4$ or Et_2AlCl alone (see Table 1). The molecular weight of the polybutadiene obtained in the polymerization catalyzed by PS-$TiCl_4$/Et_2AlCl was higher than that catalyzed by PDPAS-$TiCl_4$/Et_2AlCl or by the unsupported catalyst $TiCl_4$/Et_2AlCl.   It is   especially interesting that the butadiene conversion and the molecular weight obtained from the polymerization catalyzed by PDPAS-$TiCl_4$/Et_2AlCl showed a linear relationship versus reaction time (see Figure 5). This is consistent with a **living polymerization**, similar to the polymerization of isoprene catalyzed by PS-$TiCl_4$/Et_2AlCl reported previously.[12]

4. Influence of the Al/Ti Ratio on the Polymerization. The polymerization rate and yield depended on the concentration of monomer, amount of catalyst and mole ratio of Al/Ti in the catalyst. In the polymer-supported Ziegler-Natta catalysts, the amount of titanium complexed with polymer was a constant. The amount of aluminum, however, is easy to control by addition of Et_2AlCl just before the catalysts' use.   The polymerization of isoprene catalyzed by PS-$TiCl_4$/Et_2AlCl, followed the rate law, $Rp=k[M]^{0.30}[Ti]^{0.41}[Al]^{1.28}$.   Thus, the polymerization rate depended directly upon the concentration of aluminum in the catalyst system.[12]   Similarly, in the polymerization of butadiene catalyzed by the polymer-supported catalysts, PS-$TiCl_4$/Et_2AlCl and PDPAS-$TiCl_4$/Et_2AlCl, the yield also depended directly on the mole ratio of Al/Ti (see Figure 6). It is evident from Figure 6 that effect of the Al/Ti ratio in PS-$TiCl_4$/Et_2AlCl on the catalytic yield is more pronounced than in PDPAS-$TiCl_4$/Et_2AlCl. Thus, increasing Al/Ti in the supported catalysts is a simple and effective method to get a higher catalytic yield.

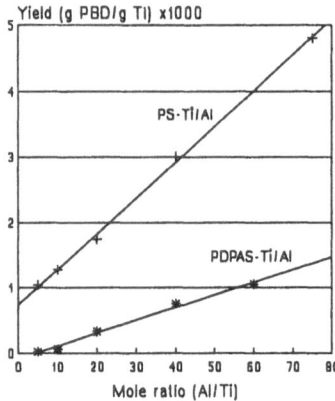

Figure 6. Dependence of polybutadiene yields at 6 hours on the mole ratio of Al/Ti in the polymer-supported catalysts.

Polymerization of Isobutylene Catalyzed by the Polymer-Supported Ziegler-Natta Catalysts

Polymers and copolymers of isobutylene are used as adhesives, sealants and insulating oils etc. They are commercially prepared by cationic polymerization. Herein, the polymerization of isobutylene, catalyzed by the polymer-supported Ziegler-Natta catalysts, was carried out at atmospheric pressure at 0 °C. In this polymerization system the concentration of isobutylene was higher than that of butadiene because isobutylene has a higher boiling point (-6.9 °C) and better solubility in benzene. The results of polymerizations catalyzed by different catalysts are summarized in Table 2.

It is evident from Table 2 that the polymer-supported catalysts had high catalytic activities for isobutylene polymerization.   In the polymerization catalyzed by PS-TiCl$_4$/Et$_2$AlCl the conversion of isobutylene was 98% in 6 h and the productivity of the catalyst was over 2.6 Kg PIB/g Ti in 6 h and over 3.7 Kg PIB/g Ti for 20 h. Based on the loss rate of the original supported TiCl$_4$, over 29.3 Kg PIB could be produced from each gram of supported titanium in PS-TiCl$_4$/Et$_2$AlCl (Al/Ti=5). Thus, PS-TiCl$_4$/Et$_2$AlCl is a highly active catalyst system for polymerization of isobutylene.

Table 2. Results of isobutylene polymerization catalyzed by various catalyst systems*

Catalyst system	Time (h)	Conversion (%)	Yield (g PIB/g Ti)	$\overline{M}w$ x1000 of PIB
PS-TiCl$_4$	22	0		
PDPAS-TiCl$_4$	22	0		
Et$_2$AlCl	22	59	(125)**	17.10
TiCl$_4$/Et$_2$AlCl (Al/Ti=5)	6	86	2717	2.17
	24	95	2310	7.27
PS-TiCl$_4$/Et$_2$AlCl (Al/Ti=5)	2	74	1935	10.26
	6	98	2658	21.73
	20	99	3702	40.10
PDPAS-TiCl$_4$/Et$_2$AlCl	2	14	56	3.86
(Al/Ti=4.2)	6	36	143	11.21
	10	55	219	18.25
	20	97	386	37.82

* Polymerization conditions: in all reactions, isobutylene (0.1 L/Min x 70 min=7 L); catalyst, 0.5 g of complex beads with Et$_2$AlCl (1.5 mL, 1.8 M in toluene); temperature 0 °C. Molecular weights of polyisobutylene (PIB) were obtained by measuring the intrinsic viscosity with an Ubbelohde viscometer in toluene at 30 °C (K=20 x10^3 mL/g, α=0.67).
** This yield was calculated based on the amount of aluminum in the catalyst.

A 36% conversion of isobutylene was achieved in 6 h (97% in 20 h) in the polymerization of isobutylene catalyzed by PDPAS-TiCl$_4$/Et$_2$AlCl (Al/Ti=4.2). The polymerization rate was much slower than that catalyzed by PS-TiCl$_4$/Et$_2$AlCl. However, the isoprene conversion increased linearly with reaction time using PDPAS-TiCl$_4$/Et$_2$AlCl (see Figure 7).   This result was similar to that observed in the polymerization of butadiene. The conversion of isobutylene reached 97% in 20 h which corresponds to 386 g PIB/g Ti.   Based on a loss of 8% of the original TiCl$_4$ in PDPAS in 60 h of use, over 4.8 Kg PIB would be produced from each gram of supported titanium in PDPAS-TiCl$_4$/Et$_2$AlCl (Al/Ti=4.2).   If the Al/Ti ratio was increased to 60, the extrapolated catalytic yield could be 2240 g PIB/g Ti in 6 h. Based on a loss of 8% of the original TiCl$_4$ in PDPAS in 60 h of use, over 280 x10^3 g PIB can be produced from a gram of PDPAS supported titanium (Al/Ti=60) before the catalyst was totally deactivated. In this case the residual titanium in the polyisobutylene obtained would be about 4 ppm.   This residue need not be removed to achieve a good quality polymer.

The polymerization of isobutylene occurs through either a cationic or a coordination mechanism.   Thus, a 59% conversion of isobutylene was obtained in 22 h using Et$_2$AlCl alone as a catalyst.   This is a much lower rate than that obtained with the polymer-supported Ziegler-Natta catalysts. But the TiCl$_4$ and the polymer-supported TiCl$_4$ Lewis acid catalysts (no Et$_2$AlCl present) did not work in this polymerization.

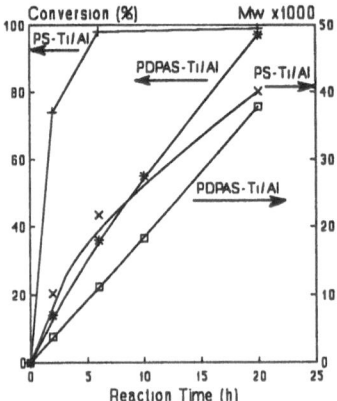

Figure 7. Dependence of isobutylene conversion and molecular weight on reaction time in various catalyst systems

It is evident from Table 2 and Figure 6 that the molecular weights of polyisobutylenes obtained over the polymer-supported catalysts, PS-TiCl$_4$/Et$_2$AlCl and PDPAS-TiCl$_4$/Et$_2$AlCl, were much higher than that obtained over the unsupported catalyst TiCl$_4$/Et$_2$AlCl. Furthermore, the molecular weights obtained over PS-TiCl$_4$/Et$_2$AlCl and PDPAS-TiCl$_4$/Et$_2$AlCl increased linearly with reaction time.   This is especially pronounced for PDPAS-TiCl$_4$/Et$_2$AlCl.   Therefore, a **living-like polymerization** process was occurring similar to polymerizations of isoprene catalyzed by PS-TiCl$_4$/Et$_2$AlCl which were reported previously.[12] Apparently, metal hydrid elimination is essentially eliminated as a route for stopping chain growth and promoting chain transfer.

Polymerization of Ethylene Catalyzed by Polymer-Supported Ziegler-Natta Catalysts

Due to the poor solubility of ethylene in benzene, the concentration of ethylene in this polymerization system was low at atmospheric pressure.   So ethylene polymerizations in benzene over the polymer-supported catalysts were difficult to carry out.   Furthermore, the PE produced covered the catalyst beads which probably retarded the rate.   The ethylene conversion was about 17.8% in 24 h using PS-TiCl$_4$/Et$_2$AlCl, and about 10.3% in 24 h employing PDPAS-TiCl$_4$/Et$_2$AlCl.   The catalytic yield was about 232 g PE/g Ti for PS-TiCl$_4$/Et$_2$AlCl (Al/Ti=5), and about 20 g PE/g Ti for PDPAS-TiCl$_4$/Et$_2$AlCl (Al/Ti=4.2). Thus, ethylene can be polymerized catalyzed by the polymer-supported Ziegler-Natta catalysts.   However,   it is inconvenient to separate the PE produced from the catalyst beads.

Copolymerization of Isobutylene with Isoprene Using Polymer-Supported Ziegler-Natta Catalysts

Butyl rubber is a copolymer of isobutylene with isoprene (1.5-4.5% mole%) which is commercially used for inner tubes, engine mounts and springs, chemical tank linings, protective clothing, hoses, gaskets, electrical insulation, etc. The butyl rubber can be prepared by cationic polymerization using Lewis acid catalyst systems at temperatures of -40 °C or lower. Recently, the copolymerization of isobutylene and isoprene was carried out in the presence of polymer-supported Ziegler-Natta catalysts at atmospheric pressure and 0 °C. Isobutylene gas (0.1 L/min. flow rate) was passed into a solution of isoprene for 70 min. The mole ratio of isobutylene to isoprene was 1:1 in the resulting solution. Various copolymerizations are summarized in Table 3. The PS-TiCl$_4$/Et$_2$AlCl catalyst system had exhibited the highest catalytic activity in butadiene polymerizations (Table 1). This catalyst system was also the most active one for isobutylene/isoprene copolymerizations (Table 3). The catalytic yield using PS-TiCl$_4$/Et$_2$AlCl (Al/Ti=5, for 24 h) reached 3.3 Kg PII/g Ti (PII: copolymer of isobutylene and isoprene). Based on the rate of titanium loss in this supported catalyst, 9.8 Kg PII can be produced from each gram of Ti. When the Al/Ti ratio increased to 60, the conversion of monomer was 97.2% in 2 h, and the catalytic yield was 3.4 Kg PII/g Ti. Based on the rate of loss of the supported titanium, the catalytic yield of PS-TiCl$_4$/Et$_2$AlCl (Al/Ti=60) could reach 119 Kg PII/g Ti. The catalytic yield of PDPAS-TiCl$_4$Et$_2$AlCl (Al/Ti=4.2, for 24 h) was 772 g PII/g Ti. Based on the rate of loss of the supported Ti (8% loss for 60 h) its catalytic yield could be 32.1 Kg PII/g Ti.

Table 3. Copolymerization of isobutylene and isoprene catalyzed by various catalyst systems*

Catalyst system	Time (h)	Conversion (%)	Yield (g PII/g Ti)	$\overline{M}w$ x1000 of PII
PS-TiCl$_4$	24	0		
PDPAS-TiCl$_4$	24	0		
Et$_2$AlCl	24	70	(274)**	8.29
TiCl$_4$/Et$_2$AlCl (Al/Ti=5)	24	70.8	2380	12.48
PS-TiCl$_4$/Et$_2$AlCl (Al/Ti=5)	2	32.1	782	7.45
	8	59.4	1448	11.87
	24	96.7	3367	13.18
PDPAS-TiCl$_4$/Et$_2$AlCl (Al/Ti=4.2)	2	10.2	90	1.22
	8	34.3	303	4.18
	24	87.4	772	11.67

* Polymerization conditions: in all reactions, isobutylene (0.1 L/Min x 70 min=7 L) was added into isoprene (21.3 g); catalyst, 0.5 g of complex beads with Et$_2$AlCl (1.5 mL, 1.8 M in toluene); temperature 0 °C. Molecular weights of polyisobutylene (PII) were obtained by employing [η]=KM^α where [η] was obtained using an Ubbelohde viscometer in toluene at 30 °C (K=21.4 x10^3 mL/g, α=0.678).
** This yield was calculated based on the amount of aluminum in the catalyst.

It is interesting that in isobutylene/isoprene copolymerizations catalyzed by PDPAS-TiCl$_4$/Et$_2$AlCl, both conversions and molecular weight increased linearly with reaction time (see Figure 8). This is consistent with a **living polymerization** similar to that found for butadiene and isobutylene.

The microstructure of the isobutylene/isoprene copolymer obtained was random. The isoprene units were almost entirely those with the *cis*-1,4 structure (^{1}H NMR). At low conversion the isobutylene/isoprene mole ratio was higher than that obtained at higher conversions. The rate of isobutylene incorporation into the polymer was faster than that of isoprene in the copolymerizations catalyzed by both of the polymer-supported Ziegler-Natta catalyst systems, PS-TiCl$_4$/Et$_2$AlCl and PDPAS-TiCl$_4$/Et$_2$AlCl.

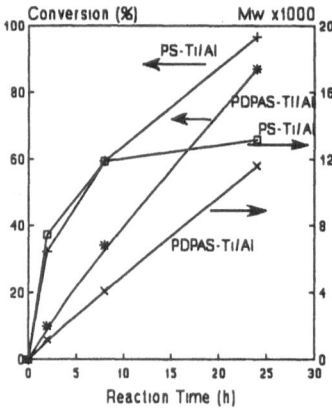

Figure 8. Dependence of conversion and molecular weight on reaction time in copolymerization of isoprene with isobutylene catalyzed by various catalytic systems

CONCLUSIONS

The polymer-supported Ziegler-Natta catalysts, PS-TiCl$_4$/Et$_2$AlCl and PDPAS-TiCl$_4$/Et$_2$AlCl, were prepared by reacting the polymer-supported Lewis acid catalysts, polystyrene-TiCl$_4$, or polybiphenylaminomethylstyrene-TiCl$_4$, with Et$_2$AlCl. These catalysts showed somewhat greater catalytic activity than the unsupported Ziegler-Natta catalyst for the polymerization of 1,3-butadiene, isobutylene, ethylene and for the copolymerization of isobutylene with isoprene. They also can be stored more easily, reused and regenerated. The overall catalytic yield of α-olefins was ca. 20-40 Kg polymer/g Ti for PS-TiCl$_4$/Et$_2$AlCl (Al/Ti=5), and 2-5 Kg polymer/g Ti for PDPAS-TiCl$_4$/Et$_2$AlCl (Al/Ti=4.2), respectively. The catalytic yield increased sharply as the mole ratio of Al/Ti was increased. Based on the rate at which titanium is lost from the catalysts, over 200 Kg polymer can be produced from each gram of the supported Ti. Therefore, the residual Ti in the polymer obtained would be as low as 4 ppm. The catalyst PS-TiCl$_4$/Et$_2$AlCl had the higher catalytic activity while PDPAS-TiCl$_4$/Et$_2$AlCl had the greater stability. Living polymerizations appeared to be occuring with all three monomers when the PDPAS-TiCl$_4$/Et$_2$AlCl catalyst was used.

ACKNOWLEDGEMENTS

Some of this work was done at Bowling Green State University, Ohio. The authors gratefully acknowledge Dr. Douglas C. Neckers for his support and guidance.

REFERENCES AND NOTES

1. (a). R. Ran, S. Jiang, and J. Shen, *Chin. J. Appl. Chem.*, **2**: 29 (1985); *Chem. Abstr.* **103**: 27939v (1985);

(b). *J. Macromol. Sci.-Chem.*, A24(6): 669-679 (1987); *Chem. Abstr.* **105**: 60145x (1987).

2. R. Ran, W. Pei, X. Jia, J. Shen, and S. Jiang, *Chem. J. Chin. Univ.*, **7**(7): 645-50 (1986); *Chem. Abstr.*, **106**(2): 6086m (1987).

3. (a). R. Ran, W. Pei, X. Jia, J. Shen, and S. Jiang, *Chem. J. Chin. Univ.*, **7**(7): 645-50 (1986); *Chem. Abstr.*, **106**(2): 6086m (1987);

   (b). *Polymer Commun.*, **5**: 379-383 (1986); *Chem. Abstr.*, **108**(3): 21080q(1988);

   (c). *Polym. Commun.*, **6**: 453-457 (1986); *Chem. Abstr.*, **108**(3): 21082s (1988);

   (d). *Sci. Bull.*, **32**(6): 388-394 (1987); *Chem. Abstr.*, **108**(5): 36846t(1988).

4. (a). R. Ran, X. Jia, W. Pei, and S. Jiang, *Acta Sci. Nat. Univ. Pekin.*, **6**: 29-35 (1986); *Chem. Abstr.*, **108**(3): 21088y (1988);

   (b). *J. Mol. Catal.*, **2**(2): 112-118 (1988); *Chem. Abstr.*, **111** (1): 7011y (1989).

5. (a). R. Ran, W. Pei, X. Jia, and X. Wu, *J. Org. Chem.*, **4**: 286-272 (1987); *Chem. Abstr.*, **108**(17): 150013u (1988);

   (b). *Acta Polym. Sinica*, **6**: 67-70 (1988); *Chem. Abstr.*, **110**(1): 7170d (1989).

6. R. Ran and J. Shen, *J. Macromol. Sci. Chem.*, A25(8): 923-933 (1988); *Chem. Abstr.*, **109**: 189478v (1988).

7. R. Ran and G. Mao, *J. Macromol. Sci. Chem.*, A27(2): 125-136 (1990); *Chem. Abstr.*, **113**(2): 6953m (1990).

8. R. Ran and D. Fu, *J. Macromol. Sci. Chem.*, A27(5): 625-636 (1990).

9. R. Ran, D. Fu, J. Shen, and Q. Wang, *J. Polym. Sci. Part A: Polym Chem.*, **31**: 2915-2921 (1993).

10. R. Ran, X. Jia, and S. Jiang, *Petrochem. Tech.*, **17**(1): 15-20 (1988); *Chem. Abstr.*, **108**(20): 168034m (1988).

11. R. Ran, X. Jia, M. Li, and S. Jiang, *J. Macromol. Sci. Chem.*, A25(8): 907-22 (1988); *Chem. Abstr.*, **109**: 150080j (1988).

12. R. Ran, *J. Polym. Sci. Part A: Polym. Chem.*, **31**: 1561-1569 (1993).

13. R. Fierri, and J. C. W. Chien, *J. Polym. Sci. Part A: Polym. Chem.*, **32**: 661-673 (1994).

14. S. Lee, K. W. S. Brian, M. P. Ripplinger, J. J. Wooster, et al, *US Patent 5231151 A* 930727.

15. K. E. Mitchell, D. C. Miller, D. W. Godbehere, and G. R. Hawley, *US Patent 5,235,011 A* 930831.

16. D. Hara, M. Sato, and M. Mori, *Eur. Patent EP 530814 A1* 930310.

17. S. I. Woo, and I. I. Kim, *US patent 5192729 A* 930309.

18. A. Sano, K. Suzuki, et al, *Eur. Patent EP 507504 A2* 921007.

19. J. C. A. Bailly, P. Behue, *Eur. Patent EP 437080 A1* 910717.

20. D. E. Gessell, D. P. Hosman, *US Patent 4945142 A* 900731.

21. R. Quijada, A. M. R. Wanderley, *Stud. Surf. Sci. Catal.*, **25**: 419-29 (1986).

22. Y. Li and D. Jun, *J. Macromol. Sci. -Chem.*, A24: 227 (1987).

23. G. M. Chemenko, E. I. Tinyakov, Ts. V. Kakuliya, L. M. Khananashvili, Yu. V. Novikov, and M. E. Volipin, *Vysokomol. Soedin. Ser. B*, **25**: 919 (1983).

24. S. A. Bedell, W. M. Coleman, and W. R. Howell, Jr., *U. S. Pat.*, **4,623,707** (1980).

25. J. Collomb, D. C. Duran, L. Havas, and F. R. M. Morerol, *Eur. Pat. Appl.*, *EP 211,624* (1987).

26. Dow Chemcal Co., *Neth. Appl. NL 85 02,580* (1987).

27. T. Yano, T. Inoue, S. Ikai, M. Shimizu, Y. Kai, and M. Tamura, *J. Polym. Sci. Polym. Chem. Ed.*, **26**: 457-467, 477-490 (1988).

28. R. Spitz, L. Duvanel, and A. Guyot, *Makromol. Chem.*, **189**: 549 (1988).

29. J. C. W. Chien and Y. L. Hu, *J. Polym. Sci. Part A: Polym. Chem.*, **25**: 2847, 2881 (1987).

PREARRANGED POLY–4–VINYLPYRIDINE NICKEL COMPLEXES AS CATALYSTS FOR THE HYDROGENATION OF ALLYL ALCOHOL

Aiaz A. Efendiev,[1] Jumshud J. Orujev,[2]
Elmar B. Amanov,[2] and Yusif M. Sultanov.[2]

Azerbaijan Academy of Sciences
[1]Institute of Polymer Materials,
Sumgait, 373204
[2]Institute of Theoretical Problems
of Chemical Technology,
Baku, 370143
Azerbaijan Republic

INTRODUCTION

In the early fifties a new type of catalysts – complexes of transition metals began to develop intensively. These were mainly soluble homogeneous complexes which had great advantages compared to traditional heterogeneous catalysts[1].

First, all molecules of the soluble complexes are easily available to the substrate, hence high activity is exhibited by homogeneous complex catalysts. Second, active centres of such catalysts are uniform, hence high selectivity is given by catalytic processes. Finally, the soluble complexes can be easily studied and characterised using conventional physical–chemical methods, including spectrometric ones and this permits obtaining adequate information about active centre structure and therefore about the mechanism of catalytic processes.

After 15–20 years of intensive study of such complexes great success has been achieved and hope appeared that they would dominate in the industrial catalytic processes. As we know, this hope has not come up to our expectations but the role of homogeneous transition metal complexes in industrial catalytic processes is increasing rapidly, especially for α–olefin polymerization. In spite of obvious advantages – high activity and selectivity and mild conditions of the reactions the homogeneous complexes have a number of shortcomings which restrict their application on an industrial scale. One of the main problems is separation of homogeneous complexes from the reaction media. In most cases it is very difficult and expensive and in some cases it is even practically impossible. Solutions of homogeneous complexes are often corrosive and in many cases they are not stable enough. From this point of view traditional heterogeneous catalysts seem to have more practical application. They are usually stable and can be easily separated from the reaction media. On the other hand they are not active and selective enough because the active centres are not uniform and not easily available for the substrate. It is much more difficult to study them using standard methods compared with homogeneous complexes. That is why the interest in supported complexes has largely grown in the last two decades.[2]

Such catalysts can combine advantages of conventional heterogeneous catalysts, such as high stability and easy separation from the reaction media with advantages of homogeneous metal complexes, such as high activity and selectivity and possibility to obtain more exact information about the structure of active centres. Use of complexing polymers as solid supports opens new possibilities for varying surrounding ligands and regulation of catalytic properties of the complexes.[3-5]

Earlier we developed a preparation of selective complexing polymer sorbents based on the use of "memory" by the polymer composition. This consists of conformational prearrangement of noncrosslinked complexing polymers favourable for ion uptake. This is followed by fixation of these optimum conformations for the ion uptake by means of intermolecular crosslinking.[6-10] It has been shown that crosslinked macromolecules might "remember" conformations advantageous for complexing metal and this leads to an essential improvement of basic sorption characteristics of the sorbent. It has been also shown that prearranged complexes are efficient catalysts for various chemical reactions due to formation of more uniform structure of the complexes in comparison with nonprearranged polymers.[11,12]

An attempt has been made to further increase the activity and selectivity of metal polymer complex catalysts by means of formation and subsequent fixation of the structure of active centres specially prearranged for hydrocarbon substrate.[13,14]

The principle consists of interacting noncrosslinked metal polymer complexes with hydrocarbon substrates or intermediate products of chemical transformations in solution, i.e. under conditions when macromolecules are still mobile enough. The second step is fixation of the structure of the active centres by intermolecular crosslinking. The third step is removal of the template from the crosslinked system. It should be done in such a way so as to prevent catastrophic destruction of the catalyst structure. Crosslinked macromolecules might "keep in mind" conformations which are advantageous for the template substrate and this results in significant increase in activity and selectivity of metal polymer complex catalysts.

In this paper prearranged complexes used as catalysts are described.

RESULTS AND DISCUSSION

Poly–4–vinylpyridine (PVP) was prepared by polymerization of 4–vinylpyridine ("Merck"). Conventional 4–vinylpyridine was purified by double distillation in vacuum (b.p. 46–47°C/10^{-4} torr). Polymerization was carried out at 60°C. A 30% mass solution of vinylpyridine in toluene in the presence of 1% mass cumene hydroperoxide was evacuated in sealed test tube at 10^{-4}torr. The polymerization reaction was substantially complete in 8 hours. The polymer precipitated from toluene solution in the form of white flakes, non–soluble in hydrocarbons and water and soluble in methanol and ethanol.

$$
CH=CH_2 \qquad \longrightarrow \qquad \left[CH-CH_2 \right]_n
$$

The resulting polymer was dissolved in ethanol and precipitated dropwise in excess of diethyl ether. After two reprecipitations the polymer was dried in a vacuum desiccator at slightly elevated temperature.

Intrinsic viscosity of PVP determined in a capillary viscosimeter at 30°C using 0,1 M LiCl ethanol solution as a solvent was 0.95.

An ethanol solution of PVP (1 mol) was mixed with an ethanol solution of $NiCl_2$ (0.5 mol) to give PVP nickel complexes. Freeze drying was carried out by rapidly freezing the mixture in liquid nitrogen followed by vacuum treatment at 10^{-4} torr and −30°C.

To obtain PVP nickel complexes prearranged for allyl and propyl alcohols ethanol solutions of each complex were mixed with allyl and propyl alcohol respectively and then freeze dried applying a procedure similar to that used for nonprearranged complexes. Dry nonprearranged and prearranged polymers were crosslinked using the following procedure. The complexes were mixed with N,N'–methylenediacrylamide and the mixtures were ground in a vibrating ball mill with a ball diameter of 0.8 cm. Tablets with a thickness of 0.25 cm and a diameter 0.8 cm were prepared from the mixture by pressing at 10^{-4} torr and heating in sealed test tubes at 150°C for 5 hours. The crosslinked tablets were ground again and particles of 150–200 mesh were used as catalysts.

In Figures 1 and 2 preparation schemes are given for conventional nonprearranged metal polymer complexes and complexes with active centres prearranged for catalysed substrate.

The conventional method for preparing is shown in Figure 1. Macromolecules containing active centres are crosslinked and after that used as catalysts. Such catalysts do not exhibit high selectivity with respect to desired substrate and, as follows from the scheme, different substrates might be sorbed on active centres with small degrees of preference. Noncrosslinked macromolecules containing active centres react with substrate in solution and these systems are prearranged to a minimum free energy. The catalysed substrate is built–in (Figure 2). In other words, active centres

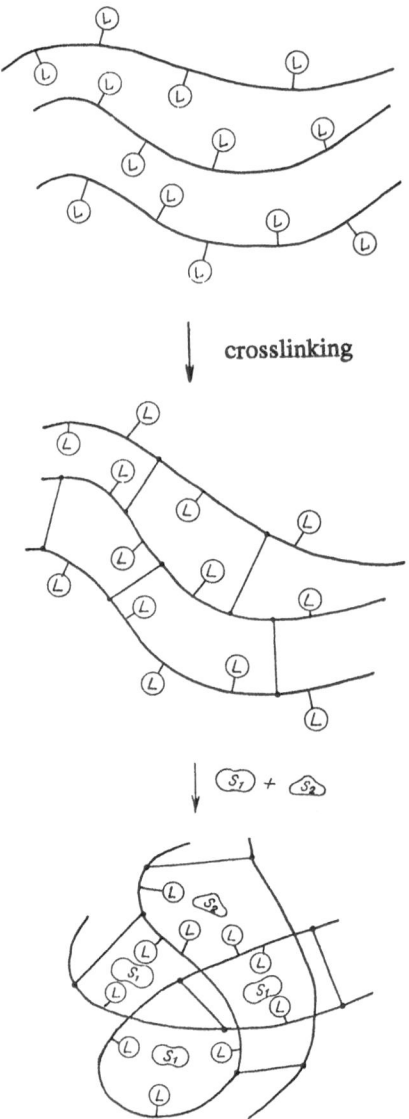

Figure 1. Scheme of preparation of nonprearranged polymer comlexes

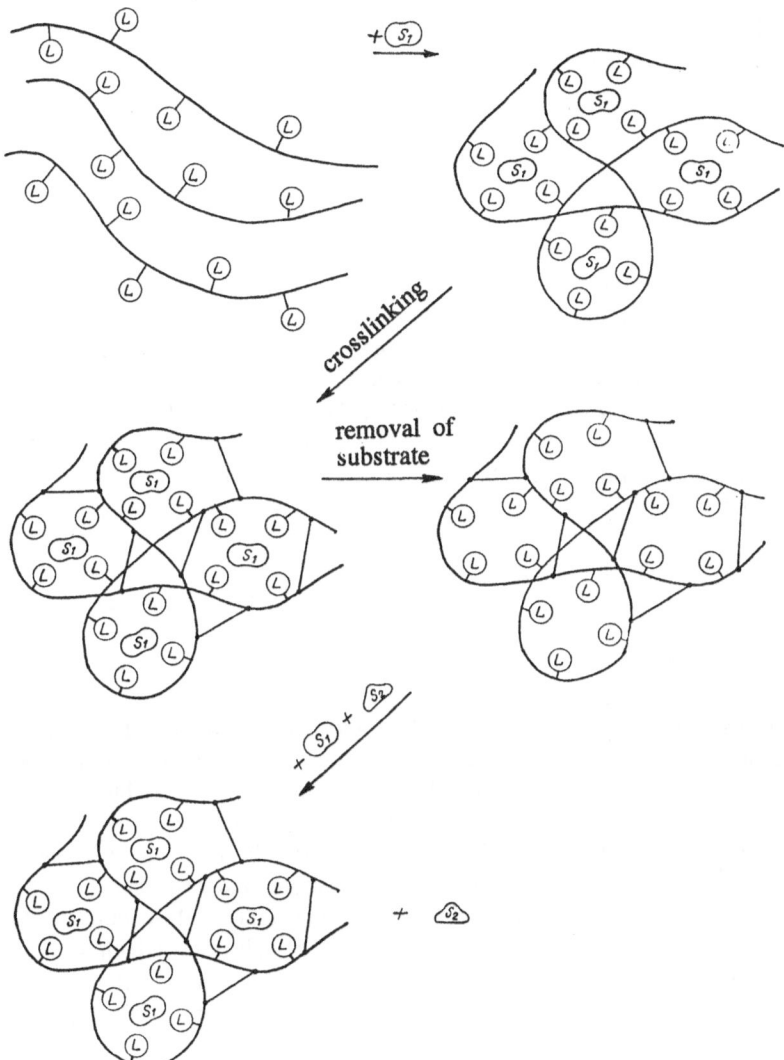

Figure 2. Scheme of preparation of prearranged polymer complexes

are formed where the built-in substrate has been already taken into account as the element of the structure. Such optimum active centre structures for this substrate are fixed by means of intermolecular crosslinking. These remain after removal of the substrate from the crosslinked system. After removal of the substrate from crosslinked polymer the system "remembers" the state when macromolecules were bound with substrate molecules as more preferable for the catalyst crosslinked in the presence of these molecules. This, in turn, decreases activation energy of substrate sorption onto active centres and improves the catalytic properties of metal polymer complexes.

Hydrogenation was carried out in a gasometric unit at $25^{\circ}C$ using gaseous hydrogen which was bubbled through the reaction medium at atmospheric pressure.

$$CH_2 = CH - CH_2OH + H_2 \rightarrow CH_3 - CH_2 - CH_2OH$$

Kinetic curves for the hydrogenation of allyl alcohol to propyl alcohol in the presence of both nonprearranged and prearranged PVP with different degrees of crosslinking are given in Figures 3–5.

It can be seen from the Figures that the rate of hydrogenation in the presence of complexes, prearranged for allyl alcohol, is much higher in comparison with nonprearranged complexes. On the contrary, the rate of reaction in the presence of complexes prearranged for propyl alcohol is lower than that in the presence of nonprearranged complexes due to blocking of active centres from complexation with allyl alcohol.

The reaction rate increases sharply with rise in the degree of crosslinking from 5 to 15% using complexes prearranged for allyl alcohol. Further increase in the crosslinking degree from 15 to 25% leads to a slight decrease of reaction rate. Presumably, an increase of crosslinking leads to a more rigid fixation of active centres structure. Further increases in crosslinking together with fixation hinders diffusion of substrate to active centres.

In case of complexes prearranged for the reaction product, i.e. propyl alcohol the reaction rate monotonously decreases with the increase of the crosslinking degree. Weakly crosslinked complexes exhibit hydrogenation rates comparable with that in the presence of nonprearranged complexes. Increases in crosslinking of the complexes, prearranged for propyl alcohol, tends to more rigidly block active centres from complex formation with allyl alcohol. Thus the reaction rate decreases.

During complex formation between substrate and catalyst, bonds both in the substrate and in the catalyst become more rigid. The substrates which are bound less strong are less rigid and therefore less reactive compared to those with stronger bonds.

Specificity might influence activation entropy of transition complex because the conformation of the substrate–catalyst complex upon sorption is close to conformation of catalyst within the activated complex. This possibility is stipulated by rigid valence angles and bonds and also restriction of rotation.

In complexes with "good" substrates, i.e. those strongly bound and easily reacted the conformation is fixed very close to that in transition state. The situation is different in the case of "bad" substrates. Rotation around the critical bond is free and favourable conformations occur less often. As a result of metal polymer complexes prearrangement, substrate stereocorrespondence like that in enzymic reaction takes place.

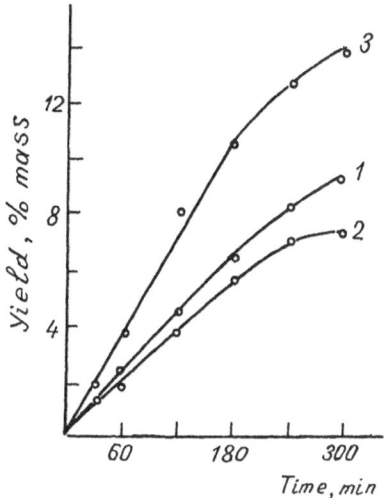

Figure 3. Kinetic curves of hydrogenation of allyl alcohol into propyl alcohol in the presence of nonprearranged (1) and prearranged for propyl alcohol (2) and allyl alcohol (3) complexes. Crosslinking degree - 5%.

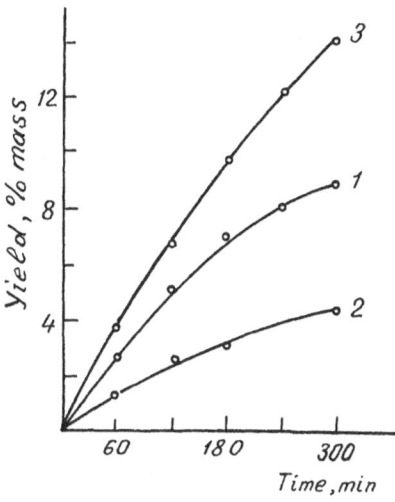

Figure 4. Kinetic curves of hydrogenation of allyl alcohol into propyl alcohol in the presence of nonprearranged (1) and prearranged for propyl alcohol (2) and allyl alcohol (3) complexes. Crosslinking degree - 15%.

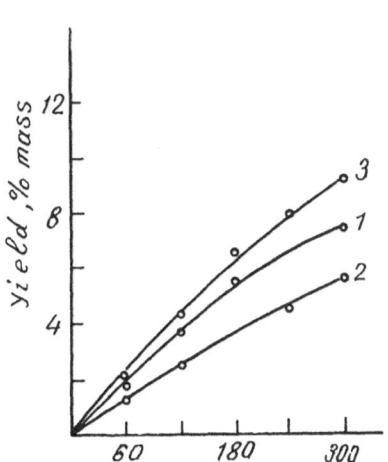

Figure 5. Kinetic curves of hydrogenation of allyl alcohol into propyl alcohol in the presence of nonprearranged (1) and prearranged for propyl alcohol (2) and allyl alcohol (3) complexes. Crosslinking degree - 25%.

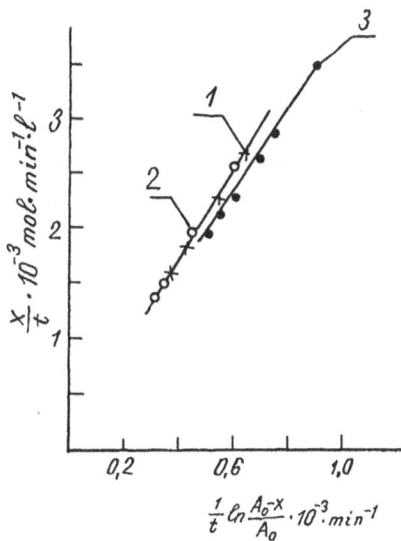

Figure 6. Kinetic curves of hydrogenation of allyl alcohol into propyl alcohol in the presence of nonprearranged (1) and prearranged for propyl alcohol (2) and allyl alcohol (3) complexes in terms of Henri equation. Crosslinking degree - 15%.

An attempt was made to determine Michaelis constants for metal polymer complex catalysts based on the simple mechanism of enzymic reactions.[15]

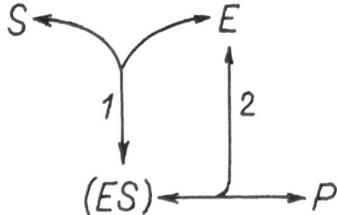

where E – enzyme; S – substrate; P – product; ES – enzyme–substrate complex.

$$S + E \underset{k_{-1}}{\overset{k_{+1}}{\rightleftharpoons}} ES \xrightarrow{k_2} E + \sum P$$

$$K_m = \frac{k_{-1} + k_2}{k_{+1}}$$

Kinetic data from the hydrogenation of allyl alcohol in the presence of nonprearranged and prearranged complexes were treated in terms of Henri equation.[16]

Results of treating this kinetic data using Henri equation are given in Figure 6. It can be seen from the Figure that the Michaelis constant for the catalyst prearranged for allyl alcohol is smaller.

However, for general case it is impossible, using K_m, to find both individual elementary constants k_{-1} and k_{+1} and their ratio – substrate constant K_S. For determination of k_{+1} the kinetics of the nonstationary reaction at the initial stages should be studied. This is technically very difficult for very rapid catalytic reactions and is often restricted by diffusion. That is why K_m is usually regarded as some efficient value being sum of elementary constants. Obviously, at $k_2 \ll k_{-1}$ the stationary concentration of complex ES only slightly differs from that at equilibrium. Thus:

$$K_m = \frac{k_{-1} + k_2}{k_{+1}} \approx \frac{k_{-1}}{k_{+1}} = K_S$$

For numerous enzymic reactions separate determination of K_m and K_S demonstrated that these values do not differ significantly. Thus, smaller value of K_m which is about the same as K_S demonstrates once more that allyl alcohol is "good" substrate for catalysts with active centres prearranged for allyl alcohol. It is not only bound stronger with catalyst but also reacts easily. Thus, the rate of hydrogenation reaction using this catalyst is higher.

The data in Figure 6 for catalysts nonprearranged and prearranged for propyl alcohol lie along straight line with a higher slope higher than that for the catalyst prearranged for allyl alcohol. This demonstrates that allyl alcohol is equally "bad" for both catalysts.

This line passes through the origin of the coordinates in Figure 6, whereas in case of the catalyst prearranged for allyl alcohol this line intersects the ordinate axis in a point of 0.3 mol $\cdot$ min$^-$1 $\cdot$ l^{-1} which corresponds to the theoretical ultimate reaction rate at infinite concentration of the substrate.

This fact confirms that the reaction forming the intermediate catalyst substrate complex takes place first and after that the hydrogenation reaction proceeds when using the catalyst imprinted by allyl alcohol.

CONCLUSIONS

From these results it can be deduced that special prearrangement of active centres of poly–4–vinylpyridine nickel complexes for substrate, i.e. allyl alcohol, leads to higher rates of hydrogenation compared with nonprearranged complexes. In case of complexes which had been prearranged for the reaction product, i.e. propyl alcohol, the reaction rate decreases because complexation with allyl alcohol of the active centres was blocked.

ACKNOWLEDGEMENT

This work was supported by International Science Foundation (Grant No RXM000).

The authors wish to thank Professor V.A.Kabanov of Moscow State University for valuable advices.

REFERENCES

1. K.Masters. "Homogeneous Catalysis by Transition Metals," Mir, Moscow (1983).
2. Yu.I.Yermakov, V.A.Zakharov and B.I.Kusnetsov. "Supported Complexes in Catalysis," Nauka, Moscow (1980).
3. "Speciality of Polymers." N.Ise and I.Tabusi, eds., Mir, Moscow (1983).
4. "Polymer–Supported Reactions in Organic Synthesis." P.Hodge and D.Sherrington, eds., Mir, Moscow (1983).
5. "Polymeric Reagents and Catalysts." Warren T.Ford, ed., American Chemical Society, Washington DC
6. V.A.Kabanov, A.A.Efendiev and J.J.Orujev, Auth. Cert. USSR 502907 (1974), Bull. Inv. 6 (1976), 58.
7. V.A.Kabanov, A.A.Efendiev, J.J.Orujev and N.M.Samedova, Synthesis and investigation of complexing sorbent with macromolecules "prearranged" for sorption, Dokl. Akad. Nauk SSSR. 238:356 (1978).
8. A.A.Efendiev, J.J.Orujev and V.A.Kabanov, Preparation of complexing polymeric sorbents with macromolecules "prearranged" for sorbing ion, Vysokomol. Soyedin. B 19:91 (1977).
9. V.A.Kabanov, A.A.Efendiev and J.J.Orujev, Complex–forming polymeric sorbents with macromolecular arrangement favorable for ion sorption, J.Appl. Pol.Sci. 24:259 (1979).

10. A.A.Efendiev and V.A.Kabanov, Selective polymer complexons prearranged for metal ion sorption, Pure & Appl. Chem. 54:2077 (1982).

11. A.A.Efendiev, T.N.Shakhtakhtinsky, L.F.Mustafaeva and H.L.Shick, Liquid–phase oxidation of ethyl–benzene over cobalt complexes, Ind.Eng.Chem.Prod., Res,.Dev. 19:75 (1980).

12. V.A.Kabanov, L.S.Molochnikov, S.A.Ilyichev, O.N.Babkin, Yu.M. Sultanov, J.J.Orujev and A.A.Efendiev, An EPR study of Cu (II) complexes with structured polymeric sorbents, Vysokomol. Soyedin. A28:2459 (1986).

13. A.A.Efendiev, J.J.Orujev, T.N.Shakhtakhtinskii and V.A.Kabanov, Structural setting of active centres of metallopolymeric complex catalysts for hydrocarbon substrate, Kinetika i kataliz. 27:451 (1986).

14. A.A.Efendiev, J.J.Orujev, T.N.Shakhtakhtinsky and V.A.Kabanov, Liquid phase oxidation of ethylbenzene in the presence of metal–polymer complex catalysts with specially prearranged structures of active centres, in: "Homogeneous and Heterogeneous Catalysis," VNU Science Press, Utrecht (1986).

15. I.V.Beresin. "Fundamental of Physical Chemistry of Enzymic Catalysis," Visshaya Shkola, Moscow (1977).

16. J.Westley. "Enzymic Catalysis," Harper and Row, New York, Evanston and London (1969).

POLYMERS WITH LIGATED PEROXOTUNGSTIC UNITS: ORGANOPHOSPHORYL MACROLIGANDS FOR THE CATALYTIC EPOXIDATION OF ALKENES.

Georges Gelbard,[1]* David C. Sherrington,[2] François Breton,[1] Mohamed Benelmoudeni,[1] Marie-Thérèse Charreyre,[1] and Doan Dong[1]

[1]Institut de Recherches sur la Catalyse-CNRS, 2 avenue Albert Einstein, 69626 Villeurbanne Cedex, France.
[2]Department of Pure & Applied Chemistry, University of Strathclyde, 295 Cathedral street, Glasgow G1 1XL, UK.

INTRODUCTION.

Epoxidation of olefins is an important way to functionalize hydrocarbons for intermediates in fine chemicals industry [1]. Organic peracids are still used as stoichiometric reagents, but, the use of hydrogen peroxide or organic hydroperoxides appears more convenient and cost effective. These reagents require a transition metal complexes as catalysts; most of them are titanium, vanadium, molybdenum or tungsten based [2].

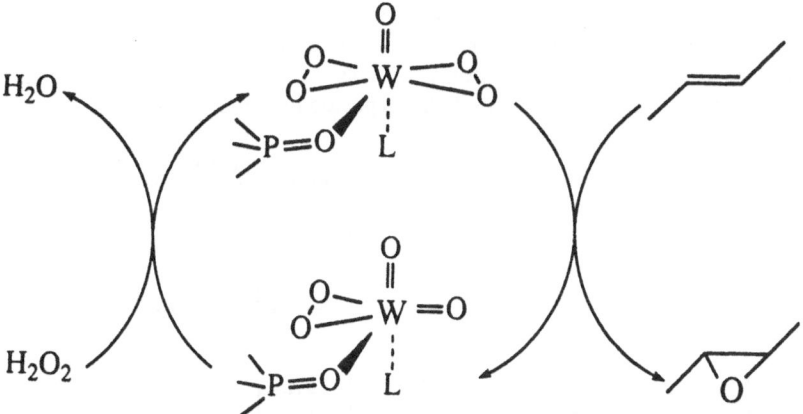

Figure 1. Schematic catalytic cycle of the epoxidation with organophospho-peroxotungstic complexes

Epoxidation with hydrogen peroxide is most attractive due to its particuarly low cost and the absence of pollution by effluents [3]. In this case, tungsten complexes are the best catalysts, we have shown that very selective and efficient epoxidations can be performed when complexing the intermediate peroxotungstic species with phosphorus(V) derived ligands which contain the phosphoryl subunit O=P<- (figure 1)[4].

To develop a process from such catalytic systems, it appears necessary to recycle the homogeneous [WO$_5$-O=P<-] system and to avoid any contamination of the epoxides with traces of metallic species. A possible way of "immobilizing" the catalyst in a reactor is to use biphasic solid-liquid systems where the metal is complexed to ligands which are covalently bound to insoluble and porous polymers according to the following principle: as soon as a ligand L=O is involved in the coordination sphere of a metal M, it becomes possible to bind the corresponding complex onto an organic polymeric matrix (P) by means of covalent links (-/\/\/\/\-) between the polymeric chains and the complexing ligand.

$$L=O-M \quad \longrightarrow \quad (P)-/\/\/\/\-L=O-M$$

This procedure for supporting homogeneous catalysis is very well documented [5]; but, a limited number of examples are known in the epoxidation using hydrogen peroxide [6]. The first part of this report will deal with non-ligated metal since free peroxotungstic anions can be immobilized on anion-exchange resins.

ANION EXCHANGE RESINS SUPPORTED PEROXOTUNGSTIC ANIONS.

The first relevant report concerned the epoxidation and hydrolysis of maleic acid into tartaric acid. A commercially available polystyrene-based anion-exchange resin was treated with a solution of sodium hydrogen tungstate NaHWO$_4$ to serve as the catalyst [7]. The polymethacrylates, which are more oxidation-resistant supports, were preferred.

POLYMETHACRYLATE BASED QUATERNARY AMMONIUM SALTS.

A series of macroporous and gel-type exchange resins were obtained by the copolymerization of glycidyl methacrylate (GMA) followed by ring opening amination with secondary amines [8] and finally by quaternization.
Methyl or butyl methacrylate (MMA, BMA) was used as comonomers, and, various amounts of ethyleneglycoldimethacrylate (EGDMA) were used as cross-linking agent. A mixture of cyclohexanol and decanol was used as porogen when a macroporous structure was required (figure 2). The resins were treated with a 0.1N solution of peroxotungstic acid, H$_2$W$_2$O$_{11}$. The composition of the supported catalysts are reported in Table 1:

Table 1. Polymethacrylates supported ammonium peroxotungstates.

Catalyst	% GMA	% BMA	% EGDMA	R^1	R^2	mmol W/g
1a[a]	-	-	-	Me	Me	1.72
1b[b]	53.4	0	46.6	Me	nPr	1.75
1c[b]	53.4	0	46.6	Me	HO-Et[c]	1.54
1d[d]	52	44.8	2.8	Me	Me	1.43
1e[d]	52	44.8	2.8	nBu	nBu	1.30

a) gel-type Amberlyst IRA958, composition not available.   b) macroporous resin, cyclohexanol / decanol as porogeneous agent. c) 2-hydroxyethyl. d) gel-type resins

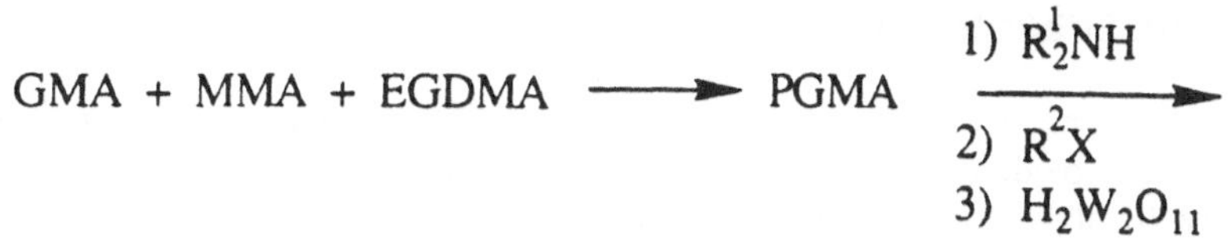

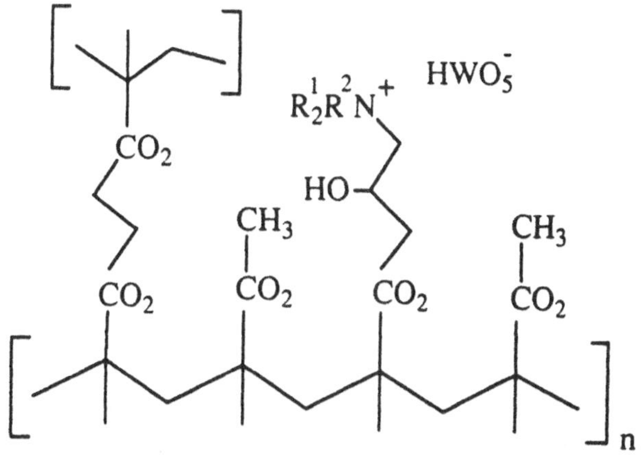

GMA + MMA + EGDMA $\longrightarrow$ PGMA $\xrightarrow[\begin{array}{l}2)\ R^2X\\3)\ H_2W_2O_{11}\end{array}]{1)\ R_2^1NH}$

Figure 2. Polyglycidylmethacrylate supported peroxotungstic catalysts.

The epoxidations were performed at 70°C with a 5 fold excess of alkene vs. H_2O_2 (cyclohexene as a typical substrate) and a 70% solution of H_2O_2. Dioxane was added to get homogeneous liquid phase. The results, reported in Table 2, show that a contribution of lipophilicity is beneficial near the metal. Interestingly, the dibutyl substituted ammonium cation **1e** exhibited higher turnover frequency (197) than was observed in homogeneous solution (TOF = 68).

Table 2. Turnover frequency (TOF) in the epoxidation of cyclohexene by H_2O_2 in the presence of catalysts 13.

Catalyst	Time (h)	Conversion % [a]	Selectivity [b]	TOF h^{-1}
1a	21	72	22	9
1b	4	89	65	34
1c	4	82	68	46
1d	4	76	49	41
1e	3	>99	54	197

a) of hydrogen peroxide.
b) in epoxycyclohexane.

POLYSTYRENE SUPPORTED PYRIDINE LIGANDS.

Pyridine nitrogen atoms can be associated with peroxotungstic units either as neutral comple-xing ligands in the N-oxide form or as pyridinium cations. The polystyrene-grafted pyridines, **2** and **3**, obtained by alkylation of 2-aminomethylpyridines with chloromethylated polystyrene, were either converted into the peroxotungstic salts, **2a** and **3a**, or first transformed into their corresponding N-oxide derivatives by treatment with hydrogen peroxide in the presence of acetic acid and then complexed to give **2b** and **3b**.

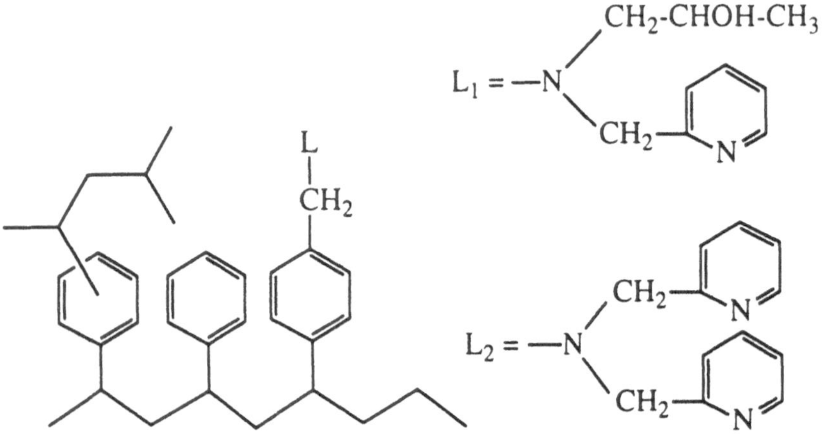

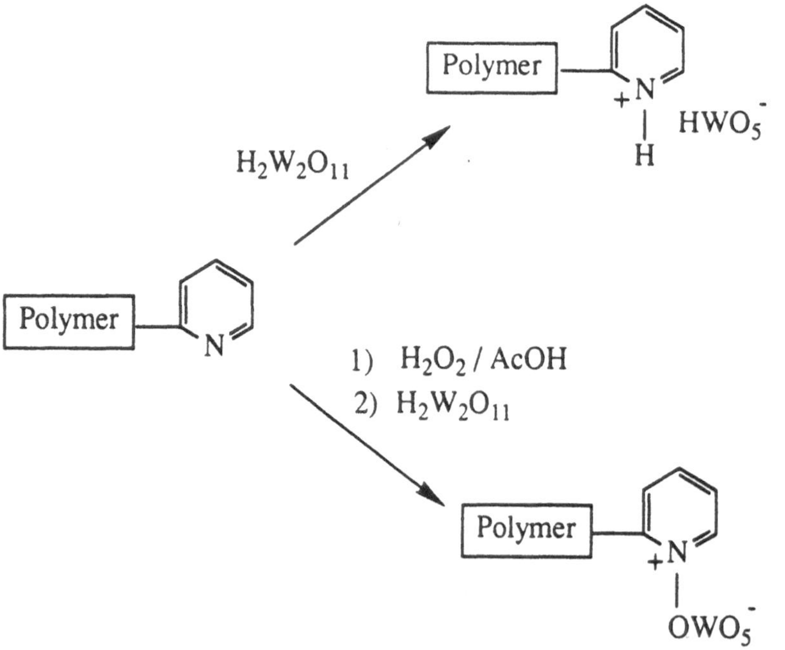

Figure 3. Supported peroxotungstic pyridinium salts **2a** and **3a** and betaines **2b** and **3b**.

Elemental microanalyses showed that the pyridine nitrogens were transformed into their N-oxide derivatives. The neutral complexes, **2b** and **3b,** may have a betainic form.

Table 3. Polymeric pyridine-peroxotungstic acid resins.

Ligand[a]	Capacity[b]	Catalyst	N/W	Capacity[c]
2	2.6	**2a**	1.37	1.93
2	2.6	**2b**	0.82	1.36
3	4.7	**3a**	2.57	1.57
3	4.7	**3b**	1.35	1.50

a) **3** and **4**: polymers XFS 43084 and XFS 4195 respectively from Dow Chemical.
b) meqN/g from the elemental analysis of the starting material. c) meqW/g.

Acidic salts of the general formula $>NH^+ HW_2O_{11}^-$ would be formed from free amine and pyridine units, but, the N/W ratios were greater than 1/2 in all cases. Thus, many basic units remain free even after treatment with an excess of peroxotungstic acid. Pyridine N-oxides were expected to give the betainic complexes $>N^+O-WO_5^-$. The N/W ratios in **2a** and **2b** were close to unity; but the N/W ratios, in compounds **3a** and **3b,** showed less than 40% of the N-oxide ligands uncomplexed.

The epoxidation of cyclohexene were performed at 70°C with a molar ratio of alkene/H_2O_2 = 5/1; 0.6 molar % of catalyst and an equal volume of dioxane (table 3).

Table 4. Epoxidation of cyclohexene with H_2O_2 using peroxotungstic acids on pyridinium and pyridine oxide resins as catalysts.

Catalyst	Conversion[a]	Yield[b]	Epoxide[c]	Diol[d]	Allylics[e]	TOF h^{-1}[f]
2a	91	84 (22)	41	43	7	6 4
2b	88	94 (22)	46	48	5	7 1
3a	72	48 (24)	43	5	35	3 3
3b	81	85 (5)	50	35	5	35

a) % of H_2O_2. b) % of all epoxidation products (time h). c) % of epoxycyclohexane
d) % of cis+trans 1,2-cyclohexanediol. e) % of 2-cyclohexenol and 2-cyclohexenone
f) average value.

Though the yields of epoxidation products are high, these catalysts give significant amounts of diol which results from a residual acidity. The remaining free pyridine and N-oxide units in the polymer appear unsufficient to buffer the acidity of the peroxotungstic units sufficiently. The high content of metal, relative to the high amount of grafted ligands, was shown later to be unfavourable for good selectivity [8].

POLYMER SUPPORTED PHOSPHONYL AND PHOSPHORYL LIGANDS.

The choice of such ligands resulted from preliminary studies [4]: Very significant improvements were made in the epoxidation of simple alkenes and allylic alcohols when the usual tungstate and peroxotungstate salts were complexed with organophosphorus ligands containing the phosphoryl unit [O=P<-] in for example hexamethylphosphotriamide (HMPA) and trioctylphosphine (TOPO).

Several cross-linked polymers containing ligands related to the phosphine oxide, phosphona-mide and the phosphoramide series were prepared

$$O=P\begin{matrix}R\\-R\\R\end{matrix} \qquad O=P\begin{matrix}R\\-R(or\ N)\\N\end{matrix} \qquad O=P\begin{matrix}N\\-N\\N\end{matrix}$$

to get the immobilized catalyst with the general structure: (P)-/\/\/\->P=O-WO$_5$.

POLYSTYRENE SUPPORTED PHOSPHINE OXIDE LIGANDS.

Polymer grafted chelating diphosphine dioxides were used for the epoxidation with a molyb-denum / TBHP system [9]. Here simple ligands were used. The phosphine oxides were grafted on several macroporous and gel-type chloromethylated polystyrenes 4. The gel poly-mers were 1% DVB Merrifield type resins from BioRad™. The macroporous XAD 4 (725 m^2/g) and XAD 16 (860 m^2/g) were supplied by Rohm and Haas™. The latter were chloro-methylated using the SOCl$_2$/dimethoxymethane system [10] and treated with a disubstituted phosphine oxide in basic media [11]:

$$[\text{Polystyrene}]\text{-CH}_2\text{-Cl} \quad + \quad O=P(H)<RR \quad + \quad NaOH \xrightarrow{\hspace{2cm}}$$
$$4$$

$$[\text{Polystyrene}]\text{-CH}_2\text{-}\underset{R}{\overset{R}{P}}=O \xrightarrow{\hspace{2cm}} [\text{Polystyrene}]\text{-CH}_2\text{-}\underset{R}{\overset{R}{P}}=O\text{-WO}_5$$
$$5 \qquad\qquad\qquad\qquad\qquad\qquad 6$$

Figure 4. Polystyrene-grafted phosphine oxides.

Table 5. Peroxotungstic acid on grafted polystyrene phosphine oxides.

Catalyst	Type	meq Cl/g[a]	meq P/g[b]	R	meqW/g[c]	P/W
6a[d]	gel SX1	1.25	1	C$_6$H$_5$-	0.17	6.37
6b	gel SX1	4.15	1.68	C$_8$H$_{17}$-	0.82	1.70
6c	gel SX1	4.15	1.93	C$_8$H$_{17}$-	0.80	1.87
6d[e]	XAD4	3.3	0.81	C$_8$H$_{17}$-	0.77	0.90
6e	XAD4	3.1	1.03	C$_6$H$_5$-	0.21	4.92
6f[f]	XAD16	2.5	1.12	C$_6$H$_5$-	0.30	3.67

a) in starting resin 4. b) in the ligand 5. c) in the final catalyst 6. d) gel-type BioBeadsSX1 from Biorad. e) and f) macroporous polytyrene from Rhom & Haas™.

These macroligands 5 showed a typical band at 1165 cm-1 for v(P=O) in the FTIR spectrum. The ^{31}P CP/MAS NMR gave broad signals at +90-95 ppm corrresponding to the phosphine oxide groups when R= C$_8$H$_{17}$ and at +75 ppm when R= C$_6$H$_5$.
Complexations were performed by suspending the phosphine oxide polymers in hot dioxane in the presence of an excess of peroxotungstic acid H$_2$W$_2$O$_{11}$. The complexed polymers 6 were characterized by elemental microanalysis and FTIR; the typical bands of the per-oxotungstic species were found at 978-968, v(W=O), 816-814 v(O-O), 544-543, v(W-O-O sym.) and 617-587, v(W-O-O asym.) cm^{-1}

The overlapping of these with bands related to the polystyrene backbone made difficult an accurate assignment of the $v(P=O)$ related to complexed phosphines oxides. The characteristics of the catalysts are given in Table 5.

These supported catalysts were checked as above in the epoxidation of cyclohexene with hydrogen peroxide. The results are reported in Table 6.

Table 6. Epoxidation of cyclohexene with peroxotungstic acid on phosphine oxide resins.

Catalyst	Conversion [a]	Selectivity [b]	TOF h[-1] [c]
soluble[d]	93.5 (1.5)	87.5	122 (83)
6a	92 (1)	59	168 (82)
6b	84 (1)	67	194 (82)
6c	82 (0.5)	39	112 (90)
6d	83 (3)	53	53 (24)
6d[e]	29 (6)	47	6.5 (4)
6e	97 (3)	62	54 (42)
6f	88 (4)	88	132 (39)

a) % of consumed H_2O_2, (time h) . b) % of epoxycyclohexane. c) at the beginning and (averaged). d) $H_2W_2O_{11}$ with 2 eq. of TOPO. e) recycling of the same sample of catalyst **6d**.

Most of these supported catalysts gave similar or even better TOF than that of the soluble catalyst, but, the instablity of the supported complex does not allow repeated uses.

POLYSTYRENE SUPPORTED PHOSPHONAMIDE LIGANDS.

The use of phenylphosphonic acid as complexing ligand for WO_5 [4][12] are described. The behaviour of the corresponding polystyrene supported dimethylamide was checked. The required polystyrene-phosphonamide was obtained according to the following sequence:

[Polystyrene] ————> [PS]-C$_6$H$_4$-PO$_2$H$_2$ ————> [PS]-C$_6$H$_4$-PO$_3$H$_2$
                         7   **8**

————> [PS]-C$_6$H$_4$-P(=O)(NRR)(NRR) ————> [PS]-C$_6$H$_4$-P(=O)(NRR)(NRR)-WO$_5$
              9   **10**

The Friedel-Crafts phosphorylation of polystyrene with $PCl_3/AlCl_3$ followed by hydrolosis afforded the phosphinic acid resins 7 which was oxidized with H_2O_2 in acetic acid to phosphonic acid **8**. The complete oxidation of P(III) into P(V) was acertained by ^{31}P CP/MAS solid state NMR spectroscopy: typical signals at +3.5 and at +18 ppm were observed for phosphinic acid in 7 and phosphonic acid groups in 8 respectively.

The resins **8** were treated first with $SOCl_2$ and then with excess dimethylamine to give the polymer ligands **9**. Analytical results are reported in Table 7.

Table 7. Polystyrenephosphonamide derived catalysts.

Ligand	N/P	Catalyst	meq W/g	P/W
9a[a]	1.42	**10a**	1.53	1.54
9b[b]	1.39	**10b**	2.03	0.79

a) amidification of the macroporous phosphonic resin Duolite ES63 from Rohm & Haas. b) phosphonylation and amidification of BioBeads SX1, a gel-type species from BioRad™.

The epoxidations with H_2O_2 were performed three times with the same sample of a macroporous and twice with a gel-type catalyst. The results are reported in Table 8.

Table 8. Reuse of catalysts in the epoxidation of cyclohexene.

Catalyst[a]	Conversion (time)	Epoxide	Diol	O_2	TOF h^{-1} (average)
10a-1	86 (3)	64	5	30	112 (91)
10a-2	91 (4)	59	15	25	31 (25)
10a-3	90 (4)	60	25	14	57 (35)
10b-1	90 (90 mn)	34	45	20	175 (67)
10b-2	82 (3)	54	15	25	147 (24)

a) run n°.

Some decomposition of H_2O_2 occurred with these catalysts. Recyclings are possible with the gel-type polymer support. The macroporous resins appear stable after the second run.

POLYBENZIMIDAZOLE SUPPORTED LIGANDS.

Polybenzimidazole (PBI) is a heat and oxidation resistant polymer [13]. It is a good candidate for the grafting of catalysts when oxidation reactions are performed. Due to apparent chemical difficulties, there are a limited number of examples of chemical modifications of this support [14]. Some epoxidations have been performed with t-butylhydroperoxide and PBI grafted Mo complexes [15].

The starting material was Celazole™, a non-porous polymer from Hoechst-Celanese™. The beads were crushed and sieved; the 100-200-mesh fraction was used. The required organo-phosphoryl groups were grafted under strongly basic conditions at the secondary H-N of imidazole ring in order to get the amide which can then be alkylated with strong electrophiles such as diphenylphosphoryl chloride or bis(dimethylamino)phophoryl chloride (figure 6).

In the diphenylphosphonamide **11** and the phosphotriamide **12** macroligands, the $\upsilon(P=O)$ appeared at 1190 and at 1180 cm^{-1} respectively in the FTIR spectrum; after complexation, υ (W=O) appears at 960 cm^{-1}. Elemental microanalysis data before and after complexation with $H_2W_2O_{11}$ are given in Table 9.

Table 9. Polybenzimidazole supported catalysts.

Polymer	%P (meq /g)[a]	%W (meq /g)[b]
11-WO$_5$	1.6 (0.5)	12.8 (0.7)
12-WO$_5$	6.0	24.6 (1.3)

a) for the macroligands **11** and **12**.
b) after complexation with $H_2W_2O_{11}$.

272

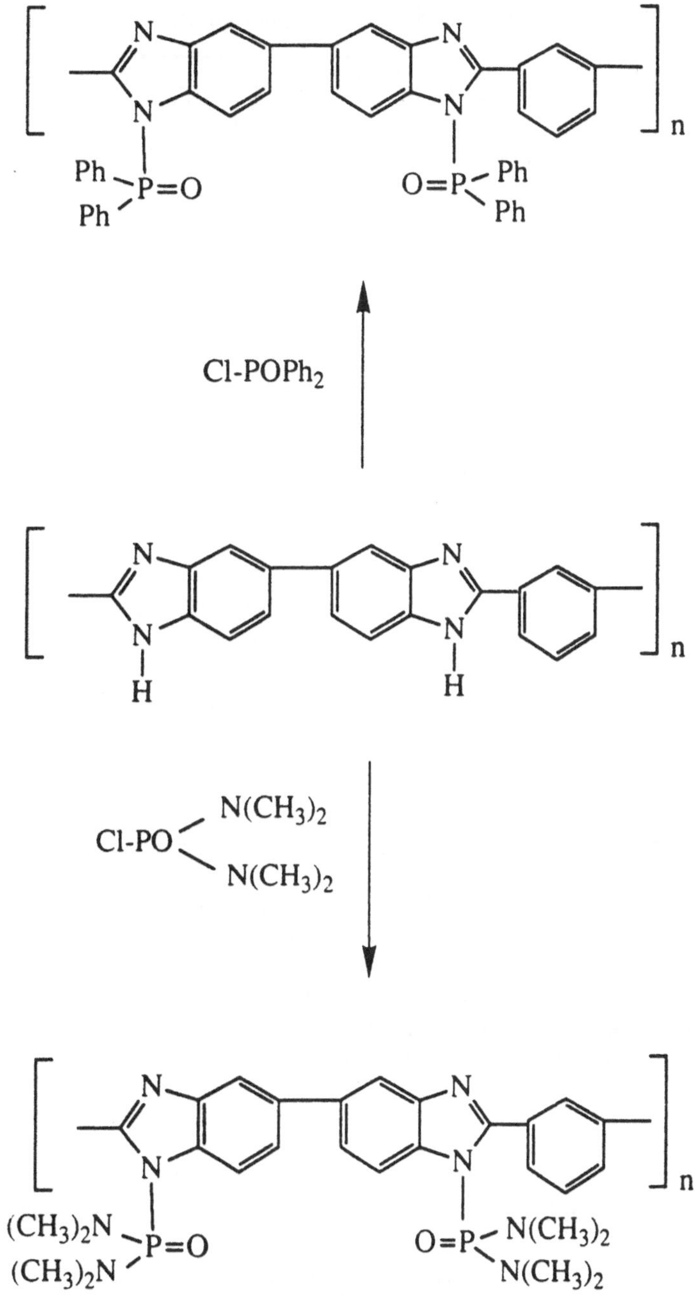

11

12

Figure 6. Phosphonamide **11** phosphotriamide **12** from polybenzimidazole.

Repeated epoxidations of cyclohexene were performed with the same sample of catalyst which was recovered by centrifugation. Experiments were run with catalyst **12**-WO$_5$ which appeared better than **11**-WO$_5$

Table 10. Recycling of polybenzimidazole-supported catalysts.

Catalysts	% Epoxide	% Diol	Total yield (TOF)
11-WO$_5$	27	17[a)	44 (6.4)
12-WO$_5$-1	42	32[b)	74
12-WO$_5$-2	49	29	78
12-WO$_5$-3	46	6	52
12-WO$_5$-4	47	13	60
12-WO$_5$-5	56	26	82

a) with 45% of allylics. b) no allylics.

This supported catalytic system is perfectly stable and can be reused five time without any loss of reactivity and modification in the selectivity. The leakage of metal in solution was below the threshold of detection (< 2 ppm), confirming earlier results [15].

CONCLUSIONS

Epoxidation of alkenes with hydrogen peroxide can be performed when peroxotungstic units are complexed to functionalized polymers. Complexing ligands, such as phosphoramides and phosphonamides, appeared more promising than amine or phosphine oxides and yielde catalysts which displayed TOF data higher than that of analogous soluble catalysts . When stable complexes are available, it becomes possible to recycle the catalyst and to perform epoxidation in a continuous-flow reactor.

ACKNOWLEDGMENTS.

The authors acknowledge with many thanks the Centre National de la Recherche Scientifique, Rhone-Poulenc, Chemoxal-l'Air Liquide and Roussel-Uclaf for financial support. We thank Rohm and Haas, Dow Chemicals and Hoechst-Celanese for providing samples of polymers and Atochem for a gift of concentrated hydrogen peroxide.
Special thanks are due to Drs. L. Krumenacker (Rhone-Poulenc), H. Ledon (l'Air Liquide) and J. Buendia (Roussel-Uclaf) for their interest in this work.

REFERENCES

1- Kirk-Othmer Encyclopedia of Chemical Technology, Interscience, New York (1985) vol 7, p 239,263.
2- a) K.A. Jørgensen, Transition-metal catalyzed epoxidations, Chem.Rev. 89:431 (1989)
   b) R.A. Sheldon, Metal-catalysed epoxidation of olefins with hydroperoxides, in "Aspects of Homogeneous Catalysis," R. Ugo, ed., vol. 4, D. Reidel, Dordrecht (1981) p.3.

3- a) W.A. Herrmann,"Organic Peroxygen Chemistry," Topics in Current Chemistry, vol. 164, Springer Verlag, Berlin (1993).
b) G. Strukul, " Catalytic Oxidations with Hydrogen Peroxide as Oxidant," Kluwer Academic Publishers, Dordrecht (1992).

4- M. Quenard, V. Bonmarin, and G. Gelbard, Epoxidation of olefins by hydrogen peroxide catalysed by phosphotungstic complexes, Tetrahedron Lett., 28:2237 (1987).

5- a) F.R. Hartley, "Supported Metal Complexes," D. Reidel, Dordrecht (1985).
b) D.C. Sherrington, Polymer-supported metal complex oxidation, Pure & Appl. Chem., 60:401 (1988).

6- a) G.G. Allan, and A. N. Neogi, Macromolecular organometallic catalysis: Epoxidation, J. Catal., 16:197 (1970).
b) Kamaludsin, and H.V. Singh, J. Catal., 137:510 (1992).

7- E. Kàlalovà, Z. Radovà, F. Švec, and J. Kàlal, Reactive polymers VII, Europ. Polym. J, 13:293 (1977).

8- G. Gelbard, F. Breton, M.T. Charreyre, and D. Dong, Polypyridine-based catalysts, Makromol. Chem. Macromol. Symp., 59:353 (1992).

9- C.C Su, C.H. Ueng, W.H. Lin, M.J. Gi, K.H. Lii, J.S. Ting, S.H. Jan, and K.O. Hodgson, Homogeneneous epoxidation catalyzed by molybdenum complexes containing neutral ligands, Proc. Natl. Sci. Counc. B ROC, 6:45 (1982).

10- L. Galeazzi, Montedison, german patent n° 24 55 946 (1975).

11- T.H. N'Guyen, and S. Boileau, Phase-transfer catalysis in the modification of polymers. Tetrahedron Lett. 2651 (1979).

12- G. Gelbard, F. Raison, E. Roditi-Lachter, L. Ouahab, and D. Granjean, Epoxidation of allylic alcohols by hydrogen peroxide in the presence of complexd peroxotungstic species, J. Mol. Cat. to appear (1995).

13- P.E. Cassidy. "Thermally Stable Polymers," Marcel Dekker, New York.

14- a) M. Hu, E.M. Pearce, and T.K. Kwei, Modification of polybenzimidazole, J. Polym Sc. A, Polym. Chem. 31:553 (1993).
b) H.G. Tang, and D.C. Sherrington, Polymer-supported Pd(II) Wacker-type catalysts, Polymer 34:2821 (1993).
c) M. Chanda, and G.L. Rempel, Polybenzimidazole resin-based new chelating agents. Reactive Polym. 11:165 (1989).

15- M.M. Miller, and D.C. Sherrington, Polybenzimidazole-supported molybdenum propene epoxidation catalyst, J. Chem. Soc., Chem. Comm. 55 (1994); J. Chem Soc Perkin II, 2091 (1994).

COBALT DICARBOLLIDE CONTAINING POLYMER RESINS FOR CESIUM AND STRONTIUM UPTAKE

W. P. Steckle, Jr., J. R. Duke, Jr., and B. S. Jorgensen

Materials Science and Technology Division
Los Alamos National Laboratory
MST-7, MS E549
Los Alamos, NM 87545

INTRODUCTION

Metals have been incorporated into polymers in a variety of ways. In this study cobalt will be indirectly attached to a polymer matrix via cobalt diacarbollide. Cobalt dicarbollide, $[(C_2B_9H_{11})_2Co]^-M^+$ (CoB2), is a sandwich molecule analagous to ferrocene where Co^{3+} is sandwiched between two carbollide ($C_2B_9H_{11}^=$) cages as depicted in Figure 1. Cobalt dicarbollide has been chosen for its demonstrated ability to selectively extract cesium and strontium from solutions containing a mixture of metals. Unfortunately hazardous organic solvents such as nitrobenzene or ethyl acetate are utilized in the extraction process. It is hoped that by incorporating CoB2 into a polymer matrix that it will act as an ion exchange resin selective towards cesium and strontium, without the need of any organic solvents.

Separation of cesium and strontium from nuclear waste presents a problem to both the Department of Energy and the commercial nuclear power industry. Over the past several decades large quantities of high level radioactive waste have been generated by the atomic industry. $^{137}Cs^+$ and $^{90}Sr^{2+}$ occur in nuclear waste as the result of fission in reactor fuels. One of the reasons for removing these particular isotopes is that they contribute upwards of 97% of the thermal energy to the storage tanks.[1] Wastes have been generated in such a manner that they contain a wide variety of metals and isotopes along with a variety of flocculating agents. Prior treatment methods have employed the use of phosphotungstic acid, ferro- and ferricyanides and tetraphenylborate to precipitate these metals. In multi-element streams this leads to a long tedious process where many precipitates are formed adding to the quantity of waste being processed. Macrocyclic crown ethers have also been employed for removal of Cs and Sr. Selectivity can be achieved by selecting the size of the cavity of the crown ether in which the metal cation fits. The drawbacks to crown ethers are their cost and the inability to remove the metal from the crown ether once it has been extracted.[2] Amberlite resins which have crown ethers sorbed on them are currently commercially available for the removal of strontium.

Research efforts towards the remediation of Savanah River and Hanford currently encorporates the use of resorcinol/formaldehyde and phenol/formaldehyde resins.[3,4] These resins are used for the alkaline side of the waste recovery process and would not be stable on the acid side, particularly those streams containing nitric acid. Supports that have exhibited selectivity towards cesium include crytstalline silcotitanates, zeolites, and pillared clays. Sodium titanate is also being evaluated as an ion exchanger/precipitant for strontium. The sorbed crown ethers previously mentioned have been found to work only in acidic media and are fouled by the presence of barium.[5]

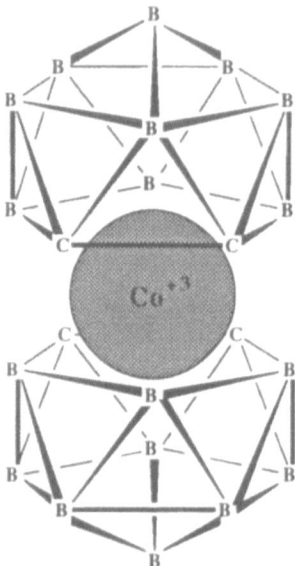

Figure 1. Structure of cobalt dicarbollide. Hydrogen atoms attached to each boron and carbon have been omitted for ease of illustration.

Although cobalt dicarbollide was first made in the 1960's, it was not until the 1970's that its use for Cs and Sr extraction was investigated by Russians and Czech researchers[6,7]. CoB_2 shows high selectivity and extraction into an organic phase. Nitrobenzene exhibited the highest distribution coefficient of the solvents investigated, three times that of ethyl acetate.[8] Due to the carcinogenic and toxic nature of this solvent coupled with more stringent environmental regulations, technologies are being developed that do not require the use of volatile organic compounds.

Prior research in the area of polymer supported metal catalysts has led to the attachment of 1,2-dicarbadodecaborane, a precursor of CoB_2, to polystyrene (PS). Chandrasekaran et al.[9] prepared a rhodium complex for use as a hydrogenation catalyst in this manner. An anion of the carborane cage was prepared by the reaction of the decaborane with n-buLi and was then subsequently reacted with a poly(chloromethylstyrene) (PCMS).

RESULTS

Experimental

Two different polymeric supports were used in this study. Poly(chloromethyl-stryene) was chosen because it can be readily modified with a variety of functional groups and is commercially available. Polybenzimidazole (PBI) was chosen for its hydrophilic nature along with its high chemical, thermal, and radiolytic stability. PCMS resins used were commercially available from Polysciences and macroporous resins were made at MST-7. Designations for the resins which were prepared by us are: X-Y-Z, where X denotes the mole % vinylbenzylchloride, Y denotes the mole % styrene, and Z denotes the mole % divinylbenzene. By scanning electron microscopy,SEM, stress cracking could be seen in the commercially available resins. Upon sulfonation these resins would further degrade to a fine powder. BET surface area measurements for the commercial resins versus those we prepared were 2 m^2/g and 21 m^2/g respectively. In order to obtain maximum dicarbollide loadings for K_d experiments, the materials made in house were chosen. The PBI resin was Aurorez microporous PBI resin beads made by Hoechst-Celanese were used as supplied.

Cyclohexane, toluene, and tetrahydrofuran (THF) used for the grafting reactions were freshly distilled from sodium under an argon purge. n-BuLi, hexamethyldisilazane, methanol, and ethanol was purchased from Aldrich and used as received. Carborane, purchased from Alfa, was purified by sublimation prior to use. Lithium carbollide and CoB_2 was prepared by CST-11at Los Alamos National Laboratory.

FTIR analysis was performed using either a Perkin Elmer 1600 FTIR or a Nicolet 730 FTIR equipped with a Nic-Plan IR microscope. Diamond windows were used in order to compress the resin beads. Beads were looked at both in transmission as well as reflectance modes. Solid state ^{13}C and ^{11}B NMR of these resins was done using a Bruker ASX 300. Surface area analysis was performed using a Micromeritics ASAP 2010 surface area analyser. Elemental analysis was performed by Desert Analytics.

Synthesis

Two reaction routes were envisioned for grafting CoB_2 onto a resin. One involved directly grafting CoB_2 to the PCMS if that were possible. The other involved attaching carborane to the polymer followed by dehydroborylation to form CoB_2. Steric hindrance and the fact that each cage on CoB_2 itself is a dianion needs to be taken into consideration. The first route still would be easier on the resin than having to expose the resin to the harsh conditions under which dehydroborylation occurs. Dehydroborylation is typically achieved by refluxing carborane with potassium hydroxide in ethanol. The second route was not pursued.

In this study a series of cross-linked resins was used. Different solvents were investigated in order to determine the rate of reaction and whether or not the degree of swelling effected the substitution. The degree of swelling not only affects the rate of reaction but also the uniformity of the grafting throughout the resin. Once it was determined that the carborane cage was actually grafted onto the resin and not merely adsorbed onto the surface, a series was prepared incorporating lithium carbollide [$(C_2B_9H_{11})^-Li^+$] and cobalt dicarbollide onto a resin. Effects of having a charged cage, counterions, and steric hindrance could be determined by grafting these different species to the resin. Whether the dicarbollide itself could be grafted directly onto the resin or if it had to be made in situ on the resin after the carborane had been grafted on had to be determined. Tetramethylammonium (TMA), sodium, and cesium salts of cobalt dicarbollide were used in this study. This reaction scheme along with the sulfonation of the resin is given in Figure 2. Reasons for sulfonation will be discussed later.

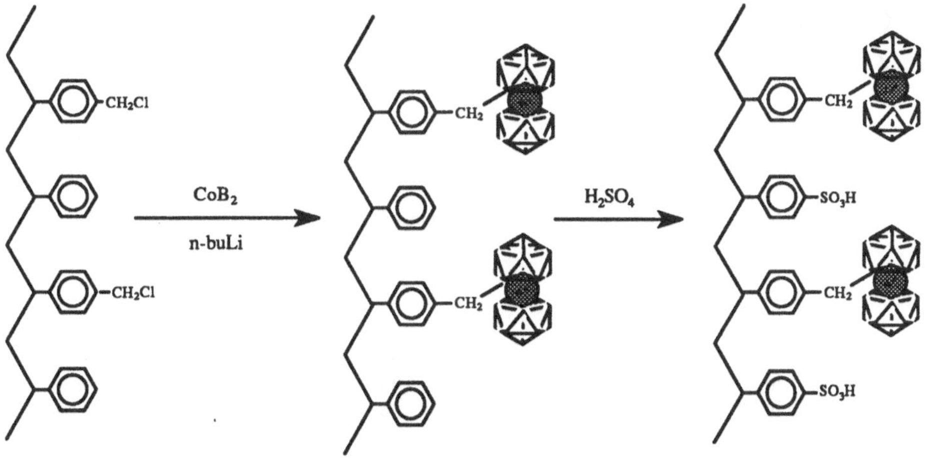

Figure 2. Anionic grafting of cobalt dicarbollide onto polystyrene and subsequent sulfonation.

Reaction conditions for both the grafting of the carborane series as well as the preparation of the PBI prior to the CoB_2 reaction were as previously cited.[7,8] In order to insure that the carboranes were not merely adsorbed to the surface of the resins an exhaustive study was performed. All samples were washed three times with methanol before soxhlet extraction for 48-72 hours in ethanol. Both methanol and ethanol are excellent solvents for the species being grafted in this study. Samples with adsorbed carborane or CoB_2 that were subjected to similar work up were found to be free of either the carborane or CoB_2.

Typical grafting reactions were done using the following procedure. Reactions were carried out in round bottom flasks equipped with a magnetic stirbar and a septum. The resin and the carborane were weighed into separate flasks. This was carried out in a glove box under an argon atmosphere. Freshly distilled solvent was transferred via syringe to the flask containing the carborane. This mixture was cooled to 0°C while stirring prior to the addition of the n-buLi. Once this mixture came to room temperature it was tranfered via syringe to the flask containing the resin. After 24 hours the anion was quenched by the addition of methanol. The sample was then filtered using a scintered glass funnel and washed three times with 50 ml of methanol. The resin was then placed in a soxhlet extractor with ethanol for 24 hours before drying to a constant weight in a vacuum oven.

As previously mentioned, the other resin investigated was polybenzimidazole. Since PBI does not contain a reactive chlorine site, epichlorohydrin was grafted on as reported by Chanda et al.[10-12] The resulting hydroxyl group from the opening of the epoxide was then protected by hexamethyldisilazane. Figure 3 shows these reactions along with the grafting of the dicarbollide onto the PBI. Attachment of CoB_2 to PBI was verified as described for PCMS.

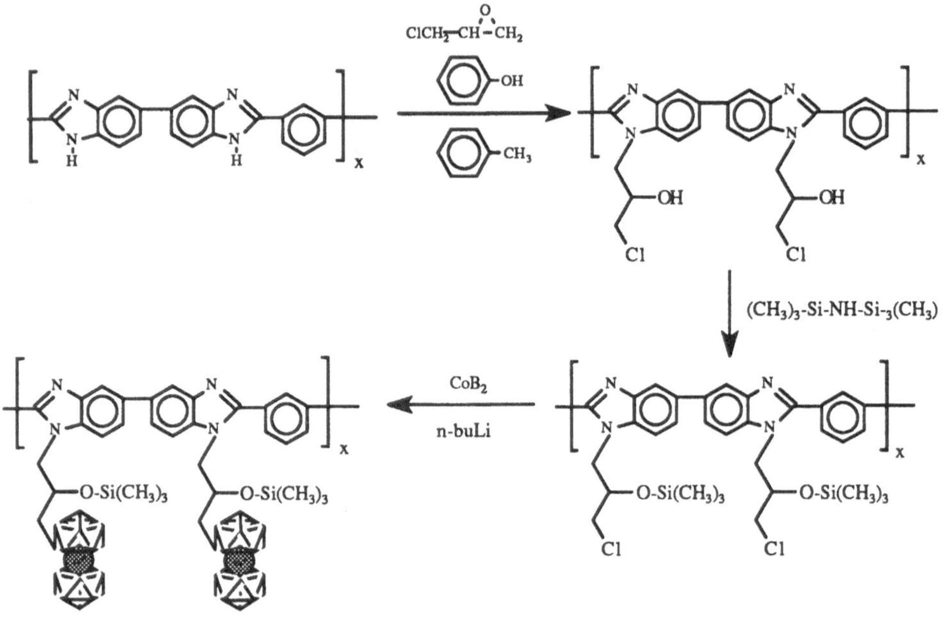

Figure 3. Reaction scheme for preparing PBI with covalently bonded dicarbollide.

DISCUSSION

Since carborane had previously been shown to react with PCMS[9] in benzene, this reaction was optimized using toluene before trying to graft the carbollides. This reaction can be followed using IR spectroscopy by monitoring the disappearance of the CH_2-Cl stretch, ca. 1265 cm^{-1}, and the growth of the strong B-H absorbance ca. 2580 cm^{-1}. FTIR spectra of PCMS, CoB_2-PS, and sulfonated CoB_2-PS can be found in Figure 3. Not only is the B-H absorbance dependent on the type of cage, it can been seen to shift 5 - 10 cm^{-1} upon grafting. The second cage in the series to be attached was the lithium carbollide. Somewhat less sterically hindered than CoB_2, this too could be readily attached to PCMS. Being an anionic species apparently did not prevent the formation of the carbanion. Surprisingly the bulky CoB_2 was readily incorporated into the resin.

Another consideration was whether or not the extent of swelling effected the reaction rate and whether the carborane was reacting uniformly throughout the resin. THF and toluene were chosen as swelling solvents and cyclohexane as the non-swelling solvent. This

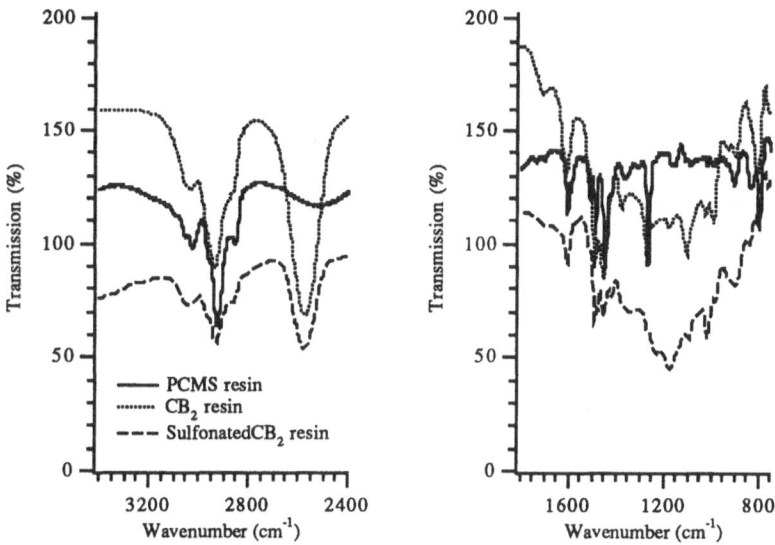

Figure 3. FTIR spectra of PCMS, CoB$_2$-PS, and sulfonated CoB$_2$-PS.

rate of reaction can be followed as a function of the normalized intensity of the B-H stretch. Samples were taken off via syringe and quenched with methanol. They were then washed several times with methanol to insure the removal of any unreacted carborane. Values of the absorbance were normalized against a C=C aromatic stretch @ 1600 cm^{-1}. By plotting the normalized B-H absorbance vs. time as seen in Figure 4, the extent of dicarbollide incorporation can be monitored. There is also a corresponding decrease in the absorbance of the CH$_2$-Cl stretch. The reaction in THF goes to completion in under 15 minutes whereas it takes over 4 hours in toluene. After 72 hours in cyclohexane the reaction had considerably slowed down at a conversion of approximately 60%. The faster reaction rate in THF is due to better solvation of both the n-buLi and carborane carbanion ion pairs in THF. The slow and incomplete reaction in cyclohexane is due to both poor solubility and low swelling of the resin. Since optimum reaction conditions were obtained in THF all subsequent reactions were performed in THF.

Uniformity of the grafting process could be readily monitored visually and confirmed by FTIR. As the white resin is swollen with the carbollide carbanion it turns a brilliant yellow. Pieces of the resin were removed from the reaction mixture and were sectioned. After reacting for an hour the yellow color was uniform throughout the resin. The quenching of the carbanion led to a uniform reversal of the color of the resin. In THF the values for the normalized B-H absorbance were consistent at 2 mm intervals from the surface to the center of a 25 mm resin cylinder. Using the optimum conditions as determined by carborane experiments, CoB$_2$ was likewise incorporated into the resin.

Elemental analysis shows that the incorporation of various salts of CoB$_2$ onto the resins in increasing order is: Na < Cs < TMA, as shown in Table 1. These values correspond to 5%, 10% and 30% incorporation based on the number of chloromethyl sites available. Loading values are low since a control sample of the TMA CoB$_2$ itself tested low for boron: calc. - 43%; found - 27%. The inherent stability of the dicarbollide could pose a problem for the complete digestion of the sample for boron analysis. Improved digestion methods are currently being investigated. Samples for elemental analysis were taken at 24 hours reaction time. Figure 4 also shows a comparison of the grafting kinetics for the TMA and Cs salts of CoB$_2$. After 24 hours the amount of TMA CoB$_2$ incorporated has reached a maximum whereas this level of incorporation was not achieved for the Cs salt until 72 hours.

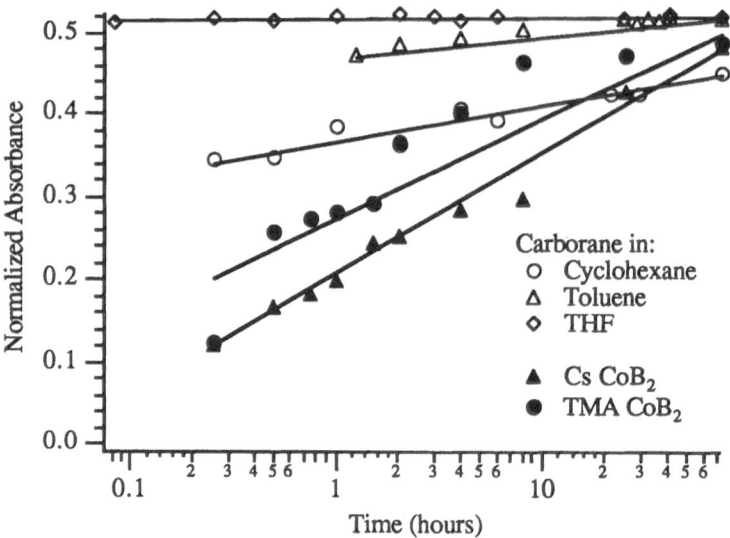

Figure 4. Following the reaction as a function of normalised B-H absorbance. The dicarbollide reactions were run in THF.

Table 1. Elemental Analysis for Cobalt Dicarbollide Containing Polymers.

Sample	C	H	N	Cl	B	S
untreated PCMS						
calc.	74.6	6.2		19.2		
found	75.2	6.7		16.1		
PS-TMACoB2						
calc.	43.1	7.6	2.6	0.0	35.8	
found	67.2	7.4	1.2	10.0	10.4	
PS-CsCoB2						
calc.	30.9	4.9		0.0	32.3	
found	70.9	6.8		11.5	3.9	
PS-NaCoB2						
calc.	37.8	6.0		0.0	39.5	
found	74.4	6.2		13.1	1.4	
SPS-TMACoB2						
calc.a	42.3	7.5	2.5	0.0	35.2	0.7
found	57.1	6.2	0.9	3.5	6.7	4.9
PBI-TMACoB2						
calc.	42.4	6.2	6.2	0.0	28.6	
found	65.9	5.3	9.9	3.2	2.5	

a) Calculations were based on an assumed sulfonation level of 5 mole%

    ^{13}C and ^{11}B solids NMR has been used to confirm the presence of carborane and CoB_2 on both the PS and PBI resins. The sharp peak shown in Fig. 5 for the dicarbollide is due to the tetramethyl ammonium counterion and not the carbons in the cage. The presence of this peak in the CoB_2-PS spectrum is further evidence of the incorporation of the dicarbollide onto the resin. ^{11}B proton decoupled MAS has only been able to provide additional confirmation of the presence of boron in the system. Fortunately carborane and CoB_2 are the only sources of boron in the system. Unfortunately quadrupolar interactions of ^{11}B leads to a significant amount of broadening of the peaks. Even the ^{13}C spectrum of carborane has only one weak fairly broad peak. When Co^{3+} is introduced into the system this broadening significantly worsens. Line broadening on the order of tens of ppm makes quantitative analysis difficult. This broadening is due more to the paramagnetic nature of the cobalt than quadrupolar interactions. Carborane itself gives rise to three sharp peaks in the ^{11}B spectrum, while the CoB_2 has peaks which are 25 ppm wide at half height. When attached to a polymer these peaks significantly broaden. This is also true for the already broad peaks of the CoB_2. Apparently tethering the carborane cage to the polymer has led to a change in the relaxation of the borons in the cage. This phenomenon warrants further investigation.

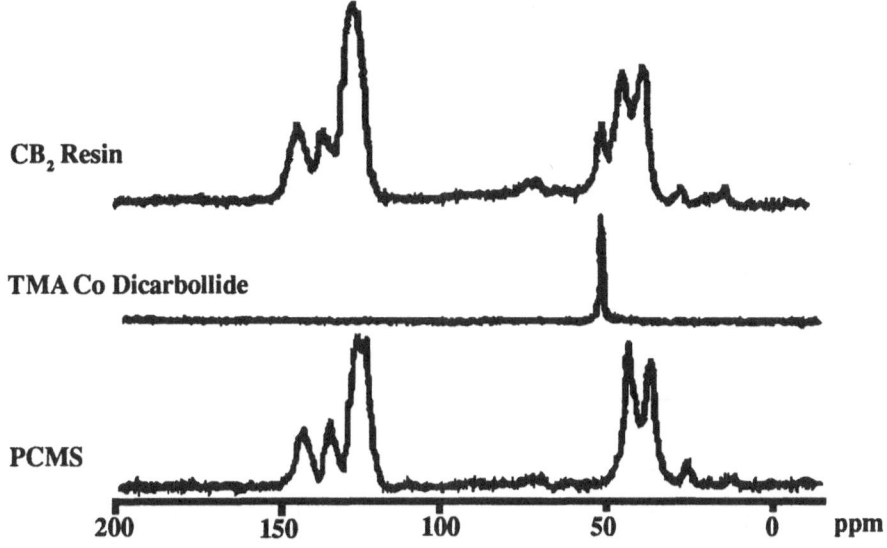

Figure 5. ^{13}C solid state NMR spectra of the starting materials and the CoB_2-PS resin.

K_d values for PS, SPS, and resins functionalized with the dicarbollide are given in Table 2. K_d values for the CoB_2-SPS, both grafted and absorbed are zero. This in part is due to the hydrophobic nature of PS. When the PS resins are placed in an aqueous environment there is no intimate contact between the resin and the dissolved metals. In order to get better contact between the CoB_2 on the resin and the solution, the support need to be more hydrophilic in nature. Since styrenics can be readily sulfonated, these resins were subsequently sulfonated as previously depicted in Figure 2. Elemental analysis of the sulfonated resins gave high values of sulfonation loadings. Even samples with higher PS content did not have sulfonation levels significantly higher than those that were highly substituted. Sulfonation appears to be occuring on phenyl rings that have other substituents. Although sulfonating the PS improves the hydophilicity of the resin, SPS is a commercially available broad based ion exchange resin.

As expected, the K_d values for SPS are high. It was disconcerting to see that the K_d values for the derivatized resins were actually lower. In order to determine the selectivity of the resins, the uptake of ^{22}Na from 0.01M HNO_3 was compared to ^{137}Cs uptake. K_d values of 424 and 918 ml/g, respectively, indicate that SPS has no inherent selectivity (Cs/Na = 918/424 = 2.2). Similar tests on CoB_2-SPS yields a selectivity of Cs/Na = 329/25 = 13.2. These initial results show that even though the total amount of cesium and strontium removed from solution by the CoB_2-SPS is less than that of SPS. the derivatized resins are six times more selective.

PBI being more hydrophilic than PS was subsequently chosen as a support for the CoB_2. Although CoB_2 could be incorporated onto the resin, the K_d values were actually negative for this resin. The dry resin was found to absorb water from the system, actually serving to concentrate the system thereby giving negative K_d values. Elemental analysis of the CoB_2-PBI samples show that only 10% of the theoretical dicarbollide was incorporated into the system. Scanning electron microscopy (SEM) shows that the surface of these resins have very few exposed pores. A cross section of these resins shows radial pores encased in a shell which is nominally 4 μm thick. The presence of this shell hinders the grafting reaction as well as preventing the uptake of Cs and Sr.

Table 2. Cesium and Strontium K_d Values Under Acidic and Basic Conditions.

Resin	0.1 M HNO$_3$		0.1 M NaOH	
	Cs	Sr	Cs	Sr
PS	0	0	0	0
SPS	104	325	239	1,374
in 0.01 M HNO$_3$	401	7,154		
Adsorbed CoB$_2$ on PS	0	0	0	0
Grafted CoB$_2$ on PS	0	0	0	0
SPS-CsCoB$_2$ 80-0-20	73	154	219	662
SPS-CsCoB$_2$ 70-10-20	86	159	307	701
SPS-CsCoB$_2$ 60-20-20	64	99	282	520

CONCLUSIONS

Cobalt dicarbollide can be covalently bonded to both PS and PBI resins. THF was found to be the most efficient solvent for the grafting both the carborane and cobalt dicarbollide onto these resins. Even when cobalt dicarbollide is tethered directly to a polymeric support it provides a material that is selective towards removing cesium and strontium from aqueous solutions without the use of any organic solvents. Resins containing CoB$_2$ have demonstrated selectivity six times greater that of sulfonated polystyrene. Since sulfonates generally do not exhibit any selectivity between various cations, the level of sulfonation needed to impart hydrophilicity but not reduce the selectivity needs to be determined. Further work is being done to determine the optimum cobalt dicarbollide loading and degree of sulfonation to obtain the optimum selectivity.

Acknowledgments

Funding was provided by the DOE Office of Environmental Management under the Efficient Separations & Processing Integrated Programs. We would also like to thank P. Hurlburt, R. Miller, T. Foreman, K. Abney, and S. Kinkead of CST for providing all the carborane derivatives for this work. K_d experiments were performed by K. Abney. SEM's were run by V. Gurule of MST-7.

REFERENCES

1. U.S. Dept. of Energy, " Department of Energy Plan for Recovery and Utilization of Nuclear Byproducts from Defense Wastes", DOE/DP-0013, Vol.2, (August 1983).
2. W. Schulz and L. Bray, Solvent extraction recovery byproduct 137Cs and 90Sr from HNO3 solutions - a technology review and assessment, *Sep. Sci. Tech.* 22:191 (1987).
3. C.D. Carlson, L.A. Bray, G.N. Brown, R.J. Elovich, K.J. Carson, Ion exchange removal of cesium from alkaline Hanford tank wastes, *NORM 94 Program and Abstracts* 56 (June 1994).
4. S.A. Bryan, C.D. Carlson, S.R. Adami, M.R. Telander, Radiation testing of organic ion exchange resins *NORM 94 Program and Abstracts* 57 (June 1994).
5. D.W. Wester, L.A. Bray, G.N. Brown, S.F. Yates, T.M. Kafka, New directions in removal of cesium and strontium from aqueous solutions *NORM 94 Program and Abstracts* 55 (June 1994).

6. J. Rais, P. Selucky, and M. Kyrs, Extraction of alkali metals into nitrobenzene in the presence of univalent polyhedral borate anions *J. Inorg. Nucl. Chem.* 38:1376 (1976).

7. V. Scasnar and V. Koprda, Extracton of 137Cs inot nitrobenzene by cobalt dicarbollide *Radiochem. Radioanal. Letters* 34:23 (1978).

8. M. Hawthorne and T. Andrews, Carborane analogues of cobalticinium ion *Chem. Commun.* 443 (1965).

9. E.S. Chandrasekaran; D.A. Thompson; and R.W. Rudolph, Attachment of 1,2-dicarbadodecaborane(12) to polystyrene. Catalysis by a polymer-bound rhodium complex *Inorg. Chem.* 17:760 (1978).

10. M. Chanda, K.F. O'Driscoll, and G.L. Rempel, Polybenzimidazole resin based new chelating agents. Ferric ion selectivity of resins with immobilized oligoamines *Reactive Polymers* 9:277 (1988).

11. M. Chanda and G.L. Rempel, Polybenzimidazole resin based new chelating agents. Uranyl and ferric ion selectivity of resins with anchored dimethylglyoxime *Reactive Polymers* 11:165 (1989).

12. M. Chanda and G.L. Rempel, Removal of uranium from acid sulfate by ion exchange on poly(4-vinylpyridine) and polybenzimidazole in protonated sulfate form *Reactive Polymers* 17:159 (1992).

MAGNETIC INVESTIGATIONS ON CU(II)- AND FE(III)-CONTAINING LIQUID CRYSTALLINE METALLOPOLYMERS

W. Haase[1], K. Griesar[1], E.A. Soto-Bustamante,[1] and Yu.G. Galyametdinov[2]

[1]Technische Hochschule Darmstadt
Institut für Physikalische Chemie, Petersenstr. 20
64287 Darmstadt, Germany
[2]Permanent address: Russian Academy of Sciences
Kazan Physico Technical Institute, Sibirsky Tract 10/7
420029 Kazan, Russia

INTRODUCTION

In the last years, we focused our interest on specific properties of paramagnetic metallopolymers[1,2] and metallomesogens[3-8] which may occur as a result of their structural features. The close-range order in these materials may provide special interactions, which are not present in related crystalline compounds. Moreover, temperature dependent magnetic susceptibility measurements can be generally used as a valuable tool to examine structural features.

The detection of intramolecular exchange interactions realized by superexchange mechanism is possible via an analysis of the magnetic susceptibility data. In general, the strength of such interactions indicates specific types of bridging ligands and geometrical features in polynuclear units. Therefore, it is possible to establish the existence of e.g. dimeric metal cores in metallomesogens (example: μ-O-iron(III)-dimer structure in ref. 3) or metallopolymers (example: μ-O-copper(II)-dimer structure in ref. 2) by means of this method.

In addition, models for the arrangement of spin-coupled ions in the crystalline and mesogenic phases of some metallomesogens have been developed based on the magnetic data. For example, antiferromagnetic 1-D-Heisenberg behaviours were reported for crystalline phases of different metallomesogens.[6,7,9]

Beside this limited number of examples for an occurence of 1-D-magnetic ordering, no 3-D-cooperative behaviours in the crystalline or in the liquid-crystalline phase of low-mass mesogenic compounds were observed, although theoretical considerations[10] suggest this.

In some cases, paramagnetic polymers with discrete transition metal complexes attached on their main chain offer pronounced deviations from the Curie-Law behaviour of their corresponding non-polymeric model compounds.[11,12,13] Although no detailed structural analysis was carried out, it was suggested that superexchange mechanisms via atoms situated in the polymer chain are essentially the reason for such deviations. To apply a more systematic approach, our investigations on polymethacrylates containing copper(II) N-salicylidene- or N-pyridoxalidene moieties were combined with structural and magnetic examinations on crystalline model compounds.[2] Here, a comparison of exchange interactions observed in the investigated polymers with those in the model compounds yields information about the arrangement in spin-centers within the polymers.

In this contribution, we present magnetic investigations on materials which essentially offer liquid crystalline and polymeric properties. Our approach to combine the magnetic behaviour of transition metal complexes possessing unpaired spins with liquid-crystalline and polymeric properties consists of using a suitable modified polymethacrylate with Schiff-Base units in the side chain for the incorporation of Cu(II) and Fe(III). The general structure of the polymer P[L$_p$] used for complexation is given in Figure 1:

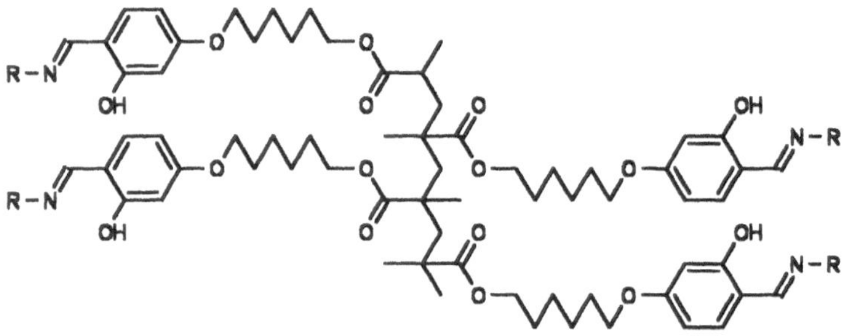

R=-Ph-OC$_8$H$_{17}$

Figure 1. Schematic representation of the chemical structure of polymer P[L$_p$] used for complexation.

Very recently, we reported on the magnetic properties of mesogenic Cu(II)- and rare-earth containing polymers derived from polymer P[L$_p$] or a similar polymer.[14] The Cu(II)-containing compound was prepared using Cu(OAc)$_2$ as starting material. In this contribution, we describe the magnetic behaviour of an analogous copper(II)-metallopolymer derived from CuCl$_2$ and the magnetic properties of 2 iron(III)-metallopolymers.

EXPERIMENTAL PART

Synthesis

The preparation of mesogenic metallopolymers with discrete transition metal complexes in their side chain was carried out using P[L$_p$] as starting materials. This compound can be used for complexation of different metal ions, since it offers N,O-bidentate pendent groups in the side chain. The (idealized) composition of the resulting metallopolymers is given in Table 1.

Table 1. Schematic representation of the idealized compositions of the investigated metallopolymers. In this scheme L_P refers to a repeating unit of the polymers, containing one N,O-chelating group.

Compound	Composition
P1	$Cu(L_P)_2$
P2	$Fe(L_P)_2Cl$
P3	$(L_P)_2Fe-O-Fe(L_P)_2$

The materials presented in Table 1 are soluble in different organic solvents like Toluene or Tetrahydrofuran. Details of synthesis as well as elemental analysis and other analytical results will be given in a separate paper.[15] The method applied for the formation of Fe-O-Fe-dimers in P3 was described in ref. 3.

As mentioned above, we have already presented the magnetic data of a copper(II) metallopolymer with identical composition to P1.[14] Along this line, P1 was prepared using $CuCl_2$ instead of $Cu(OAc)_2$ in order to test the influence of the counter ions on the reaction and their possible occurrence in the resulting polymer structure. Beside a slight difference in the degree of polymerization, the physical properties of P1 and the previous presented polymer are essentially the same.

Liquid Crystalline Behaviour

The liquid crystalline properties were examined by polarizing microscopy, Differential Scanning Calorimetry and X-ray diffraction techniques.[15]

Magnetic Susceptibility Measurements

The temperature-dependent susceptibilities in the range 4.2-300 K of the samples were recorded using a Faraday-type magnetometer. The measurements presented were done using a computer controlled Cahn D-200 microbalance and a Bruker B-MN 200/60 power supply.[16] The applied field was $\approx$ 1.5 T. The magnetic susceptibilities of P1 in the temperature range of 300-500 K were measured in the heating mode of an apparatus described in ref. 17, modified for this range. The molar susceptibilities were corrected for the underlying diamagnetism applying Pascal's scheme.[18] The $\chi(T)$-fits were performed with programs developed in our group.

RESULTS AND DISCUSSION

Liquid-Crystalline Behaviour

The phase transition temperatures of the metallopolymers P1 and P3 are summarized in Table 2.

Table 2. Transition Temperature (°C) for the mesogenic metallopolymers detemined by differential scanning calorimetry.

Compound	Phase behaviour
P1	G - 76.4 - S_A - 184.1 - I
P3	G - 82 - S_A - 147.5 - I

G = Glassy state
S_A = Smectic A phase
I = Isotropic phase

Magnetic Susceptibility Measurements

Copper(II) metallopolymer.

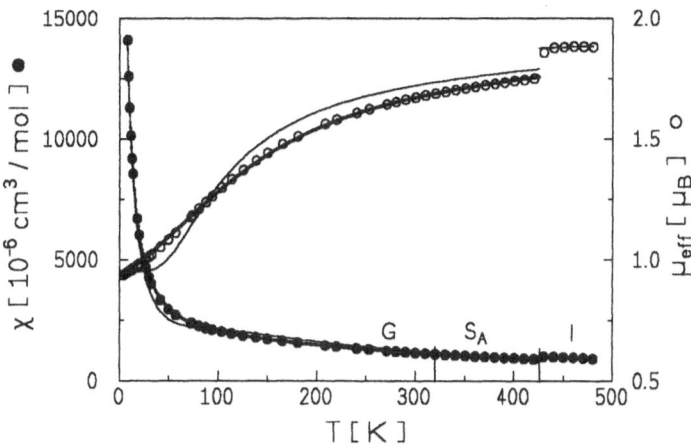

Figure 2. Magnetic susceptibilities and effective magnetic moments for P1. In the glassy and smectic phase, the wide lines based on a fit using the chain model (eq. (1)) whereas the narrow lines referred to a fit using the dimer model (eq. (2)). A fit to the Curie-Weiss law was carried out for the data in the isotropic phase (for details see text).

Figure 2 shows the temperature dependent magnetic susceptibility data of metallopolymer P1 in the temperature range 4.4-480 K. The data in the glassy and smectic phase are in agreement with a strong antiferromagnetic coupling between copper centers (1.67 μ_B at 300 K). The relative high values of μ_{eff} at 4.2 K (0.94 μ_B) is due to a remaining part of monomeric (i.e. non-coupled) copper sites, which follows a Curie- or Curie-Weiss law. In contrast to the reduced effective magnetic moment in the glassy and smectic phase, we observed normal magnetic behaviour in the isotropic phase (μ_{eff}= 1.85 μ_B at 430 K).

The temperature dependency $\chi(T)$ for P1 is in good agreement to the magnetic behaviour already reported for the analogous Cu(II)-polymer derived from Cu(OAc)$_2$.[14] Consequently, the counterion is not involved in the strucutre and the composition of P1 can be considered as Cu(L$_p$)$_2$ (see Figure 3).

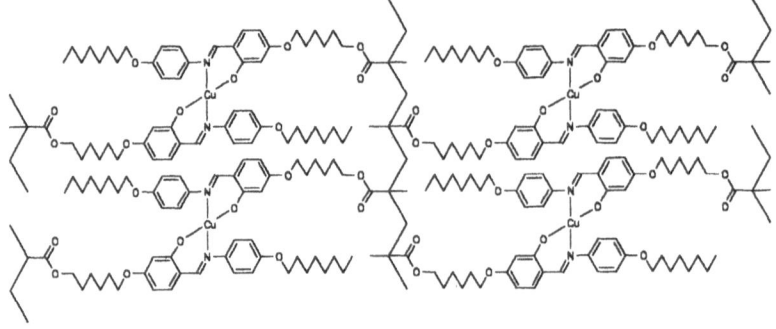

Figure 3. Schematic representation of the chemical structure Cu(Lp)₂ of P1.

The structural proposal as displayed in Figure 3 is based on the assumption of square-planar coordinated monomeric copper ions. As described in our previous publication,[14] the observed antiferromagnetic exchange interactions in our Cu(II) metallopolymers can be explained if we consider more extended arrangements of spin centers, which may realize efficient superexchange pathways:

a) dimeric units (Figure 4a)
b) linerar chain motive (Figure 4b).

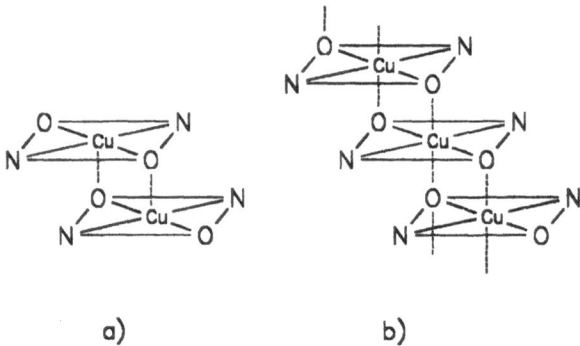

a)   b)

Figure 4. Schematic representation of different structural arrangements in the polymer: a) dimer; b) 1-D-Heisenberg linear-chain model.

Additionally, an adequate description of the magnetic behaviour of an amorphous metallopolymer has to take into account the random polymer backbone conformation and some degree of structural disorder. Therefore, the appropiate model involves a mixture of non-interacting (molar fraction x_P) and antiferromagnetic coupled (molar fraction $1-x_P$; coupling strenght J) copper species. Accordingly, we have fitted the magnetic data to Eq. (1) (dimer model) and Eq. (2) (1-D-Heisenberg linear-chain model).

$$\chi = (1-x_p)\,\chi^{II} + x_p\,\chi^{I} + N_\alpha \qquad (1)$$

$$\chi = (1 - x_p) \, \chi^{III} + x_p \, \chi^I + N_\alpha \qquad (2)$$

using

$$\chi^I = \frac{N_L \, g^2 \, \mu_B^2}{3kT} \, S(S+1) \qquad (3)$$

$$\chi^{II} = \frac{N_L g^2 \mu_B^2}{kT} \cdot \frac{\exp{(2J/kT)}}{1 + 3 \exp{(2J/kT)}} \qquad (4)$$

$$\chi^{III} = \frac{N_L g^2 \mu_B^2}{kT} \cdot \frac{0.25 + 0.14995 x + 0.30049 x^2}{1 + 1.9862 x + 0.68854 x^2 + 6.0626 x^3} \qquad (5)$$

with $x = J/kT$.

Here, N_α is the temperature independent paramagnetism. The obtained fit parameters for P1 are summarized in Table 3:

Table 3. Results of fitting the susceptibility data in the glassy and smectic phase for P1 according to dimeric (Eq. 1) and 1-D-Heisenberg (Eq. 2) model.

Model	g	$J(cm^{-1})$	$x_p(\%)$
dimer model	2.12	-79.2	26.5
chain model	2.12	-72.4	24.2

The magnetic properties of the Cu-polymer presented in this paper are nearly identical with those of the compound already reported. In both cases we yield the better fit over the complete temperature range with the chain model (Eq. (2)).

A refined fitting procedure, which allows the coexistence of copper centers occupying monomeric (molar fraction x_p), dimeric (molar fraction x_{dim}; coupling strenght J_{dim}) and chain (coupling strength J_{chain}) sites yield g=2.15; J_{chain}=-71.3 cm^{-1}, J_{dim}=-78 cm^{-1}, x_{dim}=8.6% and x_p=24.8%.

The specific arrangement of the copper(II)-ions which leads to the 1-D-Heisenberg behaviour is connected with a combination of polymeric and liquid-crystalline properties: In the isotropic phase, these interactions vanish and we have found a normal Curie-Weiss behaviour (g=2.12; θ=-7 K). For a monomeric non-mesogenic model compound, we observed a Curie-Weiss law (g=2.12, θ=-1.7 K). [14]

Cooling sample P1 down from the isotropic phase to the smectic phase leads to increased susceptibilities compared to the initial values recorded by heating (see Figure 5).

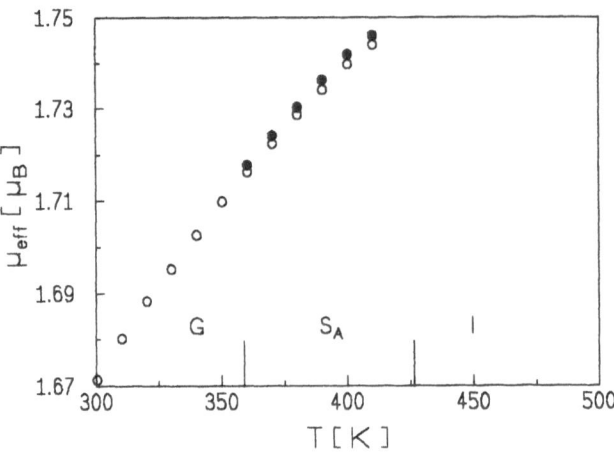

Figure 5. Effective magnetic moments for P2 in the temperature range 300-480K; (O): heating process; (●): cooling process.

This behaviour can be referred to a magnetic field-induced orientation in the liquid-crystalline phase of a magnetically anisotropic sample with its axis of maximum susceptibility parallel to the field. The orientation of a mesogenic phase leading to increased susceptibilities can be observed rather by cooling from the isotropic phase than heating up from room-temperature. The principles of such an orientation were well established for metal-free mesogenic compounds (see for example ref. 19). Here, the anisotropy is generally produced by diamagnetic phenyl groups (i.e. axis of maximum susceptibility parallel to the director); the molecules tend to orient with the director parallel to the magnetic field H. In a separate paper, we present a systematic summary of our EPR- and magnetic susceptibility investigations on Cu(II) and VO(IV) containing mesogens.[5] The way of orientation (director parallel or perdendicular to H) was detected by angular-dependent EPR-measurements. Generally for metallomesogens which contain paramagnetic ions, the total magnetic susceptibility anisotropy results from the supercomposition of the paramagnetic anisotropy of the ion and the diamagnetic anisotropy of the phenyl groups. For mesogenic square-planar coordinated Cu(II)-complexes, a competition of paramagnetic and diamagnetic anisotropy occcurs, since the axis of maximal paramagnetic susceptibility is situated perpendicular to the director. Therefore, in this case the resulting anisotropy is influenced mainly by the number of phenyl groups per copper ion. If the number of phenyl groups per copper is greater than 4, the diamagnetic anisotropy exceeds the paramagnetic one and one can observe an orientation of the director parallel to H.

In the case of our mesogenic Cu(II)-polymer P1 (4 phenyl groups per copper), we can assume that the molecular units orient with their molecular long axis perpendicular to the magnetic field. A calculation according to the scheme developed in ref. 5 leads to a theroretical value of $\Delta\chi^{i \rightarrow m} = \chi^m - \chi^i = 3.4 \cdot 10^{-6}$ cm^3/mol for the increase of susceptibility by going from the isotropic (χ^i) to the mesogenic (χ^m) phase. In the frame of this estimation, we consider a unique composition of the polymer which consist of 4 phenyl groups per square-planar coordinated Cu(II) core (see figure 3). The assembling of Cu(II)-ions in different arrangements (monomers, dimers , chain...) should not modify the orientation behaviour expected for a monomeric unit: In such arrangements as displayed in Figure 3, the coordinations planes are oriented parallel and the average anisotropy is approximately the same as for an isolated chelate unit. Further, we assume that one Cu(II)-ion exhibits a reduced susceptibility of $905 \cdot 10^{-6}$ cm^3/mol at the phase transition smectic to isotropic (426 K) due to the averaging of antiferromagnetically and non-coupled copper centers.

The difference of the observed value for $\Delta \chi^{i \to m} = \chi^m - \chi^i = 2.3 \cdot 10^{-6}$ cm^3/mol and the calculated one relates mainly to the random distribution of molecular long axis over the director which can be described by an order parameter $P \approx 0.6$.

Iron (III)-compounds. Our work concerning the mesogenic iron(III) polymers P2 and P3 (see Table 1; phase transitions see Table 2) was stimulated by previous investigations on their corresponding low-molecular mass model compounds M1 and M2 (see Figure 6a and b).

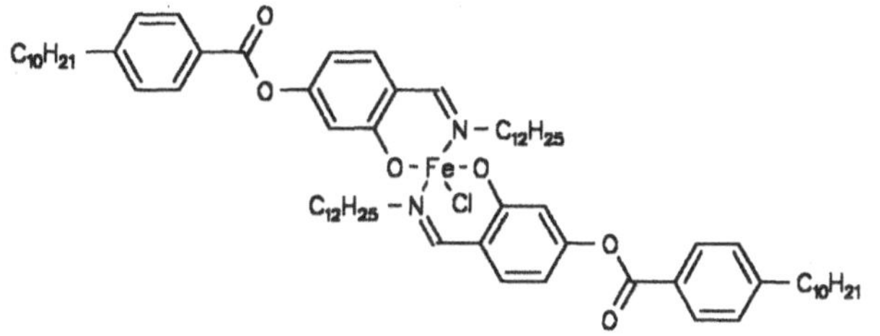

Figure 6a. Chemical structure of M1.

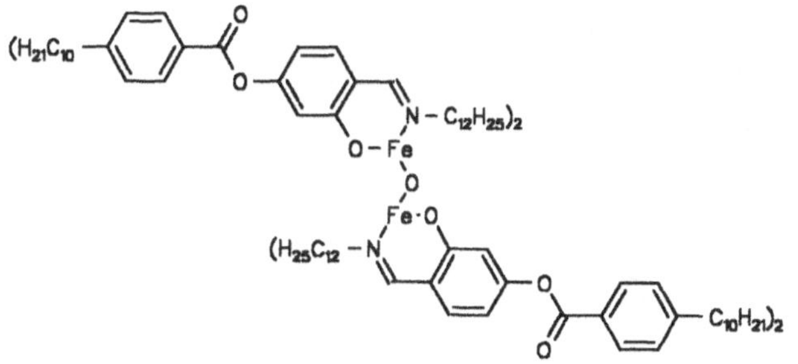

Figure 6b. Chemical structure of M2.

M2 was obtained from the reaction of M1 with oxygen as described in ref. 3. The oxygen-bridge which connects the two iron(III) centers leads to a significant variation of the magnetic properties: M1 follows a Curie-Weiss law with $\theta = -0.5$ K [2], whereas M2 exhibits a strong antiferromagnetic exchange interaction between the iron(III) centers (J=-91.8 cm^{-1}; ref. 3)

The basic idea was to transfer the well established formation of µ-oxo-bridged iron(III) centers which is accompanied with drastic changes in the magnetic behaviour to a polymer structure. The application of this reaction starting from P2 (Fe(L$_p$)$_2$Cl) leads to a polymer which offers a composition given by P3 $((L_P)_2Fe\text{-}O\text{-}Fe(L_P)_2)$, if 100% of all iron centers react. This reaction can be regarded as an additional crosslinking process - besides the crosslinking of polymer chains through iron(III) via complexation, which is still present in P2. Consequently, a detailed analysis of the magnetic data gives informations about the degree of oxygen-linking in the network of polymer P3.

Figure 7 displays the experimental temperature dependent magnetic susceptibility data and effective magnetic moments for P2 (Figure 7a) and P3 (Figure 7b) in the temperature range 4.2-300 K.

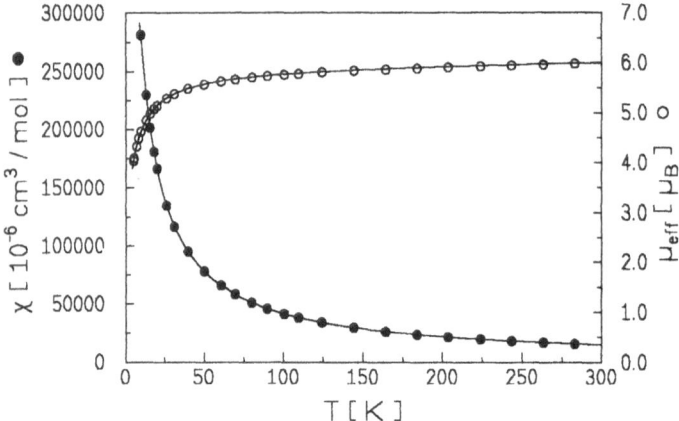

Figure 7a. Magnetic susceptibilities and effective magnetic moments for P2 in the temperature range 4.2-300 K.

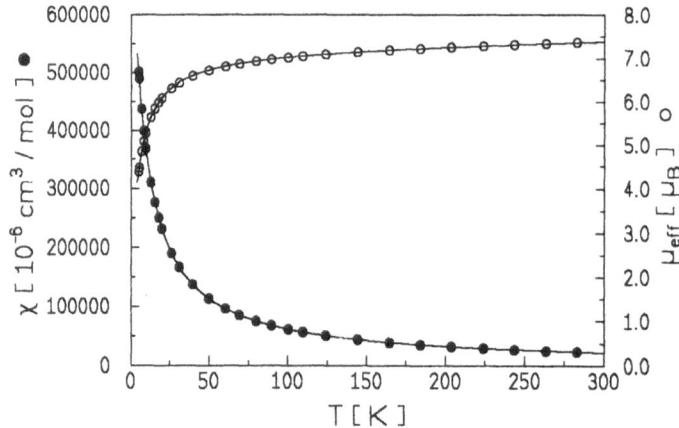

Figure 7b. Magnetic susceptibilities and effective magnetic moments per dimer for P3 in the temperature range 4.2-300 K.

296

We observed a relative drastic decrease of the effective magnetic moment at room temperature by going from P2 (μ_{eff}=5.98 μ_B per iron at 282.7 K) to P3 (μ_{eff}=7.37 μ_B per dimer; i.e. 5.21 μ_B per iron at 282.7 K). In both polymers, the effective magnetic moment decreases with lower temperatures, indicating the presence of some antiferromagnetic exchange interaction. For P3, this variation is more pronounced (μ_{eff}=4.38 μ_B per dimer; i.e. 3.10 μ_B per iron at 4.8 K) than for P2 (μ_{eff}=4.04 μ_B per iron at 4.8 K). Due to the thermal instability of the compounds P2 and P3, it was not possible to examine the magnetic behaviour in their liquid crystalline phases.

In accordance with our previous magnetic susceptibility studies on the related models M1 and M2, we can assume that this change is caused by the formation of strong antiferromagnetically coupled Fe-(μ-O)-Fe-units. To compare the fundamental changes in magnetic behaviour by going from P2 to P3 and decribe this crosslinking process using a limited number of parameters, we have fitted the temperature dependent susceptibility data of P2 and P3, applying the simplest model allowing a discussion:

The magnetic behaviour of P2 can be phenomenologically described as a mixture of non-coupled Fe(III) (molar fraction x_p) and interacting Fe(III)-ions; the interactions are represented by a Curie-Weiss law. The best fit-parameters obtained were g=2.0 (fixed), x_p=15% and θ=-8.1 K.

The relative large θ-value compared with the corresponding value calculated for the model compound M1 reflects the existence of some antiferromagnetic pathways within the polymeric structure, which are not present in the monomer. These interactions are most probably due to linking of some part of Fe(III) sites via chlorine or phenolate. However, the resulting variation of μ_{eff} versus temperature is not so pronounced as to justify a more detailed description.

As described below, we have obtained metallopolymer P3 from P2 using the reaction which transfer M1 to M2. If we use a polymer as starting material, we can expect that only a limited amount of Fe(L$_P$)$_2$Cl centers react to give (L$_P$)$_2$Fe-O-Fe(L$_P$) bridges.

Therefore, we consider that in P3 a molar fraction (1-x_p) of Fe(III) occupying Fe-(μ-O)-Fe-units; whereas the remaining part (x_p) can be regarded as obeying a Curie-Weiss law (see Eq. 6).

$$\chi = (1 - x_p)\, \chi^{II} + x_p\, \chi^I + N_\alpha \tag{6}$$

$$\chi^I = \frac{C}{3(T - \theta)} S(S+1) \tag{7}$$

$$\chi^{II} = \frac{C}{T} \cdot \frac{2\exp(2x) + 10\exp(6x) + 28\exp(12x) + 60\exp(20x) + 110\exp(30x)}{1 + 3\exp(2x) + 5\exp(6x) + 7\exp(12x) + 9\exp(20x) + 11\exp(30x)} \tag{8}$$

using x=J/KT, $C = \dfrac{N_L g^2 \mu_B^{\,2}}{k}$

To avoid a correlation of parameters, our fit is based on the reasonable approximation that the dimeric units exhibit the same coupling strength J=-91.8 cm^{-1} as observed for the dimeric model compound M2. The fitting procedure yield g=2.00 (fixed), J=-91.8 cm^{-1} x_p=75% and θ=-8.3 K i.e. 25% of the possible crosslinking centers did react. Again, the relative high θ-value may indicate the occurrence of some chlorine or phenolate-bridges. The value of x_p reflects the sterical hindrance for a complete Fe-O-Fe-crosslinking in the polymer.

Figure 8 gives a schematic view of the suggested network, including dimeric and monomeric iron(III) cores.

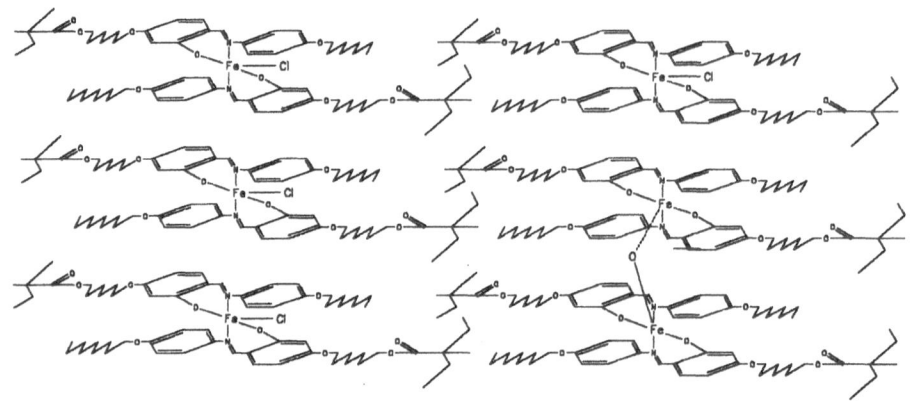

Figure 8. Suggested structure of P3.

CONCLUSION

We report on the magnetic investigations of modified methacrylates with different Cu(II)- and Fe(III)-Schiff-base complexes in their side chain.

The copper(II) containing compound exhibits strong antiferromagnetic exchange coupling in the glassy and smectic phase due to axial interactions between adjacent copper-ions. The existence of these interactions can be refered to the combination of liquid-crytalline and polymeric properties, since they are not present in the isotropic phase of the investigated polymer and in monomeric model compounds of similar structure.

The crosslinking of a iron(III) Schiff-base polymer via oxygen bridges leads to a formation of antiferromagnetic coupled Fe-(μ-O)-Fe-units. Based on the magnetic behaviour in comparison with these of monomeric Fe(III)-model compounds we can elucidate the structure of the resulting polymer.

Acknowledgement

We gratefully acknowledge financial support by Deutsche Bundespost TELEKOM, and Deutsche Forschungsgemeinschaft (DFG).

REFERENCES

1. W. Haase, K. Griesar and C. Erasmus-Buhr, in Electrical, Optical, and Magnetic Properties of Organic Solid State Materials, edited by L. Y. Chiang, A. F. Garito, D. J. Sandman (MRS Symposium Proceedings V.247, 1992), p. 455.
2. W. Haase, K. Griesar, M. F. Iskander and Y. Galyametdinov Mol. Materials, 3, 115 (1993).
3. Y. Galyametdinov, G. Ivanova, K. Griesar, A. Prosvirin, I. Ovchinnikov and W. Haase, Advanced Mater., 4, 739(1992).
4. K. Griesar, Y. Galyametdinov, M. Athanassopoulou, I. Ovchinnikov and W. Haase, Adv. Mater., 6, 381 (1994).
5. I. Bikchantaev, Y. Galyametdinov, A. Prosvirin, K. Griesar, E. A. Soto Bustamante and W. Haase, Liq. Cryst., in print.
6. V. Salas Reyes, G. Soto Garrido, C. Aguilera, K. Griesar, M. Athanassopoulou, H. Finkelmann and W. Haase, Mol. Materials, in print.
7. W. Haase, V. Salas Reyes, M. Athanassopoulou, G. Soto Garrido, K. Griesar, E. Schuhmacher, Molecular Engineering, in print.
8. R. Paschke, S. Diele, I. Letko, A. Wiegeleben, G. Pelzl, K. Griesar, M. Athanassopoulou and W. Haase, Liquid Crystals, in print.
9. M.P. Eastman, M.L. Horng, B. Freiha, K.W. Sheu, Liq. Cryt., 2, 223.
10. M. Buivydas, Phys. Status Solid B, 168, 577 (1991).
11. K.M. More, S.S. Eaton and G.R. Eaton, J. Am. Chem. Soc., 103, 1087 (1981).
12. X.S. Cheng, M. Kamachi, W. Mori and M. Kishita, Polymer Journal, 22, 39 (1990).
13. M. Kamachi, M. Shibasaka, A. Kajiwara, W. Mori and M. Kishita, Bull.Chem. Soc. Jpn., 62, 2465 (1989).
14. W. Haase, E.A. Soto-Bustamante, K. Griesar and Y. Galyametdinov, Mol. Cry. Liq. Cryst., submitted.
15. E.A. Soto-Bustamante, Y. Galyametdinov, K. Griesar, E.Schuhmacher and W. Haase, in preparation.
16. S.Gehring, P. Fleischhauer, H. Paulus and W. Haase, Inorg. Chem., 32, 54 (1993).
17. L. Merz and W.Haase, J. Chem. Soc. Dalton Trans., 875 (1980).
18. A. Weiss, and H. Witte, Magnetochemie (Verlag Chemie. Weinheim, FRG, 1973).
19. I.. Ibrahim and W.Haase, J. de Phys., Paris, C3, 40, 164 (1979).

NOVEL PHOTONIC MATERIALS CONTAINING PORPHYRIN RINGS

Zhenan Bao, Luping Yu*

Department of Chemistry
University of Chicago
5735 S. Ellis Ave.
Chicago, IL 60637

INTRODUCTION

Recently, interests in soluble, well-defined polymers containing transition metals are increasing because of the demands for advanced materials in molecular electronics, electro-optics, and catalytical processes.[1-3] Our major research effort is focused upon the synthesis of new polymers exhibiting multifunctional properties, such as photorefractive polymers which combine the photoconductivity and optical nonlinearity to manifest photorefractivity.[4] Polymers containing metalloporphyrin arouse our interest for these purposes due to their unique electronic structures and optical properties. It is known that many porphyrin and metalloporphyrin compounds are responsible for charge separation in natural photosynthetic centers and in many photosynthetic model compounds.[5] These properties are relevant to the design of photorefractive polymers.

Several examples of porphyrin-containing polymers are known. They can be classified into three categories: (a) main-chain porphyrin polymers; (b) side-chain porphyrin polymers; and (c) face-to-face porphyrin polymers. The target function for these different forms of

Metal-Containing Polymeric Materials
Edited by C.U. Pittman, Jr., *et al.*, Plenum Press, New York, 1996

polymers vary. For example, the side-chain polymers were synthesized for studying the role of porphyrin or metalloporphyrins in the biological, biomimetic reaction system and their catalytic effects on the photoreductions.[6-8] On the other hand, main-chain porphyrin polymers were synthesized for the purpose of enhancing physical properties, such as electric conductivity and third-order nonlinear optical properties. There are only very few reports of main-chain porphyrin-containing polymers. The first example was a rigid rod oligmer with four porphyrin units fused together by coplanar aromatic systems.[9] The synthesis of this extended porphyrin system involved a seven-step sequence. This conjugated porphyrin oligmer resembles a molecular wire because of its rigid-rod nature and electrical conducting property. Another example reported by Bradley et al was a conjugated zinc-porphyrin polymer linked with acetylene units.[10] This polymer was synthesized by Glaser-Hay coupling of the meso-diethynyl zinc porphyrin. The use of ethynyl groups as spacers insured that conjugation can be extended through the porphyrin units. Recently, we succeeded in synthesizing a series of porphyrin-containing polymers. This paper describes our detailed experimental results.

Our research goal, however, is not limited to synthesizing porphyrin-containing polymers but also to study the general approaches to the synthesis of these kinds of polymers. The Heck reaction attracted our attention due to its mild reaction condition.[11] This reaction has been successfully applied for synthesizing several phenylene vinylene polymers.[12-14] We also reported the utilization of the Heck coupling reaction for the synthesis of substituted polyphenylenevinylene conjugated polymers, which demonstrated an interesting liquid crystallinity.[15] Our recent experiments indicated that the Heck reaction can provide us with a useful polymerization method under mild conditions to synthesize new polymers containing porphyrin moieties. Different metals can easily be incorporated into these polymers and their effects on the physical properties can be studied. In this chapter, we will first report the detailed synthesis and characterizations of several conjugated metalloporphyrin containing polymers. Further examples using the same approach to the preparation of other types of novel porphyrin containing polymers are illustrated.

EXPERIMENTAL PROCEDURES

Synthesis

Monomer **2** was synthesized according to a literature procedure.[15] Monomer **1A** was a new compound and was synthesized in 68% yield from a dipyrromethane **3**.[16]

Metallation of monomer **1A**, by refluxing with transition metal acetate in chloroform/methanol (3:1) solution, yielded monomers **1B-1D** quantitatively.

The following is a typical polymerization procedure: Tri-n-butylamine (0.15 ml) was added to a mixture of metalloporphyrin monomer (0.12 mmol), 1,4-diiodo-2,5-dinonoxybenzene (compound **2**, 0.0718g, 0.12mmol), palladium acetate (0.0011g, 0.0047mmol), and tri-o-tolyphosphine (0.0036g, 0.012mmol) in DMF (5 ml). The reaction mixture was heated in an oil bath at 100°C under a nitrogen atmosphere until the polymer precipitated out of the reaction mixture. It was then poured into acetone (75 ml) and the precipitate was collected. To further purify the polymer, it was redissolved in chloroform or THF and the solution was filtered to remove any insoluble palladium residual. The filtrate was concentrated and precipitated into acetone. The resulting polymer was collected and extracted with methanol for 24 hours in a Soxhlet extractor.

Photoconductivity Measurements

An Ar^+ laser was used as the light source for the photoconductivity measurements. The photocurrent was determined by measuring the voltage across a resistor with a lock-in amplifier in conjunction with a mechanical chopper. The power of the laser light incident on the sample was 143 mW/cm^2 throughout the experiment. The positive voltage was applied to the illuminated ITO or Au electrode.

Fluorescence Measurements

All measurements were done in air-equilibrated chloroform solution. Steady-state fluorescence spectra were taken with Spex Fluorolog spectrofluorimeter. The fluroscence quantum yields were determined by the optically dilute method using Rhodamine 6G in ethylene glycol as reference whose quantum yield is 97%. Fluorescence decay profiles were recorded at room temperature in a time-correlated single-photon counting apparatus. An intracavity dumped Rhodamine 6G dye laser (575 nm , 3.8 MHz) was used as the exciting light source. The fluorescence at 636 nm was collected through a polarizer oriented at 54.7° (magic angle) relative to the vertically polarized excitation. The instrumental response function had a full width at the half maximum of 130 ps.

RESULTS AND DISCUSSIONS

Photoresponsive Materials

In order to study the feasibility of the Heck reaction for synthesis of porphyrin-containing polymers, we first conducted the reactions shown in scheme 1. The polymerizations were carried out smoothly in DMF. Due to the mild reaction conditions of the Heck reaction, the metallated porphyrin was incorporated into the polymer chain without affecting the metal chelation. Table 1 summarizes the results from GPC measurements (THF as solvent and poly-styrenes as standards) and elemental analysis. The structures of these polymers were also confirmed by spectroscopic studies.

Table 1. Elemental analysis and molecular weights of polymer I-A - I-C (determined by GPC using polystyrenes as standards).

Sample	cal.			found			M_w	M_n
	C%	H%	N%	C%	H%	N%		
polymer **I-A**, Fraction 1	79.63	8.80	4.22	78.36	8.89	4.15	4.6×10^4	3.0×10^4
Fraction 2							5.3×10^4	4.5×10^3
polymer **I-B**	79.74	8.82	4.23	78.36	8.91	4.17	8.3×10^3	4.2×10^3
polymer **I-C**	80.04	8.85	4.24	78.69	8.88	4.23	1.3×10^4	6.6×10^3

The ^{1}H NMR spectrum of polymer **I-A** (in CDCl$_3$) was dominated by the chemical shifts of the side alkyl chains at the region between 0.80 and 2.55 ppm. The meso proton of the porphyrin appeared at 10.10 ppm, the -OCH$_2$- protons at 4.19 ppm. The proton from the dialkoxy phenyl ring appeared at 7.34 ppm, while the aromatic proton at the phenyl ring linked directly to the porphyrin unit were found at 7.92 and 8.10 ppm. The vinyl protons were not equivalent and appeared at 7.73 and 7.82 ppm. The chemical shift of the vinyl benzene end group was visible in the spectrum as a result of the oligomeric fraction. Polymer **I-C** gave similar ^{1}H NMR spectral features. The paramagnetic property of Cu(II) severely broadened the spectrum of polymer **I-B**. However, after removing the copper ions from polymer **I-B** by treating the chloroform solution with concentrated hydrochloric acid followed by neutralizing with NaHCO$_3$, a similar spectrum to polymer **I-A** was obtained, except an additional chemical shift at -2.35 ppm was seen due to the NH proton.

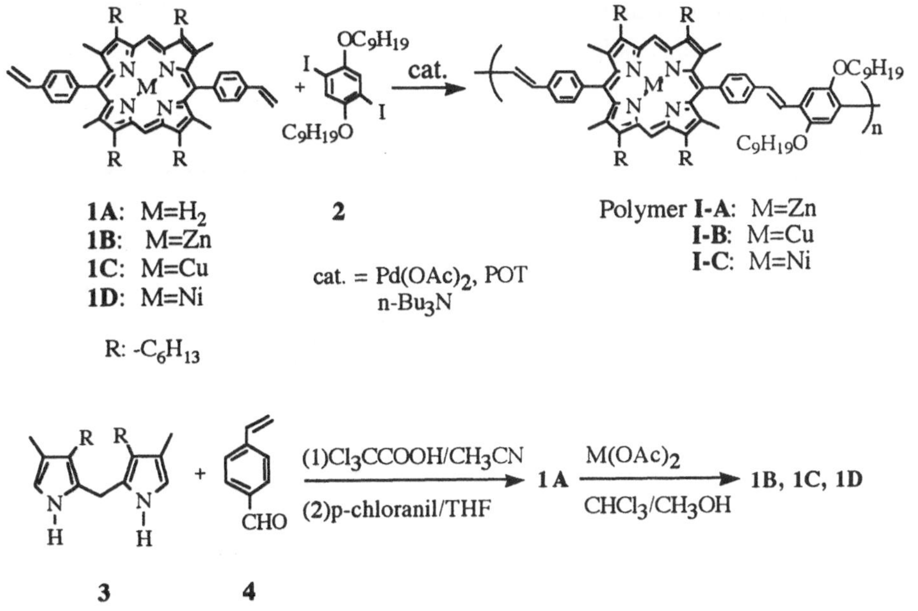

Scheme 1. Synthesis of monomers and polymers for photoactive porphyrin-containing polymers.

FTIR spectra of all the polymers showed typical absorption of the ether linkage at 1208 cm^{-1} after polymerization due to the introduction of the dialkoxybenzene units. The appearance of the absorption at 965 cm^{-1} due to the out-of-plane mode of the =C-H of the double bond indicated that their conformations were in the trans-form.

All of the polymers were soluble in common organic solvents, such as THF, chloroform and chlorobenzene. Uniform, optical-quality films were obtained by either casting or spin-coating on glass slides. The TGA studies indicated that the polymers were stable up to 400 °C under a nitrogen atmosphere. There was no melting or glass transition observed in the DSC thermograms between room temperature and 400 °C.

The UV/vis spectra of the polymers were similar with their corresponding metalloporphyrin monomers. (see Figure 1) A strong soret band at 408 nm and Q bands between 500 nm and 600 nm could be clearly identified. The emission spectrum (Figure 1 inset) of polymers **I-A** also resembled its corresponding zinc porphyrin monomer. For the THF solution of polymer **I-A**, the excitation at Q (1,0) band (550 nm) resulted in an emission spectrum at 585 nm and 640 nm. These results indicated that the metalloporphyrin moieties in the polymer backbone were kept in tact throughout the polymerization and the electron delocalization was limited. No fluorescence was observed for either monomers or polymers with the Cu and Ni porphyrin moieties.

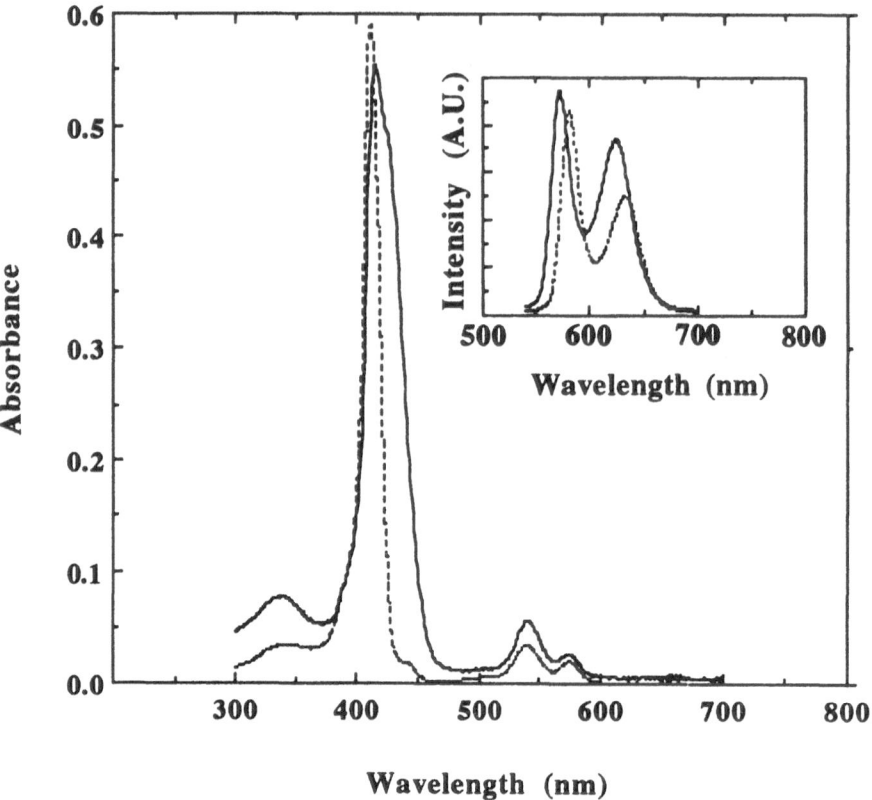

Figure 1. UV-vis and fluorescence (inset) spectra of polymer **I-A** (———) and monomer **1B** (--------).

The redox properties of these polymers were studied by cyclic voltametry (CV) in THF or chloroform solution. The results showed that polymerization did not significantly affect the redox properties of the porphyrin moieties compared to their corresponding monomers. For example, two reversible oxidation processes and a reversible reduction process for polymer **I-A** were observed at 1.01V, 1.22V, and -1.66V, respectively, in THF solution vs. SCE reference electrode. The corresponding monomer **1B** showed similar behavior at 0.98V, 1.18V, and -1.66V. This result was expected due to the limited conjugation in these systems.

The photoconducting property of these polymers was studied by using an Ar⁺ laser as the light source. It was noted that the photocurrent (I_{ph}) increased steadily under a low electric field, and increased dramatically at a high electric field (Figure 2). The dark-current was small, in the range of 10^{-9} Å measured by a Keithley 617 electrometer. A relatively high quantum yield for photocharge generation was observed, for example, a quantum yield of 2.3 % for polymer **I-B** at 514 nm with the film thickness of 1.65 µm under an electric field of 620 kV/cm, indicating a promising feature for the design of photorefractive polymers.

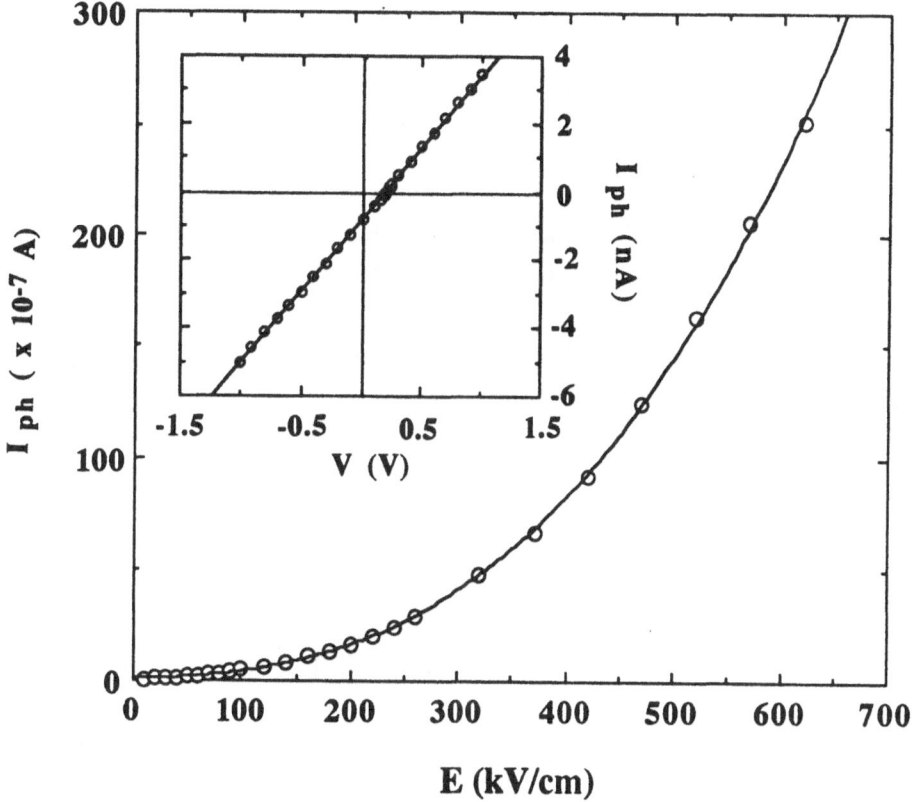

Figure 2. Photocurrent of polymer **I-B** as a function of the external field. The inset shows the photovoltage effect. The wavelength for measurements is 514 nm, and the laser intensity was 137 mW/cm².

Interestingly, all of the photo-cells exhibited photovoltage regardless of the type of electrode used, e.g. ITO/polymer/Au or Au/polymer/Au. The illuminated electrode always charged negatively. The dependence of the photocurrent on the applied voltage (Figure 2 inset) showed a displacement from the origin by $V_0 = V_{oc}$ (open-circuit voltage). When the ITO/polymer/Au cells of polymers **I-A, B**, and **C** were illuminated with 514.5 nm light, V_{oc} values were observed to be -0.17V, -0.20V, and -0.27V, respectively. Under the same conditions for a Au/polymer/Au cell, V_{oc} was found to be -0.14V for polymer **I-A**. Although the photovoltage was relatively small, the phenomena were encouraging and further improvements are possible because the synthetic approach is versatile.

Polymers with Alternated metals

The flexibility in preparing different types of monomers allowed us to synthesize polymers with alternated metals. Polymer **II-A** with Cu and Zn was synthesized according to scheme 2. As a comparison, polymer **II-B** with only Zn metal was also prepared. About 60 wt% of the polymer was soluble in CHCl$_3$. The insoluble part of the polymer was the higher molecular weight fraction since it had an identical FTIR spectrum as the soluble part. ^{1}H NMR spectra of the soluble parts were consistent with the structure as proposed. Cu-Zn and Zn-Zn showed similar spectral pattern. The existence of Cu porphyrin in the polymer was confirmed by the strong ESR signal. Both polymers were stable up to 400°C under nitrogen and no melting or glass transition temperature was observed by DSC.

The UV-vis spectra of the polymers were similar to those of the monomers except for the broadening in the Soret band. This could be due to the phenylene vinylene spacer or the exciton coupling in the polymer. The steady state fluorescence quantum yield of polymer **II-A** and **II-B** was much smaller than that of zinc monomer **5** which had a similar value to polymer **I-A**.(see Table 2) This result indicated that fluorescence quenching occurred in the polymer system. In addition, the fluorescence lifetime decreased significantly from the zinc

Table 2. Fluorescent quantum yields of monomer and polymers. (Cu-porphyrin doesn't fluorescent)

sample	monomer **1B**	polymer **I-A**	polymer **II-A**	polymer **II-B**
quantum yield (%)	1.06	1.12	0.33	0.75

monomer **5** and polymer **I-A** to polymer **II-A** and **II-B**. All the solutions were made dilute enough so that no intermolecular coupling would occur. The nature of the intramolecular interaction between the Cu and Zn porphyrins or Zn and Zn porphyrins is yet to be investigated.

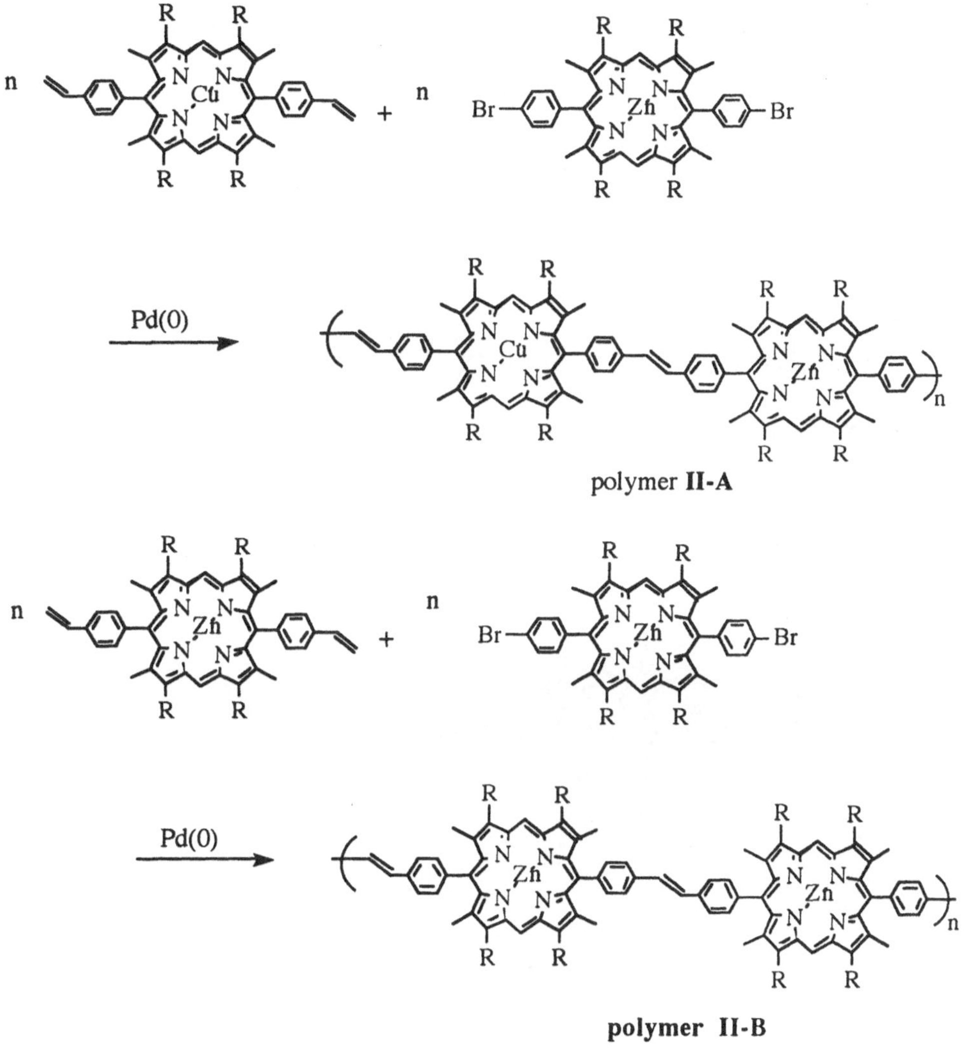

Scheme 2. Synthesis of porphyrin-containing polymers with alternated metals.

Photorefractive Polymers

Photorefractive polymers are of great interest because of their potential applications in high-density optical data storage, image processing techniques, etc. In these polymers, the large quantum yield for photocharge generation is essential in order to obtain large two-beam coupling optical gain coefficient (an important measure of photorefractive performance). Our photoconductivity measurements for polymer **I-A** - **I-C** indicated that a relatively high quantum yield for the photocharge generation can be obtained with this type of polymer backbone. If nonlinear optical chromophores are also introduced to the polymer, we expect to obtain a photorefractive polymer. Based upon the above rationale, polymer **III** was synthesized using the Heck coupling reaction. The polymer had a ratio of 1:49 between the porphyrin monomer and the chromophore **7**. The small concentration of the porphyrin unit is necessary to minimize the absorption loss for the working laser beam. This ratio can be easily changed and the effect of the amount of porphyrin can be studied.

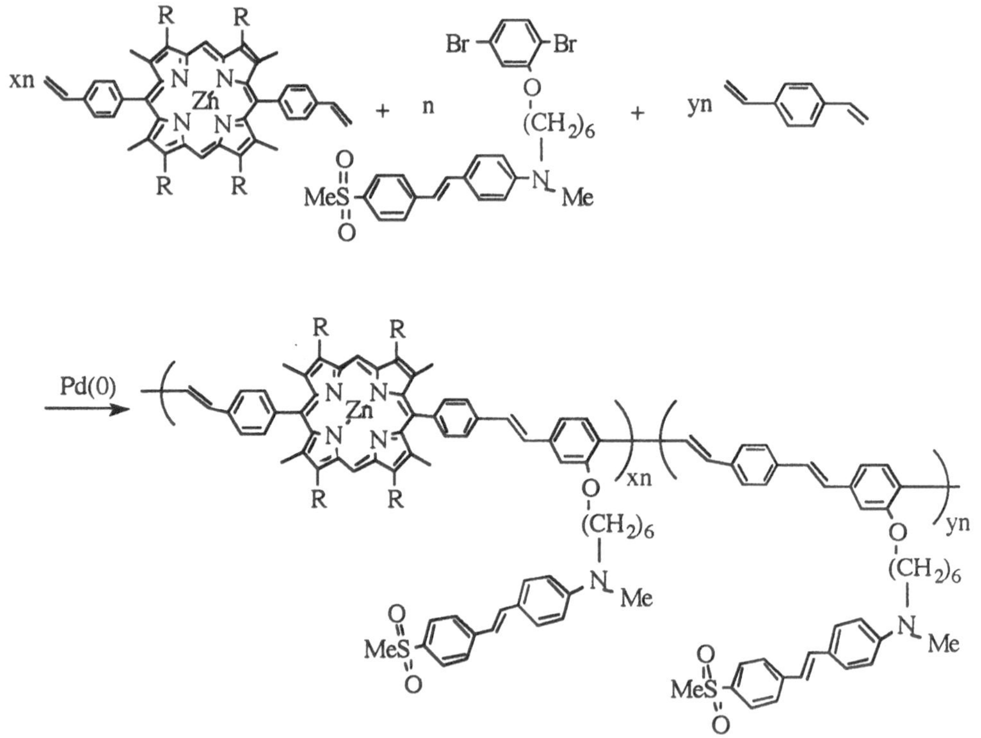

Scheme 3. Synthesis of porphyrin-containing photorefractive polymers.

Polymer **III** melted at 125°C and was soluble in different organic solvents. Optical quality films could be obtained by either casting or spin coating the polymer solution on glass slides. The film was poled by corona discharging at 100°C to align the dipoles of the chromophores, rendering it second order optical nonlinearity. Preliminary results indicated that the poled sample was photorefractive evidenced by the asymmetric energy exchange in two beam coupling experiments. Further studies are in progress.

In summary, we have demonstrated the versatility of the Heck reaction for the synthesis of polymers containing metalloporphyrins. These new polymers exhibited interesting electronic and optical properties. Further extension of the Heck reaction enabled us to synthesize multifunctional polymers-photorefractive polymers. Combining the mild reaction condition of the Heck reaction with careful structural designs, more new polymers with interesting properties can be anticipated.

ACKNOWLEDGMENT

This work was supported by the Office of Naval Research grant N00014-93-1-0092 and by the National Science Foundation (DMR-9308124), the National Science Foundation Young Investigator Program and the Arnold and Mabel Beckman Foundation (Young Investigator Award).

REFERENCES

1. M. Zeldin, K. Wynne, H.R. Allcock, Eds. "Inorganic and Organometallic Polymers" ACS Symposium Series 360, American Chemical Society, Washington, DC (1988).
2. J.E. Sheats, C.E. Carraher, C.U. Pittman, "Metal Containing Polymer Systems," Plenum, New York (1985).
3. J.M. Nelson, H. Rengel, I. Manners, Ring-opening polymerization of [2]ferrocenophanes with a hydrocarbon bridge: synthesis of poly(ferrocenylethlenes), *J. Am. Chem. Soc.* 115: 7035 (1993).
4. W.K. Chan, Y.M. Chen, Z.H. Peng, L.P. Yu, Rational designs of multifunctional polymers, *J. Am. Chem. Soc.* 115: 11735 (1993).
5. D. Dolphin, Ed. "The Porphyrin," Vol. V, Academic Press, New York (1978).
6. E. Bayer, G. Holzbach, Synthetic hemopolymers for reversible binding of molecular oxygen, *Angew. Chem. Int. Ed. Engl.* 16: 117 (1977).

7. H. Kamogawa, S. Miyama, S. Minoura, Preparation of vinyl copolymers bearing porphyrin or magnesium porphyrin dimers as pendants, *Macromolecules*, 22: 2123 (1989).

8. H. Kamogava, H. Indue, M. Nanasawa, Synthesis of vinyl polymers with pendant porphyrin dimers and their catalytic effects on photoreductions, *J. Polym. Sci.: Polym. Chem.* 12: 2317 (1974).

9. M.J. Crossley, P.L. Burn, An approach to porphyrin-based molecular wires: synthesis of a bis(porphyrin)tetraone and its conversion to a linearly conjugated tetrakisporphyrin system, *J. Chem. Soc., Chem. Commun.* 1569 (1991).

10. H.L. Anderson, S.J. Martin, D.D.C. Bradley, Synthesis and third-Order nonlinear optical properties of a conjugated porphyrin polymer, *Angew. Chem. Int. Ed. Engl.* 33: 655 (1994).

11. R.F. Heck, Palladium-catalyzed vinylation of organic halides, in: "Organic Reactions," Vol 27, John Wiley & Sons, Inc., Canada.

12. W. Heitz, W. Brugging, L. Freund, M. Gailberger, A. Greiner, H. Jung, U. Kampschulte, N. Niebner, F. Osan, H.W. Schmidt, M. Wicker, Synthesis of monomers and polymers by the Heck reaction, *Makromol. Chem.* 189: 119 (1988).

13. H.P. Weitzel, A. Bohnen, K. Mullen, Oligomeric model compounds for poly(9, 10-anthrylenevinylene), *Makromol. Chem.* 191: 2815 (1990).

14. M. Suzuki, J.C. Lim, T. Saegusa, Polycondensation catalyzed by a palladium complex. 2. Synthesis and characterization of main-chain type liquid crystalline polymers having distyrylbenzene mesogenic groups, *Macromolecules*, 23: 1574 (1990).

15. Z.N. Bao, Y.M. Chen, R.B. Cai, L.P. Yu, Conjugated liquid crystalline polymers——soluble and fusible poly(phenylenevinylene) by the Heck Coupling Reaction, *Macromolecules*, 26: 5281 (1993).

16. T. Nagata, A. Osuka, K. Maruyama, Synthesis and optical properties of conformationally constrained trimeric and pentameric porphyrin arrays, *J. Am. Chem. Soc.* 112: 3054 (1990).

METAL-CONTAINING POLYMERS AS PRECURSORS FOR THE PRODUCTION OF FERROMAGNETIC AND SUPERCONDUCTING MATERIALS

A. D. Pomogailo, A. S. Rozenberg, and G. I. Dzhardimalieva

Institute of Chemical Physics in Chernogolovka Russian Academy of Sciences, Chernogolovka 142432, Moscow Region, Russia

INTRODUCTION

Thermal transformations of metal-containing monomers (MCM) are of interest at least for two reasons: on one hand, the study of thermal decay of MCM and their transformation products makes it possible to evaluate MCM thermal stability and its role in solid state polymerization processes. Previous studies of some metal (Zn, Co, Ni etc.) acrylate thermal decompositions, both in air and in inert atmosphere,[1,2] were based on thermal analysis and provided only a qualitative picture of transformations occurring upon thermolysis. Complex kinetic studies are needed for a better understanding of these processes.

On the other hand, an investigation of MCM thermal decay is of interest in connection with the preparation of highly dispersed nano-sized metal (or metal oxide) particles stabilized in the polymer matrix. In distinction to known approaches,[3-5] the thermal transformation of transition metal acrylates ($MAcr_n$) is a unique technique, which successfully combines the processes of synthesis and chemical passivation of nano-size particles. Besides, it is an important step towards solving the problem of preparing a highly dispersed state.[6,7] Finally, the products of thermal transformation of MCM can be precursors for the production of ferromagnetic and superconducting materials.

This chapter reports the physical-chemical properties, the kinetics and the mechanism of thermal transformation of some metal acrylates: $Co(CH_2=CHCOO)_2 \cdot H_2O$ ($CoAcr_2$),[8] $Ni(CH_2=CHCOO)_2 \cdot H_2O$ ($NiAcr_2$),[8] the cluster acrylates $Cu_2(CH_2=CHCOO)_4$ ($CuAcr_2$),[8,9] $Fe_3O(OH)(CH_2=CHCOO)_6 \cdot 3H_2O$ ($FeAcr_3$),[10,11] cocrystallized $CoAcr_2$ - $FeAcr_3$ in 1:1 and 1:2 mole ratio and others ($CuAcr_2$, $BaAcr_2$, and $YAcr_3$) and bismuth ceramics as precursors for superconducting material production.

RESULTS AND DISCUSSION

The previous studies showed that copolymerization and terpolymerization of heterometallic MCM yields metallocopolymers of random structure. The thermal decay of these products results in the formation of nano-sized particles stabilized by the polymer matrix. This approach is very useful for the formation of ferromagnetic and HTSC materials.

The general method of synthesis of metal-containing monomers is the interaction of oxides, hydroxides or carbonates of metals with unsaturated acids.[12] The above-mentioned $MAcr_n$ was obtained according to Scheme 1:

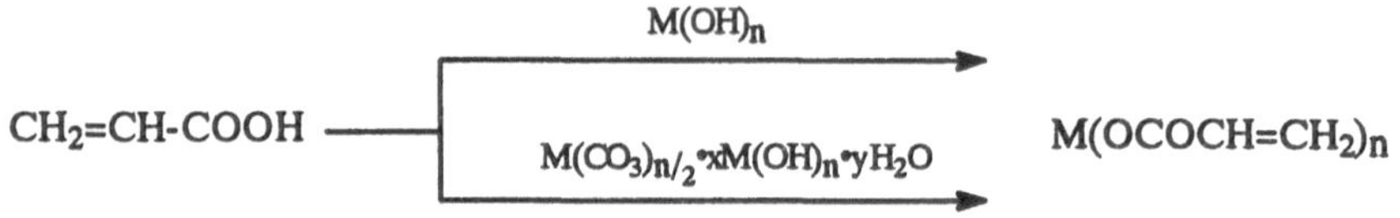

$$CH_2=CH\text{-}COOH \xrightarrow{\begin{array}{c} M(OH)_n \\ \hline M(CO_3)_{n/2} \cdot xM(OH)_n \cdot yH_2O \end{array}} M(OCOCH=CH_2)_n$$

M = Fe(II), Fe(III), Cr(III), Co(II), Ni(II) – precursors for the production of .
ferromagnetic materials

M = Y(III), Ba(II), Cu(II),
Bi(III), Ca(II), Sr(II), Pb(II), Cu(II) – precursors for the production of
superconducting materials

Scheme 1

The obtained $MAcr_n$ are complexes of mono- ($NiAcr_n$, $CoAcr_n$) and poly- ($CuAcr_n$, $FeAcr_n$) nuclear type with mono- and bidentate carboxylate ligands:

$$CH_2=CH\text{-}C \begin{array}{c} O \\ O \end{array} M \begin{array}{c} O \\ O \end{array} C\text{-}CH=CH_2$$

According to optical microscopy data, the initial samples of $CuAcr_2$ are bright green crystals, mainly in the form of plates, and of their splices with a rough surface. Needle-shaped prismatic crystals and amorphous glassy products are also formed in small proportions (up to 10 mass.%). The average crystal plate size was $(5 \times 20) \times (5 \times 50) \times (1 \times 5)$ μm^3, although the particles with dimensions less than 2 μm are also observed. The samples of $CoAcr_2$, $FeAcr_3$, FeCoAcr, and Fe_2CoAcr obtained are morphologically close to one another. They consist of glassy shapeless particles (aggregates) colored from light-orange ($CoAcr_2$) and light-brown ($FeAcr_3$, FeCoAcr, Fe_2CoAcr) (in transmitted light) to yellow (in thin layers). The average size of the aggregates is about 100-150 μm ($CoAcr_2$, $FeAcr_3$, Fe_2CoAcr) and 40-60 μm ($FeAcr_3$). The aggregates consist of particles, which have an average size of 1×5 μm ($FeAcr_3$, FeCoAcr) and 10×5 μm ($CoAcr_2$, Fe_2CoAcr). The particles of $CoAcr_2$, $FeAcr_3$, FeCoAcr, and Fe_2CoAcr do not rotate the plane of polarization of transmitted light in crossed polaroids.

Table 1. The formal kinetic regularities of gas evolution in thermal transformations of metal-containing acrylates[*]

$MAcr_n$	T_{ex}, °C	10^3 (m_o/V), g·cm^{-3}	$S°_{sp}$, m²·g^{-1}	Type of $\eta(t)$	$\eta(t)$, W_o	$k_i(T_{ex})$, $\eta_\infty(T_{ex})$, $\Delta\alpha^\Sigma_\infty(T_{ex})$, $W_o(T_{ex})$
$CuAcr_2$	190-240	2.2±0.25	14.7		$\eta(t)=$ $\eta_{1\infty}[1-\exp(-kt)]+k_2t$ up to $\eta = 0.95$ $W = \eta_{1\infty}k_1$	$k_1=9.5\cdot10^{12}\exp[-89.5/RT]$,s^{-1} $k_2=9.2\cdot10^{11}\exp[-93.5/RT]$, s^{-1} $\eta_{1\infty}=1.8\cdot10^4\exp[-27.5/(RT)]$ $W_o=1.7\cdot10^{17}\exp[-117.0/RT]$, s^{-1} $\Delta\alpha^\Sigma_\infty = 1.4(190°C)\rightarrow1.82(240°C)$
$CoAcr_2$	350-390	1.38±0.03	20.2		$\eta(t) = 1-\exp(-kt)$ $W_0=k$	$k=W_o=3\cdot10^{14}\exp[-136.0/RT]$, s^{-1} $\Delta\alpha^\Sigma_\infty \approx const = 1.54(\pm0.01)$
$FeAcr_3$	200-370	1.33-7.5	15.0		$\eta(t) = 1-\exp(-kt)$ $W_o = k$	$T_{ex} = 200\text{-}300°C$ $k = W_o = 4.2\cdot10^{21}\exp[-140.5/RT]$s^{-1} . $\Delta\alpha^\Sigma_\infty = 1.6\cdot10^{21}\exp[-14.3/RT]$ $T_{ex} = 330\text{-}370°C$ $k = W_o = 1.3\cdot10^6\exp[-73/(RT)]$, s^{-1} $\Delta\alpha^\Sigma_\infty = 1.7\cdot10^2\exp[-15.0/(RT)]$
$FeAcr_3$-$CoAcr_2$ mole ratio 1:1	340-390	0.61-1.85	9.0		$\eta(t)=\eta_{1\infty}[1-\exp(-k_1t)] +$ $(1-\eta_{1\infty})[1-\exp(-k_2t)]$ $W_o=\eta_{1\infty}k_1+(1-\eta_{1\infty})k_2$	$k_1 = 2.3\cdot10^{12}\exp[-118.5/(RT)]$, s^{-1} $k_2 = 6.0\cdot10^6\exp[-79.0/(RT)]$, s^{-1} $\Delta\alpha^\Sigma_\infty = 5.25\cdot10^2\exp[-17.9/(RT)]$ $\eta_{1\infty} = 0.65(340°C) \rightarrow 0.45(370\text{-}390°C)$
$FeAcr_3$-$CoAcr_2$ mole ratio 2:1	340-390		8.1			$k_1 = 2.6\cdot10^{12}\exp[-116.0/RT]$, s^{-1} $k_2 = 6.6\cdot10^5$ $\exp[-71.6/RT]$, s^{-1} $\Delta\alpha^\Sigma_\infty = 1.9\cdot10^2\exp[-14.3/RT]$ $\eta_{1\infty} = 0.50 (340°C)\rightarrow0.35(390°C)$

[*] m_o - the initial sample weight, V - the volume of the reaction vessel, $S°_{sp}$ - the specific of the starting sample, W_o - the starting velocity (rate) of the decomposition, $\Delta\alpha^\Sigma_\infty = \alpha^\Sigma_\infty - \alpha^\Sigma_o$

The regularities of the kinetics of MAcr$_n$ thermal decay

The thermal transformation of MAcr$_n$ is accompanied by gas evolution and the loss of weight. The reversible dehydration of CoAcr$_2$ takes place within 30-160°C. IR-analysis shows that the dehydration results in the disappearance of the following absorption bands: ν(OH) (Co-OH$_2$) 3000-3600 cm^{-1}, ρ_w(OH) + ν(Co...OH$_2$) 880 cm^{-1} and the intensity of (δ(O–H) + ν(C=C)) 1665 cm^{-1}, (ρ_r(CH$_2$) + δ(Co...OH$_2$)) 690 cm^{-1}, (δ(CH$_2$) +δ(Co...OH$_2$)) 595 cm^{-1} diminishes (the assignment is based on the data reported earlier[10,13,14,15]). There exist two ranges of evaporation. At 30-75°C evaporation is limited by the dehydrated water

$$(P_{H_2O}^I \; (T_{ex}) = 1.7 \cdot 10^7 \exp\,[-38.5/(RT)], \text{ kPa})$$

and the heat of vaporization of H$_2$O over CoAcr$_2$ (ΔH_{vap} = 38.5 kJ mol^{-1}) is close to that of pure water (ΔH_{vap} = 44 kJ mol^{-1})[16]. At 75-160°C

$$P_{H_2O}^{II} \; (T_{ex}) = 2.7 \cdot 10^4 \exp\,[-20/(RT)], \text{ kPa}$$

and evaporation is limited by the dehydration of CoAcr$_2$ (ΔH_{vap} = 20.0 kJ mol^{-1}).

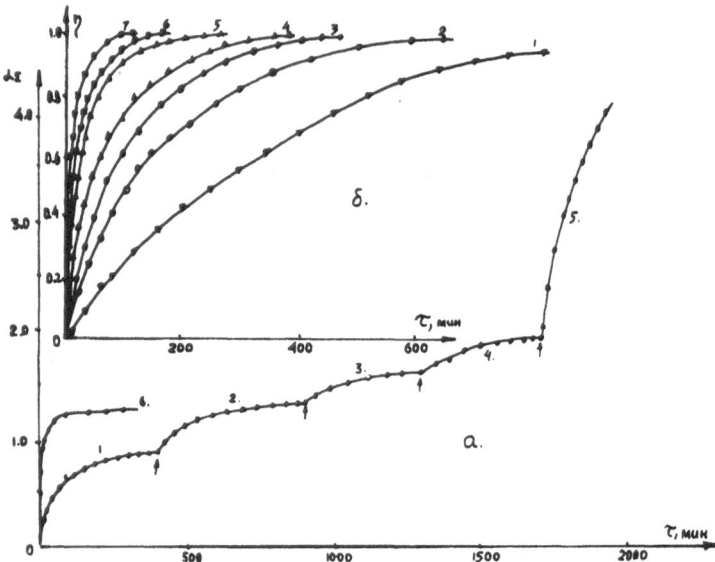

Figure 1. a) The kinetics of gas evolution from FeAcr$_3$ at T$_{ex}$, °C: 1-215, 2-250, 3-275, 4-300, 5-350, 6-240. The pointers indicate the temperature increase. b) The conversion (η) as a function of time in LT-range: 200°C(1), 205°C(2), 210°C(3), 215°C(4), 220°C(5), 230°C(6), 230°C(7) (m$_o$/V = 6.7·10^{-3}).

The kinetics of thermal decay was studied by monitoring gas evolution during MAcr$_n$ decomposition with a diaphragm zero-gage. For all MAcr$_n$, the rate of gas evolution decreases both with an increase of time and gas evolution η ($\eta = (\alpha_t^\Sigma - \alpha_o^\Sigma) / (\alpha_\infty^\Sigma - \alpha_o^\Sigma)$ where α_o^Σ, α_t^Σ and α_∞^Σ are the initial, the current and the final total number of moles of evolved gaseous products per one mole of the initial compound). The principal kinetic parameters of gas evolution during MAcr$_2$ decomposition in the autogenerated atmosphere are represented in Table 1 (static, nonisothermal conditions, the heated part of the reaction vessel was less than 0.05V, where V is the volume of the reaction vessel). It should be noted that two gas-evolution ranges were observed in the decomposition of $FeAcr_3$ (Fig.1): a low-temperature (LT, 200-300°C) range and a high-temperature one (HT, >300°C). The rates of gas evolution in both LT and HT ranges are described by the first-order equations with different temperature dependence of reaction rate constants (Table 1). For the thermolysis of cocrystallized $FeAcr_3$-$CoAcr_2$, the kinetics are described by the empirical equation, which has the form reminiscent with similar kinetic parameters (k_1, k_2, $\Delta\alpha \sum$) for FeCoAcr and Fe_2CoAcr. The level of η_{1-} is different for FeCoAcr and Fe_2CoAcr.

The product composition for thermal transformations of $MAcr_n$

Gaseous and condensed products. CO_2 is the main gaseous product of the decomposition of all of the $MAcr_n$ systems studied (IR, MS). The IR spectra exhibit a system of absorption bands characteristic[13] of CO_2, v/cm^{-1}: 3720, 3650, 3610, 2285, 670, 650 and 620. The emission of H_2 (MS), CO (IR,MS), H_2O (MS) is much weaker. The bands associated with the presence of =CH-, >C=O, -COOH, >C=C<, -OH groups are also observed, v/cm^{-1}: 3145, 3060, 3000, 2970, 1760, 1720, 1630, 1445, 1400, 1300, 1250, 1180, and 950. These bands were attributed to the absorption of acrylic acid vapor. During sample heating (in the initial stages of the transformation) the condensation of H_2O and CH_2=CHCOOH (pH ~ 3.0) proceeds on cold walls of the reaction vessel. The decomposition product is a yellow liquid condensate after evacuation at 20°C. It's IR spectra of exhibited absorption bands which are associated with the presence of H_2O and polyacrylic acid (PAcr): v OH (3480 cm^{-1}), v C=O (1705 and 1650 cm^{-1}), and a group of bands in 1450 and 600 cm^{-1} range, attributed to the deformation vibrations of δ (C-H), δ (C=O) and δ (C-C) bonds.

Only in the case of $CuAcr_2$ decomposition is C_2H_4 observed (IR): v C-H (3106, 3020, 3000, 2970 cm^{-1}); v C=C (1625 cm^{-1}); δ_{as} CH_2 (1445 cm^{-1}); δ_s CH_2 (1380 cm^{-1}); ρ_w CH_2 (950 cm^{-1})[13].

The level of gas evolution ($\alpha \sum_{\infty}$) at the end of decomposition is different for $MAcr_n$ and increases with the number of CH_2=CH-COO groups. The fractionation of gaseous products at 77K gives a proportion of uncondensed ($\alpha \sum_{<}$, H_2, CO, a small amount of CH_4 in the case of $CoAcr_2$) and condensed ($\alpha \sum_{>}$, CO_2, H_2O, CH_2=CHCOOH vapors, and C_2H_4 in the case of $CuAcr_2$) products: $\alpha \sum_{>} > \alpha \sum_{<}$, $\alpha \sum_{\infty} = \alpha \sum_{>} + \alpha \sum_{<}$. The temperature dependences at the end of decomposition are different. So, for $CuAcr_2$ $\alpha \sum_{>}$ is equal to T_{exp} 220°C, and $\alpha \sum_{>} >> \alpha \sum_{<}$. For $CoAcr_2$ $\alpha \sum_{>}$ is constant and is equal to 1.12-1.118 (350°C - 390°C), $\alpha \sum_{<}$ is equal to 0.34 (350°C) - 0.44 (390°C). In the case of $FeAcr_3$ at the end of decomposition in LT range $\alpha \sum_{>}$ is equal to 0.55 (200°C) - 0.90 (240°C), $\alpha \sum_{<}$ is equal to 0 (200-220°C) - 0.05 (240°C) and in HT range $\alpha \sum_{>}$ is equal to 3.15 (340°C) - 4.0 (370°C), $\alpha \sum_{<}$ is equal to 0.2 (340°C), - 0.7 (370°C).

The level of $\alpha \sum_{>}$ and $\alpha \sum_{<}$ for $FeAcr_3$ - $CoAcr_2$ also increases with T_{exp} at the end of decomposition: $\alpha \sum_{>} = 0.43 \cdot 10^2$ exp {-10.7/RT}, $\alpha \sum_{<} = 0.04$ (340°C) ≴ 0.35 (390°C) for FeCoAcr; $\alpha \sum_{>} = 0.40 \cdot 10^2$ exp {-18.8/RT}, $\alpha \sum_{<} = 0.15$ (340°C) ≴ 0.49 (390°C) for Fe_2CoAcr.

Solid decomposition products

At the end of the transformation the mass loss ($\Delta m/m_o$) of the sample increases monotonically with T_{exp}, but does not reach the values which could have been expected for $MAcr_n$ decay to a metal or oxide.

Both during, and at the end, of decomposition α^{Σ} changes are observed in IR spectra with the increase of the level of gas evolution. The relative intensity of the I_{rel} bands changes and a shift of the absorption frequencies takes place ($I_{rel} = D(\nu_i) / D^{max}(\nu_j)$ are the optical densities of the absorption at ν_i and ν_j frequencies corresponding to the absorption maximum in the spectra).

The IR spectra of each $MAcr_n$ share certain specific properties associated with the absorption bands of $>C=C<$ and CO.

1). At a low level of $MAcr_n$ transformation (for example, $FeAcr_3$ at $T_{exp} \sim 200°C$) the relative intensities of the absorptions associated with ν ($>C=C<$), ρ_w CH_2, ν ($=CH-C$), π ($-CH=CH_2$) decrease and even disappear. The IR spectrum of the conversion product is similar to that for the corresponding metal polyacrylate, as it can be seen from a comparison of I_{rel}.

2). The further the conversion of $MAcr_n$ proceeds, the greater the decrease of I_{rel} for ν (COO) becomes. The deformation vibrations of C-O completely disappear (in HT range for $FeAcr_3$ and $FeAcr_3$-$CoAcr_2$). This indicates a destruction of COO groups.

3). ν (C-H) is shifted to the high-frequency range, which is the evidence for the change in the nature of C-H absorption and is typical for ν (=CH-).[14]

4). The absorption between 1630-1655 cm^{-1}, associated with ν (C=C), is shifted to higher frequency (1695-1780 cm^{-1}). This shift is probably due to the appearance of conjugated C=C bonds.

5) In the case of $MAcr_n$ containing water of crystallization the value of I_{rel}, associated with valence and deformation vibrations of the coordinated H_2O molecules and M-OH_2 bonds, decreased. These changes in IR spectra are possibly due to dehydration at the initial stage of the thermal conversion.

Thus, during of $MAcr_n$ decay the transformation of metal-containing carboxyl groups and the formation of conjugated C=C bonds take place.

The value of the specific surface (S_{spec}) at the end of transformation is very peculiar. For instance, in the case of $CuAcr_2$ and $CoAcr_2$ S_{spec} increases during the transformation and its value depends on T_{exp}. S^f_{spec} of $CoAcr_2$ increases with T_{exp} (m^{2}g^{-1}): 24.1(350°), 24.5 (360°), 30.0 (370, 36.7 (380°), 42.1 (390°C). S^f_{spec} of $CuAcr_2$ increases with T_{exp} up to 220°C and then decreases (m^{2}g^{-1}): 48.0 (190°), 53.8 (200°), 59.0 (210°), 6.08 (220°), 59.4(230°), 43.7(240°C). Probably, the decrease of S^f_{spec} is due to the sintering of solid products. At the same time $S^f_{spec} > S^o_{spec}$. However during the transformation S^f_{spec} of $FeAcr_3$, $FeAcr$, Fe_2CoAcr is approximately equal to S^o_{spec} (Table 1), does not depend on T_{exp} and is 15.0, 13.6 and 11.3 m^{2}g^{-1}, respectively.

Optical microscopy of the morphology showed that solid products of transformation of $CuAcr_2$ are agglomerates of optically nontransparent dark brown particles with metal lustre (by Cu, probably) and 2.0-100 μm in size. Among the agglomerates there are small crystals with a form close to a cubic of needles of $(2-5)^3$ μm^3 size, as well as transparent prismatic needles of 1/5x20.0 μm^3 size. The fraction of the latter is 20-30 vol.% at 190-210°C, but it decreases with the T_{exp} increase and does not exceed 10 vol.%. There are also glassy-like particles, which do not rotate the plane of polarization.

Nontransparent magnetic particles (up to 40 vol.%) black colored (probably, Co or its oxides) and ~5 μm in size constitute the main part of solid products from $CoAcr_2$ decay.

Some particles reach ~60.0 μm in size. The transparent glassy-like particles were observed both in the individual state and on the surface of other opaque particles.

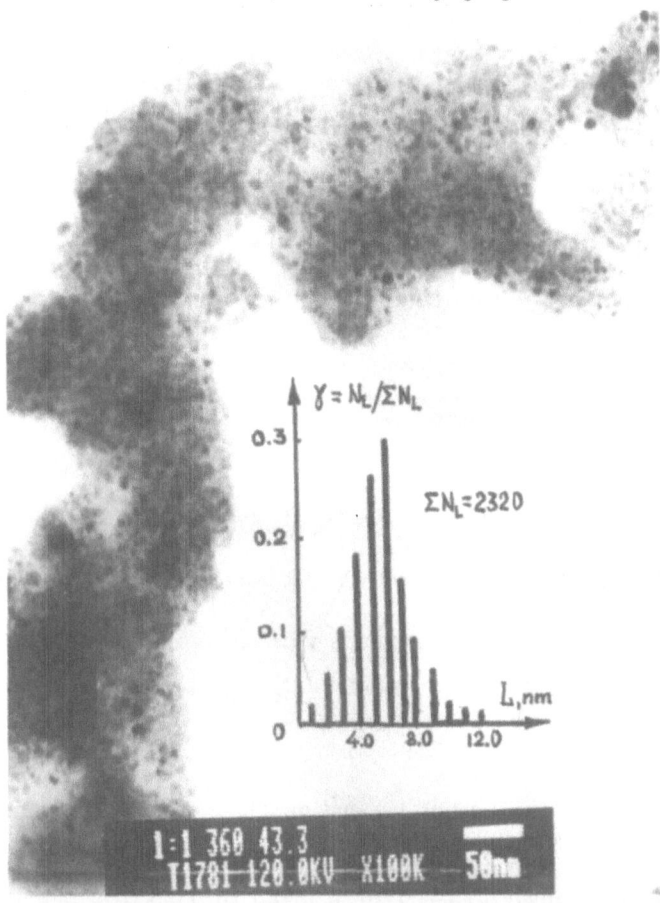

Figure 2. The electron microscopic photograph of the product of Fe_2CoAcr decay (T_{ex} = 370°C).

Optical microscopy analysis of $FeAcr_3$ and $FeAcr_3$-$CoAcr_2$ thermal conversion products suggest that the evolution of their morphology is similar. The thermal decomposition results in the growth of agglomerates (up to 100 μm), and the increase of the portion of large agglomerates in the overall mass of the product (LT range for $FeAcr_3$). When subjected to slight mechanical actions the large agglomerates break into small irregularly shaped particles of 1-3 μ size. During $FeAcr_3$ degradation in the LT range the fluidity of the sample increases, and its decomposition proceeds homogeneously in the bulk. The particles are light-yellow in transmitted light. In the HT range the sample turns to dark brown.

Electron microscopic data show the products of $FeAcr_3$, $FeCoAcr$ and Fe_2CoAcr decomposition to have the identical morphological structure. Electron-dense particles of practically spherical shape with a narrow distribution in sizes are observed. Their distribution in a less electron-dense matrix is uniform with an average separation of 8 nm. The average size of spherical particles is 6.0-13.0 nm for $FeAcr_3$, and is 6.0 nm for $FeCoAcr$ and Fe_2CoAcr. These particles are present both individually and as aggregates of 3-6 particles (50 wt.%). (Fig.2).

Possible pathways of the thermal transformation of $MAcr_n$

The thermal stability of $MAcr_n$ is defined by their weakest bonds. The structure of solvate-free $MAcr_n$ is yet unknown. However one can expect that after desolvation ($CuAcr_2$) or dehydration ($CoAcr_2$, $FeAcr_3$, Fe_2CoAcr) the structural unit is well preserved: $Cu_2(CH_2=CHCOO)_4$, $Co(CH_2=CHCOO)_2$, $[Fe_3O(CH_2=CHCOO)_6]^+$. This also concerns cocrystallized $FeAcr_3$-$CoAcr_2$. The bidentate function (typical for solvate complexes) of CH_2CHCOO groups becomes tridentate for some CH_2COO-groups. This change of coordination is known for many desolvated and dehydrated carboxylates[17]. The increase of the dentate degree for CH_2COO-groups leads to a distortion in the oxygen environment of the metal. Both the length and the strength of M-O and C-O bonds changed. Information about the strength of coordination bonds was obtained from mass spectrometric study[10] of $FeAcr_3$ carried out at various accelerating electric fields. The dependence of the $CH_2=CHCOO$ ion yield on the intensity of the accelerating electric field indicates the energetic nonequivalence of M-O bonds with the acrylic ligands.

The splitting of a weak M-O bond takes place during the thermal decay of $MAcr_n$ with the formation of the $CH_2=CHCOO$ radical. This radical reacts with the matrix (RH) to give acrylic acid.

$$CH_2=CHCOO^{\cdot} + RH \rightarrow CH_2CHCOOH + R^{\cdot} + (0\text{-}20.9 \text{ kJ mol}^{-1}) \quad (1)$$

where $R^{\cdot}$ is $^{\cdot}CHCHCOOM(CH_2CHCOO)_{n-1}$.

The analysis of other pathways which consume $CH_2=CHCOO$ radicals shows reaction (1) is preferred.

The $CH_2CHCOOH$ formed evaporates or decomposes with the formation of C_2H_4.

$$CH_2=CHCOOH \longrightarrow \left[\begin{array}{c} H_2C=C(H)\text{-}C=O \\ \underset{H\text{-----}O}{|} \quad | \end{array} \right] \longrightarrow C_2H_4 + CO_2 - (0 \div 4.18 \text{ kJ mol}^{-1})$$

$$(2)$$

Thermal effects of the reaction (1) and gas-phase reaction (2) were estimated by considering the following data and assumptions: H°_f [$CH_2CHCOOH$ (e)] = 384.6 (kJ mol^{-1})[18], H°_{vap} [$CH_2CHCOOH$ (e)] = 45.6 (kJ mol^{-1})[18], H°_f [$CH_2CH_2(g)$] = 52.3 (kJ mol^{-1})[16], H°_f [$CO_2(g)$] = 393.0 (kJ mol^{-1})[16], E [$CH_2CHC(O)O$-H] = E [$CH_3C(O)O$-H] = 461.5 (kJ mol^{-1})[19], E [R-H] = E [CH_2CH-H] = 438.5 (kJ mol^{-1})[19]. Hence H°_f [$CH_2CHC(O)O$-H g)] = 96.2 (kJ mol^{-1}).

C_2H_4 is found in $CuAcr_2$ decay, which occurs at a lower temperature as compared with other $Macr_n$ species. Reaction (2) is probably competitive with the evaporation of acrylic acid.

The radical R formed via reaction (1), can undergo the intramolecular addition to the double bond

$$(3),$$

and initiate the polymerization of the residual monomer with the formation of both linear and cross-linked polymers:

$$R_1[-CH_2-CH-]_s-R_1$$

the linear structure   (4.1)

$$R' + sM(CH_2CHCOO)_n$$

$$R_1[-CH_2-CH-]_{s/2}-R_1$$

the cross-linked structure (4.2)

which is followed by decarboxylation of metal-containing fragments:

$$R_1[-CH_2-CH-CH_2-CH-]_{s/2}-R_1 \longrightarrow R_2[-CH_2-CH=CH-CH-]_{s/2}-R_2 + (2s+2)CO_2 + (s+2)M$$

$$\quad\quad M_{1/n}O_2C \quad\quad CO_2M_{1/n}$$

(5)

where $R_2 = CH_2=CH-CH=CH-$.

The metal formed in the reaction (5) undergoes the oxidation according to the following routes:

$$M + rCO_2 \rightarrow MO_r + rCO \quad\quad (6.1)$$
$$M + rH_2O \rightarrow MO_r + rH_2 \qu\quad\quad (6.2)$$

CO and H_2 are identified by mass spectrometry and form the noncondensed (at 77K) gaseous products of decomposition.

Therefore, the solid products from the thermal transformation of $MAcr_n$ have the composition $MO_r(CH_2CHCOO)_{p-x}(CH_2CH)_x(CHCHCOO)_{q-y}(CHCH)_q$.

Using the data on the mass loss of the sample at the end of degradation, the results of low-temperature fractional distillation of the gaseous products and the material balance, one can evaluate the quantitative composition of gaseous and solid products of transformation at different T_{exp} (Fig.3).[11,20-22]

The T_{exp} dependence of the value of $\gamma = (\alpha CH_2CHCOO + \alpha CH_2CH)/(\alpha CH_2CHCOO + \alpha CHCH)$ was obtained. This quantity is a ratio of the acrylate groups depleted in hydrogen, to the amount of those within a polymeric chain, which characterizes an effective length of the formed polymer chain (2γ).

1) Polymerization and, consequently, the elimination of $CH_2CHCOOH$ proceed at temperatures lower than the observed temperature range and polymerization temperature exceeds those of dehydration. During thermal transformation of $FeCoAcr$ and Fe_2CoAcr, CH_2CHCOO radicals formed from the Fe-containing cluster monomer perform as the polymerization initiator (the thermal stability of Fe-containing monomer is lower than that of Co-containing monomer). At high temperature (HT-range) the elimination of two CH_2CHCOO radicals per one dehydrated $FeAcr_3$ cluster molecule is possible.[22]

2) Decarboxylation is a limit in gas evolution. It should be noted that decarboxylation of terminal groups proceeds much easier than the destruction of intrachain metal-containing

groups. This is consistent with the concept of a higher energy content for terminal groups on long-chain polymers.[23,24]

In FeCoAcr and Fe$_2$CoAcr copolymers the thermal stability of intrachain Co-containing carboxylate groups is lower than that of corresponding Fe-containing fragments. This is confirmed by the magnetic properties of solid products from the transformation of FeCoAcr at 360°C (Fig.4). The strong changes of σ_s, H_C and χ_σ are observed at the end of transformation when decarboxylation of Fe-containing fragments is realized.

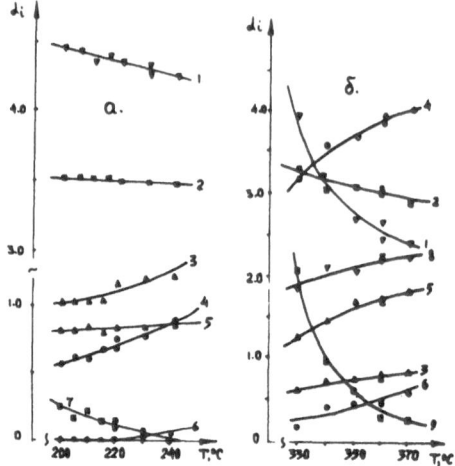

Figure 3. The yield of products of FeAcr$_3$ decay in LT-(a) and HT (b) ranges: 1 - α(CH$_2$CHCOO) + α(CH$_2$CH); 2 - α(H$_2$O); 3 - 2r (a); r (b); 4 - α(CO$_2$); 5 - α(CHCHCOO) + α(CHCH); 6 - α(H$_2$); 7 - α(CHCHCOO); 8 - α(CH$_2$CH); 9 - α(CH$_2$CHCOO).

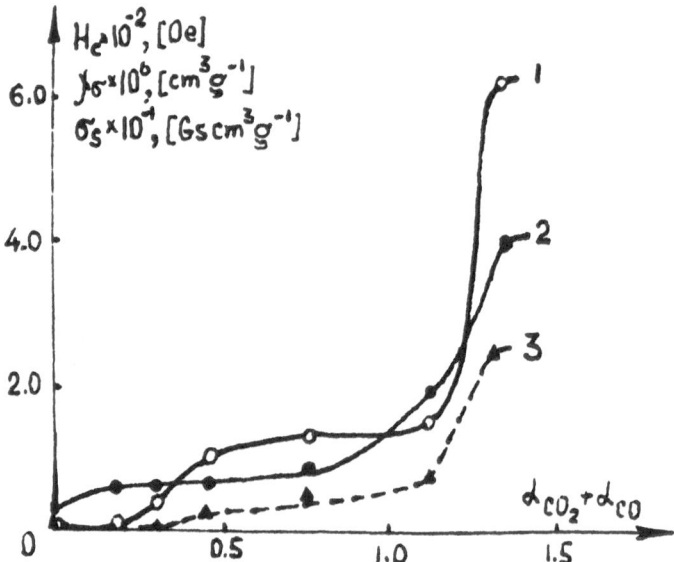

Figure 4. The dependence of σ_s, H_c, and χ_s as a function of α(CO$_2$) + α(CO) (FeCoAcr, T_{ex} = 360°C).

3). The γ value decreases with increasing T_{exp} at the end of the thermal transformation. However, the range of γ values differs for the species of $MAcr_n$ which have been studied. $FeAcr_3$ was studied decreasing from 5.75 (200°C) down to 5.00 (240°C) (LT-range) and from 2.64 (330°C) down to 1.4 (370°C) (HT-range). In the case of $FeCoAcr_3$-$CoAcr_2$ γ varies from 7.7 (340°C) to 5.0(390°C) (FeCoAcr) and from 5.5 (340°C) to 3.8 (390°C) (Fe_2CoAcr). This is the evidence for a correlation of the rates of elimination of CH_2CHCOO-radicals (the thermal stability of M-O bond) and of radical polymerization. It manifests itself most strikingly in the case of $CuAcr_2$ and $CoAcr_2$, for which a decrease of γ occurs with an increase of T_{exp}. However, $\gamma < 1$ for $CoAcr_2$ at 370°C and for $CuAcr_2$ over the whole temperature range. Probably, this is associated with a high concentration of CH_2COO- and R_1 and their recombination and decarboxylation.

Magnetic properties of $MAcr_n$ decay products.

The products of thermal decomposition of $CoAcr_n$, $FeAcr_3$, FeCoAcr and Fe_2CoAcr are ferromagnetic. Some magnetic properties of $MAcr_n$ and its solid decomposition products are given in Table 2. The specific magnetization only slightly exceeds a half of that which would be expected if all the iron was present in the products as Fe_3O_4 ferrite. Probably, Fe atoms, which do not contribute to the magnetization, are in the amorphous phase. Only the ferrite phase is detected by electron-diffraction and X-ray analysis.

Table 2. Magnetic properties of $MAcr_n$ and products of its thermolysis[1]

$MAcr_n$	M,% found	$\chi_\sigma \cdot 10^5$,[cm^{3}g^{-1}] found/calcd		σ_s,[Gs cm^{3}g^{-1}]		H_c, [Oe]		j_r	
		292K	77K	292K	77K	292K	77K	292K	77K
FeAcr$_3$	24.6	1.75/ 5.50	3.22/ 21.0	0.152	0.306	0	0	0	0
Product of FeAcr$_3$	12.4[2]	26.1	26.1	33.45	35.94	22.0	22.3	0.051	0.25
CoAcr$_2$	27.5	4.83/ 2.93	15.1/ 11.15	0.442	1.41	0	0	0	0
FeCoAcr	15.4(Fe) 13.7(Co)	2.73/ 3.17[3]	6.5/ 8.63[3]	0.258	0.792	0	0	0	0
Product of FeCoAcr		41.2	31.1	26.16	13.91	62,50	62.5	0.31	0.42

[1] Note: χ_σ - the magnetic susceptibility, σ_s - the specific magnetization, H_c - the coercive force, j - rectangularity coefficient.
[2] For Fe_3O_4. According to the data of electronografic and X-ray analysis, Fe_3O_4 is the main solid product of thermal decay of $FeAcr_3$.
[3] Calculated from experimental data χ_σ for $FeAcr_3$ and $CoAcr_2$

The ferromagnetic phase of the FeCoAcr thermolysis product has such a high magnetic anisotropy that the measuring fields of a magnetometer (±10 kOe) allow only the detection of a partial hysteresis loop in the sample. This indicates the existence of dispersed cobalt ferrite $CoFe_2O_4$. The ferromagnetic particles contain up to 50% of Fe and Co atoms. This is somewhat lower than the ferromagnetic phase of the $FeAcr_3$ transformation product which exhibits a three times lower anisotropy at 77K. The remaining Co and Fe atoms produce a weakly magnetic phase. This anisotropy probably is caused by the exchange bond in the interface between the ferromagnetic and antiferromagnetic phases. One can assume that CoO is in an antiferromagnetic phase because FeO is not found in the decomposition products from $FeAcr_3$ (the main product is Fe_3O_4).

Metal-containing polymers as precursors for the production of high-temperature superconducting ceramics

One of the basic problems in the synthesis of homogeneous high-temperature superconducting (HTSC) materials is attaining uniform metal ion distribution within the ceramic product. Such a uniformity is apparently related to the dispersion of the initial mixture of reagents and to the character of metal ion distribution.

Recent studies have appeared on the preparation of HTSC ceramics employing a preliminary polymer synthesis. This solves the dispersion problem to a great degree. In this approach a polymer matrix is prepared with a near-molecular dispersion of metal ions in the matrix. In one version of this method, polymer-analogous conversions are used to yield polymeric complexes of Y(III), Ba(II), and Cu(II) nitrates within polymethacrylic acid.[25,26] However, the immobilization of metal ions on polymer functional groups is known to proceed nonuniformly, owing to simultaneous complexation processes, a "neighbor effect", a low coefficient of diffusion through a polymer matrix, and so on. Because of these limitations, the conversion of the original polymer does not is usually exceed 30-60%. In other possible versions of HTSC ceramic synthesis, the polymer is prepared first. For example, acrylic acid was proposed to be polymerized in a mixture with Y(III) nitrate, Ba(II) acetate and Cu(II) acetate aqueous solution[27]. With this approach, however, uncertainty appears in the composition of metal-containing polymer. One cannot exclude the possibility that no chemical binding exists between metal compounds and the polymer matrix..

A much more uniform distribution of metal ions can be acheived by employing a metal-containing monomer which is then polymerized. Copolymerization of two or more monomers containing the appropriate metals should lead to the formation of a polymer matrix with a truly uniform metal ion distribution. In such a copolymerization, the polymer chain does not contain free functional groups.

In the work reported here, the synthesis of HTSC ceramic was conducted by the copolymerization of metal-containing monomers in plastic flow conditions under high pressure. We selected acrylate and acrylamide (AAm) metal nitrate complexes as the metal-containing monomers. Y(III), Ba(II), and Cu(II) acrylates were obtained by the procedures described above; these acrylates were mixed in 1:2:3 mole ratio, dissolved in a minimum volume of methanol and dried. The AAm metal nitrate complexes of metal nitrates was obtained by blending $Y(NO_3)_3 \cdot 6H_2O$, $Ba(NO_3)_2$, and $Cu(NO_3)_2 \cdot 6H_2O$ (mole ratio 1:2:3). The [AAm]/[nitrates] = 4:1 mole ratio; the water was removed under vacuum. The mixture of metal-containing monomers was processed in a Bridgman anvil type apparatus under 1 GPa pressure at room temperature; the angle of anvil rotation was 500°.

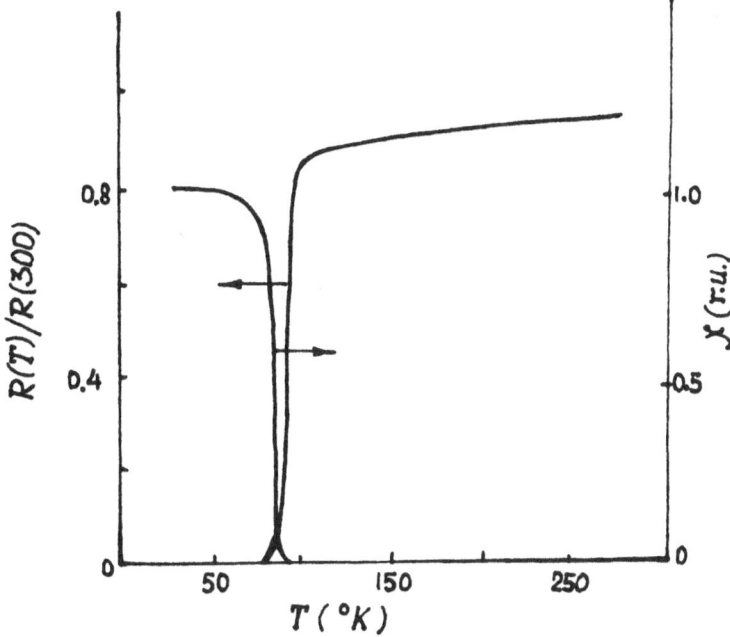

Figure 5. Temperature dependences of resistance and magnetic susceptibility of HTSC ceramic obtained by preliminary polymer synthesis from Y(III), Ba(II), and Cu(II) acrylates.

The metal-containing copolymers were heated in air for 2-5 h at 800°C in order to burn out the organic phase. The residual mixture of metal oxides was ground and pressed into tablets with a diameter of 10 mm and a thickness of 1-2 mm. These tablets were then heated at 920°C for 5 h in air. HTSC ceramics were cooled at a rate of 50°C/h. The tablets were not subjected to any additional annealing in oxygen.

The properties of HTSC ceramics are clearly illustrated in Fig.5. The temperature dependence of the resistance and magnetic susceptibility are plotted for specimens of metal acrylate-based copolymer.

The R(T) relationship demonstrates that at $T > T_c$. Thus, the ceramic behaves as a metal although the drop of the resistance in SC transition is not as great as for single crystals or for the best specimens of 1-2-3 yttrium ceramic. The quality of the normal polycrystalline state is apparently more closely related to the quality of the intergranular contacts than to the quality of the structure of the grain itself. The actual SC transition is quite sharp, and its width is no greater than 2-3 K over the range of 0.1-0.9 of the total resistance drop. Zero resistance is already reached in the sample at 87 K, which is close to the value for the best HTSC ceramics prepared under conditions of oxygen annealing.

It is known that 10-15% v/v of the SC phase is adequate for complete electrical shorting of the specimen. Therefore, the resistance method is extremely insensitive to the presence of SC phases with different T_c values. As a rule, R(T) plots show only one transition with the highest T_c for such multiphase substances. For this reason, the study of the diamagnetic shielding signal is important in analyzing the phase composition of HTSC ceramics. The plot of $\chi(T)$ indicates a sharp transition in the presence of a single SC phase. Note that the total volume of the diamagnetically shielded phase reaches 100%.

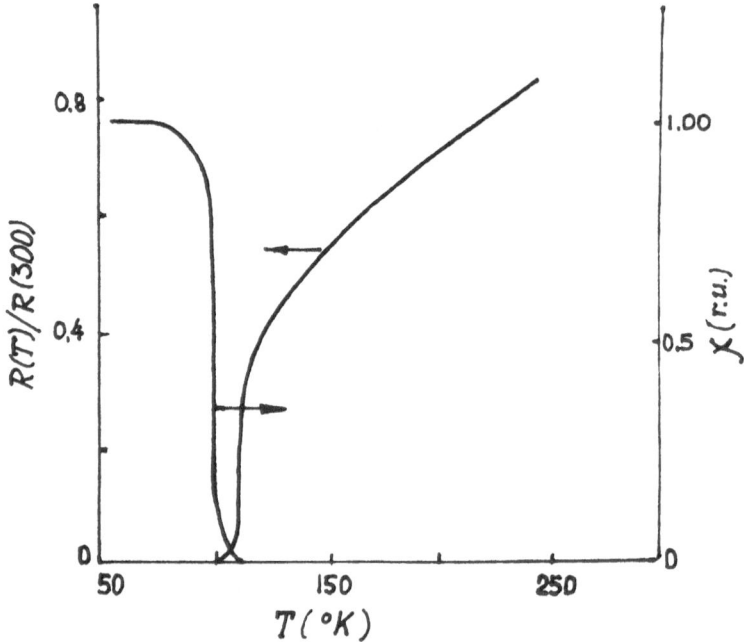

Figure 6. Temperature depedences of resistance and magnetic susceptibility of HTSC ceramic obtained from the products of spontaneous polymerization of Bi(III), Ca(II), Sr(II), Pb(II), and Cu(II)- containing monomers.

The application of polymer precursors for the synthesis of Bi(III)-containing HTSC ceramics is of great interest. It is known that the reproducible synthesis of bismuth superconducting cuprite $Bi_2Sr_2Ca_{n-1}Cu_nO_{2n+4-d}$ (n = 1-3) (it is often doped by Pb) with a high T_c (110 °K) is a complex problem.[28] As a rule, the phase with T_c = 85 K is always contained as an impurity in the samples. This is associated with the heterogeneity of the ceramic at the microlevel and the sequential phase transition 2201 → 2212 → 2223. At the same time the ceramic monophase with T_c = 110 °K was obtained by using the products of spontaneous AAM-metal nitrate polymerization as the starting reagents. The temperature dependences of the resistance and the magnetic shielding signal for specimens are given in Fig.6.

It should be noted that a large resistance drop is observed in a normal metallic state. Its value decreases twice with temperature from ambient temperature down to T_c. The conductivity of these ceramics reaches 800-1000 $\Omega^{-1}cm^{-1}$. X-ray data show the product lattice to be pseudotetragonal and its parameters to be equal to a=b = 5.410(2) Å, c=37.124 (3) Å. The temperature behavior of polymeric precursors for the bismuth ceramic is very interesting. At 420° and 870°C maximum rates of the mass loss are observed. Mass-spectrometric analysis of gaseous products evolved from the bismuth ceramic during vacuum-thermal processing was carried out. O_2^+, Bi^+ and Bi_2^+ ions compose most of gaseous products. Trace quantities of OH^+, H_2O^+, CO^+, N_2^+, NO_2^+, and CO_2^+ ions are also found at all temperatures (Fig.7). Thus, traces of residual carbon are present in the ceramic formed. Preliminary studies show that the critical current density (I_{cr}) depends slightly on the method of ceramic preparation. The density of the synthesized ceramic (r) is less than that of the initially pressed oxides. During charge sintering, vacancies are formed which increase volume. Thus, there is a relationship between the ceramic density and its critical current density:

ρ, g/cm^3	2.4	3.0	3.6	4.2	4.7
I_{cr}, A/cm^3	9.0	20	90	130	240

It is known that size reduction generates new dislocations which are flows of vacancies. Optimal sintering conditions at the final stage must be such that a branched net of dislocations should be kept up till all pores are closed.[29] The theoretical density of 2223 bismuth ceramic is equal to 6.5 g/cm^3. Therefore, one can expect the formation of the bismuth ceramic with the quite large critical current.

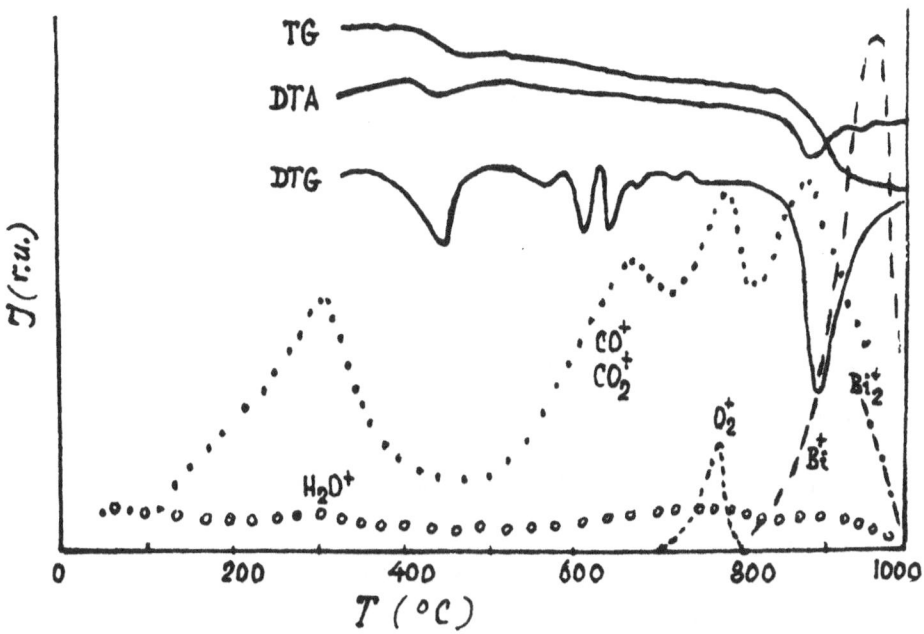

Figure 7. The thermal analysis data of Bi(III)-containing HTSC ceramic (Heating rate is 4°C, CO^+ and CO_2^+ are scaled up in fifty time)

CONCLUSION

Thermal transformations of metal-containing monomers, accompanied by their polymerization and partial or total decay of the resulting metal-containing polymers, may be a promising route for the fabrication of the materials. Kinetic features of the thermal destruction enables the formation of particles with the sizes needed and a polymer matrix. However there are still a number of the unsolved problems, among them are the following:
- the optimization of the syntheses of metal-containing precursors;
- the study of kinetics of thermal copolymerization in binary and ternary systems and the distribution of various metals along the polymer chain;
- controlling the redox processes during the formation of ferromagnetic and supersconducting materials;
- metal-metal interactions in metal-containing polymers and metal particles;
The analysis of the results achieved in this field allows one to believe that these problems will be solved in the near future.

ACKNOWLEDGMENTS

We acknowledge the International Science Foundation (NJB000) for the Support of this research.

REFERENCES

1. A.Gronowski, Z.Wojtczak, J.Thermal Anal. 26:233 (1983).
2. Z.Wojtczak, A.Gronowski, J.Thermal Anal. 36:2357 (1990).
3. S.P.Gubin, I.D.Kosobudskii, Uspekhi Khimii, 52:1350 (1983).
4. S.P.Gubin, Zh. Vsesouznogo obshchestva im.D.I.Mendeleeva 36:131 (1991).
5. R.Dageni, C & Er, 70(29):20 (1992).
6. G.V. Lisichkin, V.F.Petrunin, Zh. Vsesouznogo obshchestva im.D.I.Mendeleeva 36:131 (1991).
7. I.V.Tananaev, V.B.Fedorov, M.D.Molokhov, and L.V.Malukova, Izv. Akad. Nauk SSSR, Ser. Neorgan. Materialy 20:1026 (1984).
8. G.I.Dzhardimalieva, A.D.Pomogailo, V.I.Ponomarev, L.O.Atovmyan, Yu.M.Shulga, and A.G.Starikov, Izv. Akad. Nauk SSSR, Ser. Khim. 1525 (1988) (Bull. Acad. Sci. USSR, Div. Chem. Sci. 37:1352 (1988) (Engl. Transl.))
9. V.I.Ponomarev, L.O.Atovmyan, G.I.Dzhardimalieva, A.D.Pomogailo, and I.N.Ivleva, Koord. Khimiya 14:1537 (1988) (Sov. J. Coord. Chem., 14 (1988) (Engl. Transl.))
10. Yu.M.Shulga, O.S.Roshchupkina, G.I.Dzhardimalieva, I.V.Chernushchevich, A.F.Dodonov, Yu.V.Baldokhin, P.Ya.Kolotyrkin, A.S.Rozenberg, and A.D.Pomogailo, Izv. Akad. Nauk, Ser. Khim. 1739 (1993) (Russ. Chem. Bull., 42:1661 (1993) (Engl. Transl.)
11. A.S.Rosenberg, E.I.Aleksandrova, G.I.Dzhardimalieva, N.V.Kiryakov, P.E. Chizhov, V.I.Petinov, and A.D.Pomogailo, Izv. Akad. Nauk (5), (1995).
12. A.D.Pomogailo, V.S.Savostyanov. "Synthesis and Polymerization of Metal-Containing Monomers," CRC, Boca Raton (1994).
13. K.Nakamoto. "Infrared and Raman Spectra of Inorganic and Coordination Conpounds," Wiley, New York (1986).
14. K. Nakanishi. "Infrared Absorption Spectroscopy," Holden-Day, Inc., San Francisco (1962).
15. McCluskey, R.L.Suyder, and R.A.Condrate, J.Solid State Chem. 83:332 (1989).
16. Kratkii spravochnik fiziko-khimicheskikh velichin (Brief Handbook of Physico-Chemical Quantities) A.A.Ravdel' and A.M.Ponomareva, eds., Khimiya, Leningrad (1983) (in Russian).
17. M.A Poraj-Koshits, in: "Crystallochemistry. Advances in Science and Technology, VINITI, Moscow 15:3 (1981).
18. Khimicheskaya Entsiklopediya (Chemical Encyclopedia), Sov. Entsiklopediya, Moscow, 1 (1988) (in Russian).
19. K. Mortimer, in: "Teploty reaktsii i prochnost svyazei," Mir, Moscow (1964) (Russ. Transl.).
20. E.I.Aleksandrova, G.I.Dzhardimalieva, A.S.Rozenberg, and A.D.Pomogailo, Izv. Akad. Nauk, Ser.Khim., 303 (1993) (Russ. Chem. Bull., 42:259 (1993) (Engl. Transl.)).
21. E.I.Aleksandrova, G.I.Dzhardimalieva, A.S.Rozenberg, and A.D.Pomogailo, Izv. Akad. Nauk, Ser.Khim., 308 (1993) (Russ. Chem. Bull., 42:264 (1993) (Engl. Transl.)).

22. A.S.Rozenberg, E.I.Aleksandrova, G.I.Dzhardimalieva, A.N. Titkov, and A.D.Pomogailo, Izv. Akad. Nauk, Ser.Khim., 1743 (1993) (Russ. Chem. Bull., 42:1666 (1993) (Engl. Transl.)).

23. V.N.Kozyrenko, I.V.Kumpanenko, V.V.Loginov, I.D.Mikhailov, N.V.Chukanov, and S.G.Entelis, in: "Chislennye metody resheniya zadach matematicheskoi fiziki i teorii sistem" (Numerical Methods for the Solution of Problems of Mathematical Physics and System Theory), University of Peoples Friendship, Moscow, (1987) (in Russian).

24. N.V.Chukanov, I.V.Kumpanenko, V.V.Losev, and S.G.Entelis, Dokl. Akad. Nauk SSSR, 261:135 (1981) (Dokl. Chem., 261 (1981) (Engl. Transl.)).

25. J.W.Chien, B.M.Gong, J.M.Maadsen, and R.B.Hallock, Phys. Rev. B, 38:11853 (1988).

26. J.C.Chien, B.M.Gong, X.Mu, and Y. Yang, J. Polym. Sci., Polym. Chem., 28:1999 (1990).

27. I.Valente, C.Sanchez, M.Henry, and J.Livage, Ind. Ceram., (3):193 (1989).

28. Dubovitskii A.V., Makarov E.F., Makova M.K., Merzhanov V.A., and Topnikov V.N., Sverkhprovodimost': Fizika, khimiya, tekhnika., 4:!024 (1991).

29. Geguzin Ya.E. "Fizika spekaniya," Nauka, Moscow (1984).

BRIDGED MACROCYCLIC TRANSITION METAL-OLIGOMERS, SYNTHESIS AND ELECTRICAL PROPERTIES

Michael Hanack

Institut für Organische Chemie
Lehrstuhl für Organische Chemie II der Universität Tübingen
Auf der Morgenstelle 18
D-72076 Tübingen
Germany

Stable intrinsic semiconductors, which exhibit conductivities of 0.1 S/cm without any doping process can be easily prepared by using our "shish-kebab" approach. Hereby stacked transition metal macrocycles $[MacM(L)]_n$. with M e.g. Fe, Ru, Os and Mac = phthalocyanine (Pc) or 2,3-naphthalocyanine (2,3-Nc) are used. The bridging ligands (L) are special heteroaromatic systems, e.g. s-tetrazine (tz), substituted tetrazines, triazine (tri), substituted triazines, but also special organic dinitriles, e.g. tetrafluoroterephthalic acid dinitrile, fumarodinitrile, dicyanoacetylene, and cyanogen. We report about the preparation and the characterization of the corresponding transition metal complexes $[MacM(L)]_n$.

The intrinsic conductivities of all these compounds are a result of the low oxidation potential of the bridging ligands and due to the low lying LUMO in the corresponding bridged systems $[MacM(L)]_n$.

The electrical and physical properties, especially the UV- and Mößbauer spectra (for the corresponding iron compounds) for the bridged macrocyclic metal complexes are discussed with respect to the intrinsic semiconducting properties of these compounds.

Bridged stacked transition metal macrocycles $[MacM(L)]_n$ are synthesized in a large variety by using the so-called "shish kebab" approach[1] in which transition metal macrocycles (MacM) are reacted with bidentate bridging ligands (L). In $[MacM(L)]_n$ the transition metals are e.g. Fe, Ru, Os, Co, Rh in various oxidation states.[1] As macrocycles (Mac) the heteroaromatic systems phthalocyanine (Pc), tetrabenzoporphyrin (TBP), but also macrocycles containing more extended Ð-systems like 1,2- or 2,3-naphthalocyanine

(1,2-, 2,3-Nc), anthracenocyanine (Anc) or phenanthrenocyanine (Phc) are used.[1] As bridging ligands (L) bifunctional organic donor molecules, e.g. pyrazine (pyz), *s*-tetrazine (tz) and others (see below), for metals in the oxidation state +2, or cyanide (CN⁻), thio-cyanate (SCN⁻) or azide (N₃⁻) for transition metals in the oxidation state +3[1] are employed (Figure 1).

The coordination polymers or oligomers [MacM(L)]ₙ are practically insoluble in or-ganic solvents, but soluble oligomers have been synthesized using metallomacrocycles R₄PcM and R₈PcM, which are substituted in peripheric positions (R = *t*-bu, et, OR'; R' = C₅ - C₁₀; M = Fe, Ru).

The so obtained non-soluble and soluble coordination polymers exhibit interesting electrical and non-linear optical properties. The soluble polymers [RₓMacM(L)]ₙ form Langmuir Blodgett films and they have been recently used successfully as sensors.

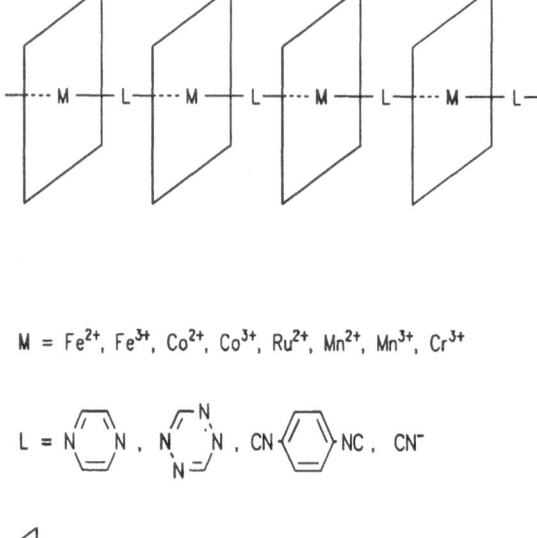

$M = Fe^{2+}, Fe^{3+}, Co^{2+}, Co^{3+}, Ru^{2+}, Mn^{2+}, Mn^{3+}, Cr^{3+}$

$L = $ ⟨pyz⟩, ⟨tz⟩ , CN⟨benzene⟩NC , CN⁻

⟨Pc⟩ $= Pc^{2-}, R_4Pc^{2-}, R_8Pc^{2-}, 1,2-Nc^{2-}, 2,3-Nc^{2-}, TBP^{2-}$

Figure 1. Schematic structure of bridged macrocyclic transition metal complexes.

In this report we concentrate on the electrical properties of these compounds, and it will be shown that the combination of a transition metal macrocycle with special bridging ligands forming the coordination polymer [MacM(L)]ₙ can lead to systems which exhibit intrinsic conductivities, in other words, which show very good semiconducting properties without external oxidative doping.

Systematic investigations are reported about the influence of the bridging ligand on the conductivity of the bridged phthalocyaninato and 2,3-naphthalocyaninato transition metal complexes [MacM(L)]ₙ. Many different bridging ligands have been used to study their special structural and electronic influence on the general behaviour of the correspond-ing coordination polymers. As bridging ligands first e.g. dabco (without double bonds), pyrazine (pyz), 1,4-diisocyanobenzene (dib), tetrazine (tz), and its derivatives e.g. 3,6-dimethyl-*s*-tetrazine (me$_2$tz) are used and the powder conductivities of the corresponding bridged coordination polymers are compared.

PcFe, PcRu, PcOs, and 2,3-NcFe react with tetrazine (tz) and dimethyltetrazine (me$_2$tz), depending on the conditions, with formation of the corresponding monomers MacM(L)$_2$ (L = tz, me$_2$tz) and the bridged systems [MacM(L)]$_n$ (L = tz, me$_2$tz). The tetrazine bridged systems, in contrast to other bridged compounds [MacM(L)]$_n$ with M = Fe, Ru or Os and L e.g pyz or dib, show already good semiconducting properties without external oxidative doping (σ_{RT} = 0.05 - 0.3 S/cm).

The dc-powder conductivities of various bridged macrocyclic transition metal oligomers in the non-doped state containing a variety of structurally different bridging ligands are listed in Table 1. The bridged complexes [MacM(L)]$_n$ with L = dabco, pyz, tz, me$_2$tz and triazine (tri), contain cofacially arranged macrocycles, which are separated by approximately the same distance (about 600 pm). Table 1 also shows the conductivity data of the corresponding monomers MacM(L)$_2$ for comparison.

A systematic investigation of the influence of the bridging ligands on the semiconducting properties in [MacM(L)]$_n$ reveal, that changing L, e.g from dabco over pyz to tz leads to a steady increase of the semiconducting properties without external oxidative doping. Powder conductivities in the order of 0.1 S/cm can be reached by using *s*-tetrazine, but also substituted tetrazine e.g. 3,6-dimethyl-*s*-tetrazine (me$_2$tz), triazine, *p*-diaminotetrazine, and others as the bridging ligands in such complexes (Table 1).

The monomeric complexes PcM(L)$_2$ (L = e.g. dabco, pyz, tz; M = Fe, Ru, Os) are prepared using the same procedures as in the case of the oligomers except that the molar ratio of the ligands L and the transition metal phthalocyanines PcM is different. They show insulating behaviour. The monomers PcM(L)$_2$ can be transformed by a simple thermal process into the oligomers [PcM(L)]$_n$, which then depending upon the bridging ligands exhibit high intrinsic conductivities (Table 1).

The following conclusion can be drawn from Table 1: dabco is a ligand containing no Ð-orbitals, which are able to interact with the metallomacrocycle (see below), the complex [PcFe(dabco)]$_n$ is an insulator. An increase in conductivity is observed for the pyrazine-bridged compounds [MacM(pyz)]$_n$, which exhibit conductivities in the low semiconducting region. However, by changing the bridging ligand from pyrazine to *s*-tetrazine the conductivity is increased by 3 to 5 orders of magnitude without external oxidative doping (Table 1). In comparison with the corresponding monomers PcM(L)$_2$ the conductivity increases even more by factors in between 10^7 - 10^9 (Table 1).

In addition to tetrazine and dimethyltetrazine other substituted tetrazines, e.g. *p*-diaminotetrazine (NH$_2$)$_2$tz have been used as the bridging ligand to react with PcRu. The corresponding oligomer [PcRu(NH$_2$)$_2$tz]$_n$ also shows intrinsic conductivity.

Triazine (tri) also forms coordination oligomers, which as shown in Table 1 show a powder conductivity in between the values of the tetrazine- and pyrazine-bridged systems. Triazine is linked to the central metal of the macrocycle (e.g. Fe or Ru) via the 1,4-nitrogens.

The observed intrinsic conductivities of this type of coordination polymers are due to the low oxidation potentials of the bridging ligands, e.g. tz and me$_2$tz. The low lying LUMO of the bridging ligand leads to a charge transfer of the metal macrocycle on to the bridging ligand thereby inducing intrinsic conductivities.

Additional bridging ligands can be used to prepare intrinsically conductive coordination polymers: *p*-tetrafluoroterephthalic acid dinitrile has been used to prepare the corresponding coordination polymer $[PcRu(CN)_2F_4C_6]_n$ which, as shown in Table 1, also shows good semiconductive properties without external oxidative doping. Using fumaric acid dinitrile NCCH=CHCN leads to the corresponding polymer $[PcRu(CN)_2(CH)_2]_n$ with powder conductivity of $\sigma_{RT} = 10^{-2}$ S/cm.

Table 1. DC-Powder electrical conductivities of monomeric and bridged macrocyclic transition metal complexes (room temperature, pressed pellets, 1 kbar).

Compound	σ_{RT} [S/cm^{-1}]
PcFe(dabco)$_2$	10^{-10}
[PcRu(dabco)$_2$ x 1.4 CHCl$_3$]$_n$	10^{-10}
PcFe(pyz)$_2$	3×10^{-12}
[PcFe(pyz)]$_n$	1×10^{-6}
PcFe(tz)$_2$	$< \quad 10^{-9}$
[PcFe(tz)]$_n$	2×10^{-2}
PcRu(tz)$_2$	$< \quad 10^{-11}$
[PcRu(tz)]$_n$	1×10^{-2}
PcOs(tz)$_2$	4×10^{-8}
[PcFe(me$_2$tz)]$_n$	4×10^{-3}
[PcRu(NH$_2$)$_2$tz]$_n$	4×10^{-3}
[PcRup-(NH$_2$)$_2$C$_6$H$_4$]$_n$	5×10^{-9}
[PcRuCl$_2$(tz)]$_n$	3×10^{-3}
[PcOs(pyz)]$_n$	1×10^{-6}
[PcOs(tz)]$_n$	1×10^{-2}
[2,3-NcFe(pyz)]$_n$	5×10^{-5}
[2,3-NcFe(tz) x 0.2 CHCl$_3$]$_n$	3×10^{-1}
[(me)$_8$PcFe(pyz)]$_n$	3×10^{-9}
[(me)$_8$PcFe(tz)]$_n$	1×10^{-2}
[PcRu(me$_2$tz)]$_n$	4×10^{-3}
[PcRu(tri)]$_2$	1×10^{-9}
[PcRu(tri)]$_n$	2×10^{-4}
[PcRu(CN)$_2$F$_4$C$_6$]$_n$	1×10^{-3}
[PcRu(CN)$_2$(CH)$_2$]	3×10^{-2}

The electrical and physical properties, especially the UV and Mößbauer spectra (for iron compounds) for the intrinsically conductive coordination polymer are discussed in respect to their intrinsic semiconducting properties. This is demonstrated with one example in the following: all the intrinsically conductive coordination polymers with group 8 transition metals and bridging ligands like tz, me$_2$tz (substituted tetrazines), triazine, tetrafluoroterephthalic acid dintrile, fumaro dinitrile exhibit physical properties which are not shown by the corresponding systems in which L = e.g. pyz, dib or dabco. All the intrinsically conductive systems show broad bands in the UV/Vis/NIR spectra, e.g. for the tetrazine-bridged systems between 1250 and 2500 nm with different maxima, e.g. for [PcFe(tz)]$_n$ at 1515 nm estimated energy gap between the HOMO of the different metallomacrocycles and the LUMO of the tetrazine in all the tetrazine-bridged systems. The broad bands observed in the absorption spectra of these complexes are assigned to charge transfer from the metallo-macrocycle to the π^*-orbital of s-tetrazine.

The results from the UV/Vis spectra obtained with $[PcFe(tz)]_n$ and the other systems mentioned are supported by UPS data of $PcFe(tz)_2$ and $[PcFe(tz)]_n$, which also show that $[PcFe(tz)]_n$ has a small band gap. With the help of XPS it could be demonstrated that for $PcFe(tz)_2$, $[PcFe(pyz)]_n$ and $[PcFe(tz)]_n$ oxygen has only a negligible influence on the conductivity of these compounds. In other words, oxygen is not a doping agent.[2] On the other hand, it should be mentioned that the conductivity is very much dependent on the substituents in the peripheric positions: $[t\text{-}Bu_4PcFe(tz)]_n$ and $[Et_4PcFe(tz)]_n$ both show CT-bands in the UV/Vis and IR spectrum. The conductivity of both compounds, however, is low in comparison with $[PcFe(tz)]_n$. The large t-butyl groups in $[t\text{-}Bu_4PcFe(tz)]_n$ hinder an electron transfer from chain to chain, thereby decreasing the macroscopic measured conductivity.

Also the Mößbauer data of s-tetrazine coordinated iron complexes can be very well correlated with an internal charge transfer from the metal macrocycle to the bridging ligand with a low oxidation potential.

Mößbauer data of s-tetrazine-coordinated iron complexes are summarized in Table 2 in comparison with the data for the iron macrocycles and the corresponding pyridine complexes. For the tetrazine-bridged complexes $[MacFe(tz)]_n$ the isomer shift δ is similar to the monomeric $PcFe(tz)_2$ whereas a clear increase of the quadrupole splitting (Δ_{EQ}) even in comparison to $[PcFe(pyz)]_n$ is measured. This is a further indication for a low band gap in the tetrazine-bridged complexes, because this effect can be explained by the thermal activation of electrons of the highest occupied band. As the contribution of such delocalized electrons to the occupation of metal centred d-orbitals is diminished, the increase in Δ_{EQ} is the result of the low band gap of the quasi one-dimensional chain structure of the tetrazine-bridged complexes.

Table 2. Mößbauer data of monomeric and bridged PcFe complexes.

Complex	δ [mm/s]	Δ_{EQ} [mm/s]
ß-PcFe	0.38	2.58
PcFe(py)$_2$	0.26	2.02
PcFe(tz)$_2$	0.15	1.79
[PcFe(tz)]$_n$	0.13	2.23
PcFe(me$_2$tz)·0.5 me$_2$tz	0.36	2.67
(Me)$_8$PcFe	0.38	2.55
(Me)$_8$PcFe(py)$_2$	0.26	1.97
[(Me)$_8$PcFe(tz)]$_n$	0.14	2.11
(MeO)$_8$PcFe	0.36	2.50
(MeO)$_8$PcFe(py)$_2$	0.25	1.95
[(MeO)$_8$PcFe(tz)]$_n$	0.15	2.19
PcFe(tri)$_2$	0.20	1.79
[PcFe(tri)]$_n$	0.27	1.91
[Et$_4$PcFe(tz)]$_n$	0.14	2.04
[Et$_4$PcFe(tri)]$_n$	0.27	1.82

The bridged macrocyclic transition metal complexes described here are one of the first stable systems which exhibit intrinsic conductivity. They are thermally quite stable and have potential technical applications.

Acknowledgement

Thanks go to my co-workers who have done the experimental work with great enthusiasm. We thank the Deutsche Forschungsgemeinschaft, the Bundesministerium für Forschung und Technologie and the Fonds der Chemischen Industrie for financial support.

REFERENCES

1. a) M. Hanack, A. Datz, R. Fay, K. Fischer, U. Keppeler, J. Koch, J. Metz, M. Mezger, O. Schneider, H.-J. Schulze, in: *Handbook of Conducting Polymers*, T.A. Skotheim, ed., Marcel Dekker, New York (1986). - b) M. Hanack, S. Deger, A. Lange, *Coord. Chem. Reviews* 83; 115 (1988). - c) H. Schultz, H. Lehmann, M. Rein, M. Hanack, in: *Structure and Bonding* 74, J. W. Buchler, ed., Springer-Verlag, Heidelberg (1990). - d) M. Hanack, M. Lang, *Adv. Mater.* 6; 819 (1994).
2. M. Dreßen, *Dissertation*, Tübingen (1992).

ENHANCEMENT OF DIMENSIONAL STABILITY IN SOLUBLE POLYIMIDES VIA LANTHANIDE(III) ADDITIVES

Robin E. Southward,[2] D. Scott Thompson,[2] David W. Thompson,[2] and Anne K. St. Clair[1]

[1]Langley Research Center, National Aeronautics and Space Administration, Hampton, VA 23665-5225

[2]Department of Chemistry and Applied Science Program, College of William and Mary, Williamsburg, VA 23187-8795

INTRODUCTION

A need exists for optically clear thin films which can endure for long periods in a space environment for applications on large-area solar collectors, inflatable antennas, solar arrays, and various space optics. Fluorine-containing polyimides such as those derived from 2,2-bis(3,4-dicarboxyphenyl)hexafluoropropane dianhydride, 6FDA, are excellent candidates for use in space due to their high optical transparency at the solar maximum wavelength of 500 nm, their outstanding resistance to radiation, and many other properties.[1-8] These polymers, however, generally display coefficients of thermal expansion (CTE) between 40 and 50 ppm. Because dimensional stability of large space structures is critical to performance and efficiency, it is desirable to lower the CTE of these polymers. The control of CTE and the ability to match the CTE of polyimides to those of inorganic materials in composite devices is of considerable challenge and interest.[9] Preliminary studies have shown that the CTE of fluorine-containing polyimides can be lowered by as much as 30 percent by the addition of modest quantities of lanthanide(III) compounds.[10,11,12]

Polyimides generally have coefficients of thermal expansion in the range of 30-60 (μm/m)/°C or ppm below their glass transition temperatures. Typically, metals and inorganic materials such as silicon, quartz, silicon carbide, alumina and other metal oxides and ceramics have CTE's less than 20 ppm. Polyimides derived from 6FDA and diamines such as 1,3-bis(3-aminophenoxy)benzene (APB), 4,4'-oxydianiline, m -phenylene diamine, 2,2-bis(4-aminophenyl)hexafluoropropane, etc. have CTE's of approximately 50 ppm, primarily due to the flexible, non-linear structure of the dianhydride unit. This work focuses on attempts to reduce the CTE of thermally cured 6FDA/APB polyimide films via the addition of soluble lanthanide(III)dopant systems which are added to a dimethylacetamide resin solution of the soluble polyimide with the expectation of maintaining visual transparency and the essential mechanical and thermal properties 6FDA/APB films. This represents a submicron composite approach to reducing the CTE of polyimides which is to be contrasted with recent attempts to reduce the CTE of polyimides by tailoring dianhydride and diamine monomer structure as summarized below.

Numata and coworkers[13-22] studied systemically the relationship of chemical structure of non-fluorinated aromatic polyimides to coefficients of linear expansion. They concluded that polymers which are both linear and rigid exhibit the lowest coefficients of thermal expansion. Trofimenko[23] and Auman[24] reported on the synthesis and characterization of a class of linear and rigid fluorinated monomers based on the 6FDA structure. These monomers are 9,9-bistrifluoromethyl-2,3,6,7-xanthene-tetracarboxylic dianhydride (6FCDA) and 9-phenyl-9-trifluoromethyl-2,3,6,7-xanthenetetracarboxylic dianhydride (3FCDA). The expectation was that the rigidity and linearity of the xanthene structure of 6FCDA and 3FCDA coupled with appropriate diamines would lead to polyimides which had the same exemplary properties of 6FDA polymers but coefficients of thermal expansion which were less than 20 ppm. Indeed, the polyimides formed from both 6FCDA and 3FCDA with both rigid and flexible diamines had lower CTE's than the corresponding polymers formed from 6FDA by approximately 20%. However, when the two xanthene-based monomers were reacted with rigid diamines such as p-phenylenediamine, polyimides with CTE's below 20 ppm were realized. Unfortunately, the linear and rigid structural features which lower the CTE also enhance charge transfer interactions which imparts color to the polyimides.

The control of the coefficient of thermal expansion has been oriented heavily toward modification of monomer structure such that linear rigid polyimides can be synthesized. Alternatively, we have shown that the addition of soluble lanthanide(III)-ligand systems to poly(amic acid) and polyimide resins can result in an increase in dimensional stability.[12] The work reported herein will focus on the modification of the imidized form of 6FDA/APB. The fact that 6FDA/APB is soluble in common solvents in the fully imidized form provides greater flexibility with regard to final thermal treatment cycles and diminishes problems associated with the release of water in curing poly(amic acid) resins. Furthermore, we have found that metal doped amic acid resins often have severely compromised molecular weights after thermal curing leading to brittle and crazed films.[25]

Most commonly, the preparation of polyimide films involves first casting a film as the soluble poly(amic acid) and then thermally curing to the polyimide. This is the usual protocol since the majority of aromatic polyimides are insoluble in aprotic polar solvents at ambient temperatures. With regard to the incorporation of class A (hard) metal ions into polyimide films, the poly(amic acid) route to doped films is often complicated because of gelation due to the coordination of the metal atoms with the amic acid carboxyl and carboxylate groups. The doping of 3,3',4,4'-benzophenone tetracarboxylic dianhydride/-4,4'-oxydianiline (BTDA/4,4'-ODA) with lanthanide(III) acetates and tris(acetylacetonato)-lanthanide(III) complexes has been found to result in a gelled poly(amic acid) resin which is intractable for film production.[25] This gelation is due primarily to the basicity of the acetate and acetylacetonate anionic ligands which abstract protons from the free carboxyl groups of the amic acid chains. These resulting polymer carboxylate groups coordinate to the metal resulting in infinite network formation. Thus, in addition to the previously mentioned advantageous properties of 6FDA/APB, we chose to examine the incorporation of lanthanide(III) compounds, solubilized in DMAc by the addition of selected ligands, into the soluble fully imidized (6FDA/APB) to circumvent gelation problems.

EXPERIMENTAL

6FDA was obtained from Hoechst Celanese and vacuum dried for 17 h at 110 °C prior to use. APB was purchased from National Starch and was used as received. The lanthanide acetates (Ce, Eu, Gd, Er, and Tm) were obtained as hydrates from Aesar and Strem at a minimum purity of 99.9% and were used without further purification. TGA established the hydrates to be essentially tetrahydrates with mole ratios of metal to water being in the range 3.9 to 4.0. Dibenzoylmethane (DBM) was purchased from Eastman and was sublimed under vacuum prior to use. Hexafluoroacetylacetone (HFA) was obtained from Aldrich, and trifluoroacetylacetone (TFA) and trifluoroacetic acid (TFAA) were purchased from Lancaster. These three ligands were distilled under nitrogen before use. Anhydrous thulium(III) chloride (99.9%) and thulium(III) nitrate pentahydrate (99.9%) were purchased from Alfa. The nitrate salt was dried under vacuum at 50 °C for 24 h, and as used thereafter had a degree of hydration of 1.5. DMAc was purchased from Aldrich as HPLC grade with water <0.03%.

Imidized 6FDA/APB powder was obtained by the addition of 6FDA (0.5% molar excess) to a dimethylacetamide (DMAc) solution of APB to first prepare the poly(amic acid) at 15 % solids (w/w). The reaction mixture was stirred at ambient temperature for 7 h. The inherent viscosity of the polyamic acid was 1.4 dL/g at 35 °C. This amic acid precursor was chemically imidized in an equal molar ratio acetic acid-pyridine solution, the pyridine and acetic acid each being three times the moles of diamine monomer. The pyridine and acetic acid each being three times the moles of diamine monomer. The polyimide was precipitated in water, washed thoroughly with deionized water, and vacuum dried at 200 °C for 20 h. The inherent viscosity of the polyimide in DMAc was 0.81 dL/g at 35 °C.

All metal doped imidized 6FDA/APB resins were prepared by first dissolving the metal salt and ligand (DBM, HFA, TFA, or TFAA) in required amount of DMAc and then adding solid imide to give a 15% solids (excluding the additives) resin. The solutions were stirred 2-4 h to dissolve all of the polyimide. The clear metal doped resins were cast as films onto soda lime glass plates using a doctor blade set to give cured films near 25 microns. The cast films were allowed to dry to a tact free state for 18 h at room temperature at 10% humidity and then were heated in a forced air oven at 100, 200, and 300 °C for 1h at each temperature (the "standard thermal cycle"). The films were removed from the plate by soaking in deionized water.

Thermogravimetric analysis (heated at 2.5 °C/min) and differential scanning calorimetry (heated at 20 °C/min) were performed in air with Seiko TG/DTA 220 and DSC 210 systems, respectively. Linear coefficients of thermal expansion were done with a Seiko TMA 100 station and are reported as the average value over the temperature range of 70 - 125 °C. The CTE samples where desiccated for 24 h before analysis. Micrographs were obtained with a Zeiss CEM 902 TEM (Virginia Institute of Marine Science). X-ray data was obtained with a Philips 3600 diffractometer. Mechanical measurements were made at ambient temperature on a Sintech Model 2000/2 table top load frame. Densities were determined using a density aqueous zinc chloride gradient column.

RESULTS AND DISCUSSION

The incorporation of soluble lanthanide(III) dopants into polyimide films offers a singular opportunity to study effects of metals in positive oxidation states on film properties. The lanthanides are a unique series of metal ions which have an exceptionally stable single tervalent oxidation state and uniformly decreasing crystal radii for coordination number of six from 117 to 84 pm.[26] The large radii allow these ions to adopt coordination numbers ranging from six to twelve with the most common being eight and nine.[27,28] Thus, we have metal ion additives which have expanded coordination spheres and are class A (hard) Lewis acids allowing enhanced binding to polymer donor atoms, particularly oxygen. Such polymer-metal coordination has been demonstrated to be important in preventing the migration (phase separation) of the metal dopant to the surface of the film during cure and in yielding a uniform homogeneous distribution of submicron metal clusters or even a mononuclear distribution of metal(III) species throughout the polymer matrix.[12,29,30]

Our goal was to prepare submicron, rather than the more typical heterogeneous "filler", composite polyimide films with inorganic particle sizes less than that required for the scattering of visible electromagnetic radiation, e.g., <200 nm. This would then retain the visual clarity of the films. To achieve this particle size it was necessary to begin with lanthanide dopants which are soluble in the polyimide resin. Lanthanide(III) nitrate salts are an obvious first choice. They are very soluble in DMAc. However, the nitrates, as will be presented later, give discolored and brittle films which may be due to the oxidizing character of the nitrate ion and/or the reactivity of the weakly DMAc solvated Lewis acidic metal(III) ion. While a few lanthanide(III) chloride salts are soluble, these also give films which are not consistently of high quality. Thus, we chose to use the stable lanthanide(III) acetates which are readily available in high purity. In this study, the lanthanide(III) acetates for the earlier metals of the 4f series (Ce, Eu, and Gd in this work) are not soluble in DMAc; the later acetates of erbium and thulium (as tetrahydrates) are soluble. The insoluble acetates were readily brought into solution (see Figure 1) by adding three equivalents of a mildly acidic β-diketonate chelating ligand derived from β-diketones such as 1,3-diphenyl-1,3-propanedione (dibenzoylmethane or DBM with $pK_a = 8$), 1,1,1,3,3,3 hexafluoro-2,4-pentanedione (hexafluoroacetylacetone or HFA with $pK_a = 5.3$), or 1,1,1-trifluoro-2,4-pentanedione (trifluoroacetylacetone or TFA with $pK_a = 6.3$) or by adding trifluoroacetic acid which is very acidic with a pK_a of 0.6. Less acidic β-diketones such as the ubiquitous 2,4-pentanedione would not undergo proton transfer to an acetate anion to bring the lanthanide species into solution as an acetylacetonate complex. We also knew that β-diketonate complexes of the lanthanides eventually undergo thermal decom-

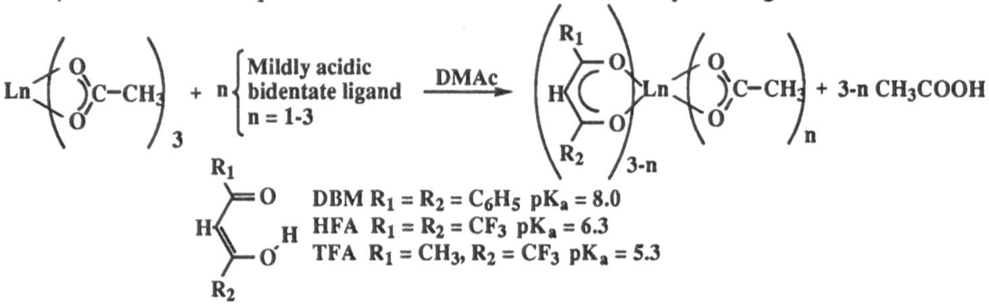

$$Ln\left[\underset{O}{\overset{O}{\bigcirc}}C-CH_3\right]_3 + n\left\{\begin{matrix}\text{Mildly acidic}\\\text{bidentate ligand}\\n = 1\text{-}3\end{matrix}\right\} \xrightarrow{\text{DMAc}} \left(\underset{R_2}{\overset{R_1}{H\bigcirc}}\overset{O}{\underset{O}{\bigcirc}}\right)_{3-n} Ln\left[\underset{O}{\overset{O}{\bigcirc}}C-CH_3\right]_n + 3\text{-}n\ CH_3COOH$$

$$\underset{R_2}{\overset{R_1}{H\bigcirc}}\underset{O'}{\overset{=O}{}}$$

DBM $R_1 = R_2 = C_6H_5$ pK_a = 8.0
HFA $R_1 = R_2 = CF_3$ pK_a = 6.3
TFA $R_1 = CH_3, R_2 = CF_3$ pK_a = 5.3

Figure 1. Solubilization of lanthanide(III) acetates with mildly acidic bidentate ligands.

position to give metal oxides.[35,36,37] Yet, having strongly coordinating ligands such as DBM in particular should provide a moderating effect on the potentially reactive Lewis acidic lanthanide(III) ions. We chose to focus on DBM as it gave the highest quality films.

Previous work has shown that lanthanide(III) acetates cause complete gelation of the poly(amic acid) precursors of 6FDA/APB and BTDA/4,4'-ODA[25]. However, in the present study we have found that the addition of three equivalents DBM to the acetate-poly(amic acid) system prevents gelation. It was further found that the addition of 6FDA/APB imide powder to lanthanide(III)-dibenzoylmethane complex solutions results in a clear and homogeneously doped resin with no apparent change in viscosity compared with a 6FDA/APB resin without metal additives. The thermally treated films were visually clear, light amber in color, homogeneous, and flexible. There was a uniform increase in the densities of the doped films.

Table I presents thermal and mechanical data on the series of lanthanide-doped films heated in a forced air oven having a polymer repeat unit to metal ion ratio of 5:1 and a metal ion to DBM ratio of 1:3. This table shows that there is only a slight, if any, increase in the glass transition temperature in the metal-containing films relative to the two metal-free control films. Such a trend in T_g was observed in the work of Stoakley and St. Clair.[10] The metal additive in all the films reduces their thermal oxidative stability with 10% weight loss by TGA occurring *ca.* 50 °C lower than the undoped control film. The coefficient of thermal expansion remains relatively unaffected with addition of the metal-ligand systems to the polymer. The addition of the metal-ligand system to 6FDA/APB produced modest changes in mechanical properties. There is a small but uniform decrease in the tensile strength and percent elongation at break for the doped films.

TEM data as illustrated for the europium-doped film in Figure 3 (air side) shows the metal particle aggregates to be in the submicron range with particle sizes less than 100 nm. Most of the aggregates are in the range of 5-60 nm. There is a relatively uniform dispersion of particles throughout the polymer matrix. There is no migration of metal complex species to the surface of the polymer which suggests from the work of Sen and coworkers[31,32,33] that there is cooordination of the polymer donor groups to the metal. This restricts migration of the metal complexes to the surface to give an apparent phase separated system. The photoelectron spectroscopic data in Table I also demonstrates that there is no metal migration. The X-ray diffraction pattern for the europium-doped film is presented in Figure 2 and is typical of those for all doped films. Consistent with the visual clarity of the films, X-ray data show the doped film to be amorphous without reflections characteristic of crystalline europium compounds with crystallite sizes large enough to scatter visible radiation. Finely divided (38 μm, 400 mesh) completely insoluble rare earth dopants, such as fluorides and oxides at the 5:1 repeat unit to metal ratio, uniformly dispersed throughout the polymer matrix, show sharp intense reflections of the inorganic material superimposed on the amorphous 6FDA/APB halo.

Table II presents data for the same five metal-polyimide systems as in Table I but thermally treated under a dry nitrogen atmosphere to ascertain if there is an important role for atmospheric oxygen with regard to polymer degradation and/or the decomposition of the tris(dibenzoylmethanato) complexes to metal oxide. The data reveal that the films cured in nitrogen are very similar in properties to those cured in air with the exception of

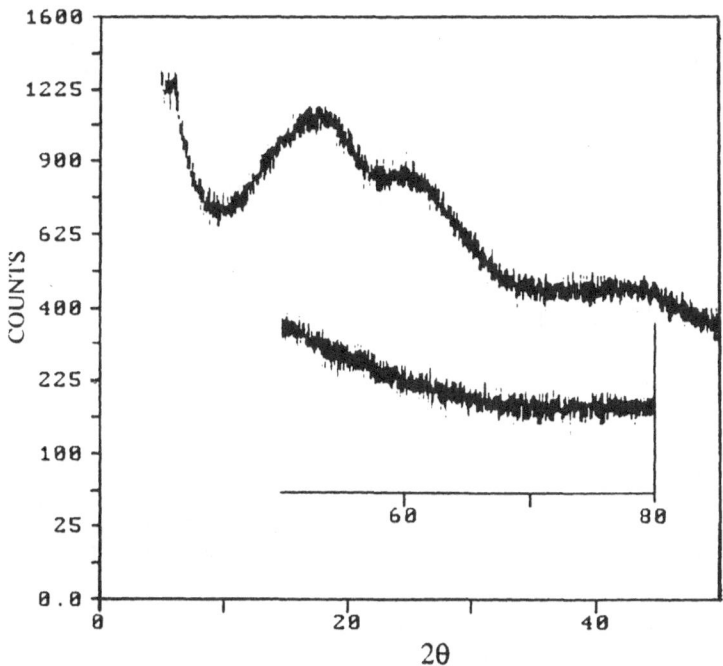

Figure 2. X-ray pattern for the europium acetate - DBM film of Table 1.

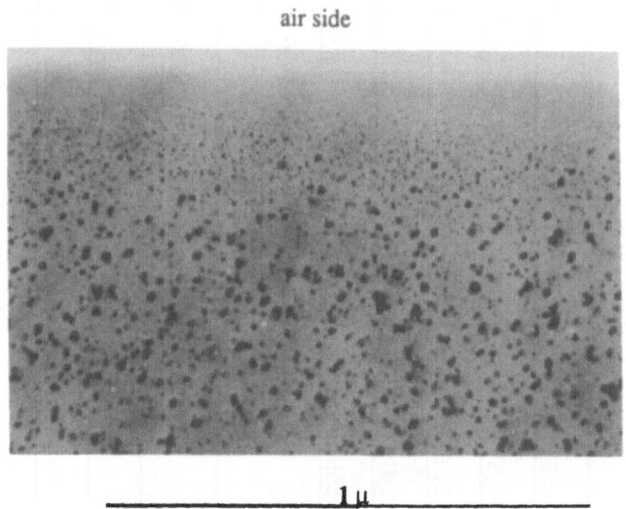

air side

1 µ

Figure 3. TEM for the europium acetate - DBM film of Table 1.

Table 1. Thermal and Mechanical Data for 6FDA/APB Films with Lanthanide (III) Acetate-Dibenzoylmethane Dopants

Additive	Theoretical % Metal	Tg by DSC (°C)	TGA Wt. Loss at 10% (°C)	CTE (ppm/°C)	Tensile Strength (Ksi)	Percent Elongation at break	Modulus (Ksi)	Density (g/cm^3)	XPS % Metal at Surface
None	Control	205	496	48.5	17.8	5.15	446	1.429	0
Ce(C$_2$H$_3$O$_2$)$_3$-DBM	3.85	210	445	46.5	14.7	4.31	412	1.468	0.17
Eu(C$_2$H$_3$O$_2$)$_3$-DBM	4.17	207	447	45.1	15.1	4.08	441	1.464	0.00
Gd(C$_2$H$_3$O$_2$)$_3$-DBM	4.38	207	439	50.2	15.8	4.04	464	1.445	0.42
Er(C$_2$H$_3$O$_2$)$_3$-DBM	4.54	207	468	47.9	15.9	4.02	489	1.459	0.08
Tm(C$_2$H$_3$O$_2$)$_3$-DBM	4.59	207	446	47.2	15.3	4.53	451	1.469	0.36

Table 2. Thermal and Mechanical Data for 6FDA/APB Films with Lanthanide (III) Acetate-Dibenzoylmethane Dopants Heated Under Nitrogen

Additive	Theoretical % Metal	Tg by DSC (°C)	TGA Wt. Loss at 10% (°C)	CTE (ppm/°C)	Tensile Strength (Ksi)	Percent Elongation at Break	Modulus (Ksi)
None	Control	205 (air)	496 (air)	48.5 (air)	17.8	5.15	446
Ce(C$_2$H$_3$O$_2$)$_3$-DBM	3.85	210	444	49.8	15.8	4.30	446
Eu(C$_2$H$_3$O$_2$)$_3$-DBM	4.17	211	433	44.0	12.8	3.11	463
Gd(C$_2$H$_3$O$_2$)$_3$-DBM	4.38	208.5	443	47.9	14.4	3.79	455
Er(C$_2$H$_3$O$_2$)$_3$-DBM	4.54	209	447	50.7	14.5	4.15	437
Tm(C$_2$H$_3$O$_2$)$_3$-DBM	4.59	210	454	48.1	13.8	4.39	402

the nitrogen cured europium film which shows the greatest lowering of the CTE, tensile strength, and percent elongation and was the only one of the ten films which was not flexible to a fingernail crease. The similarity of the air and nitrogen data would indicate that atmospheric oxygen does not play any important role in the decomposition of tris(dibenzoylmethanato)lanthanide complexes undergoing the temperature program of the standard cure cycle. Other thermogravimetric analysis work in inert atmospheres suggests that thermal degradation toward a metal oxide occurs by abstraction of oxygen from water and/or from the β-diketonate ligand.[34,35]

The europium(III) acetate/dibenzoylmethane-6FDA/APB system, which we observed in Table I to give the largest, though modest, lowering of the CTE (Table I), was examined as a function of final thermal treatment temperature. The data are contained in Table III. All films were heated at 100 °C and 200 °C for 1 h each and then at a final temperature ranging from 220 °C to 340 °C for an additional hour. From the work of Stoakley et al.[36] it was found that the intense fluorescence emission (Eu $^7F_0$ - $^5D_0$ transition) at 613 nm of tris(dibenzoylmethanato)europium(III) when dissolved in the XU-218 polyimide film and cured to 180 °C is virtually identical to that of the complex in chloroform solution. The fluorescence begins to diminish in intensity when the polymer is cured at a final temperature of 250 °C and is minimal but still visible when the polymer is cured at final temperature of 300 °C. This diminution of fluorescence intensity results from partial thermal degradation of the complex. Thus, we thought that the lowering of the CTE might track the degradation of the lanthanide-β-diketonate complex. Table III shows that there is a regular decrease in the CTE as the final cure temperature is increased from 220 °C to 340 °C at a 5:1 molar ratio. At 220 °C the elevation of the CTE above the control is due to residual solvent, which also leads to a depression in T_g. While DMAc boils at 165 °C (760 torr), it is well known that it is difficult to remove the last traces of solvent. At a final cure of 250 °C the CTE is similar to that of the controls. Increasing the temperature to 340 °C gives a significant lowering of the CTE relative to the 300 °C heated sample and relative to the control taken to 340 °C. An explanation for this reduction in the CTE is that at 340 °C the tris(dibenzoylmethanato)europium(III) complex is more completely decomposed and a europium(III) degradation product promotes light crosslinking of the polyimide. This raises T_g and increases dimensional stability. The tensile strength and percent elongation are, however, comprised, and there is the expected increase in the modulus. Table IV shows that similar but less pronounced results are obtained with the late lanthanide system erbium(III) acetate-dibenzoylmethane. Thus, higher thermal treatment temperatures appear to lead to films with lower CTE's although we would prefer to improve dimensional stability under milder conditions.

To this point we have focused only on dibenzoylmethanate as the ligand and counter ion for the lanthanide(III) ions as it readily solubilises lanthanide(III) acetates and gives clear flexible films. The influence of other anionic ligands/counter ions is demonstrated for a series of soluble thulium(III)-doped 6FDA/APB films. Data for this series is presented in Table V. All resins were prepared with a polymer repeat unit to metal molar ratio of 5:1. Thulium(III) nitrate and chloride represent simple soluble and completely dissociated lanthanide salts where the anions are neutral (i.e., derived from strong acids) and solvated thulium(III) ions dominate the solution chemistry of the metal. The thulium(III) acetate-trifluoroacetic acid system forms thulium(III) trifluoroacetate and acetic acid in solution. The trifluoroacetate anion is only weakly basic and does not coordinate strongly to the lanthanide ion. The thulium(III) nitrate film was clear, dark, and homogeneous but brittle. The films with thulium(III) chloride and thulium(III) acetate-trifluoroacetic acid were flexible and clear with high optical transparency. However, these films were bubbled. Thulium(III) acetate with hexafluoroacetylacetone and trifluoro-acetylacetone forms tris(β-diketonato) complexes in solution as dibenzoylmethane does. The film from thulium(III) acetate with HFA was severely bubbled and brittle. The analogous film system with the less acidic TFA was flexible but was occasionally bubbled and very dark. Only the 6FDA/APB films with thulium(III) acetate alone, thulium(III) acetate-DBM, and tris(acetylacetonato)thulium(III) trihydrate were clear, uniformly and lightly colored, and flexible. Of those films for which CTE's could be measured, thulium(III) chloride and tris(acetylacetonato)thulium(III) trihydrate gave significant lowerings of the CTE. These observations suggest that simple solvated lanthanide ions formed from salts with very weakly basic anions derived from the acids TFAA, HFA,

Table 3. Thermal and Mechanical Data for 6FDA/APB Films with Europium(III) Acetate-Dibenzolymethane Dopants with Non-Standard Heating Cycles in Air

Thermal Cycle (°C, 1 h at each temperature)	Repeat/ metal ratio	Theoretical % Metal	Tg by DSC (°C)	TGA Wt. Loss at 10% (°C)	CTE (ppm/°C)	Tensile Strength (Ksi)	Percent Elongation at Break	Modulus (Ksi)
100, 200, 340	Control	Zero	208	506	46.9	17.4	5.01	451
100, 200, 340	10:1	2.08	211	464	43.4	15.5	3.80	470
100, 200, 340	5:1	4.17	212	445	37.7	11.9	2.69	476
100, 200, 300	Control	Zero	207	496	48.5	17.8	5.15	446
100, 200, 300	5:1	4.17	207	447	45.1	15.1	4.08	441
100, 200, 250	5:1	4.17	202	429	47.9	15.5	3.96	461
100, 200, 220	5:1	4.17	186	397	53.2	13.8	3.65	466

Table 4. Thermal and Mechanical Data for 6FDA/APB Films with Erbium(III) Acetate-Dibenzolymethane Dopants for Comparison with Europium Data for the 340°C Thermal Treatment in Air

Cure Cycle at 100°, 200°, and 340°C	Repeat/ metal ratio	Theoretical % Metal	Tg by DSC (°C)	TGA Wt. Loss at 10% (°C)	CTE (ppm/°C)	Tensile Strength (Ksi)	Percent Elongation at Break	Modulus (Ksi)
Er(C₂H₃O₂)₃·DBM	5:1	4.54	211.3	473.	42.3	13.6	3.03	486
Er(C₂H₃O₂)₃·DBM	10:1	2.32	210.7	486	45.0	15.8	4.29	445
Control	---	Zero	208	506	46.9	17.4	5.01	451

HNO3, and HCl give less than optimum films as noted by some combination of bubbling, brittleness, and uneven or dark coloration. We suggest that in these cases the Lewis acidic

Table 5. 6FDA/APB Films with Thulium(III) Additives (Standard Heating Cycle)

Additive[a]	Tg DSC (°C)	TGA Wt. Loss @ 10%	CTE (ppm/°C)	Flexible	Appearance
Control	207	496	48.5	Yes	Colorless
$Tm(C_2H_3O_2)_3$	208	486	45.2	Yes	Light amber
$Tm(C_2H_3O_2)_3$-TFA	209	499	41.1	Yes	Brown
$Tm(C_2H_3O_2)_3$-TFAA	209	502	---b	Yes	Colorless, Bubbled
$Tm(C_2H_3O_2)_3$-HFA	207	490	---b	No	Dark, Bubbled
$Tm(C_2H_3O_2)_3$-DBM	208	446	47.2	Yes	Light amber
$TmCl_3$	209	486	42.0	Yes	Pale, Bubbled
$Tm(NO_3)_3$	210	467	37.4	No	Brown
$Tm(ACAC)_3$-$3H_2O$	210	468	35.6	Yes	Amber

a) Legend for additives.
TFA = TRIFLUOROACETYLACETONATE   TFAA = TRIFLUOROACETATE
DBM = DIBENZOYLMETHANATE $C_2H_3O_2$ = ACETATE
HFA = HEXAFLUOROACETYLACETONATE   ACAC = ACETYLACETONATE
b) Films too irregular to obtain CTE samples.

thulium(III) ion is highly reactive at high cure temperatures toward the polymer chain in a deleterious manner due to the lack of strongly coordinating ligands to moderate the metal's reactivity. In the case of the nitrate salt the oxidizing anion probably degrades the polymer. The trifluoroacetylacetonate ligand is at the borderline of basicity to give desirable films. Thus, only thulium(III) acetate and tris(acetylacetonato)thulium(III) trihydrate, the acetate and acetylcetonate anions being more basic and strongly coordinating than the anions of TFAA, HFA, TFA, HNO3, and HCl, and the thulium(III) acetate-DBM system yield films that have the essential qualities of clarity, smoothness, and homogeneity. Of these latter three systems, the tris(acetylacetonato)-complex gives a large lowering of the CTE; the modulus, tensile strength, and percent elongation being 549 Ksi, 19.9 Ksi, and 4.5 %, respectively. In these latter systems there are basic strongly coordinating ligands to moderate the reactivity of the Lewis acidic metal(III) ions.

CONCLUSION

There are only a few simple and commercially available lanthanide(III) salts which are soluble in common polyimide solvents such as DMAc. As illustrated in Figure 1 we have demonstrated that insoluble lanthanide(III) acetates, which are available in high purity at reasonable cost, are rendered soluble in polar aprotic solvents by the *in situ* addition of ligands which have mildly acidic protons and large metal-ligand formation constants. These resultant metal-ligand systems allow the further dissolution of polyimides and poly(amic acids) to occur without gelation. Thermal curing then leads to submicron composite materials. Appropriate ligands include trifluoroacetic acid and the more acidic β-diketones such as DBM, HFA, and TFA. This solubilization procedure is much more convenient than having to prepare and purify discrete isolable metal complexes. Our approach is applicable to other lanthanide(III) salts like the carbonates and additional carboxylates where the anions are of sufficient bascity to abstract acidic protons from the bidentate ligands.

The results of this study indicate that *in situ* formed tris(dibenzoylmethanate)-lanthanide(III) complexes dissolved in imidized 6FDA/APB resins on heating give films with submicron particle size which maintain essential thermal and mechanical properties and visual clarity (although lightly colored). With DBM, however, the CTE's are mostly unchanged with a final treatment temperature of 300 °C. HFA, TFA, and TFAA gave discolored and/or bubbled films. Unprotected DMAc solvated lanthanide(III) ions (from nitrate and chloride salts) gave films with compromised properties even though they lowered the CTE. The tris(acetylacetonato)thulium(III) complex gave the lowest CTE while maintaining excellent film properties. Thus, the metal complex system must be carefully tailored to avoid deleterious effects to the polymer.

The mechanism of CTE lowering is unclear at this time but may involve selective crosslinking promoted by the metal dopant and/or may involve singular properties associated with *in situ* submicron composite formation. Further studies are underway to clarify the ligand effects and the mechanism of CTE reduction.

REFERENCES

1. A. K. St. Clair, T. L. St. Clair, and K. I. Shevket, *Proceed. Am. Chem. Soc. Div. Polym. Mater. Sci. Eng.*, **51**, 62 (1984).
2. A. K. St. Clair, T. L. St. Clair, and W. S. Slemp, *Recent Advances in Polyimides: Synthesis, Characterization and Applications*, W. Weber and M. Gupta, eds., Plenum Press, New York, pp. 16-37, 1987.
3. A. K. St. Clair and W. S. Slemp, *SAMPE Journal*, **21**, 28 (1985).
4. A. K. St. Clair and W. S. Slemp, *Proceed. of the 23rd Intl. SAMPE Tech. Conf.*, **23**, 817 (1991).
5. A. K. St. Clair, T. L. St. Clair and W. P. Winfree, *Proceed. Am. Chem. Soc. Div. Polym. Mater. Sci. Eng.*, **59**, 28 (1988).
6. D. M. Stoakley, A. K. St. Clair and R. M. Baucom, *SAMPE Quarterly*, **21**, 3 (1989).
7. D. M. Stoakley and A. K. St. Clair, *Polymeric Materials for Electronics Packaging and Interconnection*, J. H. Lupinski and R. S. Moore, eds., ACS Symposium Series 407, Washington, D.C., pp. 86-92 (1989).
8. M. K. Gerber, J. R. Pratt, A. K. St. Clair, and T. L. St. Clair, *Proc. Amer. Chem., Polymer Division*, **31**, 340 (1990).
9. J. H. Lupinski and R. S. Moore, *Macromolecules*, **24**, pp. 1-24 (1989).
10. D. M. Stoakley and A. K. St. Clair, *SAMPE Quarterly*, **23**, 9 (1992).
11. D. M. Stoakley and A. K. St. Clair, U. S. Patent 5,248,519 to the National Aeronautics and Space Administration, September 1993.
12. R. E. Southward, D. W. Thompson, and A. K. St. Clair, *Proc. Am. Chem. Soc., Division of Polymeric Materials: Science and Engineering*, **71**, 725 (1994).
13. S. Numata, K. Fujisaki, D. Makino and N. Kinjo, *Proc. 2nd Int. Conf. on Polyimides*, pp. 492-510 SPE, New York, Nov. 1985.
14. S. Numata, K. Fujisaki, D. Makino and N. Kinjo, *Recent Advances in Polyimide Science and Technology*, W. D. Weber and M. R. Gupta, Eds., Society of Plastics Engineers, New York, p. 164 (1987).
15. Mikami, European Patent Application 84109054, 1, 31, 07, 1984.
16. S. Numata, and N. Kinjo, *Kobunshi Ronbunshu*, **42**(7), 443 (1985).
17. Kinjo, *J. Appl. Polym. Sci.*, **31**, 101 (1986).
18. S. Numata, K. Fujisaki and N. Kinjo, *Polymer*, **28**, 2282 (1987).
19. S. Numata and T. Miwa, *Polymer*, **30**, 1170 (1989).
20. S. Numata, N. Kinjo, *Polym. Eng. & Sci.*, **28**, 906-911, 1988.
21. S. Numata, S. Oohara, J. Imaizumi and N. Kinjo, *Polym. J.*, **17**, 981. (1985).
22. S. Numata, K. Fujisaki and N. Kinjo, *Polyimides, Synthesis, Characterization, and Applications*, (Ed. K. L. Mittal), Plenum Press, New York, 1984, Vol. 1, pp. 259-71.
23. S. Trofimenko, *Proceedings of the Fourth International Conference on Polyimides, Recent Advances in Polyimide Science and Technology*; F. Feger, et al., Eds., Society of Plastics Engineers: New York, 1993, p. 3.
24. B. C. Auman, *ibid.*, p. 15.
25. D. M. Stoakley and A. K. St. Clair, *Recent Advances in Polyimide Science and Technology*, W. D. Weber and M. R. Gupta, Eds., Society of Plastics Engineers, New York, p. 471 (1987).
26. R. D. Shannon, *Acta Crystallogr.*, **A32**, 751, 1976.
27. L. C. Thompson in "*Handbook on the Physics and Chemistry of Rare Earths*," K. A. Gschneidner, Jr., a nd L. Eyring, Eds., North-Holland, Amsterdam, 1979, Chapter 25.8
28. S. P. Sinha, *Struct. Bonding (Berlin)*, **25**, 69 (1976).

29. A. Sen, N. Manish, and J. A. Conklin, in *Metallized Plastics*, Mittal, K. L. Ed.; Plenum Press: New York, 1991; pp. 35-55.
30. J. J. Bergmeister, and L. T. Taylor, *Chem. Mater.*, *4*, 729 (1992).
31. M. Nandi, J. A. Conklin, L. Salvati, Jr., and A. Sen, *Chem. Mater.*, **2**, 772 (1990).
32. M. Nandi, J. A. Conklin, L. Saluati, Jr., and A. Sen, *Chem Mater.* **3**, 201 (1991).
33. A. Sen, M. Nandi, and J. A. Conklin, *Metallized Plastics 2*, K. L. Mittal, Ed., Plenum, New York, pp. 35-55, 1991.
34. J. V. Honene, R. G. Charles, and W. M. Hickman, *J. Phys. Chem.*, **62,** 1098 (1958).
35. K. J. Eisentraut and R. E. Sievers, *J. Inorg. Nucl. Chem.*, **29**, 1931 (1967).
36. D. M. Stoakley, D. D. Shillady, L. M. Vallarino, W. A. Gootee, and D. L. Smailes, *Recent Advances in Polyimide Science and Technology,*,W. D. Weber and M. R. Gupta, Eds., Society of Plastics Engineers, New York, p. 417 (1987).

PREPARATION OF SILVERED POLYIMIDE MIRRORS VIA SELF-METALLIZING POLY(AMIC ACID) RESINS

Robin E. Southward,[2] D. Scott Thompson,[2] David W. Thompson,[2] Maggie L. Caplan[1] and Anne K. St. Clair[1]

[1]Langley Research Center, National Aeronautics and Space Administration, Hampton, VA 23665-5225

[2]Department of Chemistry and Applied Science Program, College of William and Mary, Williamsburg, VA 23187-8795

INTRODUCTION

The use of metallized reflectors with polymeric rather than glass substrates for solar energy concentration can reduce greatly the weight and fragility of such mirrors and provide greater flexibility of packaging for subsequent deployment in space applications.[1] Silver is the metal of choice for the reflecting material. The specular reflectance of silvered mirrors is excellent, and the hemispherical reflectance of a clean silver film is greater than 97% weighted over 250-2500 nm, the range of the solar spectrum.[2] Most commonly, metallized polymeric films are fabricated in two steps. First, the polymeric film is prepared, and secondly, the metal is deposited onto the film surface by an external process such as thermal or chemical vapor deposition, sputtering, or chemical reduction from solution. Often, adhesion of the more passive metals such as copper, silver, and gold to the polymer is a problem.[3,4] In this work we report the preparation of silvered polymeric films via an especially convenient and effective single-step *in situ* self-metallization procedure involving reduction of polymer-soluble silver(I) to the metallic state. The formation of silvered mirrors, as will be described, has the potential to give improved adhesion as well as highly reflecting surfaces.

Due to the outstanding thermal and chemical stability of polyimides, we chose to investigate the formation of silvered polyimide films derived from 3,3',4,4'-benzo - phenonetetracarboxylic dianhydride and 4,4'-oxydianiline (BDTA/ODA). Poly(amic acid) resins of BTDA/ODA in dimethylacetamide (DMAc) were doped with silver(I) fluoride which was rendered soluble by the addition of hexafluoroacetylacetone. Thermally curing the silver(I)-doped resins led to metallized films which were significantly reflective. We report the characterization of these silvered films with respect to reflectivity, thermal properties, conductivity, and composition.

EXPERIMENTAL

BTDA was purchased from Allco Chemical Corporation and was dried under vacuum at 150 °C for 5 h prior to use. 4,4'-ODA was obtained from Kennedy and Klim, Inc. and was used as received. The melting points of the monomers as determined by differential thermal analysis (DTA) were 218 °C and 188 °C, respectively. DMAc (HPLC grade < 0.03% water) and silver(I) fluoride (99.9%) were purchased from Aldrich Chemical Co. and were used without further purification. Hexafluoroacetylacetone was purchased from Eastman Organic Chemicals and was redistilled under dry nitrogen.

Metal-Containing Polymeric Materials
Edited by C.U. Pittman, Jr., *et al.*, Plenum Press, New York, 1996

The BTDA/ODA poly(amic acid) resin employed in this study was prepared with equimolar amounts of monomers at 12% solids (w/w) in DMAc. The resin preparation was performed by first dissolving the diamine in DMAc in a flask flushed with dry nitrogen and then adding the dianhydride. The resin was stirred for a minimum of 5h. The inherent viscosity was 1.2 dL/g at 35° C.

Silver(I) fluoride is not soluble in DMAc. However, if one or more equivalents of HFA is added to DMAc followed by the addition of silver(I) fluoride, dissolution occurs readily and completely. Thus, silver-containing resins were prepared by first dissolving silver(I) fluoride in a small volume (*ca.* 1 mL for every 5 g of 12% BTDA/ODA resin) of DMAc containing HFA. The silver-doped resins were stirred thoroughly for 30 min before casting films. Although there is always concern for photochemical decomposition with silver systems, these doped films were not light sensitive for at least a 24 h period. Thus, the films in this study were not protected from light.

A series of films with varying concentrations of silver were made as reported in Table 1. Undoped and doped poly(amic acid) resins were cast as films onto glass plates using a doctor blade set at 17 mil to obtain cured films approximately 1 mil thick. After remaining in an atmosphere of dry slowly flowing air for 15 h, the tack-free films were thermally cured in a forced air oven. A typical cure cycle employed was 1h each at 100, 200, and 300 °C. Variations of the cure cycle did not effect the quality of metallization as long as a maximum temperature of 300 °C was attained and held for at least 1 h. Films formed metallic layers on both the air side and glass side surfaces during cure. The metallic layer that formed on the glass side during cure caused the films to detach from the glass plates except at the edges. The films were removed from the glass plates by scoring the edges with a razor blade. Samples for reflectivity measurements were made by casting or affixing films onto 2 in square glass plates.

Monomer melting points were determined by DTA at a heating rate of 20° C/min on a DuPont Thermal Analyst 2000. Inherent viscosities of the poly(amic acid) resin were obtained at a concentration of 0.5% (w/w) in DMAc at 35 °C. Thermogravimetric analyses (TGA) were obtained on the cured films in both flowing air and nitrogen (50 mL/min) at a heating rate of 2.5 °C/min using a Seiko TG/DTA 200 or TG/DTA 220 instrument. Glass transition temperatures were determined with a Seiko DSC 210 system. Surface resistivities were measured with an Alessi four point probe. Transmission electron microscopy (TEM) was done at the Virginia Institute of Marine Science, Gloucester Point, VA on a Zeiss CEM-920 transmission electron microscope. Scanning electron microscopy (SEM) was performed on a Hitachi S-510 instrument. Reflectivity measurements were made (relative to a standard silvered mirror) with a Perkin Elmer Lambda 5 UV/VIS spectrophotometer equipped with a variable angle specular reflectance accessory using a wavelength of 531 nm. X-ray data was obtained with a Philips 3600 diffractometer.

RESULTS AND DISCUSSION

All thermally cured films were metallized on both the air and glass sides. The air side was bright while the glass side was dull in appearance. Silver concentrations from 2.5% to 18.3% (calculated on the premise that after thermal curing only metallic silver and polyimide remain) gave films which were reflective on the air side. None of the films was electrically conductive on either the air or glass sides by the four point probe measurement. Many different thermal cure cycles were employed to optimize the metallization and brightness of the air side of the silvered films. Surprisingly, no effect of the thermal cure cycle on the reflectivity was observed as long as the cure cycle slowly (3 h or more from 25 °C) reached a maximum temperature of 300 °C for a least 1 h. The bulk of silver(I) is reduced in the temperature range of 170-210 °C; yet heating to 300 °C is essential to maximize the reflectivity. Post-curing at 300° C did not enhance reflectivity. The flexibility of the films was excellent with all being fingernail creaseable. All films exhibited well-adhered metals layers without loss of silver via commonly used adhesive peel tests.[5,6]

Reflectivity measurements as a function of silver concentration and angle are listed in Table 1. The values reported are an average of 3-4 readings at different points on the films. The reflectivity values seldom vary from point to point by more than a few percent. This is consistent with the scanning electron micrographs (Figure 1) of the air side film surfaces which show a uniform distribution of silver particles. While there is ample silver

Table 1. Reflectivities of silver(I) fluoride / hexafluoroacetylacetone doped films.

Percent Silver (Calculated)[a]	Repeat Unit to Silver Ratio	Equivalents of HFA	Percent Transmittance of Silvered Films (as a function of angle)		
			20°	45°	70°
2.5	8.7:1	1.1	33	25	15
5.0	4.2:1	1.1	59	48	34
7.3	2.8:1	1.0	65	51	38
8.4	2.4:1	2.0	63	50	25
9.9	2.0:1	1.0	74	61	40
12.6	1.5:1	1.0	76	65	39
18.3	1.0:1	1.2	55	45	38

a) Calculated for the silver fluoride-HFA dopant system decomposing completely to silver metal.

Table 2. Thermal data for silver(I) fluoride / hexafluoroacetylacetone doped films.

Percent Silver (Calculated)[a]	T_g by DSC	10 % Wt. Loss in Air	10 % Wt. Loss in N_2
control	271	524	540
2.5	266	393	538
5.0	268	390	548
7.3	269	385	550
8.4	268	370	523
9.9	269	378	540
12.6	268	367	515
18.3	269	371	497

a) Calculated for the silver fluoride-HFA dopant system decomposing completely to silver metal.

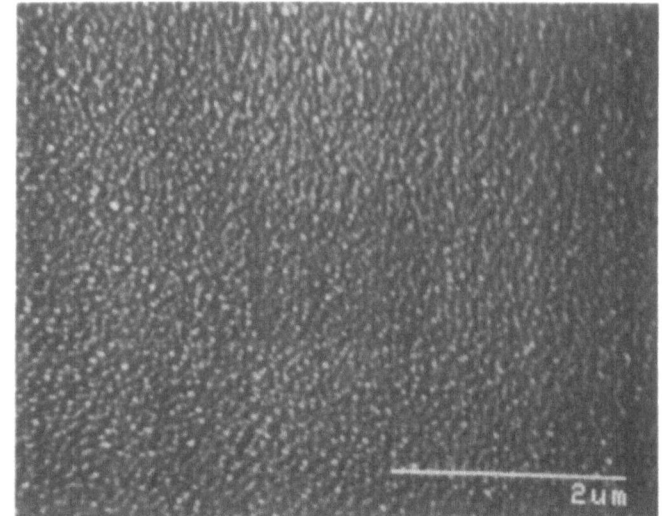

AIR SIDE

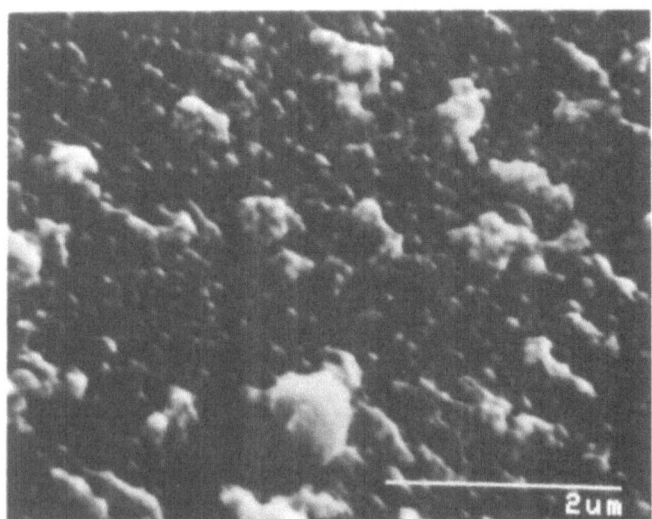

GLASS SIDE

Figure 1. Scanning electron micrograph of the 8.0% AgF/HFA BTDA/4,4'-ODA film. (See Table 1.)

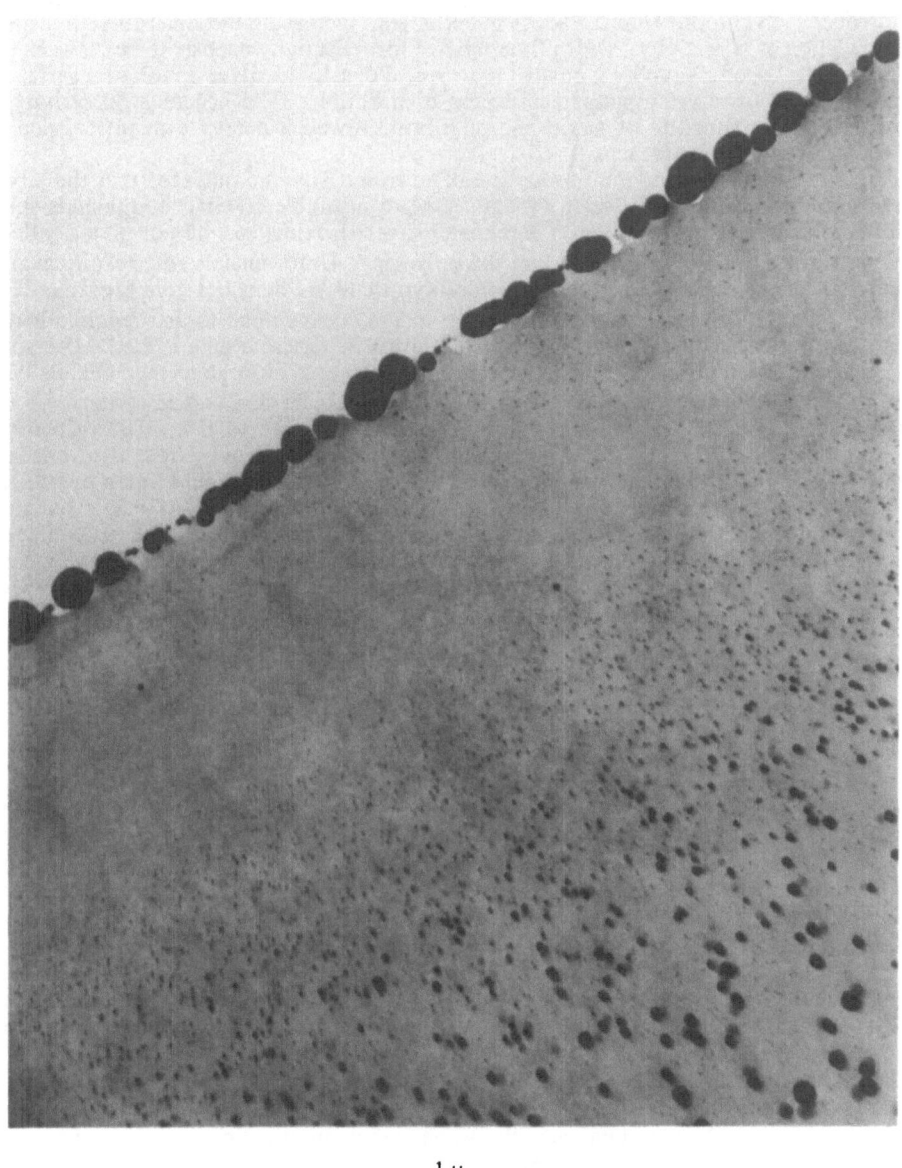

_____ 1 μ _____

Figure 2. Transmission electron micrograph of the air side of the 9.9% AgF/HFA BTDA/4,4'-ODA film. (See Table 1.)

in all of the films listed in Table 1 to form a surface mirror of *ca.* 100 nm, which is sufficient for high reflectivity,[2] we observed reflectivity to increase with increasing silver concentration, reaching a maximum at *ca.* 13% silver. Varying the HFA concentration between one and two equivalents did not affect reflectivity. The transmission electron micrograph shown in Figure 2 reveals that the surface layer of silver is *ca.* 80 nm thick with much of the metallic silver remaining in the bulk of the polymer. Clearly, migration of reduced silver to the film surface is not efficient. Increasing the silver to 18% leads to a degradation of reflectivity. The optimum range for reflectivity appears to be between 9 and 13% silver. The reflectivity is related to a critical density of silver on the film surface and to regularity of silver aggregate size and distribution. The specular reflectivity was measured as a function of angle and all films showed a decrease in reflectance with increasing angle of incidence.

It is essential for the production of mirrored silvered surfaces that the silver(I) additives be soluble in the resin system. DMAc insoluble silver(I) compounds such as silver(I) acetylacetonate, silver(I) carbonate, silver(I) oxide, etc. do not give a reflective silver surface upon thermal curing of the polymer.[7] Unfortunately, the readily available silver(I) nitrate, which has excellent solubility in DMAc, does not give highly reflective films[8-11] (*ca.* 20% reflectivity at 20° under a cure cycle similar to that reported herein). The most effective and promising silver(I) additive reported to date is the DMAc soluble organometallic complex (1,5-cyclooctadiene)(hexafluoroacetylacetonato)silver(I), [Ag(COD)(HFA)], mp 122-124 °C, studied by St. Clair, Taylor, and coworkers.[12] They found that [Ag(COD)(HFA)] dissolves in BTDA/ODA resins to give a 10 % silver cured film reflectivity of 65% at 20° which is only slightly lower than the results of this study measured under identical conditions. We thought that we might achieve an *in situ* HFA-silver(I) complex, which would behave similarly to [Ag(COD)(HFA)], by simply dissolving silver(I) fluoride in DMAc containing HFA. The driving force for β-diketonate complex formation would be the combination of the mildly acidic fluoride ion which could abstract the acidic proton from HFA and the large formation constant for the HFA-silver complex. Indeed, one equivalent of HFA renders the silver(I) fluoride soluble as the β-diketonate complex.

This approach is exceptionally convenient and allows the use of several other readily available insoluble silver(I) salts such as silver(I) acetate.[7] There does not seem to be any role for the 1,5-cyclooctadiene ligand in producing a reflective silver surface.

Thermal data for the metallized films is presented in Table 2. The glass transition temperature is only slightly lowered by the incorporation of silver into the polymer matrix. The effect of silver dispersed throughout the polymer on the polymer decomposition temperature in flowing air is pronounced. The temperature at which 10% weight loss occurs is lowered by more than 130 °C. Even low concentrations of silver promote oxidative degradation. This is not unexpected since silver metal is known to activate diatomic oxygen[13] and to interact with carbon-carbon unsaturation in organic structures[14]. Degradation in a nitrogen atmosphere is minimal, indicating that silver is promoting oxidative degradation of the BTDA/ODA. Similar thermal data trends were observed with the [Ag(COD)(HFA)]-BTDA/ODA system.[12]

Elemental analysis data for selected fully cured (in air) metallized films are presented in Table 3. The data reveal that the silver(I) fluoride and HFA additives have been substantially altered. The percent silver found is much higher than that calculated assuming that the additives remain intact. The silver found is closer to the value calculated for polymer and silver alone. For the 7.3 and 9.9 % films the silver found is higher than that calculated and suggests that during the cure cycle there has been some oxidative degradation of the BTDA/ODA. There is an increasing concentration of fluorine in the cured polymer which could be thought of as arising from residual silver(I) fluoride. However, this does not seem to be the case since a similar experiment with silver(I) acetate and HFA gives the same trend in fluorine concentration[7] and suggests that it is the HFA which is the source of fluorine. Fluorine may be incorporated into the polymer backbone via an acylation reaction of the trifluoroacetyl group formed in the thermal degradation of the silver-HFA complex.

Table 3. Elemental Analyses for AgF / Hexafluoroacetylacetone Doped Films

Metallized Film (See Table 1.)	Calculated (Wt %)						Found (Wt %)				
Elements	Ag^a	Ag^b	F	C	H	N	Ag	F	C	H	N
Control	0	0	0	71.6	2.90	5.76	0	0.065	70.4	3.05	5.58
5.0 % Ag°	5.0	4.6	0	68.7	2.79	5.53	4.43	1.42	66.9	2.65	5.37
7.3 % Ag°	7.3	6.4	0	67.3	2.73	5.41	7.93	1.73	64.4	2.72	5.19
9.9 % Ag°	9.9	8.3	0	65.9	2.67	5.30	10.6	2.17	62.4	2.57	4.99

a) Calculated for the silver fluoride-HFA dopant system decomposing completely to silver metal.
b) Calculated for the hexafluoroacetylacetonatosilver(I) complex remaining intact.

The cross-sectional view of a metallized film as presented in Figure 2 shows that much lower concentrations of the silver additive could be used if the mechanism for migration of silver to the film surface could be improved. The thickness of the air side surface layer of silver does not increase linearly with increasing silver dopant concentration in the resin. Thus, the increase in intensity with increasing additive concentration of the four most intense X-ray reflections (d = 2.36, 2.04, 1.45 and 1.23 Å) for metallic silver in the films, coupled with the elemental analysis data, make it clear that a large amount of metallic silver resides in the bulk of the film in aggregates which are less than 100 nm in diameter (Figure 2). It would appear that these silver clusters are not free to move significantly in the polymer matrix.[15] Just below the surface layer of silver there is a zone of polymer several hundred nanometers in thickness which appears to be mostly free of silver. This depletion zone was observed in the [Ag(COD)(HFA)]-BTDA/ODA system and is of unknown origin. Silver migrates to the glass side of the film as well. However, there is a much greater buildup of silver on the air side surface due, in part, to depletion of polymer (see Figure 1). The distribution of silver on the glass side is much less uniform, accounting for its dull appearance.

CONCLUSION

We have demonstrated that high levels of specular reflectance can be achieved in cured polyimide films by incorporating a silver-hexafluoroacetylacetonate complex into a poly(amic acid) resin followed by thermal curing. Furthermore, we have shown that is unnecessary to isolate HFA complexes of silver such as [Ag(COD)(HFA)] or [Ag(Me$_3$P)(HFA)] to be used as soluble metallization dopants. The efficacious HFA complex can be formed *in situ* by using a simple silver salt with a basic anion such as fluoride, acetate, carbonate, and cyanide and one or more equivalents of hexafluoroacetylacetone. We are continuing work on similar systems to understand the mechanism of reduction and to improve migration of silver to the film surface to enhance reflectivity and control conductivity.

REFERENCES

1. D. A. Gulino, R. A. Egger, and W. F. Bauholzer, "Oxidation-Resistant Reflective Surfaces for Solar Dynamic Power Generation in Near Earth Orbit," *NASA Technical Memorandum 88865*, 1986.

2. G. Jorgensen and P. Schissel in *Metallized Plastics*, K. L. Mittal and J. R. Susko, Eds., Plenum: New York, 1989, Vol. 2, pp 79-92.

3. P. F. Green and L. L. Berger, *Thin Solid Films*, **224,** 209 (1993).

4. L. J. Gerenser, *J. Vac. Sci. Technol., A,* **8,** 2897 (1988).

5. W. C. Lee, V. W. Lindberg, P. H. Wojciechowski, and F. J. Duarte in *Metallized Plastics,* K. L. Mittal and J. R. Susko, Eds., Plenum: New York, 1989, Vol. 2, pp 449-460.

6. K. L. Mittal, *Electrocomponent Sci. Technol.,* **3,** 21, (1976).

7. M. L. Caplan, R. E. Southward, D. W. Thompson, and A. K. St. Clair, *Proc. Am. Chem. Soc., Division of Polymeric Materials: Science and Engineering,* **71,** 787 (1994).

8. R. K. Boggess and L. T. Taylor in *Recent Advances in Polyimide Science and Technology*; W. D. Weber and M. R. Gupta, Eds.; Mid-Hudson Chapter SPE: New York, 1987; pp 463-70.

9. M. Linehan, D. M. Stoakley, and A. K. St. Clair, *Abstracts of Papers,* 44th Southeastern-26th Middle Atlantic Combined Regional Meeting of the American Chemical Society, Arlington, VA; American Chemical Society: Washington, DC, 1992; POLY 378.

10. A. Auerbach, *J. Electrochem. Soc.,* 937 (1984);

11. A. L. Endrey, U. S. Patent 3,073,784 (1963).

12. a) A. F. Rubira, J. D. Rancourt, M. L. Caplan, A. K. St. Clair, and L. T. Taylor, *Chem. Mater.,* in press.; b) A. F. Rubira, J. D. Rancourt, M. L. Caplan, A. K. St. Clair, and L. T. Taylor, *Proc. Am. Chem. Soc., Division of Polymeric Materials: Science and Engineering,* **71,** 509-511 (1994).

13. R. J. Madix in *Oxygen Complexes and Oxygen Activation by Transition Metals*, A. E. Martell and D. T. Sawyer, Eds., Plenum: New York, 1988; pp 253-264.

14. F. M. Hartley, *Chem. Rev.,* **163** (1973)

15. A. Foitzik and F. Faupel, *Mater. Res. Soc. Symp. Proc.,* **203,** 51 (1991).

POLYIMIDES DOPED WITH SILVER-II: SURFACE CONDUCTIVE FILMS

Adley F. Rubira,[1] James D. Rancourt and Larry T. Taylor*

Virginia Polytechnic Institute and State University
Department of Chemistry
Blacksburg, VA 24061-0212

[1]Universidade Estadual de Maringá
Departamento de Quimica, Av. Colombo 3690
87020-900 Maringá-PR
BRAZIL

INTRODUCTION

Polyimides are used for a wide range of applications in areas such as integrated electronic circuits and aerospace devices that require excellent dielectric properties, high temperature stability and chemical inertness (1). On the other hand, some applications require low electrical resistivity and high reflectivity which are characteristics that are more typical of metals. In the attempt to synthesize materials with unique combinations of properties, metal-containing polymeric composite material (2,3) have been suggested as candidates. Insulating polymers possessing desirable technological properties may be rendered conductive by mixing with conductive particles such as carbon black, metal powders, flakes or fibers and metal coated particles, but in many cases high loading levels have been necessary which spoil the polymer's properties. The approach of Taylor and co-workers (4-10) has been to dissolve additives (metal salts and organometallic complexes) into a poly(amide acid) solution. The resulting films of pre-polymer upon thermolysis undergo both imidization and metallization. Appropriate processing and the correct choice of monomers yield reflective and/or conductive films in which the polymer's properties are basically maintained (11,12). Enhanced surface reflectivity has been obtained with copper (13), gold (14), and silver (15-18, 11) compounds; while palladium, platinum (19), and tin (20) salts have improved surface electrical conductivity.

Silver-coated polyimide films because of their high solar specular reflectance, are candidates for mirrored surface applications in NASA's proposed solar dynamic power system (21). Electrically conductive polyimides can also be used in aerospace applications to dissipate charge. Currently more detailed study of silver-doped polyimides is underway. With certain monomer pairs electrically conducting rather than reflecting surfaces were produced. For example, when the dianhydride was 3,3', 4,4'-

benzophenone teracarboxylic acid dianhydride a reflective film was produced and when 4,4'-bis(3,4-dicarboxyphenoxy) diphenyl sulfide was used a conductive film resulted. Subtle changes in polyimide properties no doubt impact the nature of silver deposition. The influence of different polyimide precursors on surface electrical conductivity is reported here. In addition, the distribution of silver-containing particles in the bulk and on the surface of each film has been studied as well as the chemical composition of the silver-containing particles.

EXPERIMENTAL

Chemicals: The following monomers (Figure 1) were used in this study: (a) 3,3' 4,4'-benzophenone tetracarboxylic acid dianhydride (BTDA), Chriskev Inc. (Leawood, KS) vacuum dried overnight at 100°C; (b) 4,4'-bis(3,4-dicarboxyphenoxy)diphenyl sulfide dianhydride (BDSDA) NASA Langley Research Center (Hampton, VA), recrystallized twice from 2-butanone, and vacuum dried overnight at 120°, (c) 4,4'-oxydiphthalic anhydride (ODPA), Chriskev Inc. (Leawood, KS) vacuum dried overnight at 120°C, (d) 3,3',4,4'-diphenylsulfonetetracarboxilic acid dianhydride (DSO$_2$DA), (e) 4,4'-oxydianiline (ODA), Chriskev Inc. (Leawood, KS) sublimed at 185°C and less than 1 torr pressure then vacuum dried at 70°C, overnight; (f) 4,4'-diaminodiphenyl sulfide (ASD), Chriskev Inc. (Leawood, KS) vacuum dried at 80°C overnight; (g) 1,3-bis(aminophenoxy) benzene (APB), Mitsui-Toatsu Inc. (Tokyo, Japan) vacuum dried overnight at 70°C, (h) 3,3'-diaminodiphenylsulfone (DDSO$_2$), Chriskev Inc. (Leawood, KS) vacuum dried overnight at 70°C. N,N-Dimethylacetamide (DMAC) was obtained from Aldrich Chemical Co. (Milwaukee, WS) in a Sure-Sealed bottle under N$_2$. The silver additive (1,5 cyclooctadienehexafluoroacetylacetonato) silver(I) [Ag(COD)(HFA)] was obtained from Aldrich Chemical Co. (Milwaukee, WS) and was used as received. The dopant was stored in the dark in a refrigerator. The additive structure is also shown in Figure 1.

Composite Synthesis: Poly(amide acid) solutions were synthesized by first adding diamine (4.00 mmol) to a nitrogen-purged glass bottle with dry DMAC. Next, dianhydride (4.00 mmol) was added to the diamine with additional DMAC. The resulting solutions were stirred for 2 hours. For the silver modified polyimides, Ag(COD)(HFA) (0.5-2.00 mmol) was added as a solid with additional DMAC thereby resulting in a 10-20 wt % solids solution depending on the monomer combination. The silver containing poly(amide acid) solutions were then stirred for an additional 2 hours at room temperature. Films were prepared by spreading the silver-containing poly(amide acid) solution onto a clean, dust-free soda lime glass plate, using a doctor blade (16 or 20 mil gap), followed by heating 20 min at 80°C, then heating at 100, 200, and 300°C each for 1 hour under a dynamic (30 SCFH) atmosphere of dry breathing air. The surface of the film in contact with the soda-lime glass plate during imidization is referred to as the glass side, while that in contact with the atmosphere of the curing oven is referred to as the air side.

Measurements: Thermal analysis for the purpose of determining glass transition temperature was performed on a Perkin Elmer Model DSC-4 differential scanning calorimeter at 10°C/min heating rate under a nitrogen atmosphere. To determine the polymer decomposition temperature (PDT), dynamic (air or nitrogen) thermogravimetric analysis was performed with a Perkin Elmer Model TGS-2 thermogravimetric system at

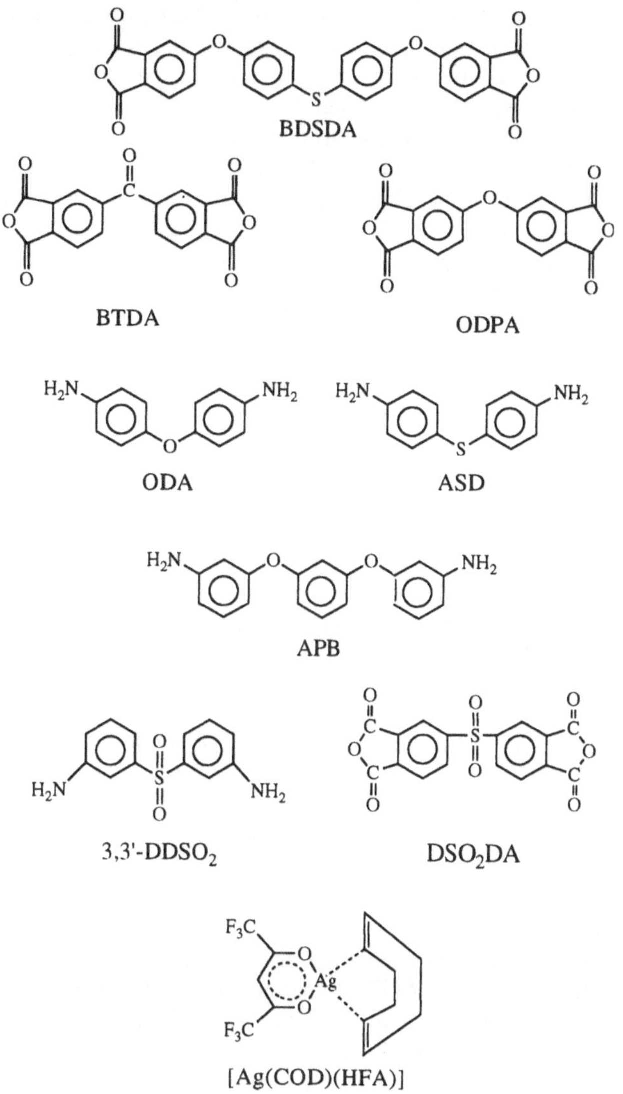

Figure 1.   Monomers and additive structures

10°C/min heating rate. Elemental analyses for silver and fluorine were obtained by Galbraith Analytical Laboratories, Knoxville, TN.

   X-ray photoelectron spectra (XPS) were obtained via a Perkin-Elmer Phi Model 5300 ESCA system. A magnesium anode (Ka = 1253.6 eV) operating at 400 W was used. The samples were attached to mounts by double-stick tape. The cited binding energies have been referenced to the aromatic carbon photopeak (C, 1s) of the polyimide backbone at 284.6 eV. Auger electron spectroscopy (AES) line scans were recorded with a Perkin

Elmer Phi Model 610 scanning microprobe system. Auger electron depth profiles were obtained via argon ion etching (~50 Å/min). Scanning electron micrographs (SEM) were taken with a ISI Model SX-40 scanning electron microscope. The samples were attached to aluminum mounts using double-stick tape and coated with a thin layer of gold in order to dissipate charge. Transmission electron microscope. Samples were imbedded in Polyscience ultra low viscosity medium and cured for 8 hours at 70°C. Using a Reichert-Jung ultra-microtome with a microstar diamond knife, cross sections of the samples were obtained between 500 and 800 Å. These sections were placed on 200 mesh copper grids prior to analysis. Surface electrical resistances at room temperature were measured using Kiethley equipment consisting of a voltage supply (model 240A), electrometer (model 610C), and a four point probe assembly (model 6105). X-ray diffraction patterns of silver-surfaced polyimide films were measured on a XDS 2000, Scintag Inc. diffractometer using CuK_α radiation generated at 40KV and 30 mA.

RESULTS AND DISCUSSION

Surface electrical resistance, measured by four point probe electrode assembly, of silver-doped BTDA, DDSO2, and ODPA-derived films was equivalent to the polyimide with no dopant ($>10^{11}$ ohms). The results on BTDA-ODA with 2:1, 4:1 and 8:1 mole ratio polymer:dopant (Table I) demonstrate the type of data obtained. After 12 minutes of argon ion sputtering a slight decrease in surface electrical resistance was observed (7.33 x 10^9 ohms). On the other hand, when the dianhydride BDSDA was used, regardless of the diamine, films with 2:1 mole ratio polymer:dopant were cloudy with surface resistance measured in the 3-10^4 ohms range. The BDSDA-ODA combination (2:1) gave the lowest resistance at 3-20 ohms. BDSDA-ODA with 4:1 mole ratio polymer:dopant yielded a surface resistance around 10^8 ohms which also had a silvery surface resembling that of doped BTDA-derived films. It is obvious from Table I that the development of a reflective or conductive surface can be drastically affected by the dopant concentration and the resin system structure.

Supporting information on chemical state and surface elemental composition of silver-doped polyimides was obtained utilizing X-ray photoelectron spectroscopy (XPS) and Auger electron spectroscopy (AES) in conjunction with argon etching. As is shown in Table II the atomic concentration of atmosphere side silver was fairly constant for non-conductive BTDA-ODA films with 2:1 and 4:1 mole ratio polymer: dopant, but for the 8:1 film the value was about 30% less. Similar trends were seen for conductive, silver modified BDSDA-ODA. To determine if intermixing between the polymer and silver was constant in each type of film, surface sputtering with argon ions for 10 minutes (5 Å/min) was done and the results are shown in Table III. The relative increase in silver concentration (upon sputtering) within the analysed region was about 40% higher for the BDSDA-derived films (24.3% to 57.2% Ag) than the BTDA-derived films (20.2% to 35.7% Ag). It is worth nothing that the carbon concentration did not change in the BTDA-derived films (58.7% vs. 58.0%) on sputtering as it changed in BDSDA (51.4% vs. 36.7%). In the BTDA case, the increase in silver concentration upon sputtering was at the expense of nitrogen and oxygen. We have shown(11) in agreement with the

Table I. Surface Resistance (ohms) for Silver-Modified Polyimide Films.

Films	Mol Ratio Polymer: Dopant	Original Surface	Rubbed Surface
BTDA-ODA	2:1	$>10^{11}$	$>10^{11}$
BTDA-ODA	4:1	$>10^{11}$	$>10^{11}$
BTDA-ODA	8:1	$>10^{11}$	$>10^{11}$
BTDA-ASD	2:1	$>10^{11}$	$>10^{11}$
BTDA-ASD	4:1	$>10^{11}$	$>10^{11}$
BTDA-ASD	8:1	$>10^{11}$	$>10^{11}$
BTDA-ODA:ASD	2:1	$>10^{11}$	$>10^{11}$
BTDA-3ODA:1ASD	2:1	$>10^{11}$	$>10^{11}$
BTDA-3,3'DDSO$_2$	4:1	$>10^{11}$	$>10^{11}$
BTDA-APB	2:1	$>10^{11}$	$>10^{11}$
BDSDA-ODA	2:1	3-20	3
BDSDA-ODA	4:1	$7.2\text{-}9.6 \times 10^8$	$>10^{11}$
BDSDA-ODA	8:1	$>10^{11}$	$>10^{11}$
BDSDA-ASD	2:1	$1.3\text{-}4.2 \times 10^4$	20
BDSDA-ASD	4:1	$2.3\text{-}6.2 \times 10^8$	$3.3\text{-}4.8 \times 10^{10}$
BDSDA-APB	2:1	$0.7\text{-}1.7 \times 10^3$	31
DSO$_2$DA-ODA	2:1	$>10^{11}$	$>10^{11}$
DSO$_2$DA-ODA	4:1	$>10^{11}$	$>10^{11}$
ODPA-ODA	2:1	$>10^{11}$	$>10^{11}$
ODPA-ASD	2:1	$>10^{11}$	$>10^{11}$
ODPA-APB	2:1	$>10^{11}$	$>10^{11}$

Table II. Relative Atomic Concentration of Silver-Doped Polyimide Films from XPS Analysis of Air Side Surface (Take-Off Angle = 45°)

Polymer	Mole Ratio Polymer: Dopant	C	O	N	S	F	Ag
	2:1	58.7	13.8	6.4	--	0.9	20.2
BTDA-ODA	4:1	60.6	10.6	6.6	--	0.5	21.7
	8:1	66.0	11.8	6.7	--	1.0	14.5
	2:1	45.5	18.6	4.1	3.4	1.2	27.2
BDSDA-ODA	4:1	43.7	18.8	2.8	3.4	2.8	28.5
	8.1	63.3	16.6	3.4	3.5	1.1	12.1

Table III. Relative Atomic Concentration of Silver-Modified Polyimide Films With 2:1 Mole Ratio Monomer: Dopant Before and After 10 min. Argon Ion Surface Sputtering. (Take-off Angle = 45°)

Films	C	O	N	S	F	Ag ·	Ag($3_{5/2}$) B.E. (eV)	Auger Peak B.E. (eV)	Ag(M4NN) K.E. (eV)[a]	Auger Parameter α (eV)[b]
							Elements			
BTDA-ODA	58.7	13.8	6.4	---	0.9	20.2	368.2	896.4	357.2	725.4
BTDA-ODA 10' sputter	58.0	3.2	3.1	---	---	35.7	368.1	896.0	357.5	725.7
BDSDA-ASD	51.4	13.3	3.8	4.3	2.9	24.3	368.6	896.8	357.0	725.6
BDSDA-ASD 10' sputter	36.7	2.6	2.3	1.2	---	57.2	368.3	896.1	357.5	725.8

[a]K.E. (eV)=(1253.6-Auger peak B.E.) [b]α=Auger (M4NN K.E.) + Ag(3d 5/2 B.E.)

Schon(22) and Larson(23) that Auger shifts are frequently more significant than XPS shifts for estimating the chemical state of silver. In Table III we show that Auger parameter after sputtering for BTDA- and BDSDA-derived films. Doped BTDA, like BDSDA, exhibited an Auger parameter after sputtering that was closer in value to that silver metal (726.0 eV). Depth profile via Auger electron spectroscopy (AES) in conjunction with argon etching was utilized to gain additional knowledge regarding the distribution of particles in both types of films (Figure 2 and 3). By comparing the depth profile of the films one can observe that the increased silver concentration as one goes deeper into the surface is higher for BDSDA-ODA films. After 12 minutes of sputtering a slight decrease in surface electrical resistance was observed for silver-doped BTDA-ODA (7.33×10^9 ohms). Measurement of surface electrical resistance on silver-doped BDSDA-ODA films after 140 minutes of argon ion etching showed some increase in surface resistance (5.75×10^8 ohms). With such a long etch period, we believe that a point has now been reached where silver concentration has decreased and carbon concentration has increased. These findings match very well with the results described earlier using both XPS and TEM (11).

To ascertain the identity of silver in these silver-doped polyimide films, X-ray measurements were made on selected reflective and/or conductive films. A typical X-ray diffractogram representative of the samples analyzed can be seen in Figure 4. From the results of X-ray diffraction analysis shown in Table IV, the main product in the silver-polyimide films composite is metallic silver for all five films investigated. We cannot rule out the possibility of the presence of a small amount of silver oxide, but if it exists, it is either in an amorphous state or as small crystallites. The X-ray results are in good agreement with results obtained on a silver metal standard and they support the Auger parameter reported earlier by us on these films (11). Similar x-ray results have been reported in polyamide-imide and poly(vinyl alcohol)/AgNO3 chelate films when treated with reducing agents (24,25). Scanning electron micrographs of the silver-doped BDSDA-ODA and BTDA-ODA are shown in Figure 5. In the BDSDA-ODA film silver particles are close enough to touch each other. On the other hand, the particles in the BTDA-ODA

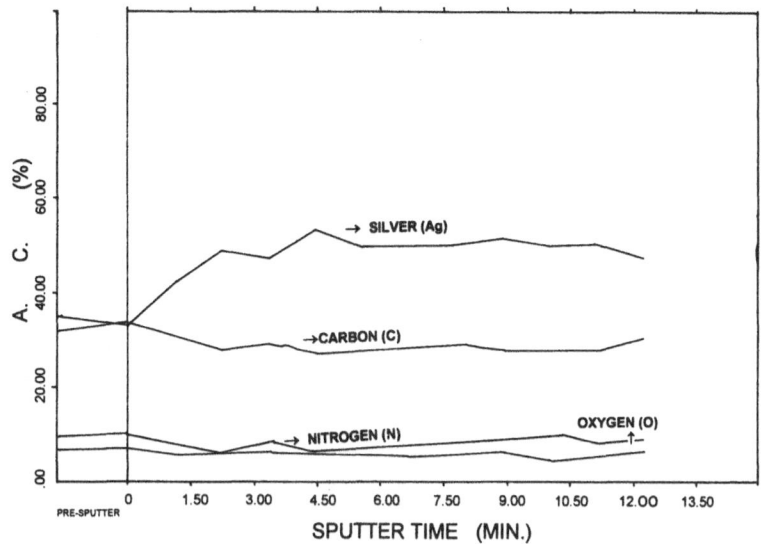

Figure 2.   Auger electron spectroscopic depth profile of air side silver-doped BTDA-ODA. Mole ratio monomer:dopant is 2:1.

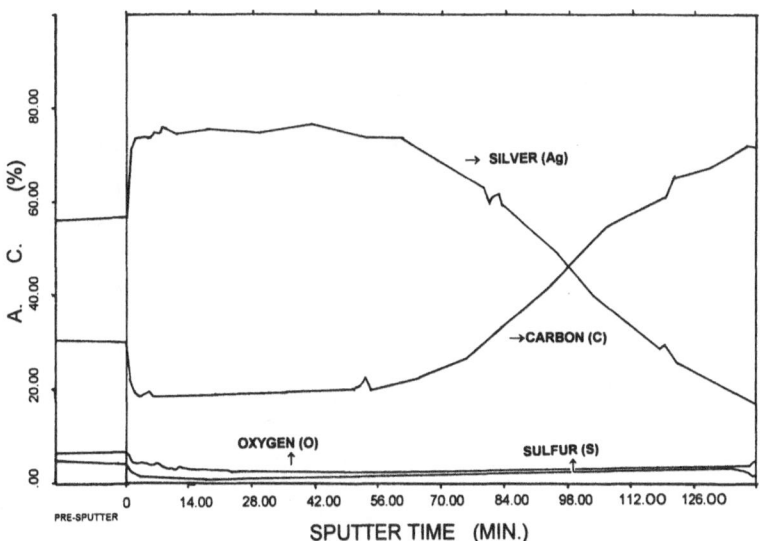

Figure 3.   Auger electron spectroscopic depth profile of air side silver-doped BDSDA-ODA. Mole ratio monomer:dopant is 2:1.

film (Figure 5b) are not in continuous contact. Similar findings can be seen in transmission electron micrographs (Figure 6). Silver-doped BDSDA-ODA has a surface layer of about 140 nm thickness and the silver particles are bonded together. The surface layer of Ag particles in BTDA-ODA, however, has about 70 nm thickness with more uniformly sized particles, but the particles are not continuous.

Table IV. Diffraction Angels (2θ) and Plane Distances (d) Corresponding to Peaks Observed in X-ray Analysis of Silver-Surfaced Polyimide Films With 2:1 Mole Ratio Polymer:Dopant. See Figure 4

Films	Peak 1		Peak 2		Peak 3		Peak 4	
	2θ	d	2θ	d	2θ	d	2θ	d
BTDA-ODA/Ag(COD(HFA)	38.17	2.356	44.33	2.042	64.50	1.444	77.43	1.232
BTDA-ASD/Ag(COD)(HFA)	38.11	2.359	44.27	2.044	64.43	1.445	77.43	1.231
BDSDA-ODA/Ag(COD)(HFA)	38.11	2.356	44.29	2.043	64.44	1.445	77.39	1.232
BDSDA-ASD/Ag(COD(HFA)	38.10	2.360	44.26	2.044	64.40	1.446	77.35	1.231
BDSDA-ASD/Ag(COD(HFA)[a]	38.06	2.362	44.23	2.046	64.41	1.445	77.33	1.233
Ag metal reference	38.14	2.359	44.33	2.043	64.50	1.445	77.61	1.230

[a]Sample with rubbed surface.

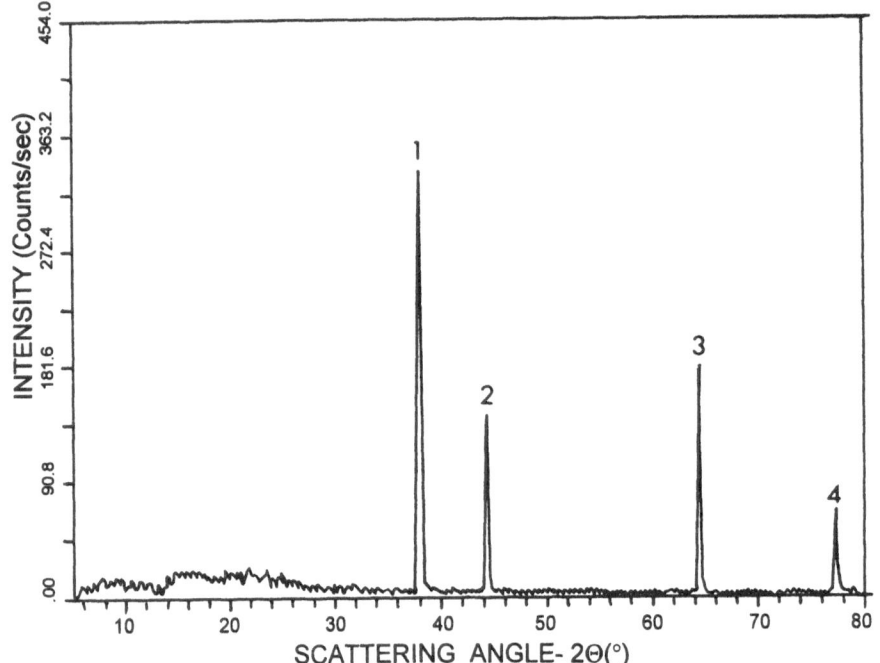

Figure 4. X-ray diffractogram pattern from a silver-doped BDSDA-ODA. Mole ratio monomer:dopant is 2:1. The broad peak at around 22° is from the polyimide matrix.

Figure 5.   Scanning electron microphotographs of silver-doped A)BDSDA-ODA and B)BTDA-ODA.
Mole ratio monomer:dopant is 2:1.

365

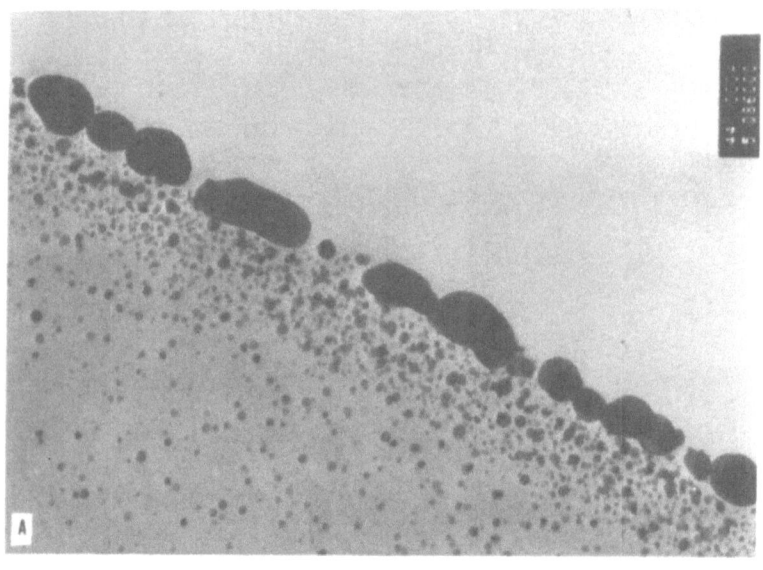

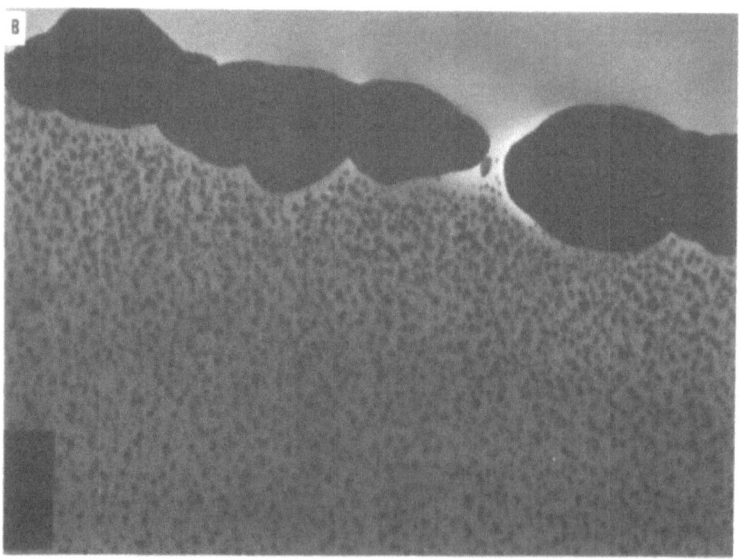

Figure 6.   Transmission electron microphotographs of silver-doped A)BTDA-ODA and B)BDSDA-ODA. Mole ratio monomer:dopant is 2:1. Magnification 112,875x.

In conclusion, the formation of the relatively thick metallic silver surface layer is responsible for the enhanced conductivity in the BDSDA-derived films. In the case of BTDA-derived films which also contain a silver surface, the greater intermixing of polyimide and silver acts to electrically shield individual silver particles. In addition, the surface of the BTDA-ODA films has a thin layer of polyimide on top of the silver which also deters surface conduction. The more uniformly sized silver particles in doped BTDA-ODA, however, lead to much more reflecting films.

ACKNOWLEDGEMENTS

The authors are grateful for a CAPES Fellowship from the Brazilian government. We also thank Steven McCartney and Frank Cromer for their assistance in obtaining some of the data described in this manuscript. The authors also gratefully acknowledge NASA for sponsoring this research.

REFERENCES

1. M. M. Koton, Polyimides Thermally Stable Polymers, Macromolecular Compounds Consultants Bureau, New York, 1988, p. 271.
2. J. E.Sheats, C. E. Carraher, and C. U. Pittman, in Metal Containing Polymer Systems; Plenum, New York, 1985.
3. H. R. Allcock, Adv. Mater., 6, 106 (1994).
4. L. T. Taylor and J. D. Rancourt, In Inorganic and Metal Containing Polymeric Materials, Sheats, J., Ed.; Plenum Press: New York, 1990, pp. 109-126.
5. S. A.Ezzell, T. A. Furtsch, E. Khor, and L. T. Taylor, J. Polym. Sci., Polym. Chem. Ed., 21, 865 (1983).
6. J. D. Rancourt, G. M. Porta, and L. T. Taylor, Thin Solid Films, 158, 189 (1988).
7. L. T. Taylor, V. C. Carver, T. A. Furtsch, and A. K. St. Clair, ACS Symp. Ser., 121, 71, (1980).
8. G. M. Porta, J. D. Rancourt, and L. T. Taylor, Chem. Mater., 1, 269 (1989).
9. R. K. Boggess and L. T. Taylor, J. Polym. Sci., Polym. Chem. Ed., 25, 685 (1987).
10. R. K. Boggess and L. T. Taylor, Recent Advances in Polyimide Science and Technology. Proceedings of the Second International Conference on Polyimides: Chemistry, Characterization, and Applications. W. D. Weber and M. R. Gupta, Eds. (Poughkeepsie, Mid-Hudson Chpt., Society of Plastics Engineers) 1987, pp. 463-70.
11. A. F. Rubira, J. D. Rancourt, M. L. Caplan, A. K. St. Clair, and L. T. Taylor, Chem. Mater., in press.
12. A. F. Rubira, J. D. Rancourt, M. L. Caplan, A. K. St. Clair, and L. T. Taylor, ACS-PMSE, 71, 509-511 (1994).
13. S. A. Ezzel, T. A. Furtsch, E. Khor, and L. T. Taylor, J. Polym. Sci. Poly. Chem. Ed., 91, 8651 (1983).
14. M. L. Caplan, D. M. Stoakley, and A. K. St. Clair, Proceed. ACS-PMSE, 69, 400-1 (1993).
15. A. Auerbach, J. Electrochem. Soc., 131, 937 (1984).
16. S. Mazur and S. Reich, J. Phys. Chem. 90, 1365 (1986).
17. S. Mazur, L. E. Manring, M. Levy, G. T. Dee, S. Reich, and C. E. Jackson, Metal. Plast. Vol. 1; K. L. Mittal and J. R. Susko, Eds., Plenum Press, New York, 1989, pp. 115-134.
18. R. K. Boggess and L. T. Taylor, in Recent Advances in Polyimide Science and Technologies, W. D. Weber and M. R. Gupta, Eds., Mid-Hudson Chapter SPE, New York, 1987, pp. 463-70.
19. T. L. Wohlford, J. Schaaf, L. T. Taylor, T. A. Furtsch, E. Khor, and A. K. St. Clair, in Conductive Polymers, R. B. Seymour, Ed., Plenum Publishing Corp., New York, 1981, pp. 7-21.
20. A. K. St. Clair, S. A. Ezzel, L. T. Taylor and H. G. Boston, "Electrically Conductive Polyimide Films, NASA Tech. Brief," August 1993, pp. 57-58.
21. D. Gaulino et al., Oxidation Resistant Reflective Surfaces for Solar Dynamic Power Generation in Near Earth Orbit, "NASA Technical Memorandum 88865", 1986.
22. G. Schon, Acta Chem. Scand., 27, 2623 (1973).
23. P. E. Larson, J. Electron Spectroscop. Relat. Phenom., 4, 213 (1974).
24. C. J. Huang, C. C. Yen, and T. C. Chang, J. Appl. Polym. Sci., 42, 2237 (1991).
25. C. J. Huang, C. C. Yen, and T. C. Chang, J. Appl. Polym. Sci., 42, 2267 (1991).

INVESTIGATION OF THE ION-BINDING PROPERTIES OF TACTIC POLY(METHACRYLIC ACIDS) USING TERBIUM(III) ION AS A FLUORESCENT PROBE

Hannia Luján-Upton and Yoshiyuki Okamoto
Department of Chemistry and the Polymer Research Institute
Polytechnic University, 6 Metrotech Center
Brooklyn, N.Y., 11201 U.S.A.

INTRODUCTION

The lanthanides have fourteen $4f$ electrons successively added to the lanthanum, La, configuration. These elements are highly electropositive and form a very stable tripositive cation, Ln^{3+}, as their primary oxidation state. The radius of these ions decrease with increasing atomic number in a trend known as the lanthanide contraction. The elements are highly electropositive because the sum of the first three ionization enthalpies is relatively low. They easily form +3 ions in solids, such as oxides, and in complexes while existing as $[Ln(H_2O)_n]^{3+}$ in aqueous solution. Other oxidation states are also possible, for example, Ce^{4+}, Sm^{2+}, Eu^{2+}, and Yb^{2+} are all stable in aqueous solutions and in solids.

The electronic configuration has a significant effect on the stability of metal ion complexes. In the case of d-type transition metal ions, their complex stabilities are the result of participation of the d electrons in the metal-ligand bond. This occurs through the overlap of hybridized metal electronic orbitals with the appropriate ligand orbitals. The $4f$ orbitals are effectively shielded from interaction with ligand orbitals by 5s and 5p electrons. If hybridization is to occur, it must involve normally unoccupied higher energy orbitals (e.g., 5d, 6s, 6p), and hybridization of this type can be expected only with the most strongly coordinating ligands. Thus, complex formation is entirely electrostatic in character, and the lanthanide complexes compare more closely with the complexes derived from calcium, strontium, and barium ions than with those derived from the d-type transition metal ions.[1-3] These cations are thus classified as class "a"[4,5] or hard acceptors.[6,7] Moreover, the fluorescence of the lanthanide ions emanate from the shielded 4f orbitals allowing them to be used as substitutional probes for Ca^{2+}, and Mg^{2+} in several biological systems.

Appreciably stable lanthanide complexes, other than the hydrated cations themselves, are obtained only when strongly chelating ligands are used and, in particular, when these ligand contain highly electronegative donor atoms such as oxygen. Lanthanide ions prefer donor atoms with the preference O>N>S and F>Cl. Complexes with nitrogen or sulfur donors are not stable in aqueous solutions due to hydrolysis.[8] The high degree of ionicity in the Ln complexes can be confirmed by the observed patterns of bond distances. It was observed that Ln—N bonds are generally longer than Ln—O bonds by as much as 0.30Å.

Terbium, Tb^{3+} has a well-defined emission and excitation spectra at room temperature and has a relatively long emission lifetime. Essentially, all emissions of aqueous Tb^{3+} complexes emanate from the $^5D_4$ excited state when an excitation less than 490nm is used. The transition regions $^5D_4 \rightarrow {}^7F_J$ (J = 0,1,2,3,4,5 or 6) all display emissions with the relative intensities being, in descending order, $^5D_4 \rightarrow {}^7F_6 > {}^7F_4 > {}^7F_3 > {}^7F_2$ (485-500nm, 580-595nm, 615-625nm, 645-655nm respectively).[9,10] All of the transition regions of Tb^{3+} show fine structuring with the sharpest structure exhibited by the $^5D_4 \rightarrow {}^7F_5$ and $^7F_3$ emissions. The $^5D_4 \rightarrow {}^7F_5$ transition at 540-555nm is also the most intense transition and remains remarkably intense under a wide variety of solution conditions, (Figure 1).[11] The transition is also extremely sensitive to the detailed nature of the ligand environment, making it the best transition to use as a fluorescence probe.[12-18] The intensity of the emission depends on three things, the degree of binding to the Tb^{3+}, the site-symmetry of the complex, and the mode of binding, whether direct or indirect excitation is involved.[19]

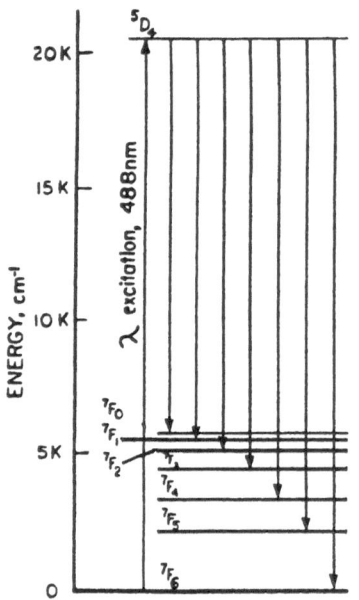 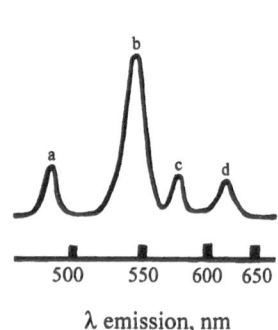

Figure 1   Spectral Transition States of Terbium
<u>Left</u>: Excitation and Emissive States of Tb^{3+}
<u>Right</u>: Emission Spectra of Terbium: a) $^5D_4 \rightarrow {}^7F_6$, b) $^5D_4 \rightarrow {}^7F_5$, c) $^5D_4 \rightarrow {}^7F_4$, d) $^5D_4 \rightarrow {}^7F_3$

Aqueous salt solutions of Tb^{3+} exist as the aquo complex $[Tb(OH_2)_n]^{3+}$ with nine water molecules (n=9), bound to the inner coordination sphere. The OH stretching mode of coordinated H_2O is effective at quenching the terbium $^5D_4$ excited state via non-radiative energy transfer. Replacement of any of these water molecules with ligands containing carboxylate anions increases both the fluorescence intensities and the lifetimes of the complexed Tb^{3+} by decreasing the concentration of quenchers in the coordination sphere.[20-23] The fluorescence intensities of the excitation spectra of Tb^{3+} are greatly enhanced upon binding to carboxy-containing polymers such as polyacrylates and polysaccharides.[24,25] The addition of simple carboxylate compounds also leads to the observation of similar effects.

Upon complexation with carboxylate compounds, a hypersensitive excitation peak at 286 nm is observed which most closely resembles the $^7F_6 \rightarrow ^5H_5$ transition.[26-28] Hypersensitive transitions are chemically more sensitive to the environment surrounding the lanthanide ion than normal f-f transitions and experience relatively large changes in shape and intensity when modifications occur in the coordination sphere of the ion.[29,30] The hypersensitive band tends to be too weak to be observed in the monomeric Tb^{3+} complexes but becomes evident and strongly allowed in certain polymeric environments.

The tacticity describes a region of ordered domains within a polymer which possess regularity in the configuration of successive pseudochiral centers. Polymers exhibiting such order are said to be "stereoregular", (Figure 2).[31] The effect of tacticity on the solution properties of poly(methacrylic acid) (PMA) has been investigated for many years. It was recognized early, for example, that isotactic PMA behaves as a weaker polyacid than syndiotactic PMA.[32] In addition, the dissociation of isotactic PMA is more endothermic than that of the syndiotactic sample.[33] The interaction with divalent metal ions such as Cu^{2+} and Mg^{2+} were also found to be different for the two polymers. Dialysis equilibria measurements indicate that the isotactic sample binds copper more strongly than syndiotactic PMA, whereas the reverse is true for the binding of magnesium and sodium.[34-36]

In the present study we investigated the binding properties of isotactic and syndiotactic PMA using Tb^{3+} as a fluorescence probe. In order to investigate the effect of microstructure on terbium fluorescence, we studied the complexation of Tb^{3+} with Kemp's triacid, (1,3,5-Trimethyl-1,3,5-cyclohexanetricarboxylic acid). Our interest in the molecule is based on its unique conformational structure which resembles a monomeric subunit of isotactic PMA. Additionally, the complexation of terbium to the isomer of Kemp's triacid, $(1\alpha,3\alpha,5\beta)$-1,3,5-Trimethyl-1,3,5-cyclohexanetricarboxylic acid, was also investigated, (Figure 3).

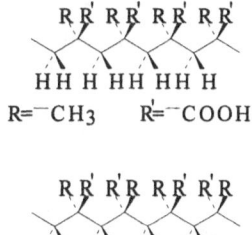

R=$^-CH_3$ R'=^-COOH

R=$^-CH_3$ R'=^-COOH

Figure 2 Structures of Tactic Poly(methacrylic acids)
Top: Isotactic PMA, Bottom: Syndiotactic PMA

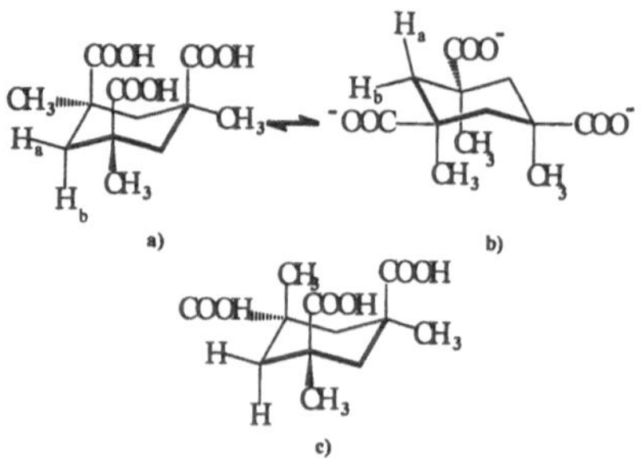

Figure 3   Kemp's Triacid Conformations: a)Undissociated Structure, b) Dissociated Structure, c) Kemp's Configurational Isomer

EXPERIMENTAL

Materials:

The isotactic and syndiotactic poly(methacrylic acids) were obtained from Professor G. Challa's laboratory, Department of Chemistry, University of Grogningen. Tacticity information was as follows: 0.97(mm), 0.03(mr), 0(rr) for isotactic PMA, and 0.01(mm), 0.07(mr), 0.92(rr) for syndiotactic PMA. The molecular weights for the polyacids were approximately 500K (iso) and 193K (syndio) as determined by solution viscometry measurements.

Kemp's Triacid and its configurational isomer were both obtained from the Aldrich Chemical Company, 99%, desiccated and used without further purification.

Preparation of Terbium Solutions

A [0.04N] stock solution of Tb^{+3} was prepared by dissolving 1.49g of hydrous terbium chloride (Rhone-Poulenc Inc., 99.9%) in 100ml of deionized water. An aqueous solution of [0.002N] of $TbCl_3$ was made from the stock and used in all fluorescence measurements. Since the degree of ionization, α, would be varied, the fluorescence intensity of terbium as a function of pH was determined at pH 4.5, 5.9, 7.1, 8.0, and 9.1.

Neutralization of Tactic Poly(methacrylic acids)

Titration of the tactic PMA's was necessary in order to determine the concentration of carboxylic acid groups per gram of polymer and also to account for any water absorption. Sodium chloride, 0.15g, [0.26M], was used as a supporting electrolyte. A titration setup protected from atmospheric CO_2 by bubbling with N_2 gas was used. All titrations were performed with 0.514N NaOH which was standardized against potassium hydrogen phthalate, using phenolthalein as an indicator. Twenty minutes were allowed to lapse before the next aliquot of base was added in order to permit equilibration. When the degree of ionization, α, was approximately 0.95, the polymers were back-titrated with standardized 0.519N HCl to a final ionization of 0.75. The HCl was also standardized by the above method using bromocresol green as an indicator.

A solution approximating 0.1N, with respect to the carboxylic acid moiety of isotactic PMA was first made by adding 0.089g of isotactic PMA to 10ml of de-ionized water. Since isotactic PMA is slightly soluble in neutral or acidic media, the polymer was stirred overnight. After 24 hrs the solution was turbid with a pH of 4.35. At this point titration was begun. The solution was allowed to stir overnight once a pH of 10.04 had been reached. The pH had dropped to 7.44 the next day, therefore the pH was adjusted to 12.4 and left stirring for 1.5hrs. Once the pH remained unchanged at 11.79, the solution was back-titrated with 0.052N HCl to a final pH of 9.89 (±0.05 pH units). Typical titration curves for the tactic PMA's are shown in Figures 4 and 5.

Polymer/terbium complexes were made by mixing two milliliters of 0.02N polymer (at various α) with two milliliters of 0.002N terbium solution. Fluorescence measurements were then conducted on these complexes.

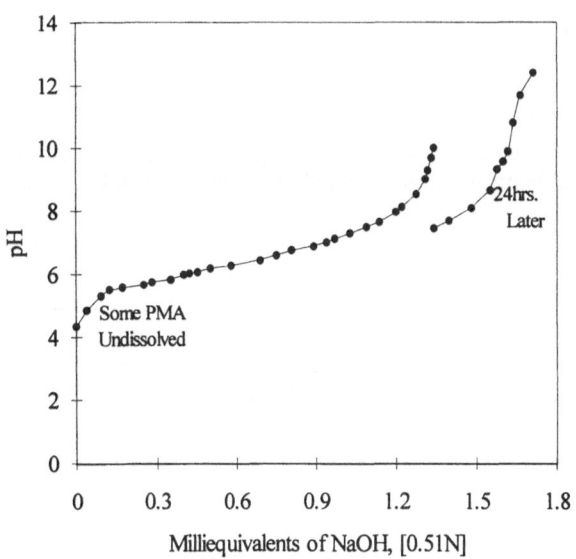

Figure 4 Titration Curve of Isotactic PMA
Isotactic PMA: 9.16meq/g in the prescence of [0.1M]NaCl at room temperature

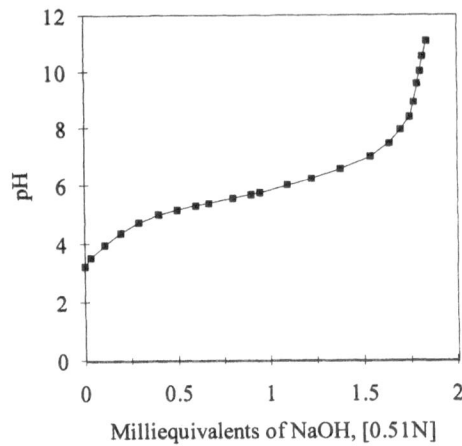

Figure 5 Titration Curve of Syndiotactic PMA
Syndiotactic PMA: 10.6meq/g in the prescence of [0.1M]NaCl at room temperature

Titration of Kemp's Triacid and its Derivatives

Kemp's triacid must first be neutralized since it is only slightly soluble in water. Titration of the acid with 0.05N NaOH gave the curve found in Figure 6. Phenolthalein was used as the indicator. The onset of the endpoint started at pH 9.0 for 0.018g of Kemp's triacid in 10ml of water (0.1M KCl) for a concentration of 0.02M in COOH. A stock solution of Kemp's triacid, 0.04N, was made by dissolving 0.172g of the acid in 10ml of de-ionized water and 40ml of 0.0512N NaOH.

Kemp's Isomer, (1α,3α,5β)-1,3,5-Trimethyl-1,3,5-cyclohexanetricarboxylic acid is the configurational isomer of Kemp's triacid and is also insoluble in water. Titration of the isomer with 0.051N NaOH gave a similar endpoint (pH=9.5) to that of Kemp's triacid. Fifty milliliters of a 0.02N stock solution was made by titrating 0.0689g of Kemp's isomer with 0.0505N NaOH in deionized water to a pH of 9.5 was attained.

Fluorescence Measurements:

Fluorescence studies were conducted on solutions of the terbium complexes for each of Kemp's derivatives and the tactic PMA's. These solutions were prepared by mixing 0.002M $TbCl_3$ with 0.02N solution of the respective ligands in order to obtain a concentration ratio of 10/1 $[COO^-]/[Tb^{3+}]$. Deionized water was used for all solutions. Excitation spectra were obtained by holding the emission wavelength fixed at 545nm, the $^5D_4 \rightarrow {}^7F_5$ Tb^{3+} band, while scanning the wavelengths between 200-400 nm. Emission was measured by fixing the absorption transition at 286nm for the tactic polymers, 240nm for Kemp's derivatives, and 230nm for aqueous Tb^{3+} while scanning the wavelengths from 400-640nm.

All experiments were done on the same day to reduce the possibility of instrumental error/deviation.

Instrumentation:

A Perkin Elmer MPF-44B and a Perkin Elmer LS50B fluorescence spectrophotometer were used to measure the fluorescence intensity of all lanthanide complexes at ambient temperature. The slits widths were set at 6nm for both the excitation and emission. A solid polymeric Eu^{3+} standard (Perkin Elmer, gauge of 0-1000 fluorescence intensity units) was used to calibrate the Perkin Elmer MPF-44B spectrophotometer. The emission wavelength was 612nm with an excitation of 396nm. A dimethyl sulfate solution of tetrakis(dibenzoyl methanato) Eu^{3+} (Perkin Elmer) was used as the standard for all fluorescence measurements.

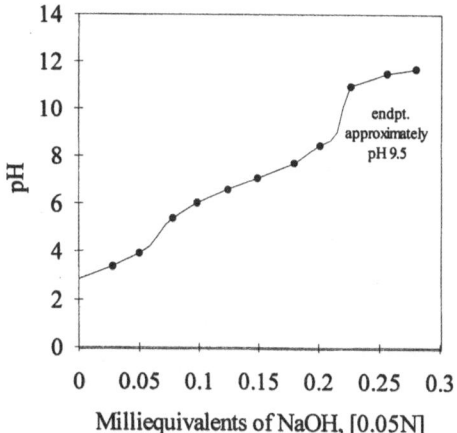

Figure 6 Titration Curve of Kemp's Triacid
Kemp's Triacid: 0.018g in the prescence of [0.1M]NaCl at room temperature

RESULTS

The excitation and emission spectra of the aqueous tactic PMA-Tb^{3+} complexes at α =0.75 (pH 6.5 for the syndiotactic PMA and 8.1 for isotactic PMA) are shown in Figures 7 and 8. The isotactic PMA-Tb^{3+} complex excitation spectrum contains a new peak near 285nm while the same peak is present only to a very small degree in the syndiotactic PMA/Tb^{3+} complex. Figure 7 shows that not only is the isotactic polymer sensitized when compared to the Tb^{3+} aquo ion, but a strong complex is formed causing a large hypersensitive transition to appear at 285nm. This peak is approximately 6-8 times larger than the corresponding peak in the syndiotactic polymer, (Figure 8).

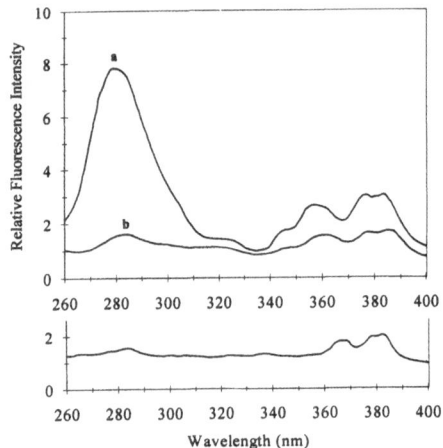

Figure 7 Excitation Spectra of Isotactic and Syndiotactic PMA/Terbium(III) Complex at α=0.75
Top: a) Isotactic PMA/Tb(III) Complex b) Syndiotactic PMA/TbIII) Complex
PMA/Tb= (10^{-2}M/10^{-3}M), λ_{em}=545nm, room temperature, Gain=1X
Bottom: Aqueous Terbium(III) Chloride, (10^{-3}M), λ_{em}=545nm, room temperature, Gain=10X

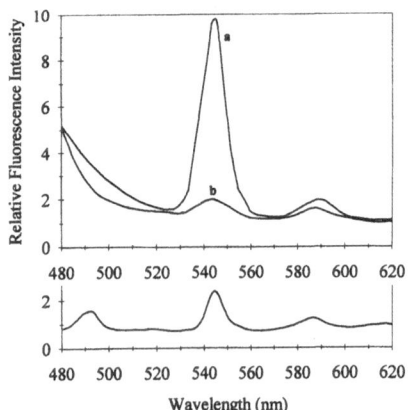

Figure 8 Emission Spectra of Isotactic and Syndiotactic PMA/Terbium(III) Complex at α=0.75
Top: a) Isotactic PMA/Tb(III) Complex b) Syndiotactic PMA/TbIII) Complex
PMA/Tb= (10^{-2}M/10^{-3}M), λ_{exc}=286nm, room temperature, Gain=1X
Bottom: Aqueous Terbium(III) Chloride, (10^{-3}M), λ_{exc}=286nm, room temperature, Gain=10X

376

The fluorescence properties for the atactic PMA-Tb^{3+} complexes were found to be similar with those of the syndiotactic PMA-Tb^{3+} complex. This is reasonable since the atactic PMA sample used possessed a great deal of syndiotactic character (>70%).

Fluorescence of Kemp's Compounds with Terbium

The two Kemp's compounds were first fully neutralized before they were complexed to Tb^{3+}. An instantaneous precipitation was evident upon complexation with Kemp's non-alkylated derivative, thereby making solution state fluorescence studies impossible. Although Kemp's triacid and its isomer also eventually precipitated from solution, their precipitation times were much longer (12-15 hours for Kemp's acid and 20 minutes for its isomer). Fluorescence measurements were carried out for the complexes of Tb^{3+} with fully neutralized Kemp's triacid, and its isomer, $(1\alpha,3\alpha,5\beta)$-1,3,5-Trimethyl-1,3,5-cyclohexanetricarboxylic acid, (Figures 9 and 10). It is apparent from both these figures that once complexed to either of Kemp's compounds the fluorescence intensity of the aqueous Tb^{3+} ion is greatly enhanced. The neutralized Kemp's acid forms a complex to Tb^{3+} which has approximately a four times higher fluorescence intensity than the complex formed with its isomer.

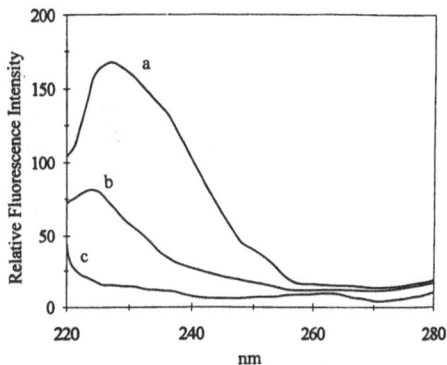

Figure 9 Excitation Spectra of Kemp's Triacid and its Isomer Complexed with Terbium (III):
a) Kemp's Trianion/Tb(III) Complex, b) Kemp's Isomer/Tb(III) Complex, [COO$^-$]/Tb(III)=(10^{-2}M/10^{-3}M), c) Aqueous Terbium(III) Chloride, (10^{-3}M), λ_{em}=545nm, room temperature.

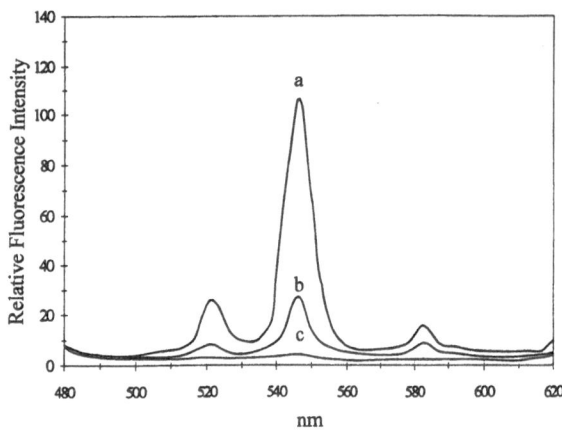

Figure 10 Emission Spectra of Kemp's Triacid and its Isomer Complexed with Terbium (III):
a) Kemp's Trianion/Tb(III) Complex, b) Kemp's Isomer/Tb(III) Complex,
$[COO^-]/Tb(III)=(10^{-2}M/10^{-3}M)$, c) Aqueous Terbium(III) Chloride, $(10^{-3}M)$,
λ_{exc}=240nm, room temperature.

DISCUSSION

This study was initiated in order to investigate the ion binding properties of tactic polyacids with Tb^{3+} ion using fluorescence spectroscopy. The difference between isotactic and syndiotactic poly(methacrylic acid) is the relative positions of the nearest neighbor ligating moieties, i.e. carboxylate, the intramolecular complexation behavior of the tactic PMAs should prove to be sensitive to tacticity. Static fluorescence measurements of the complexed terbium ion were used to probe the differences in complexation behavior between isotactic and syndiotactic PMA. Terbium(III) ion was chosen as a probe due to the well documented sensitivity of its fluorescence behavior to its complexation environment. The hypothesis was first tested on a non-polymeric configurational model, Kemp's triacid. Derivatives of this compound were also studied before the more complicated polymeric environment was investigated.

Kemp's molecule could serve as a configurational model for isotactic PMA if several conditions were met. A ring flip reversal was observed upon complexation with Tb^{3+} ion restoring the carboxylate groups to their axial positions. Also, the torsional, angular, steric and dipolar interactions among the carboxylate groups of Kemp's acid and the isotactic polymer had to be proven comparable

Menger investigated the complexation of Mg^{2+} (ionic radius 0.65Å) and Ca^{2+}(ionic radius 1.0Å) with Kemp's triacid.[37] It is known that the methylene hydrogens (H_a and H_b, Figure 3) of Kemp's compound can be used as probes for the conformational ring-flip which occurs when Kemp's acid is neutralized. In the neutralized form, Kemp's acid's once triaxial acid moieties become triequatorial. The differences in chemical shift of the methylene hydrogens, ΔCH_2, are quite different in each of these structures. The chemical shift difference, between H_a and H_b prior to the ring flip is approximately 1.44ppm. The trianionic form of Kemp's compound allows a more symmetrical environment to be experienced by these methylene hydrogens, thereby making the chemical shift differences between the doublets very small where $\Delta CH_2=0.24$ppm.

The complexation of ionized Kemp's acid with Mg^{2+} caused a reversal of the ring flip to a half chair conformation which approaches the formerly held triaxial structure ($\Delta CH_2=0.9$ppm). The energetically unfavorable half-chair conformation lies 11kcal above the chair form and is the result of a compromise between strong dipolar interactions and both angular and torsional strain. Coulombic attractions between the carboxylate groups and the cation provides enough stabilization to keep Kemp's molecule in this conformation.

Additional ^{1}HNMR data was taken on Ca^{2+}, an ion which has an ionic radius very close to that of Tb^{3+}($\sim$1.0Å). Its complexation behavior with Kemp's trianion should therefore mimic that of terbium. Although precipitation was evident shortly after mixing, due to the opening of the strained half-chair conformation to a more stable chair conformer, an NMR spectrum was observed by Menger. The NMR spectrum was obtained immediately upon mixing and a larger chemical shift difference ($\Delta CH_2=1.8$) was reported than that which had been observed for the Mg^{2+}/Kemp's complex. Hence, since ring reversals occur when Mg^{2+}, Ca^{2+}, and Al^{3+} ions complex with Kemp's molecule it is likely that a conformational ring-flip also occurs when Tb^{3+} is complexed to Kemp's triacid which allows Kemp's molecule to serve as a model for the isotactic polymer.

Since the coordination behavior of Kemp's triacid is known, the geometry of the coordinated complex is well understood. It has been found that the nearest neighbor carboxylate group distances in the preferred half-chair conformation are 5.04Å.[36] In order for Kemp's triacid to act as a model for isotactic PMA, the nearest neighbor distances between carboxyl groups in the polymer should approximate this value. These distances can be calculated, since the rotational angles and bond distances of the tactic PMA's are known.[38] The neutralized polymers are known to retain the same conformation as the acids, the repulsion between carboxylate anions can be neglected.[39] Simple geometric calculations give the nearest neighbor carboxylate distance at 2.55Å for the syndiotactic polymer and 5.16Å for the isotactic PMA.[40] The coordination site bond distances in the isotactic polymer agreed within 5% of the preferred half-chair carboxylate bond distances of Kemp's molecules. Therefore, Kemp's triacid could be used as a viable model for the complexation of the tactic PMA's with terbium ion.

Precipitation was observed after approximately 10-12hrs when the neutralized Kemp's compound was complexed with terbium. Kemp's isomer, however gave precipitation after only 20 minutes at room temperature. Menger reported precipitation after a few minutes when Ca^{+2} was complexed with Kemp's trianion in D_2O. It was postulated that a gradual conformational change occurred from a half chair to a more stable equatorial conformer. Such a conformation would be more apt to form long-range chains when complexed to metallic cations eventually precipitating out of solution. The differences in the precipitation rates of Kemp's compounds can be attributed to their structural isomerism. Kemp's isomer will always have at least one equatorial carboxylate moiety, regardless of its conformation. Therefore, it forms long-range intermolecular chains upon complexation much easily leading to a faster precipitation rate than observed for Kemp's molecule.

The relative intensities for the static fluorescence measurements of the ionized Kemp's compounds complexed to Tb^{3+} were compared. The results indicate that although both compounds complex strongly with the Tb^{3+} ion the fluorescence of Kemp's isomer is approximately four times weaker than that of the complex formed with Kemp's molecule. The contribution to the overall fluorescence intensity of excess Tb^{3+}-aquo complex is minimal due to quenching by the –OH manifold of water. The fluorescence differences between these two compounds is probably due to their configurations. This parallels the static fluorescence intensities of the tactic PMA polymers and illustratse the effect of configurational isomerism on the Tb^{3+} complex. Static fluorescence measurements on the complexes of Tb^{3+} with neutralized isotactic and syndiotactic poly(methacrylic acids) also showed a configurational effect. The fluorescence intensity of both tactic PMA/Tb^{3+} complexes showed enhancement in the hypersensitive region. The isotactic polymer was approximately eight times more fluorescent than the syndiotactic PMA under the same conditions. These data support our hypothesis concerning the importance of the relative position of the nearest neighbor ligating species.

The presence of a hypersensitive transition in the terbium/polymer complex system indicates that the terbium ion exists in a highly asymmetric environment. This stands in contrast to the fluorescence data obtained from the Tb^{3+} complexes of Kemp's compounds which did not exhibit hypersensitive transitions. The Tb^{3+} ion therefore, occupies a highly symmetric site in these complexes.

CONCLUSION

The purpose of this study was to determine whether the coordination geometry about the Tb^{3+} ion was sensitive to the distance between ligating carboxylate anions in a polymer. Isotactic and syndiotactic poly(methacrylic acids) were chosen to complex with the metal since they each possess a definite microstructure with respect to nearest neighbor ligand distance. Fluorescence measurements with Tb^{3+} ion demonstrated a six to one preference for binding with isotactic over syndiotactic poly(methacrylic acid). This preference is believed to be due to the relative distance between nearest neighbor ligating species.

In order to confirm this hypothesis Kemp's triacid was used as a monomeric model for the isotactic polymer. Calculations involving simple geometry using bond angles and lengths based on energy minimizations calculations for the isotactic polyacid and Kemp's molecule showed a common nearest neighbor ligating distance of approximately 5.0Å. The complexation behavior of Kemp's triacid could therefore be used as a model for polymeric chelation. The results of fluorescence studies on the terbium/Kemp's triacid complex and its isomeric analog corresponded to those of the tactic PMA systems. The fluorescence intensity of Kemp's triacid/terbium complex was four times as intense as that of the isomeric complex because with Kemp's triacid, Tb^{3+} can complex with three axial carboxylates while in the case of its isomer, only two axial carboxylates exist to complex to the lanthanide ion. Thus, the fluorescence intensities for the complexes of each acid was found to be different. It can be concluded that the binding efficiency of a chelating species with Tb^{3+} ion is very strongly influenced by the nearest neighbor distances of the ligating sites on the molecule.

Since the binding efficiency of a ligand is directly correlated to the number of water molecules bound to the coordination sphere of a metal ion, it might therefore prove interesting to conduct fluorescence lifetime measurements using both the polymers and their monomeric analogs in D_2O and water. Such experiments would yield information about the number of water molecules bound to the terbium ion and, by extension, the relative binding efficiencies of the various chelates studied. These investigations are being carried out in our laboratory at the present time.

ACKNOWLEDGEMENTS

The authors are greatly indebted to Professor G. Challa, and Dr. Y. Y. Tan, Department of Chemistry, University of Grogningen, the Netherlands, for providing us with tactic polymer samples. We would also like to express our gratitude to Dr. E. Donahue for his helpful calculations and and insightful discussions.

REFERENCES

1. C. K. Jørgensen, *Orbitals in Atoms and Molecules*, Academic, New York (1962), Chap. 11.
2. T. Moeller, *The Chemistry of the Lanthanides*, Reinhold, New York (1963), Chap. 3 & 4.
3. G. Schwarzenbach, *Advances in Inorganic Chemistry and Radiochemistry*, Vol. 3, H. J. Emeleus and A. G. Sharpe, eds., Academic, New York (1961), p. 265.
4. E. Nieboer and W. A. E. McBryde, *Can. J. Chem.*, **51**, 2511 (1973).
5. R. J. P. Williams and J. D. Hale, *Struct. Bonding*, **1**, 249 (1966).
6. R. G. Pearson, *J. Chem. Educ.*, **56**, 581 (1968).
7. G. Klopman, *J. Am. Chem. Soc.*, **90**, 223 (1968).
8. T. Moeller, *Inorganic Chemistry Series One*, K. W. Bagnall, ed., University Park Press, Baltimore (1972), Vol. 7, p. 275.
9. G. Blasse and A. Bril, *Phillips Technical Review*, **31**, (10), 314 (1970).
10. R. D. Peacock, *Struct. Bond.*, **22**, 84, (1975).
11. F.S. Richardson, *Chem. Rev*, **82**, 541 (1982).
12. F. S. Richardson, J. D. Saxe, S. A. Davis, and T. R. Faulkner, *Mol. Phys.*, **42**, 401 (1981).
13. J. P. Morley, J. D. Saxe, and F. S. Richardson, *Mol. Phys.*, **47**, 379, (1982).
14. J. D. Saxe, J. P. Morley, and F. S. Richardson, *Mol. Phys.*, **47**, 407, (1982).
15. F. S. Richardson, *Inorg. Chem.*, **19**, 2806, (1980).
16. F. S. Richardson, *Chem. Rev.*, **79**, 17, (1979).
17. F. S. Richardson, and T. R. Faulkner, *J. Chem. Phys.*, **76**, 1595, (1982).
18. J. D. Saxe, T. R. Faulkner, and F. S. Richardson, *J. Chem. Phys.*, **76**, 1607, (1982).
19. W. D. Horrock's, Jr., and D. R. Sudnick, *Acc. Chem. Res.*, **14**, 384, (1981).
20. W. D. Horrock's, Jr., reference 19
21. H. G. Brittain, S. P. Kelty, J. A. Peters, *J. Coord. Chem.*, **23**, 21, (1991).
22. J. Kido, Y. Okamoto, *Makromol. Chem., Macromol. Symp.*, **59**, 83 (1992).
23. J. Kido, H. G. Brittain, and Y. Okamoto, *Macromolecules*, **21**, 1872, (1988).
24. V. Crescenzi, H. G. Brittain, N. Yoshino, and Y. Okamoto, *J. Polym. Sci.: Polym. Phys. Ed.*, **23**, 437, *(1985)*.
25. I. Nagata, Y. Okamoto, *Macromolecules*, **77**, 773, (1977).
26. W. T. Carnall, P. R. Fields, K. Rajnak, *J. Chem. Phys.*, **49**, 4447, (1968).
27. W. T. Carnall, P. R. Fields, K. Rajnak, *J. Chem. Phys.*, **49**, 4412, (1968).
28. A. Rudman, S. Paoletti, and H. G. Brittain, *Inorg. Chem.*, **24**, 1283, (1985).
29. O. L. Malta, *Molecular Physics*, **42**, (1), 65, (1981).
30. R. D. Peacock, *Struct. Bond.*, **22**, 84, (1975).
31. G. Odian, Principles of Polymerization, 2nd Edition, John Wiley & Sons, New York, p568, (1970).
32. E. M. Loebl and J. J. O'Neill, *J. Polym. Sci.*, **45**, 538, (1960).
33. E. M. Loebl and J. J. O'Neill, *Polym. Prepr. Am. Chem. Soc. Div. Polym. Chem.*, **3**, 466, *(*1962).
34. Loebl, reference 31
35. J. J. O'Neill, E. M. Loebl, A. Y. Kandanian, and H. Morawetz, *J. Polym. Sci.*, **A-1,3**, 4201 (1965).
36. L. Costantino, V. Crescenzi, F. Quadrifoglio, and V. Vitagliano, *J. Polym. Sci.*, **A-1,2**, 5771, (1967).
37. M. Menger, P. A. Chicklo, and M. J. Sherrod, *Tetrahedron Letters*, **30**, 6943, (1989).

38. J. B Lando and J. Semen, *J. Macromol. Sci.,-Phys.*, **B7**(2), 297-317 (1973).

39. Y. Muroga, I., Noda, and M. Nagasawa, *Macromolecules*, **18**, 1580, (1985).

40. H. Luján-Upton, <u>Part I</u>: <u>Investigation on the Ion-Binding Properties of Carboxy-Containing Compounds using Terbium(III) Ion as a Fluorescent Probe</u>, Ph.D. Thesis, Polytechnic University, Appendix, p72, January 1995.

RATIONAL DESIGN OF NOVEL POLYELECTROLYTES: ALUMINOSILICATE/POLY(ETHYLENE GLYCOL) COPOLYMERS

Glenn C. Rawsky and Duward F. Shriver

Department of Chemistry and Materials Research Center
Northwestern University
Evanston, Illinois 60208-3113

INTRODUCTION

The field of solvent-free polymer electrolytes includes both polymer-salt complexes and more recently, polyelectrolytes. These materials comprise a class of solid ionic conductors which have been the subject of increasing attention in recent years due to their unique integration of desirable mechanical and electrochemical properties[1]. Solvent-free polymer electrolytes are the subject of much research as potential electrolytes for high energy density advanced batteries and other electrochemical devices.[2]

Considerable research has been performed with a large variety of polymer-salt complexes, and these ionic conductors have been the subject of several reviews.[3,4] After a brief introduction to these materials, we will focus on polyelectrolytes, which contain covalently bound ionizable groups and a mobile counterion. Until recently, these polymers were considered to be of little utility as ionic conductors without the addition of a high-dielectric, low viscosity solvent.[3] Rational design of molecular structure has led to new, highly conductive solvent-free polyelectrolytes in our present research.

Development of Polymer-Salt Complexes

It was known in the mid-1960's that certain salts can be dissolved in polymeric hosts containing polar groups,[5] but it was not until the work of Wright[6] in 1973 that the ionic conductivity of these materials was recognized. Despite the rather poor conductivity performance of these simple complexes of alkali-metal thiocyanate and iodide salts with poly(ethylene oxide), subsequent studies led to rapid improvements as the behavior of these systems became more well understood.

Structural analyses of pure semicrystalline PEO via vibrational spectroscopy and x-ray analyses indicate a helical conformation of the polymer chains.[7] Investigation of the cation environment in the complex $P(EO_3NaI)$ via x-ray diffraction revealed that the sodium ion was coordinated to both polyether oxygens and iodide ions.[8,9] It was postulated that ion transport

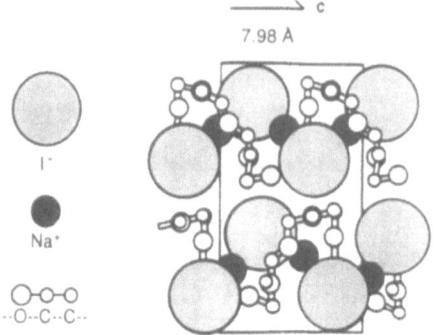

Figure 1. Crystal structure of P(EO₃NaI) polymer-salt complex.[8]

occurs via a hopping mechanism along the interior of the crystalline polymeric helix[10] of poly(ethylene oxide) (Figure 1). However, it was shown through NMR measurements[11] and experiments on amorphous polymer-salt complexes that in fact the majority of ion transport occurs within the amorphous regions of the polymer host. This realization paved the way for development of a new theory to describe ion transport in polymer electrolytes. Within the amorphous phase of a polymeric host such as PEO, macroscopic viscosities characteristic of a solid are observed, while on a microscopic regime the polymer behaves in a liquidlike fashion, due to thermal rearrangements of the polymer above its glass transition temperature. Ion transport occurs via a diffusive mechanism wherein ion motion is assisted by facile rearrangement of the polymer segments in the vicinity of the cation (Figure 2). Consistent with this model, the conductivity of polymer-salt complexes is described by the Vogel-Tamman-Fulcher (VTF) equation, which is known to apply to the viscosity behavior of supercooled liquids:

$$\sigma = \sigma_o T^{-\frac{1}{2}} e^{\frac{-B}{k_B(T-T_o)}} \qquad (1)$$

The parameter B is an apparent activation energy, T_o is a fitting parameter referred to as the "equilibrium" glass transition temperature, and σ_o is a fitting parameter related to the number of free charge carriers available. This equation may be derived by appropriate combination of the original expression describing viscosity (η)

$$\eta = C e^{\frac{B}{T-T_o}} \qquad (2)$$

with the Stokes-Einstein relation describing diffusion of ions (D) in polymer media

$$D = \frac{k_B T}{6 \pi r_i \eta_i} \qquad (3)$$

followed by substitution into the Nernst-Einstein equation, which relates conductivity to

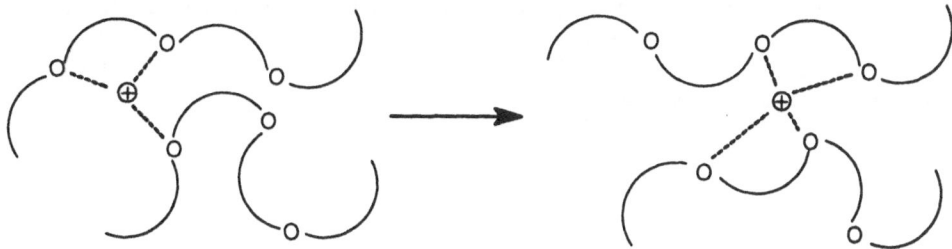

Figure 2. Cartoon of diffusive ion transport assisted by polymer chain motions.

diffusive motion of the mobile species:

$$\frac{\sigma}{D} = \frac{n q^2}{k_B T} \qquad (4)$$

The VTF description of conductivity in polymer electrolytes is now widely accepted as the proper description for almost all systems. Its validity is further supported by a variety of other treatments which can be shown to yield equivalent expressions. These include: (a) the empirically derived Williams-Landel-Ferry (WLF) equation,[12] commonly used to describe the relationship between fluidity (η^{-1}) and temperature in liquid and polymer systems; (b) the free-volume model,[13] in which local fluctuations of the molecular volume in excess of the van der Waals volume of a molecule provide "holes" into which ions may diffuse; and (c) the configurational entropy model developed by Gibbs and co-workers,[14] which describes ion transport in terms of a probability of cooperative fluctuations of the polymer and ions.

Several important points to consider in the design of polymer electrolytes arise from these descriptions. Foremost is the importance of T_g in governing the observed conductivity of a system. Instead of a simple Arrhenius-like $e^{-1/T}$ temperature dependence, conductivity is governed by an $e^{-1/(T-T_o)}$ term. The "equilibrium" glass transition temperature T_o (typically 30 - 50°C less than T_g) is the idealized temperature below which all configurational motion of polymer segments ceases, or alternatively where the free volume disappears completely. At a given temperature, conductivity will increase as T_g (and thus T_o) decreases. In addition, increasing the number of free charge carriers in the system (*i.e.* those not tightly bound to a counterion or through strong dipolar interactions with the polymer chain) will linearly increase the conductivity through the parameter σ_o. Thus the nature of the anion strongly affects ion pairing, which in turn influences the conductivity tremendously.

Polyelectrolytes

Substantially improved performance has been achieved with polymer-salt complexes (Table 1) as our understanding of these systems has grown. However, the presence of the mobile anion introduces a complicating factor in both practical application and fundamental study of these materials.

Measurements of ionic transport numbers indicate that typically the mobile anion is the dominant charge carrier; cation transport numbers ($t_+ = \sigma_+/[\sigma_+ + \sigma_-]$ in the absence of concentration gradients) are commonly only 0.2-0.6 (Table 2). This may affect practical use in, for instance, batteries consisting of a lithium metal negative electrode, polymer electrolyte, and an insertion compound as the positive electrode. A concentration polarization will arise due to the mobile anion, resulting in a limiting current

$$i_{lim} = \frac{4RT}{F} \cdot \frac{\sigma_+}{l}$$

(5)

beyond which salt depletion will be observed at the cathode/electrolyte interface. Onset of depletion will occur at a time $t \propto \sigma_-^{-1}$. Thus low t_+ results in a smaller limiting current and an earlier onset of salt depletion, leading to a less efficient system as well as reduced discharge rates.[15]

In addition, dual ion mobility complicates fundamental investigations of ion transport mechanisms. One must take into account differences in both Coulombic and steric interactions of each ion with the polymer host. The possibility of cation motion as part of a neutral cation/anion pair or low-valent ion cluster also arises, resulting in no or reduced net charge transport.[16]

For these reasons, our group[17] and others[18] have investigated polyelectrolytes with covalently bound anions such as modified poly(phosphazene)[19,20] and poly(tetraalkoxyaluminate),[21] which ensure a unity cation transport number (Table 3). However, ion pairing limits the performance of these materials.[22] One of the most successful syntheses to date involves poly(siloxane) with pendant oligoether and sterically hindered phenolate groups.[23] The use of bulky t-butyl groups surrounding the anionic site as well as incorporation of charge delocalization into the phenyl rings is designed to minimize strong electrostatic interaction between the anion and cations. We have explored the use of cryptands,[20,24] and control of both anion basicity and steric bulk to overcome the detrimental effects of strong ion pairing. Aluminosilicates are attractive as weakly basic anions that can be incorporated into a polymer, and their low basicity should reduce coulombic attraction between cation and anion, thus enhancing conductivity. In the present research, we have prepared both linear and network alkoxy aluminosilicate polyelectrolytes to examine the effects of anion basicity on conductivity. Further, we have investigated the influence of added cryptand as a cation sequestering agent which may enhance conductivity by screening electrostatic interactions between anions and the encapsulated sodium cation. In addition, a series of polymers with differing anion:etheric oxygen ratios have been prepared to determine the effect of chain length and ion concentration on conductivity and other physical properties.

Table 1. Polymer-salt complexes

Complex	σ_{312K} / S cm^{-1}	Ref.
P(EO$_{12}$LiClO$_4$)	5.6 x 10^{-6}	30
P(EO$_8$LiClO$_4$) - crosslinked	≈8 x 10^{-6}	31
P(PO$_9$LiSO$_3$CF$_3$)	2.2 x 10^{-5}	10a
Poly(ethylene succinate)$_6$LiBF$_4$	≈3 x 10^{-8}	32
MEEP$_4$LiSO$_3$CF$_3$	1 x 10^{-4}	33
aP(EO$_{50}$NaI)	9 x 10^{-5}	34

Table 2. Cation transport numbers of polymer-salt complexes

Complex	Cation transference number, t_+	Ref.
$P(EO_8LiClO_4)$	0.25	35
$P(EO_{36}NaClO_4)$	0.17	36
(Siloxane-PEO copolymer)$_n$LiClO$_4$	0.55	37
$P(EO_8LiSO_3CF_3)$	0.70	35
$P(EO_{16}LiSO_3CF_3)$	0.35 - 0.47	38

RESULTS

Experimental

Polymers were prepared according to the following procedure (Scheme 1): Poly(ethylene glycol) (MW$_{av} \approx$ 300) was reacted with excess sodium hydride, followed by allyl chloride, yielding α,ω-bis(allyloxy)poly(ethylene glycol) (1). Hydrosilylation of compound 1 with dimethylchlorosilane in the presence of Karstedt's catalyst (platinum(0) divinyltetramethyl disiloxane) formed α,ω-bis(chlorodimethylsilyl)poly(ethylene glycol) (2), which was quantitatively hydrolyzed, producing α,ω-bis(hydroxydimethylsilyl)poly(ethylene glycol) (3). Linear polyelectrolytes were then synthesized by the reaction of sodium bis(2-methoxyethoxy)aluminum hydride with a 5% excess of compound 3, to yield $[NaAl(OEOMe)_2(OSiMe_2(CH_2)_3(OE)_xO(CH_2)_3SiMe_2O)_{2/2}]_n$ [Me = CH$_3$, OE = OCH$_2$CH$_2$] (4). The network polyelectrolyte $[NaAl (OSiMe_3)(OSiMe_2(CH_2)_3(OE)_xO(CH_2)_3SiMe_2O)_{3/2}]_n$ (5) was prepared by addition of compound 3 (5% excess) to a stirred solution of triethylaluminum and sodium trimethylsilanolate. The complex of cryptand [2.2.2] (4,7,13,16,21,24-hexaoxa-1,10-diazabicyclo [8.8.8]hexacosane), with linear polyelectrolyte 4 was prepared by stirring a stoichiometric quantity (1:1 cryptand:sodium ion) of cryptand [2.2.2] with 4 in acetonitrile followed by solvent removal under high vacuum (compound 6). All polymers are transparent, golden-tan elastomeric materials. The reactions were carried out in a dry nitrogen atmosphere using standard Schlenk and glovebox techniques, and the products were stored under nitrogen.

Analyses

The identity of intermediates 1-3 were confirmed by mass spectrometry and ^{13}C/^{1}H NMR. The formation of a covalent bond between the aluminum centers and siloxy oxygens was demonstrated by progressive upfield shifts of the ^{29}Si resonance, from δ = 31ppm (2)[†,25] to δ = 15.5 (3)[12] to $\delta \approx$ 0 in the final polymer products 4 and 5 (Figure 2). Additionally, ^{27}Al NMR spectra of 4 and 5 exhibited broad signals centered around $\delta \approx$ 60-64 ppm, a value indicative of asymmetric, 4-coordinate oxo-aluminum.[26,27] All products exhibit little if any flow at room temperature. Exposure to atmospheric moisture results in slow hydrolysis of the surface of the linear polymer after 1-2 months; the network polymer remains stable for >3 months. Anal. Calc'd for 4: C 47.95, H 8.83, Si 7.29, Al 3.33, Na 2.84. Found: C 46.82, H 8.44, Si 6.99, Al 3.21, Na 3.10. Anal. Calc'd for 5: C 50.43, H 9.07, Si 7.45, Al 1.72, Na 1.47.

---

† n-Propyldimethylchlorosilane (Hüls America) ^{29}Si {^{1}H} NMR: δ 31.2 ppm (strong, R$_3$SiCl), δ 6.97 ppm (weak, (R$_3$Si)$_2$O).

Table 3: Polyelectrolytes

Polymer	σ / S cm^{-1}	T_g /°C	Ref.
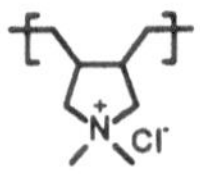 (structure: poly with N$^+$ Cl$^-$)	7×10^{-8} (26 °C)	-39	17
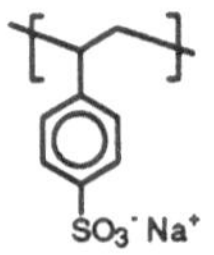 (structure: poly styrene SO$_3^-$ Na$^+$)	3×10^{-7} (26 °C)	-29	17
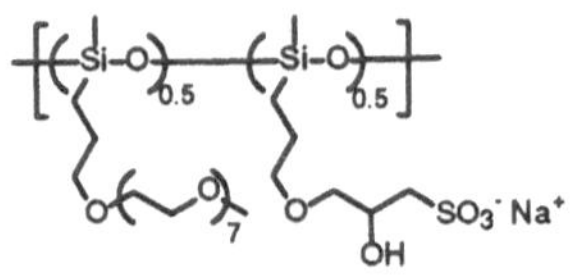 (structure: siloxane copolymer 0.5/0.5, SO$_3^-$ Na$^+$, OH)	$\sim 5 \times 10^{-7}$ (25 °C)	-65	18c
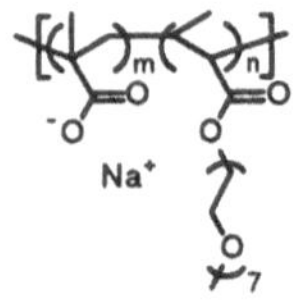 (structure: m/n copolymer, Na$^+$)	5×10^{-8} (25 °C)	-66	18g
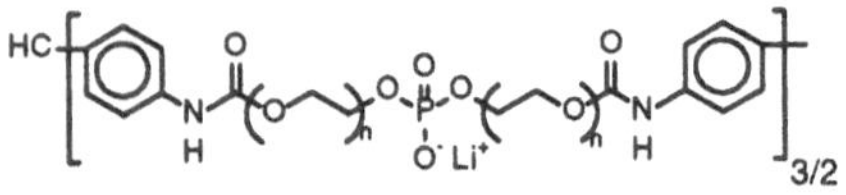 (structure: HC- aromatic urethane phosphate, O$^-$ Li$^+$)$_{3/2}$	$< 10^{-8}$ (25 °C, est.*)	-37.5	18h
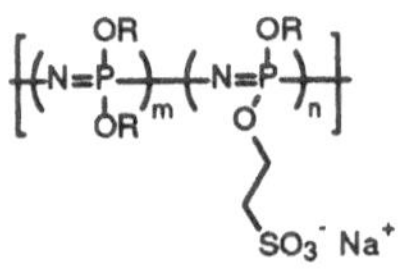 (structure: polyphosphazene OR, SO$_3^-$ Na$^+$)	$\sim 10^{-7}$ (25 °C, est.*)	-67	19
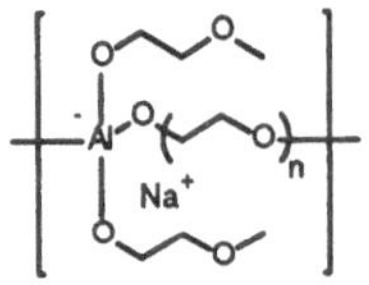 (structure: Al–O with PEO, Na$^+$)	$\sim 6 \times 10^{-8}$ (25 °C, est.*)	-31	21
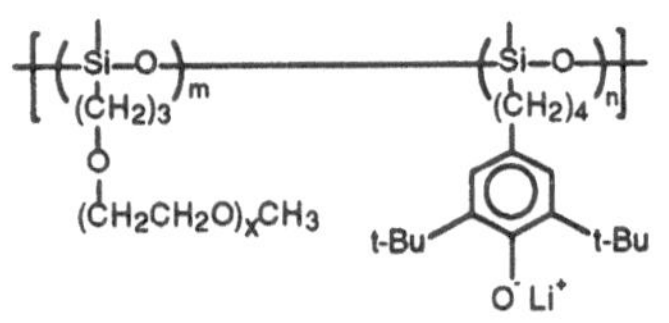 (structure: siloxane (CH$_2$)$_3$ / (CH$_2$)$_4$, (CH$_2$CH$_2$O)$_x$CH$_3$, t-Bu phenol O$^-$ Li$^+$)	5×10^{-5} (30 °C)	$\sim$ -55	23

*Values labeled as "estimates" are based on extrapolation of available data to the temperature indicated.

388

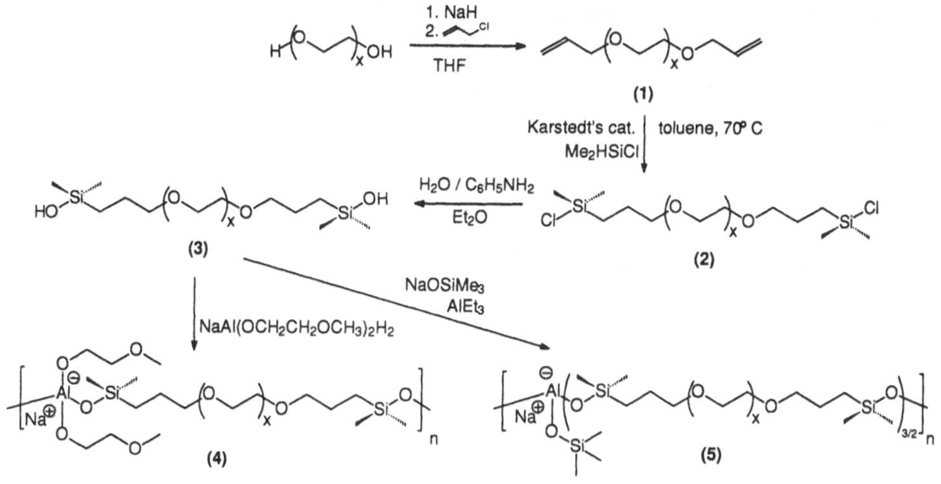

Scheme 1. Synthetic route to aluminosilicate polyelectrolytes.

Found: C 50.60, H 9.01, Si 7.26, Al 1.68, Na 1.42.[‡] Glass transitions were recorded at 40, 20 10, and 5°C/minute, then extrapolated to zero heating rate. AC impedance spectra were collected using polymer pellets approximately 0.1 cm (t) x 1.2 cm (d) sandwiched between stainless steel or tantalum blocking electrodes. Spectra were obtained for the frequency range 5 Hz - 13 MHz as a function of temperature, controlled to within 0.1°C.

DISCUSSION AND CONCLUSIONS

Glass transition temperatures and the results of fitting conductivity data to the VTF equation are given in Table 4, along with values for the related tetraalkoxyaluminate polymers **7-9** synthesized in our labs by Doan et al.[21] No evidence of melt transitions is apparent upon examination of DSC traces for polymers **4-6**, indicating completely amorphous structures. Conductivity data (Figure 3) show a progressive increase of room temperature conductivities in the series **7-4-5**, which correlates well with decreasing anion basicity, and also with decreasing glass transition temperature, T_g. The number of siloxy segments bound to the aluminate center (none in the alkoxyaluminates, two in the linear polymer, and four in the network polymer) may enhance conductivity through two mechanisms. First, by decreasing oxygen basicity, contact ion pairing is reduced.[28] Additionally, the larger Al-O-Si bond angle, longer bonds involving the Si atom, and exceptionally low rotational barrier about the Si-O bond may increase polymer flexibility near the anion. Thus a decrease in T_g is expected with added SiO groups, as observed in the sequence **7-4-5**. The change in T_g indicates an increase in the segmental motions which are essential for ion transport in polymers.[29] Comparison of the conductivities of two series of polyelectrolytes with different oligoether chain lengths to determine the effect of ion concentration and whether the behavior is further affected by a network vs. linear structure are not yet complete.

Upon addition of cryptand to the linear polyelectrolyte **4** in a 1:1 cryptand:sodium ratio, nearly an order of magnitude increase in the conductivity was observed at all temperatures. This is comparable to the increases observed previously in our labs[19,20,24] and has been attributed

---

[‡]   Analyses performed by Analytische Laboratorien GmbH, Gummersbach, Germany.

Table 4. VTF Parameters and Glass Transitions

Polymer	$\sigma_o{}^a$ $(SK^{1/2}cm^{-1})$	$B/k_B{}^a$ (K)	$T_o{}^a$ (K)	$T_g{}^b$ (K)
(4) $[NaAl(OR)_2(OSiMe_2(CH_2)_3(OE)_xO(CH_2)_3SiMe_2O)_{2/2}]_n$ $(x \approx 6.8)^c$	0.626	1048	196	224
(5) $[NaAl(OSiMe_3)(OSiMe_2(CH_2)_3(OE)_xO(CH_2)_3SiMe_2O)_{3/2}]_n$ $(x \approx 6.8)^c$	0.0621	876	181	219
(6) $[(4):[2.2.2]]_n$	2.49	1028	189	223
(7) $[NaAl(OEOMe)_2((OE)_xO)_{2/2}]_n$ $(x \approx 6.8)^c$	31.7^d	3201^d	109^d	247
(8) $[NaAl(OEOMe)_2((OE)_xO)_{2/2}]_n$ $(x \approx 9.0)^c$	0.214	1406	180	242
(9) $[(8):[2.2.2]]_n$	2.36	1107	185	---

[a] Parameters obtained from a three-variable fit to the VTF equation: $\sigma = \sigma_o\, T^{-1/2} \exp[-B/(k_B(T-T_o))]$. [b] Extrapolated to zero heating rate. [c] A value of $x \approx 6.8$ indicates $MW_{av} \approx 300$ for the poly(ethylene glycol) used in the synthesis; a value of $x \approx 9.0$ indicates $MW_{av} \approx 400$. [d] Anomalous VTF parameters are attributed to the limited amount of data available for fitting.

to a decrease in ion pairing between the bound anion and the encapsulated mobile cation.[22] The diminished magnitude of the effect observed here is attributed to the already reduced ion pairing between sodium ions and the weakly basic aluminosilicate anions. These results are supported by the calculated VTF parameters, which indicate limiting conductivities (σ_o) which are the same order of magnitude among polymers without cryptand (0.21-0.63 $SK^{1/2}cm^{-1}$), and among those with cryptand [2.2.2] (2.4-2.5 $SK^{1/2}cm^{-1}$); however, the increase upon addition of cryptand is only ≈ 0.6 orders of magnitude for the linear aluminosilicate polymer, as compared to 1.1 for the aluminate polymers. Very little change in the glass transition temperature is observed upon addition of cryptand, perhaps indicating the absence of extensive virtual crosslinking between polymer chains and cations, which would be expected to increase the microscopic viscosity. Comparison of the conductivity as a function of sodium concentration at fixed temperature for the polyelectrolytes with and without added cryptand should provide further insight as to the exact nature of the mechanism by which cryptand addition enhances conductivity.

In conclusion, Na^+ conductivity in polyelectrolytes increases in the series $Al(OR)_4^-$ to $Al(OSiR_3)_4^-$ because of decreases in basicity. We are aware of only one other polyelectrolyte with a room temperature conductivity that exceeds the performance attained by the present systems without the use of plasticizers or other additives.[23] Conductivity is enhanced significantly by a reduction in ion pairing upon addition of cryptand [2.2.2], although the effect is not as dramatic as has been observed with other less conductive systems.

ACKNOWLEDGEMENTS

This research was supported by the Army Research Office under Grant No. DAAL03-90-G-0044, and made use of MRL Central Facilities supported by the National Science Foundation, at the Materials Research Center of Northwestern University, under Award No. DMR-9120521. GCR gratefully acknowledges the Department of Defense for an NDSEG predoctoral fellowship.

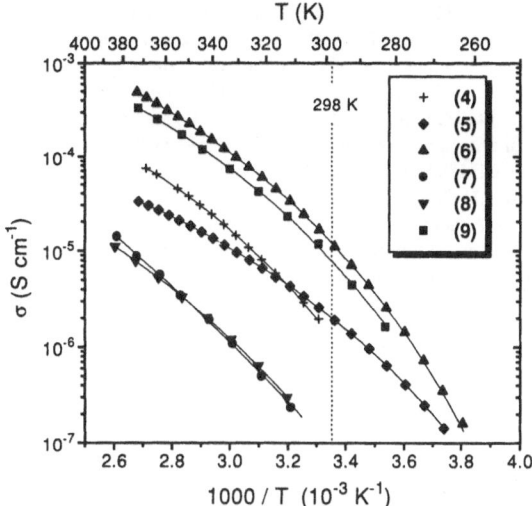

Figure 3. Temperature-dependent conductivity of aluminate (**7-9**), aluminosemi-silicate (**4,6**), and aluminosilicate (**5**) polyelectrolytes, and corresponding three-parameter VTF fits (solid lines). Best-fit VTF parameters are given in Table 4.

REFERENCES

1. D.F. Shriver and G.C. Farrington, Solid ionic conductors, *Chem. Eng. News* 63:42-57 (1985).
2. J. R. MacCallum and C. A. Vincent, "Polymer Electrolyte Reviews, Vol. 1 and 2" Elsevier, London (1987, 1989).
3. J.S. Tonge and D.F. Shriver, Polymer electrolytes, *in:* "Polymers for Electronic Applications," J.H. Lai, ed., CRC Press, Boca Raton, Florida (1989).
4. M.A. Ratner and D.F. Shriver, Ion transport in solvent-free polymers, *Chem. Rev.* 88:109-24 (1988).
5. (a) A.A. Blumberg, S.S. Pollack, and C.A.J. Hoeve, A poly(ethylene oxide)-mercuric chloride complex, *J. Polym. Sci. Part A*, 2:2499 (1964). (b) R.D. Lunberg, F.E. Bailey, and R.W. Callard, Interaction of inorganic salts with poly(ethylene oxide), *J. Polym. Sci. Part A*, 1(4):1563 (1966).
6. (a) P.V. Wright, Electrical conductivity in ionic complexes of poly(ethylene oxide), *Br. Polym. J.* 7:319 (1975). (b) D.E. Fenton, J.M. Parker, and P.V. Wright, Complexes of alkali metal ions with poly(ethylene oxide), *Polymer* 14:589 (1973).
7. Y. Takahashi and H. Tadokoro, Structural studies of polyethers (-(CH$_2$)$_m$-O-)$_n$. X. Crystal structure of PEO, *Macromolecules*, 6:672 (1973).
8. figure adapted from: S. Okamura and Y. Chatani, Crystal structure of PEO sodium iodide complex, *Polymer*, 28:1815 (1987).
9. (a) B. Papke, M.A. Ratner, R. Dupon, T. Wong, M. Brodwin, and D.F. Shriver, Structure and ion transport in polymer-salt complexes, *Solid State Ionics* 5:83 (1981). (b) B. Papke, M.A. Ratner, and D.F. Shriver, Vibrational spectroscopy and structure of polymer electrolytes, PEO complexes of alkali metal salts, *J. Phys. Chem. Solids* 42:493 (1981). (c) D. Teeters and R. Frech, Temperature dependent spectroscopic studies of PPO and PPO-inorganic salt complexes, *Solid State Ionics* 18/19:271 (1986).
10. (a) M.B. Armand, J.M. Chabagno, and M.J. Duclot, Poly-ethers as solid electrolytes, *in:* "Fast Ion Transport in Solids: Electrodes and Electrolytes," P. Vashishna, J.N. Mundy, and G.K. Shenoy, eds., North-Holland, New York, p.131 (1979). (b) J.M. Parker, P.V. Wright, and C.C. Lee, A double helical model for some alkali metal ion-poly(ethylene oxide) complexes, *Polymer* 22:1305 (1981).
11. (a) M. Minier, C. Berthier, and W. Gorecki, Thermal analysis and NMR studies of a PEO complex electrolyte: PEO(LiSO$_3$CF$_3$)$_x$, *J. Phys.* 45:739 (1984). (b) C. Berthier, W. Gorecki, M. Minier, M.B. Armand, J.M. Chabagno, and P. Rigaud, Microscopic investigation of ionic conductivity in

alkali metal salts-PEO adducts, *Solid State Ionics* 11:91 (1983).

12. M.L. Williams, R.F. Landel, and J.D. Ferry, The temperature dependence of relaxation mechanisms in amorphous polymers and other glass-forming liquids, *J. Am. Chem. Soc.* 77:3701 (1955).

13. (a) G. Grest and M.H. Cohen, Liquids, glasses, and the glass transition: a free volume approach, *Adv. Chem. Phys.* 48:455 (1981). (b) M. Watanabe and N. Ogata, Ionic conductivity of polymer electrolytes and future applications, *Br. Polym. J.* 20:181 (1988).

14. (a) J.H Gibbs and E.A. DiMarzio, Nature of the glass transition and the glassy state, *J. Chem. Phys.* 28:373 (1958). (b) J.H Gibbs, "Modern Aspects of the Vitreous State," Butterworths, London, Ch.7 (1965). (c) G. Adam and J.H. Gibbs, On the temperature dependence of cooperative relaxation properties in glass-forming liquids, *J. Chem. Phys.* 43:139 (1965). (d) B.L. Papke, M.A. Ratner, and D.F. Shriver, Conformation and ion-transport models for the structure and ionic conductivity in complexes of polyethers with alkali metal salts, *J. Electrochem. Soc.* 129:1694 (1982).

15. M. Doyle, T.F. Fuller, and J. Newman, Modeling of galvanostatic charge and discharge of the lithium/polymer/insertion cell, *J. Electrochem. Soc.* 140(6):1526-33 (1993).

16. M.C. Lonergan, Ph.D. thesis, Northwestern University, Evanston, IL (1994).

17. L. C. Hardy and D. F. Shriver, Preparation and electrical response of solid polymer electrolytes with only one mobile species, *J. Am. Chem. Soc.* 107:3823-8 (1985).

18. (a) S. Takeoka, K. Horiuchi, S. Yamagata, and E. Tsuchida, Sodium ion conduction of perfluorosulfonate ionomer/poly(oxyethylene) composite films, *Macromolecules* 24:2003-6 (1991). (b) D. J. Bannister, G. R. Davies, I. M. Ward, and J. E. McIntyre, ionic conductivities for poly(ethylene oxide) complexes with lithium salts of monobasic and dibasic acids and blends of poly(ethylene oxide) with lithium salts of anionic polymers, *Polymer* 25:1291-6 (1984) . (c) G. Zhou, I. M. Khan, and J. Smid, Cation transport polymer electrolytes. Siloxane comb polymers with pendant oligo-oxyethylene chains and sulphonate groups, *Polym. Commun.* 30:52-5 (1989). (d) T. Hamaide, C. Carré, and A. Guyot, Ionic conductivity in sulphonate end-capped poly(ethylene oxide), *in*: "Second International Symposium on Polymer Electrolytes," B. Scrosati, ed., Elsevier, New York (1990). (e) M. Watanabe, S. Nagano, K. Sanui, and N. Ogata, Estimation of Li$^+$ transport number in polymer electrolytes by the combination of complex impedance and potentiostatic polarization measurements, *Solid State Ionics* 28-30:911-7 (1988) . (f) E. A. Reitman and M. L. Kaplan, Single-ion conductivity in comblike polymers, *J. Polym. Sci.* 28:187-91 (1990). (g) E. Tsuchida, N. Kobayashi, and H. Ohno, Single-ion conduction in poly[(oligo(oxyethylene)methacrylate)-co-(alkali-metal methacrylates)], *Macromolecules* 21:96-100 (1988). (h) J. F. LeNest, A. Gandini, H. Cheradame, and J. P. Cohen-Addad, Cationic transport features of ionomeric polymer networks, *Polym. Commun.* 28:302-5 (1987).

19. S. Ganapathiappan, K. Chen, and D. F. Shriver, A new class of cation conductors: polyphosphazene sulfonates, *Macromolecules* 21:2299-2301(1988) .

20. K. Chen and D.F. Shriver, Magnesium ion conducting polymeric electrolytes, *Chem. Mater.* 3:771-2 (1991).

21. K. E. Doan, M. A. Ratner, and D. F. Shriver, Synthesis and electrical response of single-ion conducting network polymers based on sodium poly(tetraalkoxy aluminates), *Chem. Mater.* 3:418(1991).

22. K.E. Doan, B.J. Heyen, M.A. Ratner, and D.F. Shriver, Influence of cryptands and crown ethers on ion transport and vibrational spectra of polymer salt complexes, *Chem. Mater.* 2:539 (1990).

23. T. F. Yeh, H. Liu, Y. Okamoto, H. S. Lee, and T. A. Skotheim, Polyelectrolytes with sterically hindered anionic charges, *in*: "Second International Symposium on Polymer Electrolytes," B. Scrosati, ed., Elsevier, New York (1990). An ostensibly identical polymer, synthesized in our laboratory, exhibited a lower conductivity (1.4x10-6 S cm^{-1} @ 30°C vs. 5x10^{-5} S cm^{-1}): M.C. Lonergan, M.A. Ratner, and D.F. Shriver, Cryptand addition to polyelectrolytes: a means of conductivity enhancement and a probe of ionic interactions, *J. Am. Chem. Soc.* in press (1995).

24. K. Chen, S. Ganapathiappan, and D.F. Shriver, Cryptate effects on sodium-conducting phosphazene polyelectrolytes, *Chem. Mater.* 1:483-4(1989) .

25. H. Marsmann, ^{29}Si-NMR spectroscopic results, *in*: "Oxygen-17 and Silicon-29 (NMR, Basic Principles and Progress; Vol.17)," P. Diehl, E. Fluck, and R. Kosfeld, eds., Springer-Verlag, New York (1981) pp. 78, 147, 166, 194-5.

26. O. Kříž, B. Čásenský, A. Lyčka, J. Fusek, and S. Heřmánek, ^{27}Al NMR behavior of aluminum alkoxides, *J. Magn. Reson.*, 60:375-81 (1984).

27. F.J. Feher, T.A. Budzichowski, and K.J. Weller, Polyhedral aluminosilsesquioxanes: soluble organic analogues of aluminosilicates, *J. Am. Chem. Soc.*, 111:7288-9(1989).

28. H. Schmidbaur, Recent developments in the chemistry of heterosiloxanes, *Angew. Chem. Int. Ed.* 4(3):201 (1965).

29. M.A. Ratner and A. Nitzan, Conductivity in polymer ionics: dynamic disorder and correlation, *Faraday Discuss. Chem. Soc.* 88:19-42 (1989).

30. P. Ferloni, G. Chiodelli, A. Magistris, and M. Sanesi, Ion transport and thermal properties of poly(ethylene

oxide) -LiClO₄ polymer electrolyte, *Solid State Ionics* 18/19:265 (1986).

31. J.R. MacCullum, M.J. Smith, and C.A. Vincent, The effect of radiation-induced crosslinking on the conductance of LiClO₄-PEO electrolytes, *Solid State Ionics* 11:307 (1984).

32. R. Dupon, B.L. Papke, M.A. Ratner, and D.F. Shriver, Ion transport in polymer electrolytes formed between poly(ethylene succinate) and LiBF₄, *J. Electrochem. Soc.* 131:586 (1984).

33. P.M. Blonsky, D.F. Shriver, P. Austin, and H.R. Allcock, Poly-phosphazene solid electrolytes, *J. Am. Chem. Soc.* 106:6854 (1984).

34. J.P Lemmon, R.L. Kohnert, and M.L. Lerner, Characterization of a stoichiometric range of sodium salt complexes of amorphous poly[(oxymethylene)oligo(oxyethylene)] by differential scanning calorimetry and ²³Na NMR, *Macromolecules* 26:2767-70 (1993).

35. A. Bouridah, F. Dalard, D. Deroo, and M.B. Armand, Potentiometric measurements of ionic mobilities in PEO electrolytes, *Solid State Ionics* 18/19:287 (1986).

36. M. Leveque, J.F. LeNest, A. Gandini, and H. Cheradame, Ionic transport numbers in polyether networks containing different metal salts, *Makromol. Chem. Rapid Commun.* 4:497 (1982).

37. P.G. Hall, G.R. Davies, J.E. McIntyre, I.M. Ward, D.J. Bannister, and K.M.F. LeBrocq, Ion conductivity in polysiloxane comb polymers with ethylene glycol teeth, *Polym. Prepr.* 27:98 (1986).

38. P.R. Soresen, and T. Jacobson, Limiting currents in the polymer electrolyte: PEO_xLiSO₃CF₃, *Solid State Ionics* 9/10:1147 (1983).

DETERMINATION OF ACIDITY IN THE INTERIOR OF THE CROSS-LINKED POLYELECTROLYTE GRAIN BY THE USE OF pH-SENSITIVE PROBES

Leonid S.Molochnikov,[1] Elena G.Kovalyova,[1]
Igor' A.Grigor'ev,[2] Vladimir A.Reznikov[2]

[1]Ural State Wood Technology Academy Siberian highway,
37 Ekaterinburg, 620032 Russia.
[2]Institite of Organic Chemistry Siberian Branch of the Russian
Academy of Scienes Lavrent'ev's Avenu, 9 Novosibirck, 630090 Russia.

Introduction

Circumstantial evidence[1,2] suggests that the state of water in ionite micropores and the acidity of the medium significantly differ from the ones in solution contacting the sorbent. Therefore, the conditions existing in the interior of ionite grains may not be approximated by measurements in solutions, which are performed by researchers. However, it is precisely these conditions that determine complexating processes or catalysis using sorbents as catalysts. Recently, the synthesis of stable nitroxyl radicals (NR) which are sensitive to changes of medium acidity in connection with participation in reactions of proton exchange and may be used as pH-probes, has been carried out.[3,4] Using the electron spin resonance spectroscopy (ESR) method, the principal possibility to measure pH-value in a solution at the site of radicals location as well as in micropores of ionites appeared.

Results and Discussion

Studies were carried out by us on the carboxylic cationites obtained by polycondensation of salicylic acid, phenol and formaldehyde (KB-51) and by copolymerization of acrylic acid with divinyl benzene (KB-2), and containing o-oxybenzoic and carboxylic groups, respectively. Their preparation has been carried out by a known method.[5] The pH-sensitive nitroxyl radicals (NR) of imidazole (R1) and imidazolidine (R2 and R3) types (Table 1), synthesized in the Novosibirsky Institute of Organic Chemistry of the Siberian Branch of the Russian Academy of Sciences by V.V.Khramtsov[6], have been used for pH-value determination in the interior of ionite grains. The aqueous solution of radicals (with concentrations equal to 0.5 mmol/l) were titrated by HCl and NaOH solutions at $298^{0}K$, achieving the necessary pH-value.

Table 1. The ESR parameters and pKa-values of NR.

Radical	pKa ± 0.1	g ± 0.0001		a_N ± 0.006 mT	
		RH⁺	R	RH⁺	R
R1	6.1	2.0051	2.0048	1.510	1.610
R2	4.6	2.0051	2.0048	1.485	1.610
R3	3.0	2.0051	2.0048	1.485	1.620

The parameter "a" characterizing the constant of a hyperfine structure (a_N) was calculated as the distance between the first and the second components of the triplet in the radical ESR spectrum. The parameter "f" has been calculated by the formula:

$$f = I_{RH^+}/(I_{RH^+} + I_R),$$ (1)

where I_{RH^+} and I_R are peak intensities of the high field components (the third component) of the ESR signals of RH⁺ and R radicals forms respectively. The dependences of "a" on pH-value in solution at ionic strengthes μ=0.145 and 1 mol/l and "f" on pH-value at μ=0.05 mol/l are illustrated in Figure 1. The ESR spectra in the X-wave band were recorded by the ESR spectrometer RE-1307. The potentiometric titration of cationites has been carried out by the method of a number of samples[5] at the relationship solid phase (g) - solution (ml) 1:100 and μ=0.145, 1 and 3.5 mol/l for cationite KB-2x3 and μ=0.05 mol/l for cationite KB-51. After establishing equilibrium, the measurements of equilibrium solutions pH-value were determinated with the use of the pH-metre EW-74 (the accuracy in determination is 0.05) and by use of the graduation curves (Figure 1). Experiments to determine the acidity in the interior of an ionite grain was carried out as follows. The cationite grains were blotted and were transferred to the tube for measuring and pH-values were determined by ESR spectra of the NR. In the case of the NR R2 and R3, the washing of the cationite grains with a small amount of water led to a complete washing out of the NR, indicating the lack of binding probes with the sorbents. For the radical R1, washing out R1 was dependent on an equilibrium pH-value and this feature will be discussed below. The curves of the potentiometric titration of the cationites, using the pH-metre and pH-probe, are shown in Figure 2.

It was found that for the radicals R2 and R3 located in the phase of cationites the highfield component in the ESR spectrum was split into two components corresponding to the RH⁺- and the R-radical forms. For radical R1 the same splitting does not happen that, apparently, is

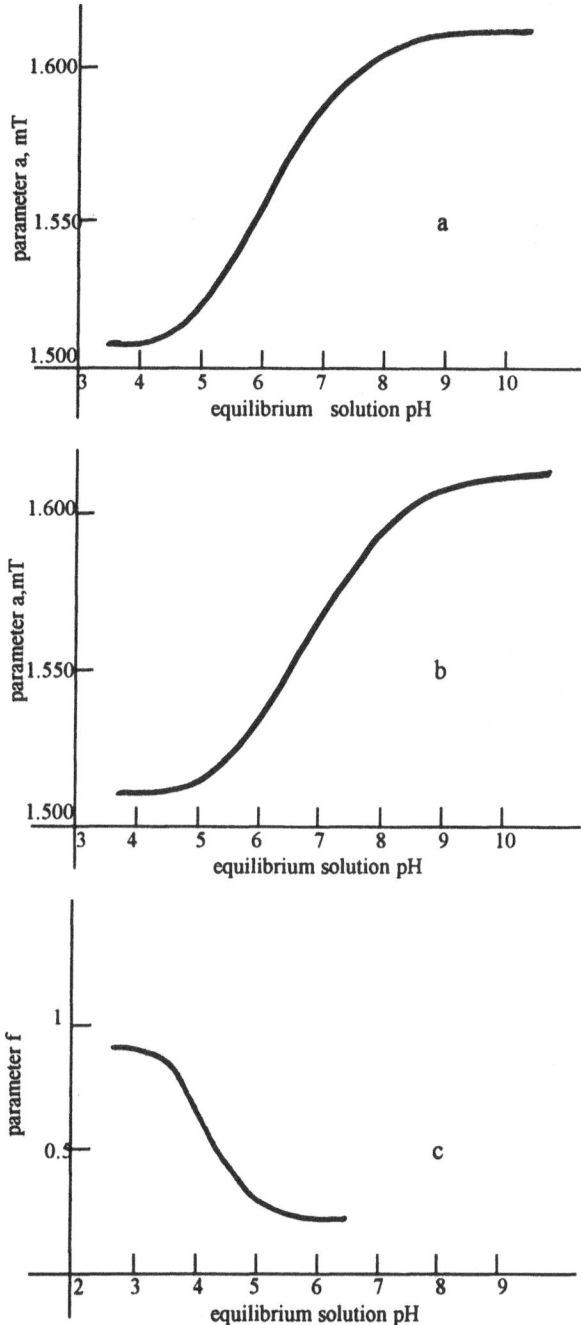

Figure 1. The graduation dependences of the parameter "a" of the radical R1 ESR spectrum on equilibrium solution pH-value at μ=0.145 (a) and μ=1 (b) and of the parameter "f" of radical R2 on equilibrium solution pH at μ=0.05 (c).

connected with transition from slow (in solution) to fast (in the interior of ionite grain) $R \rightleftharpoons RH^+$ exhange.[6] Therefore, when the principal use of the NR R2 and R3 are in research of cationite KB-51, we used the graduation relationship f = f(pH), and for the radical R1 used for research

397

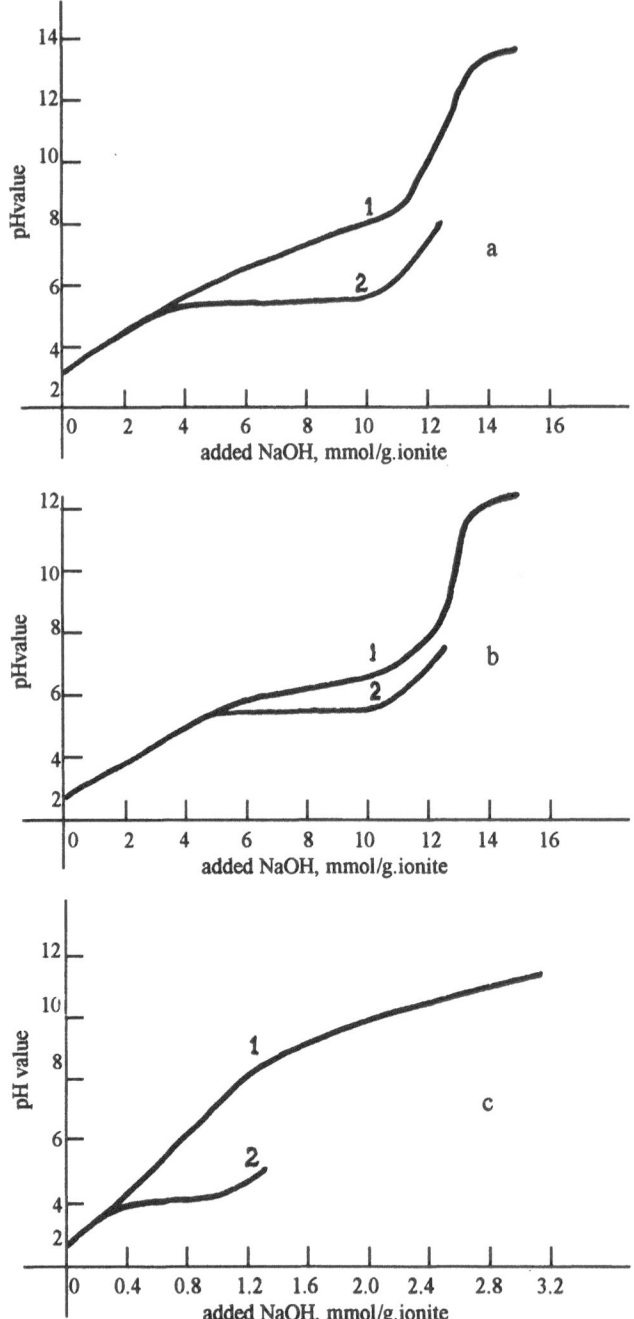

Figure 2. The curves of potentiometric titration of cationites KB-2x3 at μ=0.145 (a), 1 (b) and KB-51 at μ=0.05 (c) on evidence of pH-metre (1) and pH-probe (2).

of KB-2x3 we used the relationship a = f(pH). During the experiments to determine the acidity in the interior of the ionite grains, with the use of the pH probes, the differences of pH-values of equilibrium solutions over an ionite and in the interior of its grains have been revealed only in the range of functional groups titration of the sorbents. In this range, the lower pH-values in the phase of an ion-exchanger has been determined by us than for the corresponding equilibrium solutions (Figure 2). At the same time, the identity of the pH-metre readings in the equilibrium solution and the data pH-probe in the interior of ionite grains has been pointed out in high-acidic mediumes for the carboxylic cationites. The given identity was also noted everywhere over the range of pH-values being accessible to research with the use of the available NR for the high-acidic cationite KU-2 in the Na-form and porous chloromethylated copolymer of styrene with divinylbenzene (ion-exchange groups are lacking). The pKa-values of the carboxylic acid groups on the cationites calculated by the modified Genderson-Gasselbah equation (2), using the pH-values determined for the equilibrium solution and from the pH-probes, see Table 2.

$$pKa = pH - m \cdot lg[\ a\ /\ (1 - a)] \qquad (2)$$

The pKa- and m-values obtained with the use of equation (2) are in good agreement with the earlier determined values.[7,8] The effect of reducing the pKa by 1, with an increase of the salt

Table 2. The acid-base characteristics of used ionites

Ionite	Exchange capacity mmol/g	μ	m^1	pKa^1	$pKa \pm 0.1^2$
KB-51	1.12	0.05	2.00	5.1	4.0
		0.145	2.67	6.2	5.1
KB-2x3	13.5	1	2.25	5.3	5.2
		3.5	1.93	5.1	5.1

[1]obtained with the help of calculation by the equation (2)
[2]determined on evidence of pH-probes.

concentration in the external solution in 10, are also well known.[5] At the same time, the calculated values are cardinally different from the determined ones with the help of the pH-probes (the coincidence of pKa for KB-2x3 at μ=3.5 is accidental). The ionization constants of the carboxylic acid groups turned out to be in fact substantially beyond the determined ones according to the data from the potentiometric curves and the tendency of increase of their values with increase of ionic strength is lacking. The lack of dependence of ionic strength has been pointed also for the calculated[9] thermodynamic (true) ionization constants of carboxylic groups of methylacrylate cationite Amberlite IRC-50. The calculation was based on the development of Michaeli's and Kachalsky's ideas[10] and was carried out with regard to the Donnan's distribution and changes of system energy at the expense of the electrostatic interaction of the dissociated functional groups revealing itself in conformation reconstruction of polymeric chains. It gave pKa-values on the order of 5.1 units pH, close to the values obtained by us. However, some experiments are extremely labor-intensive and the calculations involve a number of arbitrary suppositions. On the contrary, our measurements use the standard procedure of potentiometric titration and provide the direct pKa determination of functional groups. The horizontal plot on the potentiometric curves obtained with the use of pH-probes has attracted our attention (Figure

2). These curves resemble the curves for monomers titration. They demonstrate the lack of influence of the dissociated part of polymeric chains functional groups in the cross-linked polyelectrolytes on the acid-base characteristics of groups which remained non dissociated. The shape of the curves for potentiometric titration of ionites, by which the calculation leads to dependence of pKa on dissociation extent, is explained exactly by the appearent chain charge. As a result, the parameter m being different from 1 appears in the Genderson - Gasselbah equation (2). The previously mentioned dependence of washing out of the radical R1 from an ionite on the pH-value of the equilibrium solution is illustrated in Figure 3. For values in excess of the ionization pKa of R1 and the ionite carboxylic groups, the washing out proceeds without hindrance (Figure 3, curves 1 and 2). The washing out of radicals is seriously difficult in the event that the radical exists in the RH$^+$-form, and the carboxylic groups are in the unprotonated form (Figure 3, curves 3 and 4). This effect is connected with electrostatic interaction between RCOO$^-$ groups and the protonated form of a radical. In this case the limitation of mobility of

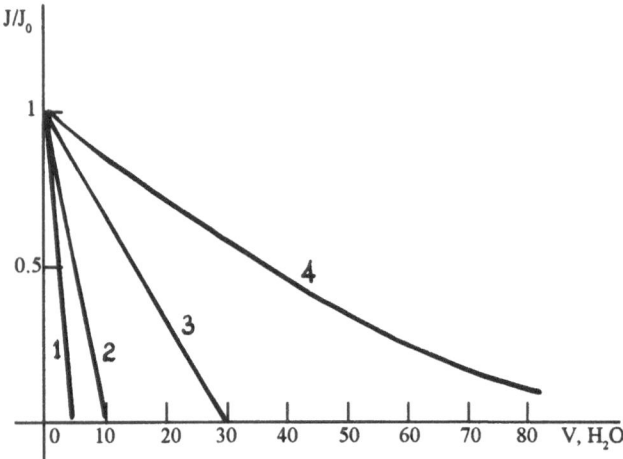

Figure 3. The dependence of relative intensity of the radical R1 ESR signal in the interior of the ionite grain ($\mu=1$) on water amount leaked through the ionite KB-2x3, where J is the intensity of the radical R1 ESR signal in phase of ionite with washing out by H$_2$O; J$_0$ is the starting intensity of radical R1 ESR signal in the phase of ionite. 1 - pH 8.65, 2 - pH 7.15, 3 - pH 6.05, 4 - pH 5.3.

the radical in the ionite phase also reveals in the spreading highfield component in the ESR spectrum showing superposition of the signals from the R- and RH$^+$-forms. However, the sorption interaction of cationite functional groups with radicals has no influence on the parameter "a" that allows determination to be made of the pH-values on the ionite interior. Thus, the offered method is rather simple, allowing direct determination of pH-value in the interior of discribed ionite grain and pKa of its functional groups. In this case there is no need to insert the empiric coefficient m in equation (2).

Conclusions

The use of the pH-sensitive radicals opens up new possibilites for studing acid-base properties of cross-linked polyelectrolytes (and polycomplexonates on their base) and for critical examination of concept previously formed in this field. But, narrowness (on the order of 1.5 units pH) of sensitivity for each of the nitroxyl radicals used and the necessity to have a large set of pH-probes with different pKa-values is a disadvantage for the described method.

Acknowledgments

The authors of the paper are indebted to B.K.Radionov for supply of the research samples.

References

1. V.V.Mank, O.D.Kurilenko. "Investigation of Intermolecular Interaction in Ion-exchange Resins by NMR Method," Naukova Dumka, Kiev (1976).
2. G.Zundel. "Hydration and Intermolecular Interaction," Academic Press, N.Y (1969).
3. J.F.W.Keana, M.J.Acarregui, S.L.M.Boyle, 2,2-Disubstituted-4,4-dimethylimidazolidinyl-3-oxy nitroxides: indicators of aqueous acidity through variation of a_N with pH, J.Am.Chem.Soc. 104:827 (1982).
4. V.V.Khramtsov, L.M.Wainer, I.A.Grigor'ev, L.B.Volodarsky, Proton exchange in stable nitroxyl radicals, Chem. Phys. Lett. 91:69 (1982).
5. N.J.Poliansky, G.B.Gorbunov, N.L.Polianskaia. "The Method of Investigations of Ion-exchangers," Chemistry, Moscow (1976).
6. V.V.Khramtsov. "Proton Exchange in Nitroxyl Radicals," Abstract of candidate's thesis. Novosibirsk (1985).
7. H.P.Gregor, L.B.Luttinger, E.Loeble, Metal-polyelectrolyte complexes, J. Phys. Chem. 59:34,366 (1955).
8. G.S.Libinson. "Physico-chemical Properties of Carboxylic Cationites," Nauka, Moscow (1969).
9. A.Chatterjee, J.A.Marinsky, Dissociation of methacrylic acid resins, J. Phys. Chem 67:41 (1963).
10. I.Michaeli, A.Katchalsky, Potentiometric titration of polyelectrolyte gels, J. Polymer Sci. 23:683 (1957).

MOLECULAR COMPLEXES OF CELLULOSE WITH METALS NEW DEVELOPMENT

Nina E.Kotelnikova,[1] Deitrich Fengel,[2] and Valery P.Kotelnikov[3]

[1]Institute of Macromolecular Compounds, Russian Academy of Sciences, St.Petersburg 199004, Russia

[2]Institute for Wood Research, University of Munich, Munich 80797, Germany

[3]"THESA" Co., Ltd, St.Petersburg 197198, Russia

INTRODUCTION

The existence of periodical substructures in cellulose has been shown in many papers[1-4]. In order to confirm this, various chemical methods were applied. In combination with advanced experimental physical procedures, particularly with electron microscopy, these methods give visualization of structural elements. Accoring to some researches, these periodical structures are formed by aggregates of cellulose molecules 30-40 A in width and with the length depending on the degree of polymerization of the sample. The mode of torsion of the macromolecules, i.e. the mode of steric arrangement of the macromolecules with respect to each other, is of importance and can vary. This determines distinction in cellulose hydrogen bonding system and, hence, its reactivity.

The reactivity of cellulose during its derivatization is usually estimated by chemical methods according to the conversion of the initial sample and to the degree of substitution of the resulting products. However, all heterogeneous chemical reactions on the fibre surface start on the morphological level and, therefore, it is the morphological fibril structure of cellulose that is the first object for attack by chemical reagents.

The observation of morphological changes on the surface of cellulose fibrils, which occur during chemical reactions, carried out by scanning electron microscope (SEM), yields valuable but mostly qualitative data. It is also known that metal ions bound to a biological system may have a structural role[5]. To our knowledge, there are no literature data confirming this fact with respect to such a biological polymer as cellulose. The aim of this paper is to clarify cellulose morphological structural elements by inserting metal ions into cellulose. These metal ions may be located between the supramolecular elements and in this way may correspond to the arrangement of periodical structures in cellulose. In this respect the present study may be regarded as fundamental. On the other hand metal-containing cellulose complexes can be applied as basic substances for the preparation of advanced new materials.

This paper concerned with the SEM (combined with EDXA) observation of modified cellulose samples containing metal ions, such as, lithium, sodium, and potassium.

EXPERIMENTAL

Microcrystalline cellulose (MCC) was used as starting material. Modified MCC samples were prepared under conditions described elsewhere[6,7]. It was established that the new type of cellulose metallic derivatives containing Li, Na, and P ions were obtained.(They will be called below MCC - Li, MCC - Na, and MCC - P). Metal ions were inserted into MCC in a complex with the solvent. In the present study the samples prepared in the ethylene diamine medium were examined.

The same samples were subjected to a repeated and thorough treatment with distilled water followed by vacuum drying at 40°C. This set of treatment was performed to destroy the chemical bonds between MCC and the metal ions bound in a complex with the solvent, to remove this complex, and in this way to isolate the regenerated initial sample. After this treatment the regenerated samples did not contain any solvent and one of them contained a few amount of potassium (Table).

Table. Effect of water treatment on the metal content in MCC samples

MCC samples containing metal ions	Metal content (wt %)	
	Before treatment	After treatment
MCC - Li	10.6	0
MCC - Na	12.3	0
MCC - P	13.2	1.7

The SEM used for observation was a Leitz AMR 1200 B. The samples were sputtered with gold-palladium.

RESULTS AND DISCUSSION

Figures 1 - 4 present the electron micrographs of the initial MCC sample (1), MCC complexes containing metal ions (2a, 3a, 4a) and the same samples after their treatment with water (2b, 3b, 4b). It is seen that MCC complexes containing metal ions show considerable changes in fiber morphological structure compared to the original MCC. There are only a few amount of fibrils at the surface of the fibers. The diameter of the fibers increases 1.5-2 times. This implies that reagents penetrate into the fiber from its ends. Selective local surface swelling in fiber parts regularly distributed along its length is also seen. The metal distribution along the fibers is completely uniform as indicated by X-ray analysis (EDXA).

The same samples after their treatment with water reveal the totally destroyed morphological fiber structure. Moreover, the deep cracks are formed along the entire fibre length and they are perpendicularly distributed to the fiber axis. Their arrangement is equally uniform for all three types of these complexes.

The uniform distribution of these cracks along the fiber length after the removal of metal ions probably indicates that they had been uniformly distributed in the fiber before the removal. The approximate quantitative evaluation of the distance between cracks gives the following data: about 1 - 3 mcm for the complexes MCC - Li and MCC - Na, and about 3 mcm for the complexes MCC - P. These facts indicate that:
- metal ions implanted into cellulose are uniformly distributed and play the structural role;
- the removal of metal ions leads to the visualization of morphological structural elements in cellulose;
- the fiber consists of structural elements formed by a helical twisted fibrillar ribbon.

Within the ribbon fibrillar bundles are located with a length of about 50 mcm and from 1 to 8 mcm in diameter. This fact implies that each fibril is about 0.2 - 0.3 mcm in diameter, which

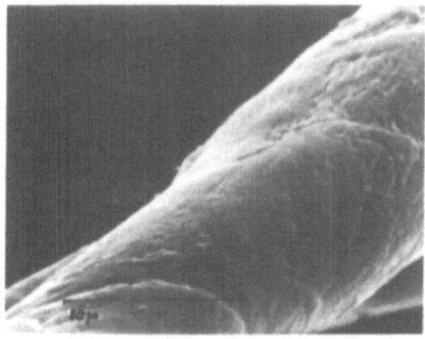

Figure 1. SEM micrographs of untreated MCC sample.

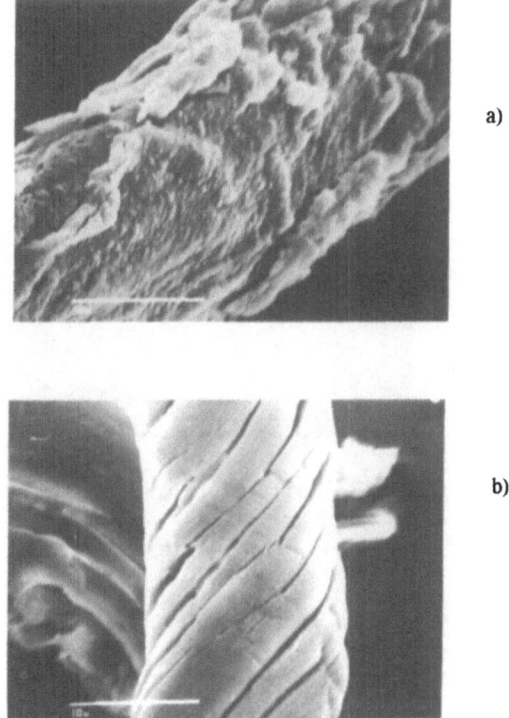

a)

b)

Figure 2. SEM micrographs of MCC - Li fiber (a) and of the same fiber after treatment with water (b).

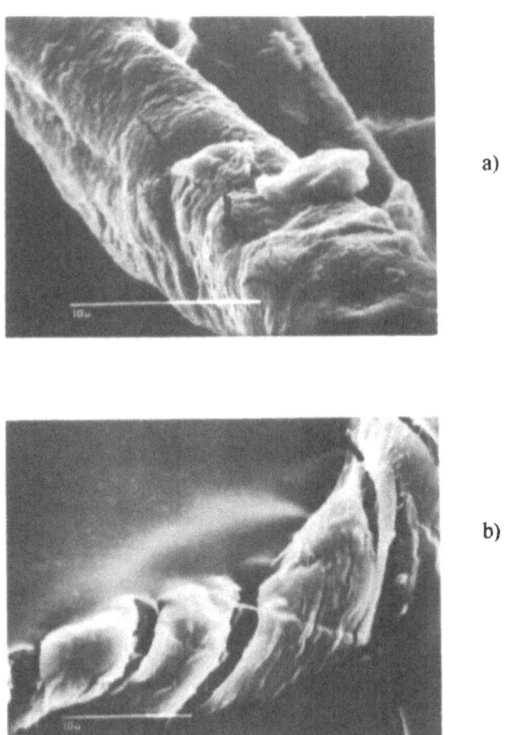

a)

b)

Figure 3. SEM micrographs of MCC - Na fiber (a) and of the same fiber after treatment with water (b).

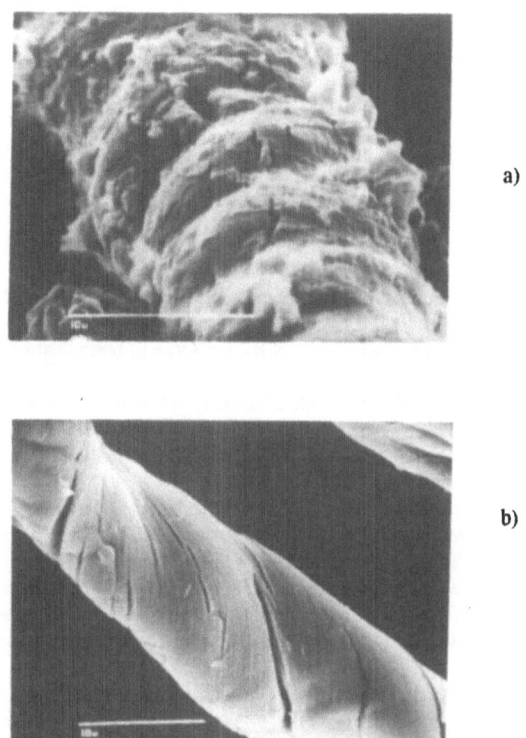

a)

b)

Figure 4. SEM micrographs of MCC - P fiber (a) and of the same fiber after treatment with water (b).

corresponds to the amount of 10 microfibrils or 30 - 60 elementary fibrils (or to the same number of crystallites).

Consequently, it became possible to show using chemical methods that it is the morphological fibrillar structure that mainly undergoes the attack by metals ions in the complex with the solvents. The bonds between morphological structural elements on the cellulose fibril surface are the most labile because they react most readily and are subsequently broken forming regular cracks parallel to fibril arrangement. The relatively regular periodicity of the arrangement of these bonds indicates that the primary morphological macro - level of celulose fiber is periodical. On passing to the micro - level the periodicity of the arrangement of structural elements (elementary fibrils and microfibrils) is known to be retained. Thus, periodicity is the specific feature of the cellulose structure on the micro - level as well as on the macro - level.

ACKNOWLEDGEMENT

The financial support of Deutsche Forschugsgemeinschaft is appreciated.

The authors thank Dr. Manfred Stoll for excellent technical assistance and for fruitful discussion.

REFERENCES

1. M. Stoll and D. Fengel, Electron microscopic visualization of individual cellulose molecules, Holzforschung, 43:7 (1989).
2. D. Fengel and M. Stoll, On the probability of the existence of periodical substructures in cellulose, Macromol. Chem., 190:2491 (1989)
3. J. Sugiyama, H. Chanzy, and G. Maret, Orientation of cellulose microcrystals by strong magnetic fields, Macromolecules, 25:4232 (1992).
4. S. Mizuta and R.M.Brown, High resolution analysis of the formation of cellulose synthesizing complexes in Vaucheria hamata, Protoplasma, 166:187 (1992).
5. J. Reedijk, Metal coordination by natural macromolecules, in: Abstracts of the 5th International Symposium on Macromolecule-Metal Complexes (MMC V), Bremen, Germany (1993).
6. N.E. Kotelnikova, D. Fengel, V.P. Kotelnikov, M. Stoll, New metallic derivatives of cellulose. Compounds of microcrystalline cellulose and potassium, J. Inorg. Organomet. Polymers, 4:315 (1994).
7. N.E.Kotelnikova, V.P. Kotelnikov, New molecular complexes of cellulose with metals, in: Abstracts of Symposium on Cellulose and Lignocellulosics Chemistry, Guangzhou, China (1991).

ELECTRON IMPACT MASS SPECTROSCOPY OF CONDENSATION METAL-CONTAINING POLYMERS

Charles E. Carraher, Jr.,
J.William Louda, Dorothy Sterling,
Earl Baker, Alberto Rivalta
and Qingmao Zhang

Department of Chemistry
Florida Atlantic University
Boca Raton, FL. 33431

INTRODUCTION

Numerous advances have been made recently in mass spectrometry. Many of these advances are related to instrumentation. Other advances are simply related to varying the conditions under which the mass spectra are obtained. Electron impact mass spectroscopy, EI-MS, is the most widely utilized mass spectrometry employed in the study of polymers and it is the emphasis of this presentation.

Fragmentation of polymeric materials occurs through at least two major routes-electron bombardment and thermally "encouraged" bond breakage. The former is controlled through variation in the energy, intensity and conditions of electron bombardment associated within the mass spectrophotometer. The latter can be controlled external (e.g. TG-EIMS) to the spectrophotometer or as part of the mass spectrophotometer. Studies carried out utilizing both internal and external sources of thermal agitation will be described.

Following is a brief description of some of the different instrumentation that exists today.

Instrumentation

ION-TRAP MS-Ion trap MS (IT-MS or FT-MS) is a technique that combines mass analysis/ion-optics and detection. While some variation exists among IT instruments, analysis and detection occurs when ions, held (trapped) in a magnetic field, absorb energy at their circulating (cyclotron) resonance frequency. The mass spectrum is a scan of rf from which the absorption of energy is processed by Fourier Transformation

The vast majority of mass analyzers are of the "dispersive" variety where the actual sorting or dispersion of ionic species occurs and is recorded. This "separation" or dispersion can be effected through variations in charge, kinetic energy, centrifugal force, frequency of a particular wave-form and combinations of the above. Following is a brief discussion of some of the dispersive mass analyzers.

Sector Mass Analyzers-utilize an electrostatic or magnetic field or both (dual sector or double focusing) to alter the preferred straight path of moving particles into a somewhat circular pathway. A magnetic (or electrostatic) sector separates and focuses ions through variation of the magnetic field strength imposed about a curved flight tube.

The combined effects of repeller, accelerator, source exit slit, electrostatic potentials and magnetic field strength are called the "ion-optics" of the sector instruments. There is typically an inverse relationship between instrument sensitivity and resolution because resolution increases result through decreases in the "widths" of the focal "windows". Resolution also decreases with increase in scan speed and cycling. This restriction with scan speed and cycling is a major reason that sector instruments are being used less with GC and LC combinations.

Quadrupole Analyzers-employ four poles, hence their name. These "poles" are rods aligned in parallel with the vector path of ions accelerated from the source towards the detector. Opposing poles form two pairs, one of DC and one of RF voltage. Ions accelerated from the source towards the detector traverse the core of this quadrupole arrangement. Application of DC and RF voltage to the pole pairs causes the ions to acquire osculations in response to the attractive or repulsive forces set up by the sign and strength of these fields. Resonance occurs when a certain m/z (or m/e) value attains a stable oscillation within the field core and is thus able to pass through the mass filter. Only the resonant ion will be "in focus' and all others, lower and higher masses, will impinge one of the charged rods. Scanning is typically accomplished by varying both the DC and RF voltages while keeping the ratio between the two constant.

Time-Of-Flight Analyzers (TOF)- In conventional dispersive and IT mass spectrometers, ions are, in essence, continually being formed and accelerated into the mass analyzer. However, by breaking this continuity of ion flux either by electronic "gateing" at the source or by temporarily spacing ionization events one can obtain an exact time of flight initiation. Once this is achieved, the precepts of kinetic energy relationships can be applied. This yields a mass filter wherein a given time-of-flight through a field free region (source to detector) will equate to an ion of only one m/z value. Thus, due to source potentials, all ions of a given charge and acceleration within a given potential will be of the same energy and will exhibit a dispersion of velocities inversely related to their masses. Simply stated, heavier ions travel slower than lighter ions of like kinetic energy.

SAMPLE INTRODUCTION

A wide variety of procedures that have been utilized to introduce the sample to be analyzed into the mass spectrophotometer. Following is a brief discussion of some of these.

Batch or Gas Sample Introduction-The most simple method for the introduction of a sample into a mass spectrometer is through controlled "bleed" of a substance in the gas phase. This is a method of choice for "pure" materials which are gases at relatively low temperatures in moderate vacuum (T<150°, P=10.4$_{torr}$) and is generally used for the introduction of helium and mass marker compounds. Helium is often used for "leak checking" the instrument. Thus helium is introduced into the system and the instrument "tuned" precisely on the peak signal at 4.00260 m/z. The helium supply is turned off and the instrument's vacuum system activated and joints and seals tested while monitoring the level of response at 4 m/z for leaks.

Many of the mass markers are perfluoro-organics. These are held in a reservoir or expansion volume within the batch manifold. Again, the instrument is "tuned" to known masses that are derived from the mass marker materials.

Batch-type systems also allow for the introduction of mass standards concurrent with the "run" of an unknown increasing the resolution of the mass spectrum.

Gas Chromatography Introduction-GC-MS is a very common instrumentation for the analysis of liquids and gasses and materials that can be converted into gasses through moderate heating. Thus, the major drawback to GC and GC-MS is the requirement that the sample be volatile at temperatures below which neither the liquid phase, typically silicone based, of the column nor the sample exhibits thermally induced decomposition. The first constraint, the upper thermal limit of the GC column, is constantly being expanded upwards and ranges of 375 to 450 are possible. The procedure makes use of the enormous separating power of GC.

Liquid Chromatography Introduction-is becoming a "routine" operation. During LC, the solvent (mobile phase) is a liquid and when introduced and vaporized into the mass spectrometer, along with the dissolved materials to be tested, creates an extremely high gas load on the system's vacuum system. Numerous techniques for the direct coupling of LC to the mass spectrometer have been used. Two of the most common are the "moving belt" or "transport interface" and "thermospray".

Direct Insertion Probe (DIP) Techniques- includes the use of thermogravimetric coupled MS (TG-MS), pyroprobe MS and programmable pyroprobe MS. Most commercial MS instruments can be equipped (or come already equipped) with a direct insertion probe. Typically, these probes can be considered as "heating tubes" where the sample is placed and the temperature raised to produce volatile ions that are "distilled" into the MS where analysis occurs.

IONIZATION TECHNIQUES

There also exists a wide variety of ionization techniques. The following is a brief description:

Electron Impact (ei or EI)-The most common ionization technique is electron impact (ei or EI). During ei, the gas phase sample encounters a beam of electrons emitted from an energized filament. The number and variety of ions generated are dependant on the ionizing energy employed. Many instruments are set-up to operate in the 70 eV range. In essence, the reaction which occurs during ei is the simple removal or addition of electron (s) from the target species, due to the impact of electrons in the beam giving positively or negatively charged ions. Most standard instruments "analyze" only the positive ions.

Chemical Ionization (ci or CI-is a variation of ei-MS where a reagent gas (such as water, methane, hydrogen, ammonia) is present in the source in large excess in comparison with the test material. This excess ensures that direct ei of the sample is rare. The reagent gas is ionized and undergoes ion-molecule collision with either another reagent gas or with the sample. Once the reagent gas has been ionized and its reactive species generated, further ion-molecule collisions involve sample molecules eventually generating sample-containing ions for MS analysis. The process is often referred to as being "soft" since fragmentation of the test materials is generally not as great as in the case of ei-MS.

Secondary Ion Mass Spectroscopy (SIMS)-Desorption-ionization during SIMS experiments results from the impact of highly energetic ions (such as Ar+, Cs+ and Xe+) into the sample which has been thinly coated onto a metallic (generally silver) direct exposure probe. Ionization mechanisms have been characterized as including cationization (in which the sample reacts with a charged metal atom (Ag+)) of the support and yields a desorbed organometallic (MAg+, electron transfer giving positive and negative molecular ion-radicals (M+, M-) and momentum transfer from the beam ions to sample salts giving ionized forms (such as MNa+). SIMS spectra can be obtained where the dominant ions are either molecular/"quasi-molecular" or fragments depending upon the energy and flux of the employed ion beam.

Fast Atom Bombardment (FAB)-Like SIMS, FAB uses inert gasses (Xe, Ar) which are ionized and accelerated from an ion-gun providing high energy particles for desorption and ionization. However, in FAB, these accelerated ions (Xe+, Ar+) are partly neutralized by electron capture, generating the Fast Atoms used for Bombardment of the sample. In FAB, the sample (with in a non-volatile liquid matrix such as glycerol, triethanolamine and 4-octyl aniline) is bombarded. Ionization, aided by the liquid matrix, and desorption occur when the fast-atom impinges the target material. Fresh sample reaches the surface by diffusion and the experiment is thereby extended in time.

Laser Desorption (LD)-LD-MS employs high energy, usually pulsed, laser radiation for the desorption and ionization of the target material. The laser light (photons) hits the target and the photon energy becomes converted to kinetic/thermal energy giving spontaneous vaporization (desorption) and ionization. Laser sampling allows a materials to be "undressed" and analyzed layer-by-layer.

Other ionization techniques include Plasma Desorption, Photoionization, Desorption Ionization and Field Desorption.

EXPERIMENTAL

Sampling techniques were chosen that would encourage larger fragments of the polymer to become analyzed. Samples were ground to a fine powder for MS analysis.

High resolution electron impact positive ion mass spectral analyses were carried out at the Midwest Center for Mass Spectroscopy, Lincoln Nebraska employing a Kratos MS-50 Mass Spectrometer. EI-MS was also conducted using a modified Dupont 21-491 and a modified Dupont 21-496 Mass Spectrometer. A direct insertion probe was typically employed unless otherwise noted. Rapid heating was accomplished using a direct insertion probe.

RESULTS AND DISCUSSION

The metal-containing polymeric materials studied in the current work can be described with the metal being part of the polymer backbone

        B
-(-A-M-)- [where M represent the metal atom and A and
        B

B represent non-metal containing (typically organic) moieties]

and where the metal is attached to an already formed polymer without acting as a cross linking agent

    -(A-B-)-
     M

or where the metal-containing moiety can act as a cross linking agent.

    -(A-B-)-
     M
    -(A-B-)-

Here we will not distinguish between coordination and condensation polymers but will not emphasize metallocene-containing materials. Following is a discussion of general results as a function of some of the important instrument variables employing EIMS.

Heating Rate

Many metal-containing materials generate few ion fragments if the heating rate is low (such as 1° C/min). While heating rates in the region of 20° C/min permit the generation of sufficient ion fragments to allow decent mass spectra, such moderate heating rates are generally insufficient to create a sizable number of metal-containing ion fragments. Rapid heating rates (ballistic heating) typically allow both the generation of sufficient "more massive" fragments but also encourages the formation of metal-containing fragments.

Through studying weight losses using TGA (for instance 1), it is obvious that the evolution of volatiles as a function of time generally decreases as the heating rate decreases. Metal-containing moieties can encourage this trend through formation of thermally induced and more thermally stable bonds as the temperature becomes sufficient for bond disruption to occur. In some special cases the metal-containing moiety acts as a combustion catalyst where thermal degradation can occur dramatically for even slow heating rates (2).

Temperature

To the 1980's it was customary to heat samples to only 150 to 250 C
*because most EI-MS were obtained on smaller molecules that were volatile in this range
*because of the limited thermal stability of the employed chromatography assemblies connected to the MS and
*because most of the commercially important and available polymers decomposed within this range giving sufficient spectra to allow for sample study and identification.

With the introduction of direct insertion probes and programmable heating stages and the interest in obtaining MS of nonvolatile and more thermally stable polymeric materials and in attempts to generate more massive ion fragments, sample heating rates and temperatures have increased.

Many of the metal-containing reactants are volatile using moderate heating rates and temperatures. Thus, good mass spectra with metal containing ion fragments are produced for tributyltin chloride and dibutyltin dichloride (heating to 200 C using moderate heating rates). For the tributyltin chloride ion fragments identified as being Sn-Bu, Sn-Cl, Bu-Sn-Cl and Bu-Sn-Bu Cl are found in abundance. For the products of tributyltin chloride and lignin and the products from dibutyltin dichloride and lignin, no metal-containing ion fragments are found when heating to 200 C using either rapid or moderate heating rates (3). In fact, for the products rapid heating to about 300 C and above is required to produce a number of tin-containing ion fragments.

This difference in the conditions needed to generate metal containing ion fragments was employed as a method to determine the purity of the lignin product. Thus, for the product formed from reaction of lignin with tributyltin chloride, tin-containing ion fragments were found appearing in good concentration at about 200 C (Table 1). As the product was purified through washing with different organic liquids, the amount of tin-containing ion fragments obtained at 200 C decreased until they disappeared (Table 2). The purity of the final product was verified through Mossbauer Spectroscopy (4). For comparison, the purified product, when rapidly heated to about 450 C gives numerous tin-containing ion fragments (for instance Table 3).

Table 1.   Relative abundances of the major ion fragments for the "uncleaned" product of lignin and tributyltin chloride heated to 200°C.

m/e	%-Rel.	Assignment	m/e	% Relative	Assignment
55	62	Bu	173	14	
56	43	Bu	175	33	SnBu
57	82	Bu	177	41	
60	44	L	209	23	
69	42	L	211	36	
71	36	L	212	45	BuSnCl
73	22	L	213	22	
81	20	L	215	20	
83	20	L	266	49	
100	36	L	267	30	
118	20	Sn	268	82	
120	22		269	56	
151	48	L	270	100	Bu$_2$SnCl
152	36		271	21	
153	24	SnCl	272	53	
155	43		274	23	
157	14				

Table 2.   Relative abundances of the major ion fragments from the products of lignin and tributyltin chloride heated to 200°C.

m/e	Bu₃Sn-Lignin (%-Rel. Abundance)	Assignment
53	70	
55	72	Bu
56	60	Bu
57	100	Bu
62	40	L
77	15	L
81	15	L
91	15	L
100	13	L
102	16	L
122	12	L
123	20	L
124	18	L
137	82	L
138	12	L
151	22	L

L = Lignin derived; Bu = Butyl derived

Table 3.   Examples of lignin associated ion fragments derived from lignin-tributyltin chloride with heating to 450°C.

m/e	Form of Tin
209-213, 242, 270, 271, 287	Sn
267-269, 288-290, 345	SnBu
399	SnBu₂
453	SnBu₃

Ion fragment abundances vary with the EI-MS operating conditions. Since most of the standard books and computer banks are based on standard conditions where moderate heating rates and temperatures were employed to gather the data, mass spectra obtained under significantly differing conditions should give different relative ion abundances. Tables 4 and 5 contain matches for Cp as a function of temperature. At low to moderate temperatures, the values are in decent agreement with the standard that was obtained at about 200°C. At high temperatures the match is poor (Tables 4 and 5).

Table 4.   Relative abundances of the major ion fragments derived from
            cyclopentadiene

m/e	66	65	39	40	38	63	67	62	31
Product(temp)									
Standard(200°C.)	100	47	32	27	8	8	6	6	5
$Cp_2TiCl_2.HQ(200C°)$	100	45	34	33	6	8	6	6	7
$Cp_2TiCl_2,HMPA(400°C)$	100	50	a	a	a	b	45	b	a
$Tp_2ZrCl_2,HMPA(400°C)$	100	45	a	a	a	8	34	5	a
$Cp_2HFCl_2,HMPA(400°C)$	100	45	a	a	a	b	61	b	a

HMPA = Hematoporphyrin IX; HQ = Hydroquinone
a. Recorded starting at m/e = 50.
b. Below recorded limit.

Table 5.   Relative abundances for the major ion fragments derived from
            cyclopentadiene for the condensation products of Acid Red 88 and
            Group IV B metallocene dichlorides.

m/e metallocene (Temp)	66	65	39	40	38	63	67	62
Standard (200°C.)	100	47	32	27	8	8	6	6
$Cp_2Ti(400°C.)$	65	29	a	a	a	b	100	10
$Cp_2Zr(200°C.)$	100	38	28	25	-	1	8	7
$Cp_2Zr(400°C.)$	31	22	a	a	a	b	100	b
$Cp_2Hf(400°C.)$	100	43	12	18	-	27	21	15

a.   Recorded starting at m/e = 50.
b.   Below recorded limit

Affect on Nature of "A" and "B"

"A" and "B" represent the organic portion of the organometallic polymer. As noted before, bond scission occurs because of at least two major factors-heat and electron impact. Major fragmentation sites typically result from either susceptibility to electron impact or thermal instability. As the heating rate and temperature increase, differences in the thermal stability of particular sites become less important because the available energy is sufficient to break most of the bonds and the rapid heating rate will decrease the importance of rearrangements. Thus, the presence of statistical abundances of ion fragments (Tables 5 and 6), the variety of ion fragments produced, and generally the number of ion fragments produced increases as the heating rate and temperature increases. (The highest temperature studied with rapid heating is about 650° C.; aka Curie-Point Pyrolysis = CuPy) The products from bis(2,2-bipyridine)ruthenium dichloride and 1,8-octanedithiol illustrates this. When heating the polymer using a moderate heating rate to about 200°C, ion fragments were obtained to only m/e= 193. Using rapid heating to 450°C gives ion fragments to about m/e=700 (5). Further, ion fragmentation patterns for A and B are temperature dependant.

Table 6 contains data for n-octane heated to 150 C. and the similar ion abundances for the product for bis(2,2-bipyridine)ruthenium dichloride and 1,8-octanedithiol. There is little

416

Table 6. Major ion fragments associated with n-octane (moderate heating rate to 150°C.) and ions derived from rapid heating to 450°C. of the product of bis(2,2-bipyridine)ruthenium dichloride and 1,8-octanedithiol.

m/e	57	43	45	41	97	55	58	44
n-octane	100[a]	28	22	8	8	7	7	2
Product	100[a]	0	0	0	17	86	0	0

a. % Relative abundance

Table 7. Fragmentation patterns of moieties of selected organometallic polymers obtained at about 200°C.

Phenyl$_2$ SnCl$_2$ and Cellulose (for Benzene)

m/e	78	51	52	50	39	77	79	76	74
Standard	100	21	20	18	14	14	7	6	5
Product	100	21	23	20	23	19	7	6	6

Py$_2$MnCl$_2$ and Terephthalic Acid (for Pyridine)

m/e	79	52	51	50	78	39
Standard	100	70	36	26	12	12
Product	100	70	36	26	12	12

Py$_2$MnCl$_2$ and Terephthalic Acid (for Benzene)

m/e	78	51	52	50	39	77	79	76	74
Standard	100	21	20	18	14	14	7	6	5
Product	100	20	18	13	18	22	5	7	4

similarity between the two. For comparison, Table 7 contains results for polymeric materials heated to about 200 C. with the ion fragments of standard materials also heated to 150 to 200 C. Here, there is good agreement.

For vinyl polymers modified through reaction with metal-containing reactants, the ion fragmentation patterns for the unmodified polymers [such as poly(vinyl alcohol), poly(vinyl amine) and Poly(acrylic acid)] are very similar to those of the modified products, with the exception of ion fragments associated with the metal-containing moiety (for instance 6-8 and Tables 7 and 8). The hydrocarbon ion fragments will generally vary by only the absence of an appropriate number of hydrogens from the pattern given by linear polyethylene or from shorter chained hydrocarbons.

Table 8. Significant ion fragments derived from the condensation product of dibutyltin dichloride and poly(acrylic acid) (from m/e > 50)

m/e	Rel. Int.	Assignment
55	20	C_4H_x
56	8	
57	17	
60	10	
67	7	
69	13	C_5H_x
70	6	
71	8	
73	10	
81	6	
83	9	C_6H_x
97	5	C_7H_x
105	1	
109	2	
110	1	C_8H_x
111	2	
129	3	
241	1	$C_3H_{13}CO_2CHCH_2$
248	0.9	$BuSnCOCHCH_2$
256	1	$BuSnCO_2CHCHC$

Combination and rearrangement can occur within the solid product prior to volatilization. For instance, for the condensation product of dibenzyl dichloride and poly(ethylene-co-sodium acrylate) significant amounts of dibenzyl are formed during analysis (7,9). The amount of dibenzyl formed is in great excess to that predicted from simple gaseous collisions.

Isotopic Abundances

Many metals are composed of several isotopes that can assist in the identification of ion fragments. To be of use, the relative isotopic abundances should be present in relative abundances typically greater than 10 %. Table 9 contains a listing of common metals that met this requirement.

Table 9.   Listing of common metals composed of several isotopes (Source: Handbook of Chemistry and Physics).

Metal	Isotopes*
Mg	24, 25, 26
Ni	58, 60
Cu	63, 60
Zn	63, 65
Ga	69, 71
Ge	70, 72, 74
Rb	85, 87
Zr	90, 91, 92, 94
Mo	92, 95, 96, 98
Ru	99, 100, 101, 102, 104
Pd	105, 106, 108, 110
Ag	107, 109
Cd	110, 111, 112, 113, 114
Sn	116, 118, 120
Sb	121, 123
Ba	137, 138
Ce	140, 142
Hf	177, 178, 179, 180
W	182, 183, 184, 186
Pt	194, 195, 196
Hg	198, 199, 200, 201, 202
Pb	206, 207, 208

*Present in relative abundances at 10% and higher.

Table 10 lists an example of an ion abundance match for tin for the seven most abundance tin isotopes. In reality, tin has ten isotopes but three (112, 114 and 115) are present in significantly small relative abundances (1% and less) and are not particularly useful. While a reasonable rule-of-thumb is to only expect reasonable matches for abundant isotopes present in relative abundances of ten and higher, Table 10 also contains matches for less abundant isotopes.

Table 10.   Relative abundance of $OSnC_2H_3$ associated ion fragments for the condensation product of poly(vinyl alcohol) and dimethyltin dichloride.

m/e	116	117	118	119	120	122	124
%-Natural Abundance	13	7.7	24	8.7	33	4.7	6.0
$OSnC_2H_3$ m/e	159	160	161	162	163	165	167
%-Relative Abundance	18	7.6	24	10	33	3.4	4.2

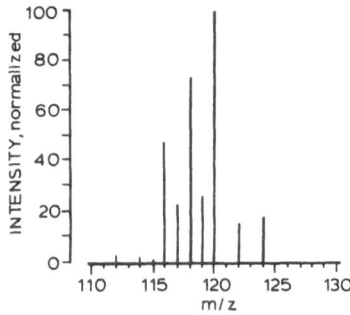

Figure 1. Normalized isotropic distribution for tin.

As expected, the more significant isotopes a metal has, the more "parcelled" out the related ion abundances will be and consequently, the more room for variance as matches are made.

Because some metals have several isotopes, it is at times useful to "draw" a scale that corresponds to the relative abundance of each of the isotopes and use this "picture" to allow you to "hunt" the mass spectrum for matches. Figure 1 contains such a "picture" for tin.

In addition to the use of metal isotopic matches, non-metals can also be composed of several isotopes that are useful in the identification of ion fragments containing them. Thus, the ratio of the relative isotopic abundance for $Cl(35)/Cl(37)$ is $3:1$. Ion fragments identical in composition, except containing one atom of chlorine, should be present in approximately a $3:1$ ratio. As an additional observation, most EI-MS instruments operate in a "positive ion" mode, rejecting negative charged ion fragments. Thus, the abundances of Cl are generally low since it is an atom that would rather form a negative ion than a positive ion. Thus, compounds, including HCl, will typically be more easily found if their relative abundances are even near that of Cl itself.

REFERENCES

1. C. Carraher, H. M. Molloy, T. Tiernan, M. Taylor and J. Schroeder, J. Macromol. Sci.-Chem., A16(1), 195 (1981).
2. J. Sheats, C. Carraher, D. Bruyer and M. Cole, Coatings and Plastics, 34(2), 474 (1974).
3. C. Carraher, D. Sterling, T. Ridgway, J. W. Louda, "Biotechnology and Polymers", Plenum, NY, 1991.
4. C. Carraher, D. Sterling, T. Ridgway, W. Reiff and B. Pandya, "Interpenetrating Polymer Networks" (D. Klempner, L. Sperling and L. Utracki, Eds.) American Chemical Society, Washington, DC, 1994.
5. C. Carraher and Q. Zhang, Polymer P., 35, 702 (1994).
6. C. Carraher, L. Reckleben and C. Butler, Polymeric Materials: Science and Engineering, 63, 704 (1990)
7. C. Carraher, F. He, D. Sterling, F. Nounou, R. Pennisi, J. W. Louda and L. Sperling, Polymer P., 31(2), 430 (1990).
8. C. Carraher, C. Butler and L. Reckleben, "Cosmetic and Pharmaceutical Applications of Polymers" (C. Gebelein, T. Cheng and V. Yang, Eds.), Plenum, NY, 1991.
9. C. Carraher, F. He, D. Sterling and J. W. Louda, unpublished results.

MODEL SYSTEMS AND STRUCTURE, FUNCTION AND REACTIVITY

RELATIONSHIPS IN TRANSITION METAL-CONTAINING BIOPOLYMERS

Geoffrey B. Jameson

Department of Chemistry and Biochemistry
Massey University
Palmerston North
New Zealand

INTRODUCTION

Metal complexes are at the heart of many biological processes. Remarkably few structural motifs are used by Nature to accomplish such diverse functions as oxidation and reduction, hydrolysis and condensation, transport of electrons and small molecules, and transformations of chemical energy into electrical and mechanical energy. One of the challenges of bioinorganic chemistry is to relate, for a particular structural motif such as a heme group, the structural features of not only the metal center but also the surroundings to the function and reactivity of the metalloprotein. While much effort has been devoted to relating the stereochemistry of the active site of metalloproteins to thermodynamic aspects of function, there is increasing interest in the stereochemical basis of kinetic aspects of protein function, such as rates of electron transfer, rates of ligand binding and reaction mechanisms for enzymes.

Three distinct solutions to the sequestration and transport of molecular oxygen (dioxygen, O_2) have evolved: the hemoglobin family in which an encapsulated heme group binds O_2 (and other small molecules); the hemocyanin family in which O_2 binds to a pair of copper ions; and the hemerythrin family in which O_2 binds to one of a pair of iron atoms. Note that hemocyanin and hemerythrin are non-heme metalloproteins: the *hem* prefix denotes in this context blood and not a porphyrinatoiron(II) species. These three distinct motifs appear in other metalloproteins that mediate a variety of other biological processes. Many aspects of bioinorganic chemistry have been comprehensively reviewed recently.[1] Following the Introduction, in the second part of this article, dioxygen (O_2) and carbon monoxide (CO) binding to myoglobin, its mutants and its models are discussed -- systems for which the basic inorganic chemistry is well characterized. Pertinent background to these first two parts are chapters on biological and synthetic dioxygen carriers,[2] and dioxygen reactions,[3] and recent reviews on synthetic heme dioxygen complexes[4] and the mechanism

Metal-Containing Polymeric Materials
Edited by C.U. Pittman, Jr., *et al.*, Plenum Press, New York, 1996

of ligand recognition in myoglobin.[5] Since the last review (1983) that tried to place in perspective structural, thermodynamic and kinetic aspects of O_2 and CO binding to the hemoglobins and their models,[6] the quantity and quality of structural data on the hemoglobins has increased by almost an order of magnitude. However, despite the proliferation of model systems since 1983, there have been only a few significant new structural results from model systems that give insight into O_2 and CO binding to hemes.[7] Finally, some of the difficulties underlying comparisons of ligand binding in different solvents (e.g. water for biological systems and toluene or benzene for many model systems) are presented.

The third part of this article, to a large extent, reports recent work by our group on the bioinorganic chemistry of metalloproteins that contain an oxo-bridged diiron moiety or a multinuclear copper complex. This area is also well-reviewed recently,[1,8-10] but in contrast to the wealth of structural, thermodynamic and kinetic data that exists for the hemoglobins and their models many aspects of the basic inorganic chemistry remain to be elucidated. Although asymmetry is a near-ubiquitous feature of dinuclear metal species in biological systems, there are few systematic studies on the consequences of asymmetry on structural, spectroscopic and chemical properties of unsymmetric dicopper and diiron model systems.

In addition to reviewing the bioinorganic chemistry surrounding dioxygen binding systems, this article seeks to provide:

(1) a broad overview of principles that determine the general functions of metalloproteins -- that is, the means by which a motif, such as the heme group, is adapted for redox as opposed to dioxygen- or electron-transport functions;

(2) a close analysis of the means by which specific functions -- for example, dioxygen transport and storage -- are modulated;

(3) a contrast of the differing roles of model systems for biological systems that are characterized in seemingly exhaustive detail -- the hemoglobins and their models -- and those systems that are relatively sparsely characterized, some by only structure-relative spectroscopic techniques -- the oxo-bridged diiron and multicopper systems.

With respect to the modulation of O_2 affinity and the mechanism of cooperativity in hemoglobins, there is a greater diversity than that enshrined in textbooks and many reviews, where discussions are generally limited to the archetypical hemoglobins, the monomeric sperm-whale myoglobin, and the tetrameric adult human hemoglobin. With respect to the role of model systems for well characterized metalloproteins, it is not inappropriate to ask the questions: *Is there a crossover point between model systems shedding light on biological systems and model systems reflecting light primarily on themselves? Are studies on model systems for hemoglobins and studies on the hemoglobins now heading on divergent paths?*

Protein Control of Function and Reactivity

Currently it is only for the hemoprotein family that extensive relationships among structure, function and reactivity exist. Processes that are mediated by the heme motif include the following, for which the porphyrin moiety is abbreviated as P, axial bases by B and exogenous ligands by L:

(1) electron transport or electron transferase activity (e.g. cytochrome c)
$$(B)_n Fe^{III}P + e^- \leftrightharpoons (B)_n Fe^{II}P \quad (n = 1, 2)$$

(2) dioxygen transport (e.g. hemoglobin) and dioxygen storage (e.g. myoglobin)
$$BPFe^{II} + O_2 \leftrightharpoons BPFe^{II}O_2 \text{ or } BPFe^{III}O_2^{-I}$$

(3) dioxygen activation for redox reactions (e.g. cytochrome $P450$)
$$O_2 + \text{R-H} + 2\,H^+ + 2\,e^- \xrightarrow{\quad BPFe^{III} \quad} \text{R-OH} + H_2O$$

(4) dismutation of hydrogen peroxide (e.g. catalase)

$$2 \text{ H}_2\text{O}_2 \xrightarrow{\text{BPFe}^{\text{III}}} \text{O}_2 + \text{H}_2\text{O}$$

and oxidation of substrates by means of peroxo species (e.g. horse radish peroxidase)

$$\text{H}_2\text{O}_2 + \text{AH}_2 \xrightarrow{\text{BPFe}^{\text{III}}} 2 \text{ H}_2\text{O} + \text{A}$$

(5) and reduction of dioxygen to water, in conjunction with copper centers (Cu_A and Cu_B) and heme groups (heme a_3 and heme a), and with electron and proton transport (cytochrome c oxidase)

$$\text{O}_2 + 4 \text{ H}^+ + 4 \text{ e}^- \xrightarrow{\text{BPFe}^{\text{II}}..\text{Cu}_B} 2 \text{ H}_2\text{O}.$$

The **general** function of a metalloprotein, in this case a hemoprotein, is determined by several factors, including:

(1) *Ligands provided by the protein to the metal center.* For example, all hemoglobins known to date have an imidazole moiety from histidine coordinated to the heme group, whereas hemoproteins that activate O_2 use a variety of axial ligands, such as cytochrome *P450* where thiolate from cysteine is coordinated.

(2) *Control of the degree of coordinative unsaturation at the metal center.* That is, there is provision (or exclusion) of sites at which exogenous ligands, such as dioxygen, may bind. By excluding exogenous ligands from the vicinity of the metal, the metal center is restricted to a structural role if the metal is redox-inactive, such as Zn^{2+}, Ca^{2+}, Mg^{2+}, or to electron transferase activity if the metal is redox active, such as $\text{Fe}^{\text{II/III}}$. For example, cytochromes c are six-coordinate low-spin species in both the Fe^{II} and Fe^{III} oxidation states, with methionine and histidine providing axial ligands. Cytochromes c' on the other hand have a lower reduction potential ($\sim$-100 mV, compared with 200 to 320 mV for the c type), and feature a five-coordinate high-spin species where the second axial position is rendered inaccessible to ligands by a non-coordinating bulky side chain, such as tryptophan. Note also that the electron transferase function of cytochromes c and c' is facilitated by the minimal structural rearrangement that accompanies electron gain or loss. Protein control of coordinative unsaturation differentiates cytochromes c from c'.[11]

(3) *Control of access of a second substrate to a first substrate.* For example, both hemoglobin and heme oxygenases, such as cytochrome *P450*, bind O_2. Control of substrate access differentiates hemoglobin, where there is access to only small molecules, from cytochrome *P450*, where O_2 and a second (organic) substrate can bind in close proximity. In addition, cysteine coordination to cytochrome *P450* facilitates the activation of the coordinated dioxygen to hydroxylate the organic second substrate, compared with histidine coordination to hemoglobins. In addition, cytochrome *P450*, peroxidases and catalases all generate the same active oxygen species, $\text{Fe}^{\text{IV}}(\text{P}^{\cdot -})=\text{O}$ (or $\text{Fe}^{\text{V}}(\text{P}^{2-})=\text{O}$).[3] In the case of cytochrome *P450*, the active oxygen is transferred directly onto the substrate. On the other hand, for peroxidases oxidation of the organic substrate at a site remote from the active oxygen appears to occur by electron transfer.

The **precise** function is determined by a subtle interplay of many factors. For example, cytochromes c share identical heme groups and axial ligands, yet the reduction potentials span over 100 mV (10 kJ mol^{-1}).[11] To be discussed in detail later, the affinity of hemoglobins for O_2 spans over five orders of magnitude ($\sim$28 kJ mol^{-1} at room temperature),[2] modulated in part by hydrogen bonding between the coordinated O_2 and nearby polar residues and in part by steric factors that reduce O_2 affinity. Moreover, the intrinsic very high affinity that isolated five-coordinate heme species have for CO is substantially diminished for the hemoglobins[2,5] -- a feature of fundamental physiological significance.

With regard to iron in biological systems, there are only four basic structural motifs: a heme motif (Figure 1(a)),[2] an iron-sulfur cluster motif based upon $Fe_2S_2(SR)_2$ (Figure1(b)),[12] a mononuclear non-heme iron motif featuring histidine and oxyanion ligands, such as phenolate from tyrosine and carboxylate from aspartic and glutamic acids (Figure 1(c)),[13] and a dinuclear non-heme iron motif based upon Fe-O(H)-Fe (Figure 1(d)).[8,9,14,15] This last motif is being identified in an increasing number of metalloproteins of diverse func-

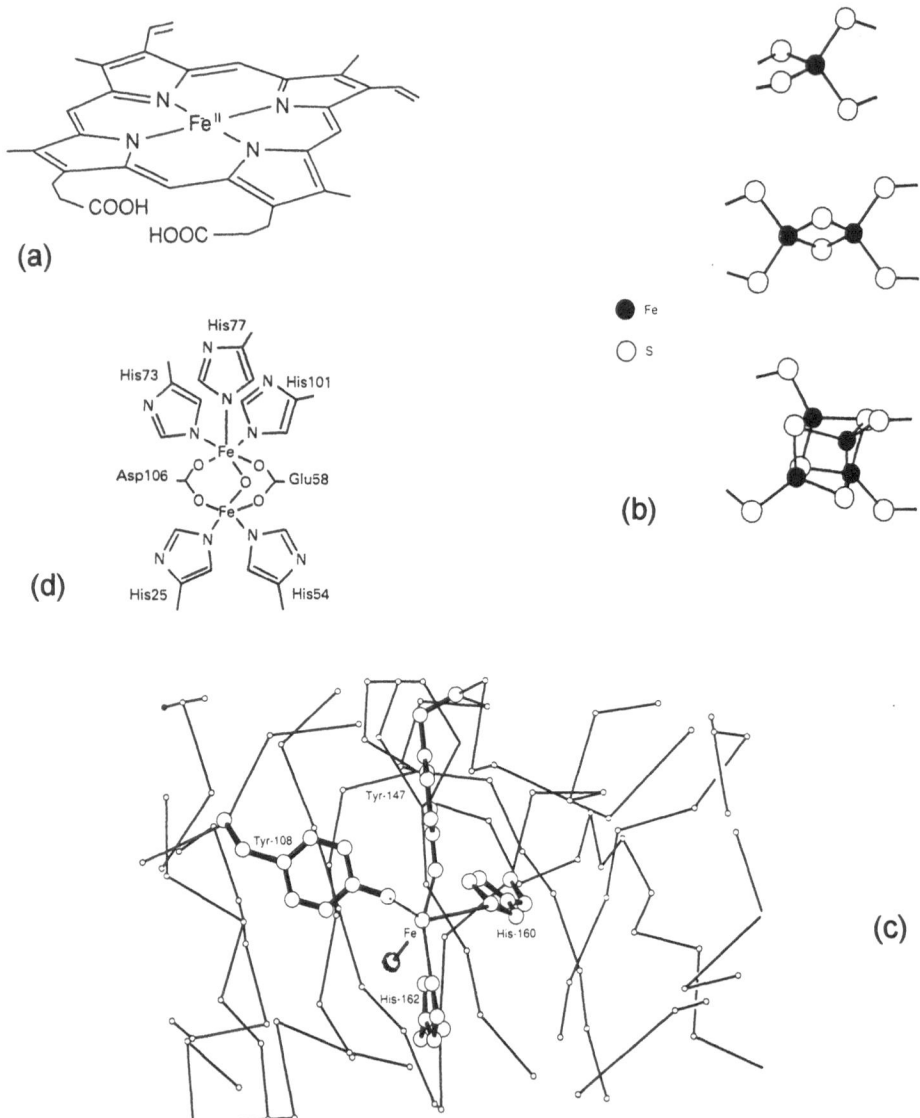

Figure 1. Structural motifs found in iron-containing metalloproteins. (a) The heme *b* group. This or related heme groups are found in hemoproteins. (b) Iron-sulfur clusters. The $Fe(SR)_4^{0/1-}$ moiety is found in rubredoxin and the $(RS)_4Fe_2S_2^{2-/3-}$ and $(RS)_4Fe_4S_4^{1-/2-/3-}$ moieties in ferrodoxins and high potential iron proteins.[12a] (c) The mononuclear non-heme non-sulfur iron complex illustrated for protocatechuate 3,4-dioxygenase.[3b] (d) The triply bridged μ-oxo diiron(III) motif illustrated for methemerythrin.[8]

tion. For example, the Fe-O(H)-Fe motif is found in an oxygen carrier, hemerythrin,[15] which shuffles between Fe(II)-OH-Fe(II) and Fe(III)-O-Fe(III) states; a hydrolase, purple acid phosphatase,[8,17] which is active from the mixed-valent state; a reductase, ribonucleotide reductase;[18] and a hydroxylase, methane monooxygenase,[19] where anionic ligands stabilise an Fe(III)-OH-Fe(III) state. Ferritin contains polymeric Fe^{III}-O-Fe^{III} units. Systems containing the Fe-O(H)-Fe moiety are discussed in more detail later.

Model Systems as Probes of Metalloprotein Structure, Function and Reactivity

The determination of complete protein structure, and, in particular, of the active-site structure and its surroundings, is an increasingly more tractable problem, as a result of the development of methods to express near-gram quantities of proteins (and mutants) from a DNA sequence and as a result of developments in the techniques of single crystal X-ray diffraction and multi-dimensional NMR spectroscopy. Notwithstanding these developments, model systems still offer invaluable insight into the spectroscopy, structure and function of metalloproteins.

First, structurally and spectroscopically well characterized non-biological or model systems provide a reference library of data for comparison with the corresponding data from biological systems. It is important that not all efforts are directed to replicating synthetically all the features of the biological system. A restricted reference library invites conclusions that the absence of evidence for a particular structural motif is evidence of that motif's absence. Conversely, evidence taken in support of a particular motif may be consistent also with a motif yet to be characterized. For the first instance, there is ample inorganic precedent in a wide variety of systems, porphyrin, non-porphyrin and organometallic, that the strongly bent Fe-C≡O moieties (120°) reported at various times for carbonmonoxy-myoglobin, MbCO,[20] are to say the least surprising. Also, the distinctive spectroscopic and magnetic properties associated with symmetric manifestations of the Fe^{III}-O-Fe^{III} motif have allowed its biological (and unsymmetric) occurrences to be identified. Fe^{III}-O-Fe^{III} species dominate the chemistry of iron when dioxygen and water are present and have been widely studied independently of biological systems.[21,22] In the second converse instance, the distinctive and unique spectroscopic features of "blue" copper proteins opened up new areas of research in copper chemistry. The moiety responsible, $(N_{his})_2$Cu(II)-SR, remains elusive to study in non-biological situations. Moreover, early efforts to interpret the uv-visible spectrum of oxyhemocyanin led to the proposal that in addition to a μ-1,2-peroxo bridge there existed a second bridge[23] -- an idea that fell on hard times when the structure of deoxy-hemocyanin[24a] revealed no potential candidates for this second bridge. Resolution of the conundrum came when it was found that dioxygen bridged the two copper centers in oxy-hemo-cyanin in a μ-η^2:η^2-peroxo manner.[24,25]

Second, with well-designed model systems and structurally well-characterized proteins, it may be possible to determine intrinsic behavior of a particular structural motif and, thence, to elucidate the role that the protein plays in modifying properties. This is the area of biomimetic chemistry, where additionally model systems may shed light on the detailed mechanism of action of certain enzymes, for example, cytochrome *P450*.

Third, notwithstanding advances in protein crystallography, model systems still provide structural data that can be more than an order of magnitude more precise than currently attainable in biological systems. Changes in bond lengths of 0.10 Å are associated with changes in ligand affinity in model systems of several orders of magnitude.[26]

Fourth, reactive intermediates of the biological system, may be isolated and characterized in appropriate model systems. For example, chemical reactions may be studied by means of model systems soluble in organic solvents, especially toluene, at low temperatures and over temperature ranges that are inaccessible for the biological system because of the freezing of water and denaturation of the protein.

Fifth, in well characterized biological systems, such as the hemoproteins, the precisely determined stereochemistry of metalloporphyrinato systems, permeates any analysis of structure, function and reactivity relationships. For example, it is now part of the lore of porphyrinatoiron(II) stereochemistry, that for five-coordinate species, with few exceptions, the iron center is high-spin and the iron atom is about 0.5 Å from the plane of the porphyrin towards to the fifth ligand.

Modelling Hemoproteins and Non-Heme Metalloproteins

The underlying challenge in model systems for the hemoglobins was to avoid the rapid autoxidation of iron(II) by dioxygen, as described in more detail in below. Immobilization of heme,[27] low temperature,[28] and built-in protection of the dioxygen-binding site[29] allowed isolation of iron-dioxygen adducts. Two of the early goals of model systems for the hemoglobins included (1) the stereochemistry of the metal-dioxygen moiety, which had been theoretically discussed for some time,[30] and (2) the elucidation of the precise structural changes occurring upon ligand binding in order to test the validity of Perutz model[31] for cooperative ligand binding to hemoglobin. Two factors facilitated the explosive development of model systems: First, it was long established that vertebrate hemoglobins contained a heme *b* moiety -- the iron(II) derivative of protoporphyrin IX (Figure 1(a)). This ligand, providing obligately tetradentate, square-planar coordination (for most metals), could be modelled by the synthetically accessible tetraphenylporphyrin (Figure 2(a)), which because of its symmetry had desirable crystallization properties, and which could be extensively modified.[4] Second, it was found that cobalt-substituted hemoglobin also bound dioxygen, generating an adduct with a distinctive EPR signature.[32] While (base)porphyrinatocobalt(II) species in aprotic solvents bind dioxygen very weakly, cobalt(II) complexes of Schiff bases (Figure 2(b)), a tetradentate non-macrocyclic ligand also offering square-planar coordination, bind dioxygen readily.[33]

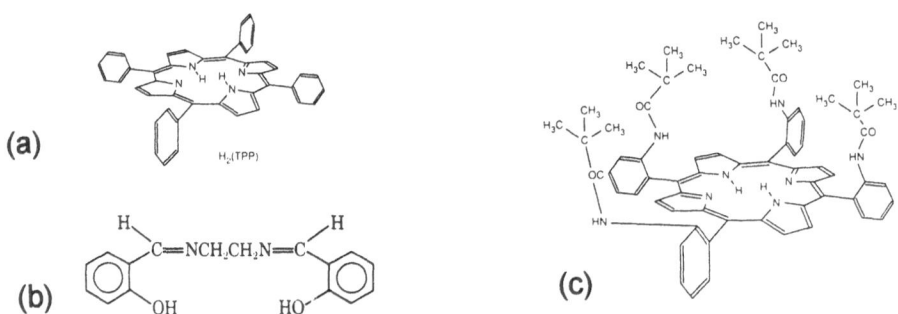

Figure 2. (a) Tetraphenyl porphyrin, widely derivatized for model systems for hemoproteins.[4] (b) The Schiff-base ligand, salicylidineimine, H_2salen, widely derivatized for model systems for cobalt-substituted hemoglobins.[33] (c) The "picket fence" porphyrin, for which stable iron-dioxygen adducts and their properties could be studied at room temperature.[29]

In 1972, the first 1:1 $Co:O_2$ species was structurally characterized using a Schiff base ligand;[34] in 1974, the first 1:1 $Fe:O_2$ complex, using a highly derivatized tetraphenylporphyrin, the "picket fence" porphyrin, illustrated in Figure 2(c).[35] In 1978, the first direct comparison of structural changes occurring upon oxygenation were determined at a reasonable level of precision for a model system again using the "picket fence" porphyrin.[36] These species and associated structural changes have served as yardsticks against which the stereochemistry of the $Fe-O_2$ moiety in oxyhemoglobin species can be compared: oxymyoglobin (1978) revealed a strongly bent Fe-O$_2$ moiety and structural changes consistent with those seen in model systems,[37] whereas oxyerythrocruorin (1979) revealed a surprisingly linear geometry and an unexpected movement of the iron atom relative to the porphyrin plane[38] The structures of both fully[39] and partially oxygenated tetrameric hemoglobins,[40] have been determined, as well as other structures of liganded hemoglobin, frozen in the T state.[41] In a crystallographic *tour de force*, the structures of unliganded myoglobin (at 85 K) and of carbonmonoxymyoglobin (at 85 K), and its photoproduct (MbCO*, at 20 K) have been determined at high resolution (1.5 Å).[42] For MbCO*, which has a lifetime of only milliseconds at ambient temperature, the photodissociated CO is found in the binding pocket, and the structure of MbCO* sits midway between MbCO and Mb.

In large part the early goals have been accomplished, although definition of the FeO$_2$ moiety to better than 0.08 Å in the O-O separation and 4° in the Fe-O-O bond angle has yet to be accomplished because of disorder in the orientation of the Fe-O-O moiety that precludes accurate determination of bond parameters, even though the nominal precision is about 0.02 Å in the O-O bond length. Functional models for myoglobin and hemoglobin exist. However, as greater understanding is sought, differences rather than similarities between model systems and the hemoglobins have become accentuated and the insights offered by model systems into structure-function-reactivity relationships of hemoglobins are increasingly more indirect. Nonetheless, even if relationships among stereochemistry, thermodynamics and dynamics of ligand binding are quantitatively deficient, there is at least some degree of qualitative predictability.[4,5]

However, for model systems the relative instability of all dioxygen adducts of non-biological systems characterized to date, the ease with which hemochromes (two axial bases to porphyrinatoiron(II)) form relative to systems with different axial ligands and the extreme difficulty with which elaborations of tetraphenylporphyrin or of various hemes crystallize in a manner suitable for single-crystal X-ray diffraction studies, results, twenty years after the first crystal structure of an FeO$_2$ species,[35] in only two synthetic FeO$_2$ crystal structures of moderate precision,[35,36] a handful of Fe-CO structures,[7,43] and only semi-quantitative definition of the Fe-O-O stereochemistry.

Creating realistic models for hemoproteins is a less formidable task than that awaiting synthetic chemists for non-heme metalloproteins, especially if the metal complex is multinuclear. For models of hemoproteins, four of the five or six ligands can be delivered in approximately the correct square-planar geometry by a variety of tetradentate ligands. On the other hand, for non-heme systems, three, four, five or even six ligands are provided by the protein, other ligands may be exogenously delivered, and the relative disposition of all the ligands with respect to one another is variable; metal ions may be singly, doubly or triply bridged. However, it has become clear that at least some structural motifs represent thermodynamic sinks -- that is they spontaneously self-assemble.[44] For example the μ-oxo-bis(μ-carboxylato) core of diiron proteins such as hemerythrin,[16] and the $(RS)_4Fe_4S_4$ moiety of ferrodoxins[12] are, at least in hindsight, readily assembled. However, as noted earlier, the $(RS)Cu^{II}(imidazole)_2$ moiety of blue copper proteins has proved elusive to create in non-biological situations. Table 1 summarizes some of the intrinsic differences between modelling the active sites of hemoproteins and of metalloproteins featuring non-sulfur dinuclear metal complexes at the active sites.

The synthesis of dinuclear and multinuclear metal complexes that are amenable to precise structure determination and that have spectroscopic properties similar to those of the metalloprotein has proved challenging, especially if unsymmetrical species are sought. A greater challenge has been to endow these model systems with functionality, such as ligand binding properties, approximating that of the targetted metalloprotein.

Table 1. Comparison of metalloproteins containing non-sulfur di-iron and di-copper complexes.

Hemoproteins	Dinuclear active sites
Fixed stereochemistry	Variable stereochemistry
Tetradentate heme	------
1 - 2 protein-supplied ligands	3 - 5 protein-supplied ligands per metal
0 - 1 exogenous ligand	1 - 3 exogenous ligands
Pseudo-2-fold symmetry	Asymmetric
Spectroscopy ↔ structure	Spectroscopy ?? structure
Structure ↔ function	Structure ?? function
Site-directed mutagenesis of non-coordinated residues	------

Terminology for Describing Symmetry in Multinuclear Metal Species

The use of the terms symmetric, unsymmetric, asymmetric and dissymmetric (with or without an "al" on the end) in discussions of dinuclear systems that lack point symmetry greater than C_1-1 is confusing and variable. In the absence of dictionary distinction in a chemical context, the following definitions and applications are suggested (the Schoenflies symbol precedes the Hermann-Mauguin or International symbol):

Symmetric: possessing point symmetry greater than C_1-1 with respect to the interchange of metal atoms; that is containing rotational symmetry of at least C_2-2, inversion symmetry (C_i - -1) or mirror symmetry (C_s-m) perpendicular to the metal atoms' internuclear axis;

Asymmetric: obligately lacking point symmetry greater than C_1-1, with respect to interchange of metal atoms;

Unsymmetric: lacking point symmetry greater than C_1-1, with respect to interchange of metal atoms, but the lack of symmetry may be accidental;

Dissymmetric: associated with or resulting from lack of symmetry as defined above.

Accordingly, dissymmetric effects are effects resulting from lack of symmetry, avoiding the ambiguous phrase, asymmetric or unsymmetric effects. This is in accord with the etymology of the prefix *dis,* that leads to word pairs such as disability and inability, disinterested and uninterested, disforested and unforested. In its scientific usage Webster's New Twentieth Century Dictionary (Unabridged) does not make a distinction among *un-*, *a-*, and *dissymmetric*, except for dissymmetric in the sense of opposite, as in D and L enantiomers. It is recommended that *asymmetric* is reserved for contexts where lack of symmetry is unavoidable or intrinsic to the ligand and its mode of coordination. Thus an asymmetric dinucleating ligand would lead to symmetrical complexes only in the event of it not dinucleating, or where higher order oligomers are formed from asymmetric dinuclear subunits that are paired by, say, two-fold symmetry.

Although, the qualifiers are non-quantitative, it is useful to specify the degree of unsymmetry by the words such as *slight, moderate* and *high*. For example, the qualified phrase "slightly unsymmetric" is appropriate when the extent of departure from effective C_{2v}-mm2 symmetry is the interchange of an amine for an imine ligand at one metal, or when one suite of ligands at one metal center is rotated slightly ($\lesssim 20°$) with respect to the identical suite of ligands at a second metal center. In the context of magnetic coupling between metal centers similar qualifiers are commonly used. The qualifier *accidental* is reserved for situations where the origins of the lack of symmetry are not obvious on steric or electronic grounds, and the qualifier *obligate* for where the lack of symmetry is unavoidable. Symmetry operations greater than C_1-1 that leave the metal centers unmoved are excluded, since there is no convenient way to represent in a word or two a complex that has, for example, one tetrahedrally and one octahedrally coordinated metal centre but with a mirror plane in which lies the intermetal vector. This situation where the metal centers are clearly chemically and spectroscopically distinct seems best summarized as being unsymmetric, in accord with current practice.

STRUCTURE, FUNCTION AND REACTIVITY RELATIONSHIPS IN HEMOGLOBINS

The study of hemoproteins, and especially the hemoglobins, has been facilitated by the diversity of species arising from natural variations (multiple-site variations between species or single-site variants), or from engineered variants (by site-directed mutagenesis). Few other classes of metalloproteins have been so intensively studied to date in so many variants and by so many techniques, including X-ray and neutron diffraction and EXAFS techniques, a battery of spectroscopic techniques from long to pico-second time-scales. By way of contrast to the wealth of structural data for hemoproteins, which is becoming increasingly more precise and accurate, there is a paucity of structural data on model systems to correlate directly with the extensive thermodynamic and dynamic data on ligand binding.

While site-directed mutagenesis on myoglobins and hemoglobins has yielded profound insights into structure, function and reactivity relationships, even with detailed structural information some changes are not readily explicable. There is, moreover, the danger of preordained changes leading to preordained conclusions, in model as well as biological systems. Recent studies on random mutants of myoglobin, revealed unexpected effects on dioxygen-binding dynamics of some changes.[45] In particular, there is now a large volume of structural, spectroscopic, kinetic and thermodynamic data on sperm whale myoglobin mutants that has been accumulated by one team under common conditions[5] -- obviating many earlier problems of comparing data that came from different laboratories, were measured under different conditions and were analyzed under different protocols.

In this section the role of non-coordinated moieties in stabilizing dioxygen adducts, in controlling absolute and relative affinity for ligands, such as O_2 and CO, and in controlling ligand-binding dynamics that lead to the rapid binding and release of ligands to porphyrinatoiron(II) species will be discussed. Vertebrate hemoglobins are the archetype of the cooperative interactions that are enabled by the association of subunits into oligomeric quaternary structures. However, not all oligomeric hemoglobins achieve cooperativity by the same mechanism. This and several other recent developments in hemoglobin structure and function that address oversimplifications found in the secondary literature, are also reviewed .

Ligand Binding to Hemes

σ-Donor, π-acceptor molecules of all sizes (e.g. CO, O_2, NO, RNC, RNO, pyridyl and imidazolyl-species) bind strongly to simple five-coordinate (base)porphyrinatoiron(II) species, where the axial base is a nitrogen-containing heterocycle, such as (substituted)-imidazolyl or pyridyl species, thereby completing an approximately octahedral stereochemistry.[2] However, species such as pyridyl, imidazolyl and alkyl isocyanides with very bulky alkyl groups do not bind to heme groups encapsulated in proteins. In the absence of protein, coordinated O_2, in particular, is prone to further reaction, in processes known as autoxidation, that lead to oxidation of iron(II) to iron(III). Largely through studies on non-biological systems, the stereochemistry of porphyrinatoiron species as a function of spin state, oxidation state and axial ligands is known in a detail and variety unparalleled for other motifs.[46] Some of this variety is summarized in Table 2. Note that of the variety of spin states and coordination numbers found in hemoproteins, only two, possibly three, are sampled by hemoglobins in their normal physiological function. The non-polar nature and diamagnetism of the Fe-CO moiety are generally well accepted -- even if the geometry of the Fe-CO moiety remains a topic of discussion. The nature of the Fe-O_2 moiety in oxyhemoglobins and models remains controversial, not least because spurious physical observations have developed a theoretical reality that persists into the most recent reviews.[47] Two examples will suffice.

Table 2. Oxidation and spin states of iron porphyrins and their biological occurrences.[2,a]

	Fe^{II}		Fe^{III}	
High spin	Fe(PF)(2-MeIm) **Hb** Cytochrome c'	5-coord	Fe(TPP)Cl **Cytochrome P450(ox)**	5-coord
$[Fe^{II}\ S=2(d^6)]$ $[Fe^{III}\ S=5/_2(d^5)]$	Fe(TPP)(THF)$_2$ **Hb(H$_2$O)?**	6-coord	$[Fe(TPP)(H_2O)_2]^+$ $[Fe(OEP)(3-ClPy)_2]^+$(293K) Fe(OEP)(Py)(NCS) **MetHb(H$_2$O)**	6-coord
Intermediate spin $[Fe^{II}\ S=1(d^6)]$ $Fe^{III}\ S=3/_2(d^5)$	Fe(TPP) no biol. occurrence	4-coord b	$[Fe(TPP)(C(CN)_3)]_n$ Fe(TPP)(OClO$_3$) ($S=3/_2, 5/_2$) **Cytochrome c'**	5-coord
Low spin $[Fe^{II}\ S=0(d^6)]$ $[Fe^{III}\ S=1/2(d^5)]$ $[(S=1/2, Fe\text{-}NO]$	Fe(TPP)(NO) **Hb(NO)·DPG(T)**	5-coord	[Fe(TPP)(Ph)]	5-coord
	Fe(PF)(2-MeIm)(O$_2$)^c Fe(TPP)(1-MeIm)(NO) Fe(TPP)(Py)(CO) Fe(TPP)(1-MeIm)$_2$ **Hb(CO), Hb(NO),**^{b} **Hb(O$_2$)**^{c} **Cytochrome b$_5$,c,c$_3$**	6-coord	$[Fe(TPP)(Im)_2]^+$ $[Fe(OEP)(3-Cl-Py)_2]^+$ (98 K) Fe(TPP)(Py)(CN) Fe(TPP)(Py)(NCS)^d **Cytochrome b$_5$, c, c$_3$** **MetHb(CN)**	6-coord

^{a} Abbreviations: PF, picket fence porphyrin dianion; TPP, 5,10,15,20-tetraphenylporphyrin dianion; OEP, 2,3,7,8,12,13,17,18-octaethylporphyrin dianion; DPG, 2,3-diphosphoglycerate; THF, tetrahydrofuran; Im, imidazole; py, pyridine; Ph, phenyl anion; see also Figure 4 for diagrams of porphyrins. ^{b} Could be placed in Fe^{III} column. ^{c} Could be placed in Fe^{III} column ($S=0$). ^{d} Non-linear Fe-NCS moiety.

First, over-interpretation of the apparently short O-O separation found in the structure of a model system for oxymyoglobin -- an ineluctable consequence of disorder[35] --- coupled with a spurious value for the O-O stretching frequency for the same model system[48a] led to the notion that O_2 bound in a singlet oxygen($^1\Delta_g(O_2)$)-like mode,[48a] an error now enshrined in one bioinorganic text. Theoretical calculations then produced a non-polar Fe-O_2 moiety for which hydrogen bonding interactions with polar groups would be repulsive[48b] -- a conclusion at odds with experimental observation.

Second, a spurious measurement in 1977 of paramagnetism for oxyhemoglobin,[48c] originally found by Pauling and Coryell to be diamagnetic,[48d] inspired theoretical justification[48e-g] that there exist low-lying paramagnetic excited states. Notwithstanding challenge by Pauling,[48h] reconfirmation of diamagnetism by several groups[48i-k] and retraction in 1985 by the original authors,[48l] this spurious observation is cited in a 1994 review[47] and compared with theoretical calculations accompanied only by an ambiguous reference to Pauling's challenge.[48h] There is presently no experimental evidence to support notions of low-lying paramgnetic excited states for either oxy- or carbonmonoxyhemoglobin. While theoretical support remains underwhelming, there is ample physical evidence that the

Fe-O$_2$ bond is strongly polarized, Fe^{III}-O$_2^{-I}$, while the Fe-CO bond is weakly polarized. This contrast is of the utmost physiological importance to most organisms using hemoglobins for dioxygen transport and storage.

Control of Autoxidation

In the absence of protein or some other superstructure surrounding the binding site for dioxygen, porphyrinatoiron(II) species (PFe^{II}) oxidize within milliseconds at room temperature by a mechanism that involves bimolecular contact of a PFe-O$_2$ species with an Fe^{II} species, as indicated in the scheme below:[49]

$$PFe^{II} + O_2 \rightleftharpoons PFe\text{-}O_2$$
$$PFe\text{-}O_2 + PFe^{II} \rightleftharpoons PFe^{III}\text{-}O_2\text{-}Fe^{III}P$$
$$PFe^{III}\text{-}O_2\text{-}Fe^{III}P \rightarrow 2\ PFe^{IV}{=}O$$
$$PFe^{IV}{=}O + PFe^{II} \rightarrow PFe^{III}\text{-}O\text{-}Fe^{III}P$$

The first reaction above is the reaction of physiological importance for hemoglobins. The dinuclear species, PFe^{III}-O$_2$-Fe^{III}P, can be isolated and studied at low temperatures.[50] Bimolecular contact can be prevented by encapsulation of the heme group in proteins, such as myoglobin, illustrated in Figure 3, or immobilization on a solid support,[27a] encapsulation in micelles[27b] or by porphyrin superstructure, a selection of which is shown in Figure 4. A recent (1994) and comprehensive listing of porphyrins used as model systems is available in the review by Reed and Momenteau.[4]

The rate of O$_2$ binding is characterized by the rate constant $k_{on}(O_2)$ (often k_1 or k_+) and the rate of O$_2$ dissociation by $k_{off}(O_2)$ (often k_{-1} or k_-). The equilibrium constant for O$_2$ binding $K_C(O_2)$ may be obtained thermodynamically as

$$K_C = [PFe\text{-}O_2]/([PFe][O_2]),$$

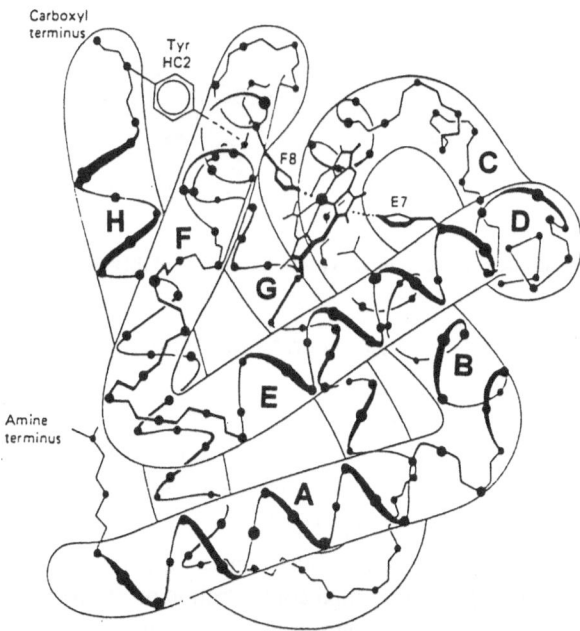

Figure 3. Structure of myoglobin showing the encapsulated heme group. Helices are labelled A - H.

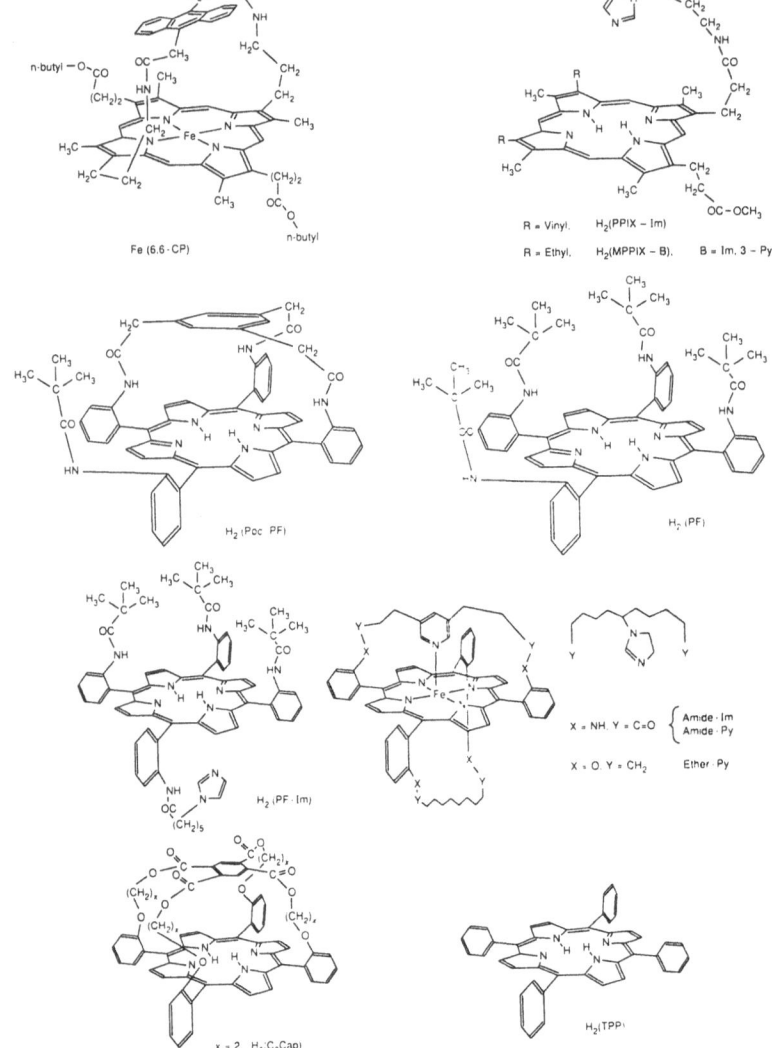

Figure 4. A selection of synthetic porphyrins used to probe the binding of dioxygen and other small molecules.[2]

(more strictly as activities), or as the partial pressure required for half-saturation ([PFe-O$_2$] = [PFe]) whence

$$P_{1/2}(O_2) = 1/K_P = 1/(K_C K_H),$$

where K_H is the Henry's law constant that relates gas solubility to partial pressure. Frequently, the equilibrium constant is determined kinetically as

$$K_C = k_{on}(O_2)/k_{off}(O_2).$$

In protic environments, where bimolecular contact of hemes is prevented, two other mechanisms for autoxidation prevail: at high concentration or partial pressure of dioxygen when essentially all dioxygen binding sites are full, proton-assisted displacement of the coordinated dioxygen as a hydrosuperoxide moiety predominates.[51] The Fe-O$_2$ moiety resembles at least formally an iron(III) superoxo species, Fe^{III}-O$_2^{-I}$:

$$PFe\text{-}O_2 + H^+ \rightarrow PFe\text{-}O_2H^+ \rightarrow PFe^{III} + HO_2^{\cdot}$$

At low concentration of O_2, interaction of O_2 with water in the binding cavity leads to the direct formation of the superoxide anion:[51]

$$PFe^{II}\cdots OH_2 + O_2 \rightarrow PFe^{III}\cdots OH_2 + O_2^{-\cdot}.$$

Control of Ligand Affinity at the Heme

Stereochemical parameters of porphyrinatoiron(II) species relevant to O_2 and CO binding are illustrated in Figure 5.[2] Values of the structural parameters shown in Figure 5 are given in Table 3 for high-spin five-coordinate and low-spin six-coordinate species.

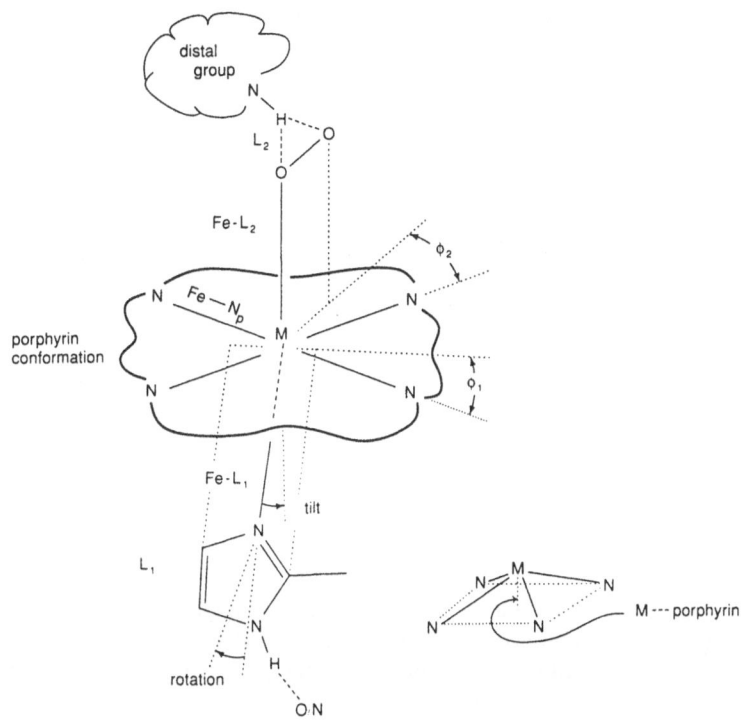

Figure 5. Structural parameters and features that determine affinity at the heme for a sixth ligand.[2]

Probable values for the intrinsically preferred orientation are provided. The intrinsic values entered in the table pertain to the hypothetical situation of the gas phase, where no interactions between the heme or its axial ligands with solvent or distal residues occur. These values are based, moreover, upon consideration of steric factors alone, since there are few reliable data to favor, for example, an eclipsed rather than bisecting conformation of the axial base or of the Fe-O-O moiety with respect to Fe-N_{porph} bonds. (See Figure 5.) Ranges of values actually found to date are also entered. Notice that a wide range of Fe-O-O and Fe-C≡O angles are apparently observed, and that the movement of the iron into the plane of the porphyrin is not always complete. Consideration of steric factors leads to the following conclusions for structural features and changes that would decrease ligand affinity relative to the hypothetical gas-phase state:[2]

   (1) *Incomplete movement of the iron atom into the heme plane (M..Por ≠ 0):* Model systems that offer restraints to the movement of the iron into the plane of the porphyrin,

such as a 2-methyl substituent on an imidazole ring, have much lower O_2 and CO affinity than model systems that employ axial bases lacking bulky substituents. Bonding of O_2 and CO is synergistic, requiring strong σ-bonding of the axial base to the iron center in order to facilitate π-back bonding from the iron into O_2 and CO π* orbitals. Note that for stronger σ-donors, such as nitric oxide (NO), the bond to the imidazole is weakened upon the binding of NO, and, in contrast to CO and O_2, NO binds strongly to porphyrinatoiron(II) species in the absence of an axial base.

Table 3. Stereochemistry at the heme: control of O_2 and CO affinity at the heme.

Parameter[a]		Intrinsic	Observed range
M..Por (Å)	5-coord	0.5	0.2 to 0.5
	6-coord	0.0	0.0 to 0.2
	change	0.5	0.3 to 0.5
ϕ(Base) (°)	5-coord	45	0 to 20
	6-coord	(45)	0 to 20
	change	0	0 to 20
τ(Base) (°)	5-coord	0	0 to 10
	6-coord	0	0 to 10
	change	0	0 to 10
Planarity	5-coord	planar	planar (?)
	6-coord	planar	planar (?)
	change	none	(??)
Fe-O=O (°)		~120	120 to 150
Fe-C≡O (°)		180	150[b] to 170

[a] See Figure 5 for definition of parameters. [b] Values less than 150° are from MbCO structure's where the CO moiety is disordered.[20]

(2) *Axial base remains tilted (τ ≠ 0):* Bonding of the axial ligand with the iron center is strongest in a symmetrical disposition. A 2-methyl substituent on an imidazole ring forces tilting to minimize steric clash of the methyl substituent with the porphyrin ring. Unliganded and liganded hemoglobin structures trapped in the low affinity T-state, show, *inter alia*, a tilted disposition of the imidazole ring of the proximal histidine.

(3) *Base adopts a more eclipsed orientation (ϕ(Base) ≈ 0):* In an eclipsing orientation, steric clash of the imidazole ring with porphyrin nitrogen atoms is maximized, hindering movement of the iron and attached imidazole group into the plane of the porphyrin on binding of the sixth ligand.

(4) *Fe-O-O adopts a more eclipsed orientation:* In an eclipsed orientation, the terminal oxygen atom clashes with porphyrinato nitrogen atoms more strongly than from a bisecting position. This effect is relatively small, as in the eclipsed orientation there is one bad contact, while in the bisecting orientation there are less severe contacts but there are two of them. Similarly, the energy difference between eclipsed and bisecting orientations of the axial base is expected to be relatively small.

(5) *Heme distorted from planarity:* For low-spin Fe^{II} systems an essentially planar conformation of the porphyrin appears to be the preferred orientation, although the energy needed to induce considerable distortions appears to be rather small, as, at least in the case of mesotetraphenylporphyrinatocopper(II), essentially planar, ruffled and saddled conformations of the porphyrin ring are observed in different crystal forms.[51] Thus, decreases in affinity associated with significantly non-planar porphyrin rings have been attributed more to increased steric clash between porphyrin and axial base, than

to the distortion itself.[7c]

(6) *Fe-CO geometry distorted from linearity:* In countless terminally coordinated metal carbonyl species, the distortion of the M-C≡O moiety from linearity amounts to only a few degrees. The geometry of the Fe-C≡O moiety in carbon monoxide adducts of hemoglobins has been extensively studied with respect to the mechanism by which hemoglobins discriminate strongly against CO, relative to the strengths of CO binding to non-biological model systems. A thorough reanalysis of this controversial topic has been recently advanced,[5] and will be presented later.

Relative to the hypothetical gas-phase situation with the iron atom centered in a planar porphyrin ring, these above factors increase the free energy of the six-coordinate species, thereby increasing the rate of ligand dissociation. Another strategy for lowering affinity, is to lower the free energy of the *five*-coordinate species, relative to the hypothetical gas phase situation of a planar porphyrin ring with the iron ~0.5 Å from the plane of the porphyrin. At one extreme, coordination of a second axial base lowers dramatically the affinity for O_2 and CO. To a lesser extent, a probable weak coordination of tyrosine to the heme in a sperm whale myoglobin mutant, where the distal histidine is replaced by tyrosine, lowers affinity for O_2 by a factor of 200 relative to the native protein (see Table 5). Such effects would tend to decrease the rate of ligand binding.

On the other hand, increased affinity results from increasing the free energy of the five-coordinate species or by decreasing the free energy of the six-coordinate species. Factors that involve only the (base)porphyrinatoiron(II) moiety include:

(1) *Fe-Base moiety in five-coordinate species pushed into heme plane:* Advancing along the reaction coordinate towards the product increases free energy of the five-coordinate species. This has recently been observed in the structure of photolyzed MbCO,[42] where the Fe-N_{His} bond is intermediate between that for coordinated MbCO structure and the relaxed ligand-free Mb structure.

(2) *Rotation of the axial base into a less eclipsed conformation:* Steric clash of axial base with the porphyrin ring is reduced by rotation from an eclipsed to a bisecting orientation upon binding the sixth ligand, or by a change from a tilted to symmetric disposition, if not sterically prevented (e.g., a 2-methyl substituent).

These, then, are some of the factors by which (axial base)porphyrinatoiron(II) stereochemistry can modulate ligand affinity and ligand binding dynamics. Most of the conformational perturbations are induced by the protein or in models by the porphyrin superstructure. Direct interaction of protein or porphyrin superstructure with bound ligands is discussed in the next subsection. Identification of factors where rates of association and dissociation are perturbed by changes in free energy of the transition state, as opposed to changes in free energy of the product and reaction species may be uncovered in linear free energy relationships between affinities and rate constants (to be discussed later).

Control of Ligand Affinity by the Surroundings

Changes in molecular volume, polarity and polarizability upon ligation of the sixth ligand, lead to changes in solvation energies for five- *versus* six-coordinate species. All other things remaining unchanged, an increase in molecular volume upon ligation of a sixth ligand, stabilizes the five-coordinate state relative to the six-coordinate -- that is affinity is lowered. Qualitatively, this has been seen for model systems, comparing ligand affinities of unsubstituted meso-tetraphenylporphyrinatoiron(II) species, where the sixth site is solvent accessible, with the "picket fence" porphyrin derivative, where the sixth site is somewhat less solvent accessible, but readily accessible by small molecules, such as O_2 and CO.[2,4] However, in this comparison, as with many other comparisons made among model systems, more than one parameter of the system is changing.

The specific interactions of coordinated sixth ligands with distal moieties has proved amenable to study in biological systems by site-directed mutagenesis. Critical, highly conserved and conservatively substituted residues in the hemoglobins include the distal histidine (position 64 in the primary structure or position 7 on helix E of the canonical myoglobin from sperm whale, abbreviated [64]E7), leucine at [29]B10, valine at [68]E11, and phenylalanine at positions [43]CD1 and [46]CD4 on the loop between the C and D helices. This is shown schematically in Figure 6. In the last several years the kinetics and thermodynamics of CO and O_2 binding to a total of 42 native and mutant myoglobins, mostly of sperm whale, have been studied by one team; eleven of these mutants have also been structurally characterized.[5] Table 4 lists the mutants; those enjoying structural characterization, mostly as the CO adduct are in bold-face type. This has permitted unprecedentedly direct correlation

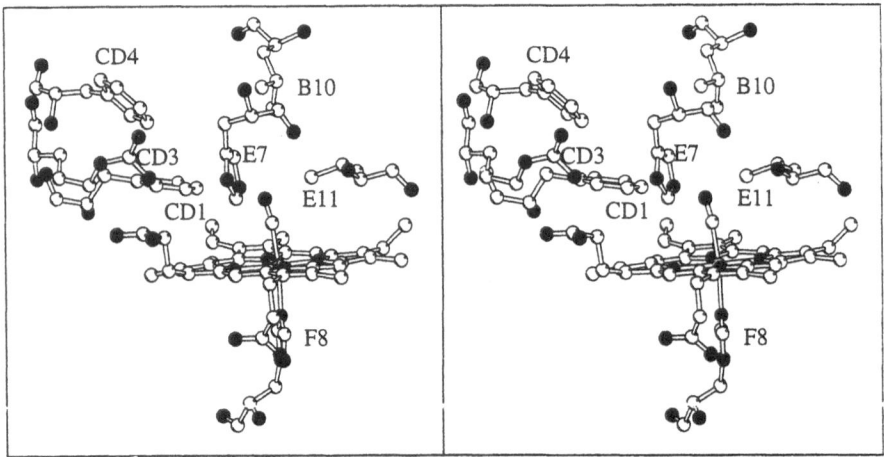

Figure 6. Distal residues surrounding the binding site for dioxygen and carbon monoxide in myoglobin.[5]

of structure with function and reactivity. Further strengthening the correlations for sperm whale myoglobin and its mutants are very similar observations from a more restricted set of human and sheep myoglobin mutants. Because the changes are largely confined to internal surfaces of the protein, in essence the consequences of a single change in residue can be followed without the concerns engendered in model systems that the structure observed in the solid state, if directly observed at all, may differ from the structure observed in solution where the thermodynamic and kinetic measurements are made. Moreover, the slow rate of autoxidation of myoglobin and of most of its mutants, facilitates in principle direct measurement of O_2 affinities, although in practice affinities for dioxygen and carbon monoxide, $K_C(O_2)$ and $K_C(CO)$, are usually determined kinetically, as the ratio of k_{on}/k_{off}. The stability of biological systems to autoxidation contrasts that of many model systems, where stability to autoxidation is brief. A selection of ligand binding parameters from sperm whale myoglobin and selected mutants is provided in Table 5. The highest and lowest values of each parameter are highlighted in bold-face type.

Distal effects that enhance affinity for dioxygen include:

(1) *Hydrogen bonding of coordinated dioxygen to distal moieties:* Relative to the gas-phase situation of ligand binding to a simple (axial base)porphyrinatoiron(II) species, BPFe, the polar $Fe-O_2$ bond is stabilized by hydrogen bonding to distal moieties. This will be discussed in more detail later. Replacement of the distal histidine([64]E7) by a

non-polar residue, such as valine, lowers O_2 affinity by at least an order of magnitude relative to the native protein; the decrease in affinity lies mainly in an increased rate of dissociation, consistent with the absence of a hydrogen bond.

Table 4. Native and mutant myoglobins studied by the groups of Olson, Sligar and Phillips.[5] Mutants subjected to crystallographic analysis are shown in bold-face type.

Myoglobin	His E7	Myoglobin	Val E11
SW, **Pig**, Human	---	SW, **Pig**, Human	---
SW	**Gln**	SW	Gln
SW	Thr	SW	Asn
SW	**Tyr**	SW, Pig	Ser
SW	Arg	**Pig**	**Thr**
SW	Asp	SW	**Phe**
SW	Trp	SW	**Ala**
SW	Met	SW	**Leu**
SW	Phe	SW, Pig	**Ile**
SW, Human	**Gly**		
SW, Human	Ala	Myoglobin	Phe CD1
SW, **Pig**, Human	**Val**		
SW, Human	**Leu**	SW, **Pig**, Human	---
SW, Human	Ile	SW	Trp
		SW	Val

Myoglobin	Leu B10	Myoglobin	Phe CD4
SW, **Pig**, Human	---	SW, **Pig**, Human	---
SW	Trp	SW	**Val**
SW	Phe	SW	**Leu**
SW, Human	Ala		
SW, Human	Ala		
SW	**Val**		
SW, Human	Ile		

(2) *Interaction of polarizable species with coordinated dioxygen:* Replacement of the non-polar leucine residue at position B10 by the more polarizable phenylalanine (L29F), enhances O_2 affinity by more than an order of magnitude. The increase arises primarily in an order of magnitude decrease in the dissociation rate constant, $k_{off}(O_2)$.

Replacement of the distal histidine by glycine, H64G, decreases O_2 affinity, but less than expected. In the space vacated by the imidazole moiety, a water molecule is observed in the mutant oxidized myoglobin, met^{H64G}Mb, and the carbonmonoxy adduct, ^{H64G}MbCO. It is proposed that this water is also present in ^{H64G}MbO$_2$ and is hydrogen-bonded to the coordinated dioxygen moiety. Autoxidation rates are inversely correlated with dioxygen affinity. Charged distal groups, for example a distal aspartate residue (H64D), lead to autoxidation rate constants more than three orders of magnitude greater than that for the native myoglobin.

Affinities for CO are rather insensitive to the nature of the distal group, in particular to replacement of the distal histidine by non-polar residues such as alanine, valine, isoleucine, phenylalanine and threonine (presumably with the polar -OH group directed away from the coordinated CO molecule. A substantially enhanced CO affinity occurs for the leucine mutant, H64L --- no satisfactory explanation for this has been advanced yet, but greater polarizability of this moiety compared to valine may be a factor. Only in the case of severe steric bulk, such as the tryptophan mutant L29W, is CO affinity greatly diminished. And in only one mutant does steric bulk, rather than electrostatic interactions,

437

Table 5. Ligand affinities of myoglobins and selected sperm whale myoglobin mutants.[5]

Myoglobin	$k_{on}(O_2)$ $\mu M^{-1} s^{-1}$	$k_{off}(O_2)$ s^{-1}	k_{ox}^a h^{-1}	$K_C(O_2)$ μM^{-1}	$k_{on}(CO)$ $\mu M^{-1} s^{-1}$	$k_{off}(CO)$ s^{-1}	$K_C(CO)$ μM^{-1}	M_c
Sperm whale[b]								
native	17	15.	0.055	1.1	0.51	0.019	27.	25.
64E7 Tyr[a]	6.7	3 200	>100.	**0.0021**	0.50	**0.092**	5.4	2 600
29B10 Phe	21.0	1.4	**0.005**	**15.0**	0.22	0.006	37.0	2.5
64E7 Gly	**140.**	1 600	44.	0.090	5.8	0.038	150	1 700
64E7 Val	110	**10 000**	33.	0.011	7.0	0.048	150	**14 000**
29B10 Trp	**0.25**	8.5	0.18	0.029	**0.0039**	0.008	**0.48**	16.
64E7 Leu	98.	4 100	10.	0.023	**26.**	0.024	**1 100**	48 000
68E11 Asn	1.9	**0.54**		3.5	0.041	**0.0096**	4.3	**1.2**
Hb *Ascaris*[c]	1.5	0.0041		370.	0.21	0.018	12.	0.032
Mb *Aplysia*[d]	15.	70.		0.22	0.49	0.02	25.	110.
LegHb[e]	156.	11		14.	13.5	0.012	1100.	80.

[a] The aspartate mutant 64E7 mutant probably has the greatest k_{ox}, as the rate of autoxidation precluded measurement of $k_{on}(O_2)$ and $k_{off}(O_2)$. [b] Reference 5; typographical errors for tyrosine entry corrected. [c] Reference 52. [d] Reference 53. [e] Reference 54.

appear to decrease CO affinity more than O_2 affinity: this occurs when the valine at [68]E11 is replaced by the bulkier isoleucine. Indeed, replacing the distal histidine by the bulky tryptophan, leaves CO affinity unchanged, while O_2 affinity decreases by more than an order of magnitude. Replacement of the non-polar valine residue at position [68]E11 by polar residues, such as the isosteric asparagine (V68N), decreases CO affinity and increases O_2 affinity. Comparing O_2 and CO binding, using the discrimination factor

$$M_C = K_C(CO)/K_C(O_2)$$

shows that sterically relatively benign substitutions lead to values of M_C spanning more than four orders of magnitude. In general, factors stabilizing O_2 binding destabilize CO binding, as shown in Figure 7; the correlation is weak.

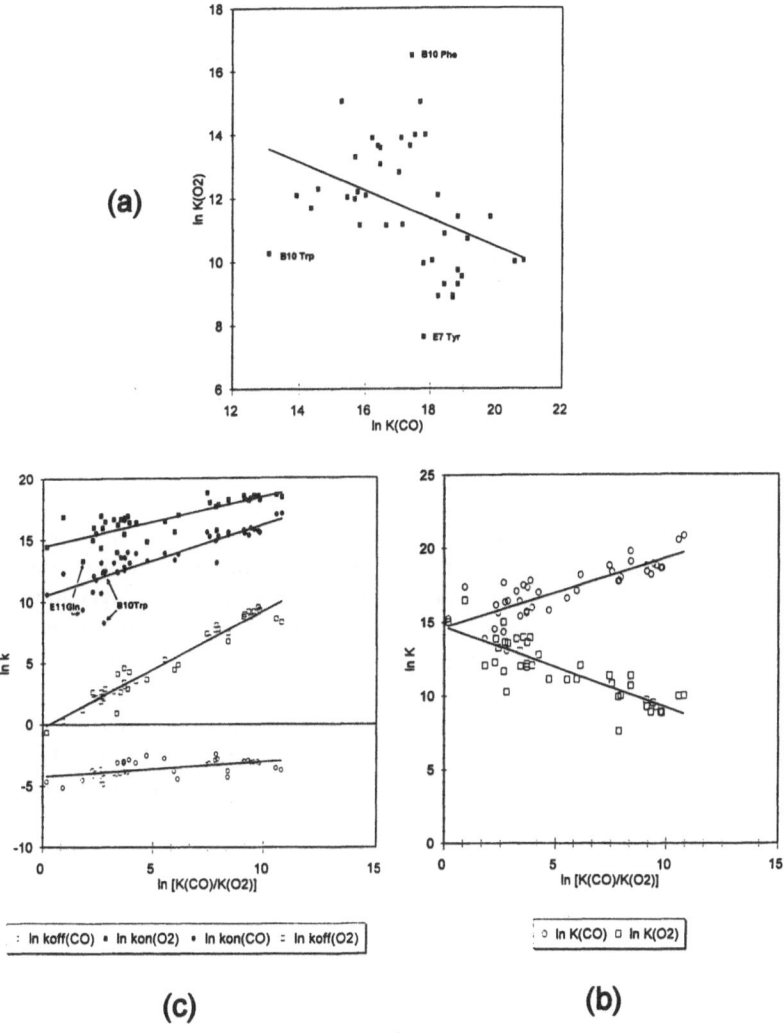

Figure 7. Comparison of O_2 and CO binding for myoglobin mutants. (a) Plot of $ln\ K_C(O_2)$ *versus* $ln\ K_C(CO)$. (b) Plot of discrimination factor, $M_C = K_C(CO)/K_C(O_2)$ *versus* $ln\ K_C(CO)$ and $ln\ K_C(O_2)$. (c) Plot of discrimination factor versus $ln\ k_{on/off}(CO)$ and $ln\ k_{on/off}(O_2)$. Note that the discrimination factor is most sensitive to $k_{off}(O_2)$. Data taken from reference 5.

439

Control of Ligand-Binding Dynamics

Table 5 also summarized the range of association and dissociation rate constants for the binding of O_2 and CO to myoglobin mutants. More detailed studies at low temperatures and at shorter time scales,[55] and the trapping of the photolyzed MbCO for crystallographic characterization[42] reveal that the ligand binding process is vastly more complex than that summarized by the equation:

$$BPFe + L \rightleftharpoons BPFeL \quad (L = O_2, CO).$$

However, the intricacies of residues moving to permit access to the ligand binding site, the progress of the ligand from solution to the vicinity of the binding site and the actual binding and dissociation processes are outside the scope of this review. Suffice to note that such studies have been made on primarily biological systems to date.

Control of Autoxidation by Hydrogen Bonding. Strong binding of dioxygen depresses the rate of autoxidation[5] -- contrary to naïve presumptions that stronger binding leads to greater charge separation in an $Fe^{III}\text{-}O_2^{-I}$ species and a greater susceptibility to proton-assisted displacement of $HO_2\cdot$. Autoxidation is facilitated in the absence of a hydrogen bond between the coordinated dioxygen and an H-X moiety (X = N<, O-). Hydrogen bonding between the coordinated dioxygen ligand and distal moieties, such as water and imidazole from histidine, not only increases O_2 affinity but also enhances the stability of the $Fe\text{-}O_2$ moiety, so that while a formally $Fe^{III}\text{-}O_2^{-I}$ species is stabilized the heterolytic cleavage shown below is disfavored:

$$Fe^{II} + O_2 \rightarrow Fe\text{-}O_2 \rightarrow Fe^{III} + O_2^{\cdot -}.$$

Control of Ligand Dissociation and Association. O_2 dissociation rate constants span four orders of magnitude. Replacement of the distal histidine by non-polar residues, leads to an increase of two to three orders of magnitude in the O_2 dissociation rate constant, due to loss of hydrogen bonding. For the glycine mutant (H64G), for which the dioxygen affinity is not as low as that of other mutants involving non-polar substitutions, the lower dissociation rate constant may be due to hydrogen bonding to a water molecule that may be present in the imidazole site -- such water is observed in the metMb crystal structure. Replacement of the non-polar aliphatic valine residue by polarizable phenylalanine and polar residues, such as glutamine and asparagine, (and also isoleucine), decrease the dissociation rate constant by as much as an order of magnitude. Consistent with the proposal of a specific interaction between coordinated dioxygen and this phenylalanine moiety[5] is the complementary observation that replacement of the distal phenylalanine residue at [46]CD4 by aliphatic moieties leaves CO affinity essentially unchanged, but decreases O_2 affinity by more than an order of magnitude, primarily by an increase in the dissociation rate constant.

In contrast to the wide range of dissociation rate constants for O_2, those for CO vary by little more than one order of magnitude. Substitutions at the distal histidine site, whether non-polar, polar (tyrosine) or charged (but not arginine, which may be oriented out into the solution), all show moderate increases both in the $k_{off}(CO)$ and $k_{off}(O_2)$, and the $k_{on}(CO)$ and $k_{on}(O_2)$, with the exception of tyrosine and tryptophan. At least for the non-polar residues, the presentation of a non-polar surface to the aqueous solution appears to function as a phase-transfer catalyst for CO and O_2 from the polar aqueous environment to the less polar internal environment of the ligand-binding pocket. CO and O_2 are both about an order of magnitude more soluble in aromatic solvents, such as toluene, than in water.

Whereas $k_{on}(CO)$ span nearly four orders of magnitude, for the same selection of mutants, $k_{on}(O_2)$ span just over two orders of magnitude. Non-polar mutants at His E7 increase $k_{on}(O_2)$, and most mutants at Val E11 site decrease $k_{on}(O_2)$. Bulky distal residues, such as tryptophan at the leucine B10 site, depress $k_{on}(CO)$ and $k_{on}(O2)$; the polarizable phenylalanine residue stabilizes both the CO and O_2 adducts against dissociation.

Linear Free Energy Relationships. LFERs can offer insight into the possible resemblance of transition-state species to the reactant or product species. When ligand binding data from all 42 myoglobin mutants are examined, strong correlations are not found, especially for O_2 binding (Figure 8(a) and (b)). For the binding of CO, $k_{off}(CO)$ is essentially independent of K(CO), but between *ln* $k_{on}(CO)$ and *ln* $K_C(CO)$, the slope ≈ 1 -- a value that has been interpreted in model systems showing these trends for O_2 binding as being consistent with "a binding mechanism involving a large conformational change of the encumbered porphyrin macrocycle taking place before binding occurs and being retained in the

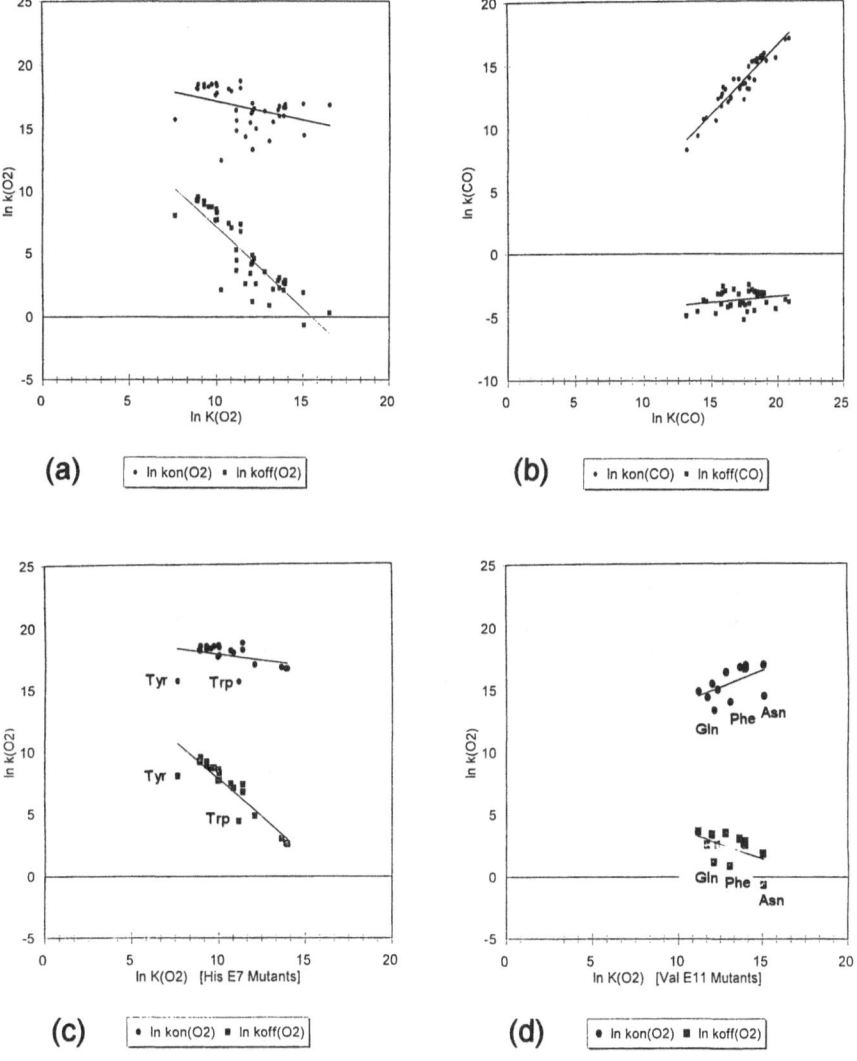

Figure 8. Linear free energy relationships (*ln* K_C *vs ln* k_{on} and *ln* K_C *vs ln* k_{off}) for ligand binding to selected vertebrate myoglobins and their mutants. Data taken from reference 5. (a) LFER for O_2 binding to 42 mutants. (b) LFER for CO binding to 42 mutants. (c) LFER for O_2 binding to 20 mutants, involving the distal histidine (E7). (d) LFER for O_2 binding to 10 mutants, involving the distal valine (E11).

bound state." There is structural evidence contrary to such a mechanism for myoglobin: in the photodissociated MbCO structure, for which the dissociated CO lies uncoordinated above the heme, the $Fe-N_{His}$ bond is intermediate between those for the fully bound MbCO and fully dissociated Mb structures[42] -- on the assumption that the photodissociated species represents an intermediate on the ligand dissociation (and association) path, and not a kinetic blind alley. Moreover, as described in more detail later, the binding of CO to Mb appears to be accompanied by expulsion of water from the binding pocket.

For O_2 binding to the myoglobin mutants essentially opposite correlations are observed compared to CO binding. Restricting the set to the distal histidine mutants, improves the degree of correlation Figure 8(c), especially if the points belonging to the very bulky tryptophan and the possibly coordinated tyrosine mutants are excluded. The correlation is similar to that seen for O_2 binding to flat "open" porphyrins. There, a repulsive interaction between O_2 and the five-coordinate porphyrinatoiron(II) species in the transition state was proposed.[4] For ligand binding to myoglobins, one might expect this effect to obtain also for CO, but it does not. A repulsive interaction of O_2 with water in the binding pocket, which devolves into a favorable hydrogen bonding of coordinated O_2 seems more appropriate. The strongly negative slope for O_2 dissociation is consistent with hydrogen bonding to the $Fe-O_2$, an explanation not available to explain the strong negative slopes for $ln\ K_C(O_2)$ versus $ln\ k_{off}$ for the flat "open" porphyrins.

Restricting the set to the valine E11 mutants reveals, but not strongly, that the factor(s) depressing the association rate constant also tends to enhance the dissociation rate constant (Figure 8(d)). Note that the curve relating $ln\ K_C(O_2)$ with $ln\ k_{on}(O_2)$ for the E11 mutants has opposite slope to that for the E7 mutants. For the E11 mutants, separating out the residues with hydrogen bonding potential (glutamine, asparagine and phenylalanine), leads to a good correlation for the remaining seven points. The correlation here resembles that for some sterically encumbered "picket fence" porphyrin compounds offering amide moieties towards the coordinated dioxygen. Note that phenylalanine, typically considered hydrophobic, is also found making specific dipolar contacts with polar species, as observed in the structure of the dioxygen derivative of the Leu ^{29}B10 Phe mutant -- the sperm-whale myoglobin mutant showing the highest affinity for O_2.[5c]

In summary, linear free energy relationships do not reveal microscopic mechanisms -- mechanisms that seem reasonable for model systems are inapplicable to myoglobin mutants, which then raises doubts as to the uniqueness of the interpretation for the model systems in the absence of accompanying structural data. There is, further, a real temptation with LFER's to indulge in "data snooping": from the original 42 mutants, it is possible to select a group of 12 mutants, where $ln\ K_C(O_2)$ versus $ln\ k_{on}(O_2)$ (Figure 8(a)) are excellently correlated with a slope close to unity -- opposite to the trends separated out above.

Ligand Binding in Model Systems

Comparing model systems. A vast and imaginative array of different model systems have been synthesized. Comparison among model systems is fraught with difficulties; direct comparison between model and biological systems is hazardous. Beyond macroscopic features, it is becoming increasingly apparent that the nature of ligand binding to model and biological systems rarely share a common mechanism. At the same time, it must be stressed that a complete understanding of ligand binding to porphyrinato-iron(II) species requires that the nature of ligand binding in non-biological systems also be studied and understood, since it is from the *differences* between systems, and the different mechanisms by which accidentally similar parameters obtain that insight into ligand binding has been obtained. Two examples will be discussed later: the importance of hydrogen bonding, and the stereochemistry of the Fe-C≡O moiety.

Some of the difficulties in comparing model systems with one another and with biological systems include:

(1) *Role of solvent:* The role of solvation in determining ligand-binding properties is generally unknown when comparing one system in one solvent with another system in a second solvent.

(2) *Porphyrin-substituent induced perturbations:* Porphyrin substituents, e.g. caps over the ligand-binding site, distort the porphyrin ring from planarity. The extent of distortion may differ between liganded and unliganded derivatives and the conformational energy involved is usually unknown.

(3) *Indirect structural comparisons:* There is dearth of structural information on many model systems. In others, structural data obtained on one system (e.g., the "picket fence" porphyrin) is applied to give a structural perspective on kinetic and thermodynamic data on other ostensibly closely related systems (e.g., a "picket fence" porphyrin derivative where the axial base is covalently delivered to the iron by replacing one of the "picket" substituents).[2,4,6]

(4) *Structural studies in the crystalline state applied to kinetic and thermodynamic studies in the solution state:* While there is ample evidence that the structure determined by X-ray diffraction methods in the crystalline state for both proteins and smaller molecules is a good approximation to the structure in solution, the structural perspective accorded to thermodynamic and kinetic data is, then, necessarily indirect. EXAFS and NMR methods offer a bridge between the crystalline or solid state and the solution state.

(5) *Kinetic data are reported in molar units; equilibrium data measured directly are frequently reported in units of gas pressure.* The solubility of dioxygen and carbon monoxide in water differs by almost an order of magnitude from that in non-protic solvents, while in a given solvent, for example water or toluene, the solubility of CO and O_2 are similar to within about 20 %. Moreover, solubility data for gases dissolved in solutions of other substances, may be different to those data for pure solvents. Reliable solubility data for O_2 and CO are scarce.

(6) *Limited extent of comparison:* Only rarely can more than two model systems be compared. For this reason, the correlation between the presumed cause and the effect measured may be accidental rather than actual. Linear free energy relationships between affinity and rate constants for association and dissociation of ligands may help in grouping compounds where ligand binding shares a similar mechanism. Linear free energy relationships allow limited conclusions to be made regarding the similarity of the transition state to either reactant or product species.

Relating Ligand Binding as K_P to K_C. Further to items (5) and (6) above, for equilibrium measurements of ligand affinity, measurement of O_2 affinity as

$$P_{1/2}(O_2) = 1/K_p = [BPFe] P(O_2) / [BPFeO_2],$$

where $P_{1/2}(O_2)$ is the partial pressure of dioxygen required to saturate half of the binding sites, has the advantage of being solvent independent. At equilibrium, the chemical potential of O_2 or CO in the gas phase must equal that in the solution phase, and to a reasonable approximation the activity coefficients for the liganded and unliganded (axial base)porphyrinatoiron(II) species are the same for a given solvent, leaving the concentrations of these species to cancel out when $[BPFe] = [BPFeO_2]$. This facilitates comparison of relative CO and O_2 affinities among different solvents. So long as the solubilities of O_2 and CO in a given solvent are similar

$$K_H(O_2) \approx K_H(CO),$$

443

where $K_H(X)$ is the Henry's law constant (often in units, mol L^{-1} mm-Hg^{-1}), then the discrimination parameter expressed in concentration units,

$$M_C = K_c(CO)/K_C(O_2),$$

is approximately the same as that expressed in pressure units:

$$M_P = K_P(CO)/K_P(O_2) = P_{1/2}(O_2)/P_{1/2}(CO),$$

since

$$P_{1/2}(O_2) = 1/K_P = 1/(K_H(O_2) K_C(O_2))$$

and

$$P_{1/2}(CO) = 1/K_P = 1/(K_H(CO) K_C(CO)),$$

which leads to

$$M_P \approx M_C K_H(CO)/K_H(O_2) \approx M_C.$$

In water $K_H(CO)$ and $K_H(O_2)$ are essentially identical; in toluene they differ by $\sim 20\%$.

Relating Ligand Binding in Different Solvents. The comparison of rate constants and equilibrium constants derived as the ratio of association and dissociation rate constants requires an attention to solvent that is usually overlooked, if data from one system in one solvent are to be validly compared with corresponding data from another system in a second solvent. Equilibrium constants measured thermodynamically in terms of partial pressure of the gas are, as detailed above, a good approximation to the true thermodynamic values where activities (or fugacities) are used. However, rate constants, and hence equilibrium constants derived from rate constants, are expressed and derived in terms of concentrations, not activities. Henry's law permits approximate comparison of equilibrium constants and, with some assumptions, association rate constants determined in *different* solvents. For convenience, suppose that in two solvents denoted [1] and [2], two systems have identical affinity as expressed by $P_{1/2}$ values:

$$P_{1/2}{}^{(1)} = P_{1/2}{}^{(2)},$$

then in terms of concentrations

$$K_C{}^{(1)} \approx K_C{}^{(2)} K_H{}^{(2)}/K_H{}^{(1)}.$$

Thus ligand affinity measured as K_C in one solvent can be compared to ligand affinity in a second solvent.

In terms of rate constants

$$K_C{}^{(1)} = k_{on}{}^{(1)}/k_{off}{}^{(1)} \text{ and } K_C{}^{(2)} = k_{on}{}^{(2)}/k_{off}{}^{(2)}$$

which leads to

$$k_{on}{}^{(1)}/k_{off}{}^{(1)} \approx (k_{on}{}^{(2)}/k_{off}{}^{(2)}) (K_H{}^{(2)}/K_H{}^{(1)}).$$

Now since k_{off} describes a first order reaction, and on the assumption that ligand dissociation is primarily a bond scission process that is solvent independent -- a reasonable assumption if the two systems being compared are identical in all respects except solvent but of less certainty otherwise -- then

$$k_{off}{}^{(1)} \approx k_{off}{}^{(2)}$$

and the expression above reduces to

$$k_{on}{}^{(1)} \approx k_{on}{}^{(2)} (K_H{}^{(2)}/K_H{}^{(1)}).$$

Therefore, *association rate constants can be expected to intrinsically differ by the ratio of the solubility of the gaseous ligand in the two solvents.* In the case of toluene and water this difference amounts to a factor of about 10. To a crude approximation, then, association rate constants measured in water are intrinsically an order of magnitude greater than association rate constants measured in toluene. The reasonableness of the assumption is justified by noting that $k_{off}(CO)$ for myoglobin mutants are insensitive to polar or non-polar distal residues, as models for solvent. Further and pertinent to item (6) above, the insertion of ligand-binding parameters measured in water for myoglobin onto a linear free energy plot for model systems in toluene or other aprotic, relatively non-polar solvents[4] is unjustified, as the effect of different gas solubility in different solvents on rate constants is ignored.

Temperature Dependence of Ligand Affinity. Comparison of ligand affinities measured at a single temperature runs the risk of leading to temperature-dependent conclusions, since

$$\Delta G_C = -RT \ln K_C = \Delta H - T \Delta S.$$

Unless ΔS is more or less constant in different systems, ligand affinities at one temperature will not parallel ligand affinities at other temperatures. The problem is exacerbated if ΔS for one system is opposite to that for another system, as apparently is observed in the hemocyanin family where for some hemocyanins ΔS is substantially positive, while for some other hemocyanins it is negative.[2] However, in all porphyrinatoiron(II) systems studied to date, biological and non-biological, the ΔS for dioxygen binding are negative, with a value of $\sim$-100 J mol^{-1} K^{-1}.[2] Thus differences in ligand affinities for porphyrinato-iron(II) systems are largely, and fortunately, confined to differences in the enthalpy contribution to the Gibbs free energy of binding, which is essentially temperature independent.

Stabilization of Fe-O$_2$ Moieties by Hydrogen Bonding. In both protein and model systems the importance of hydrogen bonding is now unequivocal.[4,5,56] This was not always so and the apparent absence of a hydrogen bond for the "picket fence" porphyrin O$_2$ adduct led to theoretical calculations that O$_2$ binding was *destabilized* by hydrogen bonding.[48b] While the FeO$_2$ ..HN(amide) distance of this model system is much longer than that for a typical hydrogen bond, there remains significant electrostatic interaction, since hydrogen-bonds diminish in energy as only $1/r$.[56] In an elegant model system to demonstrate the effects of hydrogen bonding on dioxygen affinity, an ether strap, that provided a protected site for O$_2$ binding, was replaced by an amide strap, that additionally offers hydrogen bonding capability to coordinated O$_2$ (Figure 9). The effects are summarized in Table 6.[57,58] A clear enhancement of O$_2$ binding, primarily by decrease of $k_{off}(O_2)$ was found, consistent with stabilization of the dioxygen adduct by hydrogen bonding. There are spectroscopic data

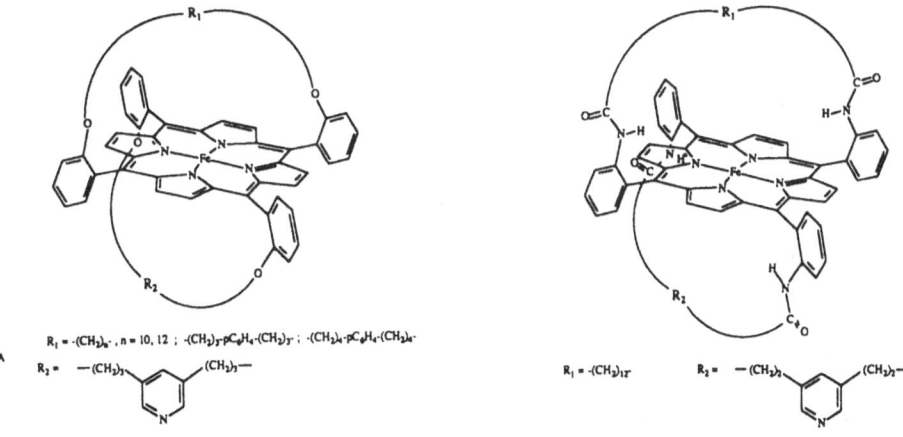

Figure 9. Schematic diagram of ether- and amide-strapped "basket handle" porphyrins.[58]

Table 5. Ligand binding to the iron center of "basket-handle" porphyrins of Figure 9.[58]

	Amide strap	Ether strap
k_{on} (O$_2$) (μM^{-1} s^{-1})	360	300
k_{off} (O$_2$) (s^{-1})	5,000	40,000
K_C (O$_2$) (μM^{-1})	0.072	0.0075
k_{on} (CO) (μM^{-1} s^{-1})	1150	990
k_{off} (CO) (s^{-1})	35	68
K_C (CO) (μM^{-1})	33	15
M_C	460	2,000

in support of a hydrogen bond, but no full structural characterization by X-ray diffraction techniques. After correction for the difference in ligand solubility between water and benzene (see above), the binding constants resemble those for native myoglobin, but the rate constants for both association and dissociation are orders of magnitude greater than those for native myoglobin.

From studies of met- and deoxymyoglobins, it is now fairly well-established that the binding pocket contains a water molecule hydrogen-bonded to a polar distal residue, histidine in the case of sperm whale deoxymyoglobin. From the decrease in dioxygen affinity when the distal histidine was replaced by glycine, the strength of the FeO$_2$--H-N(imidazole) hydrogen bond was estimated to be ~8 kJ mol^{-1}.[59] This represents a lower limit, as the coordinated dioxygen molecule is probably hydrogen bonded to a water molecule occupying the space vacated by the histidine.

For MbCO, the Fe-CO moiety is only weakly polarized. Thus, dipolar interactions of the coordinated CO molecule with distal residues are much weaker than the FeO$_2$··H-N< hydrogen bond. In further confirmation of the non-polar nature of the Fe-CO bond, the binding pocket of MbCO species is devoid of water, except for the glycine mutant.[5] Thus, for biological systems, the following processes, illustrated in Figure 10, occur for O$_2$ and CO binding when histidine, or asparagine, is the distal residue at ^{64}E7:

$$\text{Fe H}_2\text{O}\cdot\cdot\text{His} + \text{O}_2 \rightleftharpoons \text{FeO}_2\cdot\cdot\text{His(H}_2\text{O})$$

$$\text{Fe H}_2\text{O}\cdot\cdot\text{His} + \text{CO} \rightleftharpoons \text{FeCO} + \text{H}_2\text{O}.$$

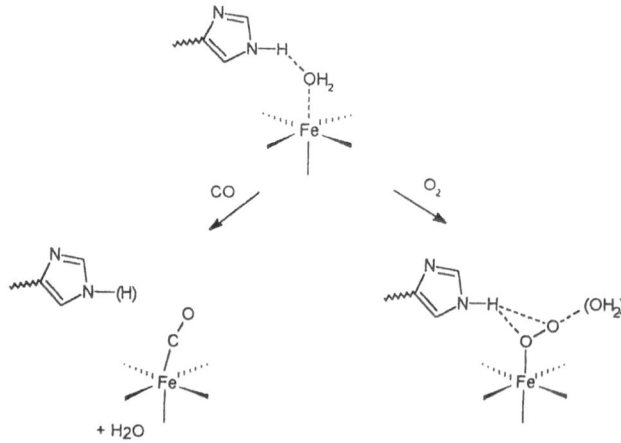

Figure 10. Schematic diagram contrasting O$_2$ and CO binding to sperm whale myoglobin.

In binding O_2 to sperm-whale myoglobin (and other vertebrate myoglobins), there is a hydrogen-bond compensation, where as one hydrogen bond is broken, another and stronger hydrogen bond is formed. This is not found in model systems, nor does hydrogen-bond compensation occur for carbon monoxide binding, in either biological or model systems. In general, in binding CO water is expelled from the binding pocket (except for the $H^{64}E7G$ mutant). Thus the water molecule hydrogen-bonded to the distal histidine in unliganded Mb stabilizes the five-coordinate species, relative to the six-coordinate, and, hence, lowers the ligand association rate constant, decreasing CO affinity. For mutants where the distal histidine is replaced by non-polar moieties bulkier than glycine, the binding pocket of unliganded Mb is devoid of water, leading to higher CO affinity, lower O_2 affinity and increased association and dissociation rate constants for both CO and O_2 binding. Tyrosine and tryptophan mutants behave differently, because of coordination to the heme and extreme steric bulk, respectively.

For model systems ligand-binding processes differ from those for native myoglobins, as expulsion of water, or other species, from the ligand-binding pocket does not appear to occur, although hydrogen bonding of the coordinated O_2 molecule to a polar moiety, H-X, may occur. The ligand-binding processes in model systems are summarized below:

$$Fe + O_2 \rightleftharpoons FeO_2 \cdots (H\text{-}X)$$

$$Fe + CO \rightleftharpoons FeCO$$

Neither model systems nor myoglobin mutants approach the extraordinarily high affinity for O_2 shown by the hemoglobins from *Ascaris*.[52] The origin of the high affinity lies primarily in a very slow rate of dissociation of 0.0041 s^{-1} -- more than two orders of magnitude slower than any of the mutants yet characterized. On the other hand, binding of CO to hemoglobin *Ascaris* resembles native vertebrate myoglobins in both kinetic and thermodynamic aspects. Thus, for hemoglobin *Ascaris* enhanced affinity for O_2 is not concomitant with diminished affinity for CO, as in the case of most myoglobin mutants. Structure analysis has begun,[52c] and it is tempting to speculate that the structures of the O_2 and CO derivatives are fundamentally different, with additional hydrogen-bonding capability for the O_2 derivative, compared to vertebrate hemoglobins, and extrusion of at least one hydrogen bonding moiety (protein-derived or water) into the solvent for the CO derivative. A similar rationalization was advanced for the anomalously high CO affinity of an arginine mutant of sperm whale myoglobin,[5] while in the structure of fluoromethemoglobin of *Aplysia* an arginine residue, normally directed out into solution, folds back into the ligand-binding pocket to hydrogen-bond to the coordinated fluoro ligand,[60] as there is no distal histidine or other polar residue.

In contrast to hemoglobin *Ascaris*, hemoglobins from the nitrogen-fixing nodules of leguminous plants, legHb, exhibit affinities for both O_2 and CO that are an order of magnitude higher than those of any vertebrate myoglobin or its mutants yet characterized.[54] This is achieved by enhanced rate constants for ligand binding, similar to the high rate constants for the sperm whale E7 His→Gly mutant, concomitant with ligand dissociation rate constants similar to native sperm whale myoglobin. The importance of structural data as a foundation for the microscopic interpretation of essentially macroscopic ligand-binding parameters, such as k_{on} and k_{off} and K_C can not be over-emphasized, as the myoglobins show a rich and not always expected variety of mechanisms for controlling ligand binding.

CO Binding and Fe-C≡O Stereochemistry

For myoglobin, O_2 affinity can be decreased while CO affinity can be increased by replacing the polar distal histidine by non-polar moieties. The discrimination factor,

$$M_C = K_C(CO)/K_C(O_2),$$

increases by up to three orders of magnitude. To a lesser extent, additional polar or polarizable moieties can enhance O_2 affinity at the expense of CO affinity, decreasing the discrimination factor by more an order of magnitude. It had been proposed[61] and widely expounded[2,4] that the distal histidine had a dual role of providing a hydrogen bond to the coordinated dioxygen -- now unequivocally established by direct observation of oxymyoglobin structures[37-40] and by spectroscopic methods -- and of decreasing CO affinity by preventing CO binding in a linear and perpendicular manner.[61] This proposal was set against an early backdrop of some surprisingly bent Fe-C≡O moieties reported for MbCO.[20] The contribution of steric effects to the discrimination factor has been extensively studied in a variety of model systems that seek to prevent the Fe-C≡O moiety from achieving its preferred linear and perpendicular geometry.[2,4,7,62] A significant but small degree of bending and tilting has been observed in several model systems coupled with an increase in discrimination against CO binding by three orders of magnitude. However, similarly and more bent and tilted Fe-C≡O moieties are now found in a variety of sperm whale Mb mutants, as well as in the native Mb, including the His E7 Gly mutant where there is no steric hindrance to a perpendicular and linear Fe-CO stereochemistry.[5,63] The direction of tilt of the Fe-CO group in MbCO structures is not invariant.

No explanation has been advanced, yet, for the observation of bent and tilted Fe-C≡O moieties in MbCO structures, but it may be noted that in model systems where the group is linear, the symmetry approximates fourfold. In myoglobin, the asymmetry of the heme and the asymmetry of the environment around the coordinated carbonyl moiety are such that degeneracy of the π^* orbitals is lifted, possibly leading to the observed bending and tilting.

It is tempting to try and relate structural and spectroscopic features of Fe-C≡O moieties to ligand-binding properties. Table 7 summarizes a variety of parameters relevant to CO binding to porphyrinatoiron(II) species. For MbCO structures there is no correlation of Fe-C≡O bond angle with affinity, as might be expected if steric effects alone caused bending, nor of the C≡O stretching frequency with Fe-C≡O angle. There is a reasonable correlation of the C≡O stretching frequency with affinity and with k_{on} rate constants. There is interestingly a reasonable correlation, not previously noted, with the discrimination factor, possibly the result of the cancellation of steric factors common to both CO and O_2 binding, leaving primarily polarity factors. The correlations are illustrated in Figure 11.

Table 7. CO affinity, CO stretching and Fe-CO geometry. Data taken from references 4, 5 and 63.

	$\nu(CO)$ (cm^{-1})	M_C[c]	$k_{on}(CO)$ (μM^{-1} s^{-1})	$K_C(CO)$ (μM^{-1})	Fe-CO[d] (°)
Fe(PF-Im)[a]	1965*		27,000	4,600	180
Fe(C$_2$Cap)(1-MeIm)[a]	2002		4200	19	173, 175
Fe(PocPF)(Im)[a]	1965		240	67	173
Mb (E7 Leu)[b]	1966		48,000	26	1,100	156
Mb (E7 Gly)[b]	1965		1,700	5.8	150	159
Mb (E7 Gln)[b]	1945		460	1.0	82	171
Mb[b]	1941	25	0.51	27	169
Mb (E11 Phe)[b]	1938		29	0.25	14	162
Mb (E11 Ile)[b]	1938		9.5	0.050	2.1	--
Mb (B10 Phe)[b]	1932		2.5	0.22	37	157
Mb (E11 Asn)	1916		1.2	0.041	4.3	--

[a] In toluene or benzene solvent; see Figure 4 for diagrams of porphyrins. [b] In water at pH 7. [c] $M_C =:$ $K_C(CO)/K_C(O_2)$. [d] MbCO structures determined from diffraction data to better than 2.0 Å; Fe-CO angles are reliable to $\pm5°$.

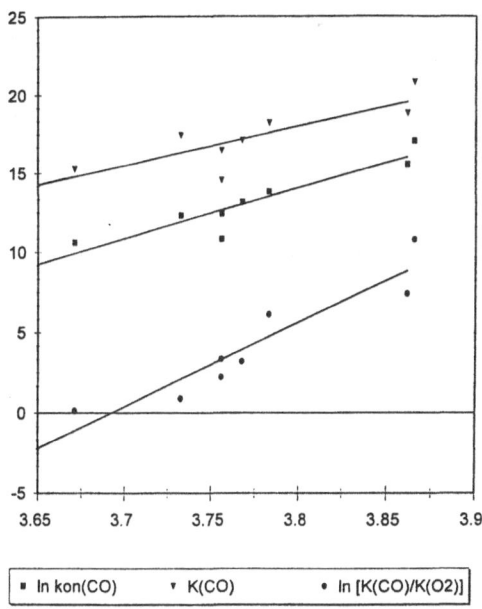

Figure 11. Correlations of the square of the CO stretching frequency ($1 \times 10^{-6} \, v^2 \, (cm^{-2})$) with the natural logarithms of the association rate constant, the binding constant and the discrimination factor. Data taken from reference 5b and 63.

While a low stretching frequency (energy) represents greater back-bonding into π^* orbitals of carbon monoxide and a stronger Fe-C bond, the strength of the Fe-C bond need not be correlated with either $K_C(CO)$, $k_{on}(CO)$ or $k_{off}(CO)$, since (i) $K_C(CO)$ represents the difference in free energy between five and six-coordinate species, and not the energy of the six-coordinate species with the Fe-C bond, and (ii) $k_{on}(CO)$, which is well-correlated with $K_C(CO)$, represents the difference in energy between the five-coordinate reactant and the transition-state species. This leaves $k_{off}(CO)$ for possible correlation with $v(CO)$; even though bond scission is probably the primary contributor to the overall value for $k_{off}(CO)$, differences in bond strength, as evidenced by $v(CO)$ are not manifest in the dissociation rate constant.

Misconceptions, Oversimplifications and Myths on Hemoglobins

A reading of biochemistry texts, and the anthropomorphic bias of most reviews would lead to the presumption that the behavior seen in vertebrate hemoglobins is general to all hemoglobins. That is, the structure and properties of sperm whale myoglobin, now studied in recombinant form from organisms whose objections to sacrifice have not found a human voice, represent monomeric hemoglobins, and the $(\alpha\beta)_2$ tetrameric structure of adult hemoglobin (HbA), readily available without sacrifice of the organism, represents the structure and properties, especially the mechanism of cooperativity, of all hemoglobins. In the last few years, a number of important results not only on vertebrate hemoglobins but also on invertebrate hemoglobins have appeared that challenge over-simplistic pictures of hemoglobin structure, function and reactivity. In particular, the mechanism of cooperative ligand binding is increasingly complex, as more data are accumulated.

Alternative Associations of Myoglobin-Like Subunits. That some invertebrate hemoglobins form high order oligomers (up to 192 subunits in size) that bind O_2 cooperatively has long been known.[64] The arrangement of subunits remains to be characterized. For lower-order oligomers, such as the vertebrate hemoglobins characterized to date, the subunits are arranged with the heme groups well separated and not in direct contact with one another (Figure 12).

β_2 β_1

C   G

F

α_2 α_1

Figure 12. Structure of tetrameric $(\alpha\beta)_2$ vertebrate hemoglobin, showing well-separated hemes.

However, the recent structure determination of a homodimeric, cooperatively binding hemoglobin from *Scapharca* reveals an association of subunits unlike that of any of the subunits of canonical tetrameric hemoglobins (Figure 13).[65] Direct heme-heme contact is observed, and the subunit interface is between the E and F helices, not the G and H of human hemoglobin. Relative to the canonical myoglobin structure (see Figure 3), there is an additional helix (pre-A), a longer F helix (the helix containing the proximal or coordinated histidine) and poor superposition of F and H helices. Further, in contrast to the minimal rearrangement of distal residues to accommodate the sixth ligand that is observed for vertebrate hemoglobins and myoglobins and most, but not all (see above) mutants, for this dimeric hemoglobin a large movement of a phenylalanine residue is required to accommodate the ligand in the ligand-binding pocket.

A homotetrameric hemoglobin from *Urechis caupo*, that shows non-cooperative binding, possibly because the allosteric effector remains to be found, also shows subunit contacts different to adult hemoglobin and to the dimeric hemoglobin from *Scapharca*. Compared with sperm whale myoglobin, the G and H helices of *Urechis caupo* hemoglobin are on the outside surface of the tetramer, with subunit contacts between the GH turn and E helix, and between the E helix and AB turn.[66]

Intermediate Quaternary States of Human Hemoglobins. To a first approx-imation, ligand binding data fit a two-state model, often referred to as the MWC (Monod, Changeux, Wyman) model of allostery.[69] Much crystallographic evidence has accumulated in support of this model[70] since the seminal paper by Perutz[31] and its elaboration by Baldwin and Chotia.[71] Vertebrate hemoglobins exist in equilibrium between two distinct quaternary structures: One structure designated, designated T, has a relatively low affinity for O_2 and CO, and is exemplified by unliganded hemoglobin. The other structure, designated R, has relatively high affinity for ligands, and is exemplified by fully liganded hemoglobin, e.g. oxyhemoglobin. From structures where both liganded and unliganded forms are frozen in

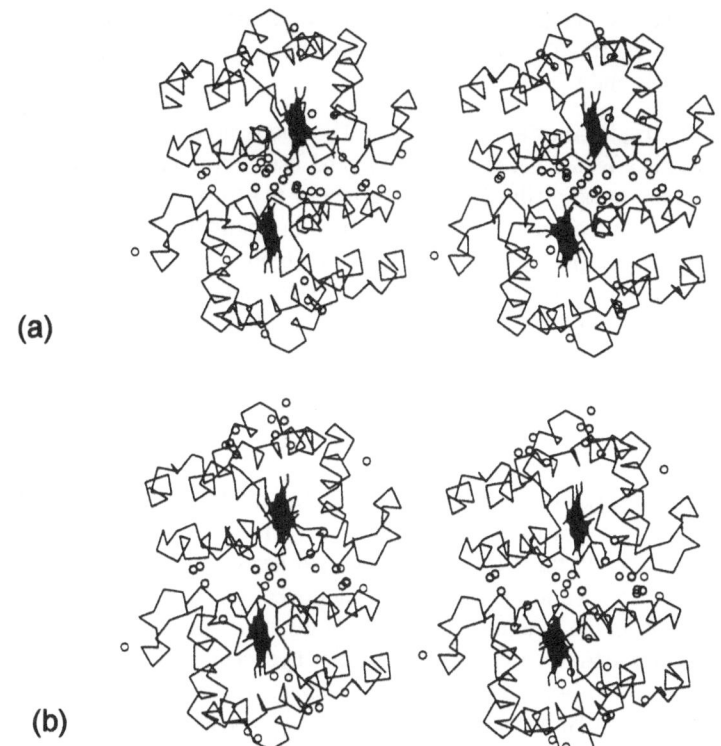

Figure 13. Stereodiagram of the dimeric hemoglobin from *Scapharca* in (a) deoxy and (b) carbonmonoxy forms, showing heme..heme contact.[65] Note the rearrangement of water molecules in the interface region upon CO binding.

the low affinity T state, it is found that ligand binding is accompanied by small changes in the immediate heme environment that are not propagated to the surface and hence to other subunits.[40,41,70] It is also possible for the β subunits of human hemoglobin to self-assemble into a homotetramer, for which the structures of the fully liganded carbonmonoxy *and* of the unliganded β_4 hemoglobin resemble the R quaternary structure.[72] On the other hand, when ligand binding switches the quaternary structure from T to R states, substantial changes in the immediate heme environment are observed, as well as changes remote to the heme associated with the change in quaternary structure.[40,41,70]

In addition to the T and R quaternary states, there is now good thermodynamic evidence for a third quaternary structure intermediate in energy between the R and T states and frequented by partially liganded hemoglobin.[73] Recently, a third quaternary structure (R2) was crystallographically characterized for hemoglobin, which may be relevant to the species thermodynamically characterized.[74] However, for mutant hemoglobins, the situation is less clear. For hemoglobin Ypsilanti from thermodynamic studies it was concluded that "..quaternary Y may represent the intermediate cooperativity state of normal hemoglobin that binds the last oxygen."[75] However, crystallographic characterization of hemoglobin Ypsilanti[76a] led to the conclusion[76b] that the structure of the CO derivative was better described as a modified R state, rather than as an intermediate on the path from T to R states.

While human adult hemoglobin is not a simple two-state allosteric protein, the structural features of the intermediate third state are not unequivocally defined. Moreover, for other tetrameric hemoglobins, mutants or other species, the quaternary states may not be identical in number and essential structure to native adult hemoglobin. Finally, the free energy of cooperativity ($\sim$ 15 kJ mol^{-1} -- the difference in ligand affinity between the T and R states) appears to lie in part with the R $\leftrightarrow$ T quaternary structure change and in part with sequential cooperativity within the R and within the T states.[77]

Alternative Heterotropic Allosteric Effectors. The cooperative binding of ligands to oligomeric hemoglobins is affected not only by the ligand itself (homotropic allostery), but also by other species (heterotropic allostery). The effect of pH (protons) -- the Bohr effect and in its more extreme manifestation, the Root effect -- on ligand binding is well-known, as is the effect of organic phosphates on vertebrate hemoglobins. For human hemoglobin, 2,3-diphosphoglycerate modifies the O_2 binding curve by stabilizing the low affinity T-state form of hemoglobin, sufficiently to ensure the transplacental transfer of O_2.[67] Other hemoglobins use inositol hexaphosphate. However, the use of organic phosphates as modulators of oxygen affinity and stabilizers of low affinity structures is not ubiquitous: recently, it was found that zinc ions performed functions similar to organic phosphates for earthworm hemoglobin.[68]

Hydrophilicity of the "Hydrophobic Pocket." Despite the long-standing observation of water in the binding pocket of deoxyhemoglobin, and water remaining in the binding pocket of oxymyoglobin,[5,37,38] there is still frequent and near-automatic inclusion of the word "hydrophobic" in reference to the ligand-binding pocket together with neglect of the water molecule(s) present in the binding pocket. While the entire pocket that encapsulates the heme is mostly hydrophobic (or lined with non-polar hydrocarbon residues), there is a hydrophilic space in the ligand-binding pocket in the vicinity of the iron center, due to the distal histidine in the case of sperm-whale myoglobin and many vertebrate hemoglobins (but not elephant myoglobin where a polar glutamine residue occurs). Recently, it has been shown that this water is central to the mechanisms by which hemoglobins discriminate against the binding of CO (see above and below).[5] For hemoglobins lacking a distal histidine or other polar residue in the binding pocket, other mechanisms are employed to stabilize polarized (e.g. O_2) or anionic ligands coordinated to the heme. In the case of hemoglobin from *Aplysia* an arginine, normally projected into solution for vertebrate hemoglobins, enters the binding pocket, along with three water molecules, to hydrogen-bond to polar ligands, as in fluorometMb derivative.[60]

Differentiation of Ligand Binding. The following points are reiterated:

(1) Reports of strongly bent Fe-C$\equiv$O moieties, with angles more acute than 150°, are incorrect.

(2) In hemoglobin, myoglobin and their mutants, there is significant but small bending and tilting of the Fe-C$\equiv$O moiety relative to the perpendicular to the heme plane. The extent is independent of affinity, C$\equiv$O stretching frequency, and ligand-binding kinetics.

(3) Hydrogen bonding stabilizes coordinated dioxygen by at least 8 kJ mol^{-1} .

(4) The relative affinity of myoglobin for CO and O_2 is determined primarily by binding-pocket polarity, water in the binding pocket, and hydrogen bonding effects of distal groups to water and the polarized Fe-O_2 moiety, and secondarily by steric effects.

Kinetic Measurements for Thermodynamic Affinity. It has generally been assumed that equilibrium constants for CO and O_2 binding determined kinetically as the ratio k_{on}/k_{off} are equivalent to those measured thermodynamically. Recent work suggests that these may differ by as much as a factor of five.[78] The reaction summarized by

$$Mb + O_2 \leftrightarrow MbO_2$$

is not an elementary reaction and the overall rate constants, k_{on} and k_{off}, are a composite of rate constants for the bumpy path from solvent to binding site. Indeed biphasic binding is observed in some myoglobin mutants.[5] That the equilibrium constants determined kinetically differ so little from those measured thermodynamically is fortuitous. In any event, comparison of thermodynamically- and kinetically-derived equilibrium constants should be made with caution, as noted earlier[6a] and recently reiterated.[4,78] However, there are now data from less heterogeneous sources and systems than before that invite closer and more reliable comparisons.[5,45]

TOWARDS STRUCTURE, FUNCTION AND REACTIVITY RELATIONSHIPS IN NON-HEME METALLOPROTEINS

To a large extent, the basic inorganic chemistry associated with a heme group and its axial ligands is well-defined in terms of structure, chemistry, spectroscopy and magnetic properties. It is from this foundation that model systems that actually incorporate functions of the biological systems have evolved, as described in the previous section. The questions being addressed have evolved in complexity from basic chemistry -- *what are the basic requirements for reversible ligand binding?* -- to basic stereochemistry -- *what is the geometry of the Fe-O$_2$ moiety?* -- to functional aspects -- *what factors control ligand binding?* -- to the design of physiologically functional blood substitutes. A perspective on future directions of hemoglobin research is deferred to the final section. For metalloproteins containing diiron and dicopper species, such as hemerythrin and hemocyanin, respectively, there still remain basic questions of how to assemble dinuclear complexes that in the first instance resemble spectroscopically and magnetically the biological systems -- that is establishing relationships among structure and spectroscopy. The subtleties of functional models remain in large part to be addressed. More so than for hemoproteins, there is the problem of establishing that the species mediating interesting redox chemistry in solution resemble the more or less stable species that are crystallographically characterized, since there is usually no stereochemically invariant moiety comparable to the heme group of hemoproteins for which only the identity but not the relative stereochemistry of exogenous ligands have to be identified.

The rapid progress of models for the hemoglobins was described earlier. The contrasts between the hemoprotein family and metalloproteins lacking the stereochemically restrictive heme group are well-illustrated by the slow genesis of models for oxyhemocyanin. In deoxyhemocyanin the protein constrains the two copper(I) atoms to close proximity, although no bridging group links them. Hemocyanin has been known for some time to bind dioxygen symmetrically between two copper atoms to give an effective dicopper(II)-peroxo moiety.[79] The O-O stretching frequency is anomalously low by comparison with peroxo-type ligands binding in either a triangular η-1,2 mode to a single metal or in a μ-1,2 mode between two metals. A variety of symmetrical bridging modes for dioxygen are possible, and a variety of μ-1,2-peroxo species have been characterized at low temperature[80] -- low temperature being needed to overcome an unfavorable entropy change for dioxygen binding that results from not having the copper atoms predisposed to coordinate dioxygen, as they are in hemocyanin. Eventually, from a rather simple ligand system, a μ-η^2:η^2-peroxo species was thoroughly characterized (see Figure 14) and found to be spectroscopically congruent with oxyhemocyanin.[24] This geometry had been mostly overlooked as a possible geometry, as the only precedents were two peroxo complexes of the oxophilic lanthanum and uranium ions.[81] Close on the heels of the model system came the structure of oxyhemocyanin, which confirmed the prediction of the μ-η^2:η^2-peroxo geometry.[25] The structure of the first dicopper-dioxygen model congruent with oxyhemocyanin occurred nearly twenty years after the first iron-dioxygen model congruent in structure with oxyhemoglobin. And from the prediction of a symmetrical dicopper(II) peroxo moiety in

1976, sixteen years elapsed before the distinctive spectroscopic and magnetic properties of oxyhemocyanin were fully elucidated -- that is relationships among structure, spectroscopy and magnetism established. In the intervening years, the fundamental coordination chemistry of copper(I) and copper(II) in dinuclear systems has been extensively explored,[82] so that now some of the factors which determine the mode of dioxygen binding to unconstrained model systems have been identified.[83]

Arthropodan hemocyanin and all functional hemocyanin models are symmetric. Molluscan hemocyanin and tyrosinase, and copper proteins with trinuclear motifs appear to be unsymmetric:[84] synthetic challenges remain for the synthesis of asymmetric dicopper complexes, especially tricopper species.

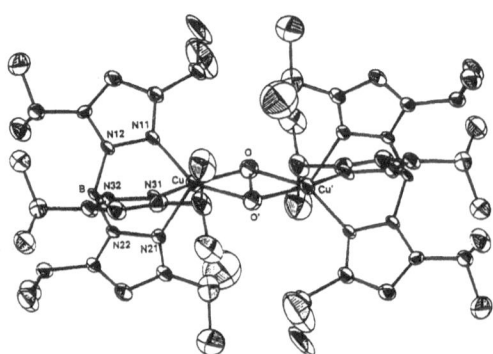

Figure 14. The μ-η^2:η^2-peroxo-dicopper(II) model for oxyhemocyanin.[24]

Further, in contrast to the hemoglobins, the roles of the histidine stereochemistry around the copper centers of hemocyanin and of the non-coordinating residues near the ligand-binding site in controlling ligand affinity and dynamics remain uncharacterized, either by comparison of binding data from other species, or by site-directed mutagenesis. At least part of the problem lies in the large minimal size of a functional agglomeration of subunits and the large size of the subunit itself, compared to hemoglobins, and perhaps also in the apparently lesser importance ascribed to this motif to human physiology, as compared to hemoglobins and cytochrome c oxidase.

Hemoglobin and hemocyanin comprise two of the three known solutions that have evolved for the transport of O_2. The third solution, hemerythrin, is a member of a family of proteins that feature in their oxidized forms a diiron(III) μ-oxo or μ-hydroxo core that is additionally bridged by carboxylato species. In contrast to the symmetric bridging mode that O_2 adopts in oxyhemocyanin, in oxyhemerythrin O_2 binds to only one of the pair of iron atoms, leading to a terminally coordinated peroxo moiety $Fe^{III}..Fe^{III}$-O_2H^{-II} , a mode of dioxygen coordination that remains crystallographically characterized only in the biological system.[15] The first model systems that reproduced the distinctive spectroscopic and magnetic properties of metazidohemerythrin, specifically the thermodynamically stable triply bridged μ-oxo-bis(μ-carboxylato) core, were characterized in 1984 (see Figure 15).[85] Notwithstanding the intrinsically asymmetric diiron complex of hemerythrin, symmetric model systems have provided considerable insight into the basic structural, spectroscopic and magnetic properties. Asymmetry is shared by other members of the μ-oxo diiron family of metalloproteins, which include ribonucleotide reductase, methane monooxygenase, rubrerythrin and purple acid phosphatase (where asymmetry is especially pronounced).[8,9,17]

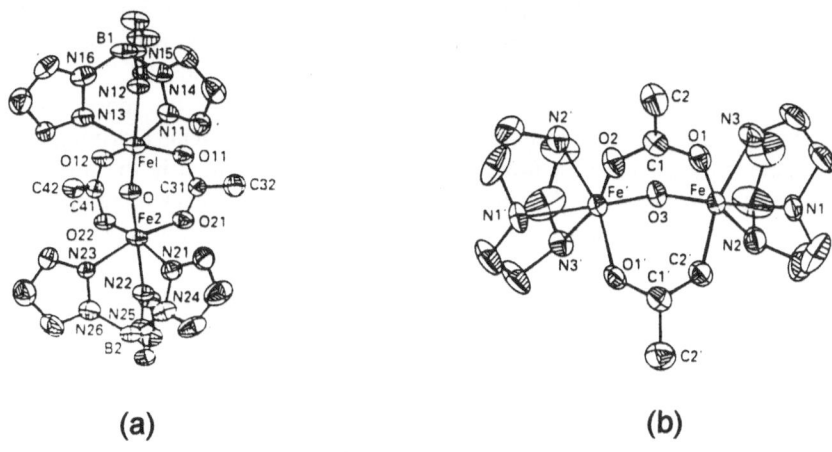

(a)	(b)

Figure 15. The first model systems for metazidohemerythrin, containing the μ-oxo-bis(μ-carboxylato) core. (a) Coordination at each iron is completed by tris(pyrazolyl)borate, HB(pz)$_3^-$.[85a] (b) Coordination at each iron is completed by 1,3,5-triazacyclononane.[85b]

The spectroscopic signatures of heme groups are thoroughly characterized, while those of μ-oxodiiron(III) species are of more recent and less complete determination.[21a] To date an oxo bridge *uniquely* mediates strong antiferromagnetic coupling between a pair of iron(III) centers, and usually leads to an enhanced symmetric Fe-O-Fe mode in the resonance Raman spectrum. The spectroscopic signatures of multiply-bridged μ-1,2-carboxy-lato-μ-oxo species have also been defined. However, spectroscopic signatures uniquely diagnostic of μ-hydroxo moieties, as occur in reduced forms of hemerythrin, in the diiron(III) state of methane monooxygenase,[14] but not the diiron(III) state of the similarly coordinated ribonucleotide reductase,[18] and as probably occur in purple acid phosphatases,[17] remain to be defined. Systems of weak analogy to metalloproteins serve as a valuable reference library to eliminate possible structures for the active site of incompletely characterized metalloproteins. And on the other hand, the distinctive metal complexes at the heart of many metalloproteins have inspired new areas of inorganic chemistry, such as O_2 binding to iron derivatives of non-porphyrin lacunary macrocycles,[86a] and fruitful analogies between metalloporphyrinato systems and totally inorganic heteropoly anions.[86b]

Myriads of symmetrical dicopper and oxo-bridged diiron species are known. The induction of asymmetry remains a key challenge in the area of dinuclear μ-(hydr)oxodiiron, and di- and trinuclear copper systems, since the perturbations in spectroscopic and magnetic behavior that may result from asymmetry remain incompletely defined.[87,88] Moreover, in order to generate functional models in a deliberate manner requires, in general, the creation of unsymmetric ligand systems, where the ligands present and coordinative unsaturation at a pair of metal centers can be controlled. The efforts of various groups, including the author's, towards unsymmetric and obligately asymmetric systems are described below.

Asymmetry in Metalloproteins with Multinuclear Metal Complexes

Asymmetry is a near ubiquitous feature of multinuclear metal complexes in biological systems, with the exceptions of arthropodan hemocyanin and some iron-sulfur clusters. Reasons why this should be so remain unaddressed. The following rationale is advanced:

(1) *Protein backbone intrinsically asymmetric:* The directionality of the backbone

chain of a protein molecule renders asymmetry in resulting metal complexes difficult to avoid. Indeed of the dinuclear metalloproteins, hemocyanin is the oddity with its effectively symmetric binding site and symmetric dioxygen adduct.

(2) *Directionality of electron flow:* In the case of electron transport, asymmetric systems, or the asymmetric assembly of metal complexes, provide a driving force and a directionality for electron flow. This is elegantly illustrated in the protein nickel hydrogenase, where an Fe_4S_4 cluster with one histidine supplementing the otherwise ubiquitous cysteine ligand sits near one surface and a string of Fe_3S_4 clusters span the protein to the opposite surface.[89] Cytochrome c oxidase offers also an asymmetric assembly of metal complexes to move electrons from cytochrome c to a highly asymmetric mixed-metal site where dioxygen is coordinated and reduced.[90]

(3) *Symmetric coordination disfavored:* An asymmetric dinuclear complex disfavors the symmetric bidentate coordination of potentially bidentate substrates, with resultant formation of stable intermediates. In asymmetric complexes, ligand affinity at one metal differs to that at the second. In this regard the asymmetry of tyrosinase, where dioxygen binds and proceeds to react with an organic substrate *via* a series of asymmetric intermediates,[24] may be contrasted with the symmetry of hemocyanin, where a stable symmetrically and reversibly coordinated dicopper-peroxo adduct results.[25]

(4) *Different coordinative unsaturation at metal sites:* Compared to mononuclear systems where at most two adjacent ligand sites are available for substrates (with fewer ligands the metal would bind to the protein with low affinity), in dinuclear systems, three or more sites are potentially available in the case of diiron systems, allowing for the coordination of more than one substrate. Asymmetry permits substrates to be preferentially directed to a site. In the case of purple acid phosphatases, the substrates are phosphate esters and water. In the case of a single substrate binding, a dinuclear system offers immediately two reducing equivalents to the substrate. For example dioxygen is effectively reduced to the peroxo state on coordination to dinuclear systems, whereas for hemoglobins, only a formal one-electron reduction occurs.

The occurrences of asymmetry in multinuclear metal complexes are reviewed next, and then recent results of the author's group on asymmetric dinuclear complexes are reported.

Asymmetry in Biological Di- and Trinuclear Copper Complexes. Figure 16a shows the asymmetric dicopper complex of a proposed intermediate in the catalytic cycle of tyrosinase[24] and Figure 16b the asymmetric trinuclear complex of ascorbate oxidase. Ascorbate oxidase appears spectroscopically as a mixture of a mononuclear Type II copper(II) species, showing EPR and uv-visible spectra typical of simple mononuclear copper(II) species, and a Type III, EPR-silent dinuclear copper(II) species.[91] Of the three copper atoms, the pair that is coupled appears to change with exogenous ligand.[92] Few unsymmetric synthetic dicopper(II) complexes are known, and the range of structural, spectroscopic and magnetic properties that may be associated with asymmetric di- and tricopper complexes remains largely uncharacterized.

Asymmetry in Cytochrome c Oxidase. After many years of study, recent investigations by means of site-directed mutagenesis have provided a clearer picture of the architecture of cytochrome c oxidase, and, in particular, of the ligands offered to the metal centers.[90] One six-coordinate heme functions to receive from cytochrome c electrons, which are passed through a helix to a second, coordinatively unsaturated heme that lies in close proximity to a copper site. Dioxygen binds between these last two metals. A second copper site, possibly a dinuclear copper complex, and also the first copper, facilitate proton movement, since in addition to the protons consumed in the reduction of O_2 to water

$$O_2 + 4\,H^+ + 4\,e^- \rightarrow 2\,H_2O$$

another four protons are pumped across the mitochondrial cell membrane. A current picture of cytochrome c oxidase is shown in Figure 17. The nature of bridging ligands that give rise to the distinctive magnetic properties of the coupled heme..copper moiety of cytochrome c oxidase is the object of intense study by means of model systems.[3a,93]

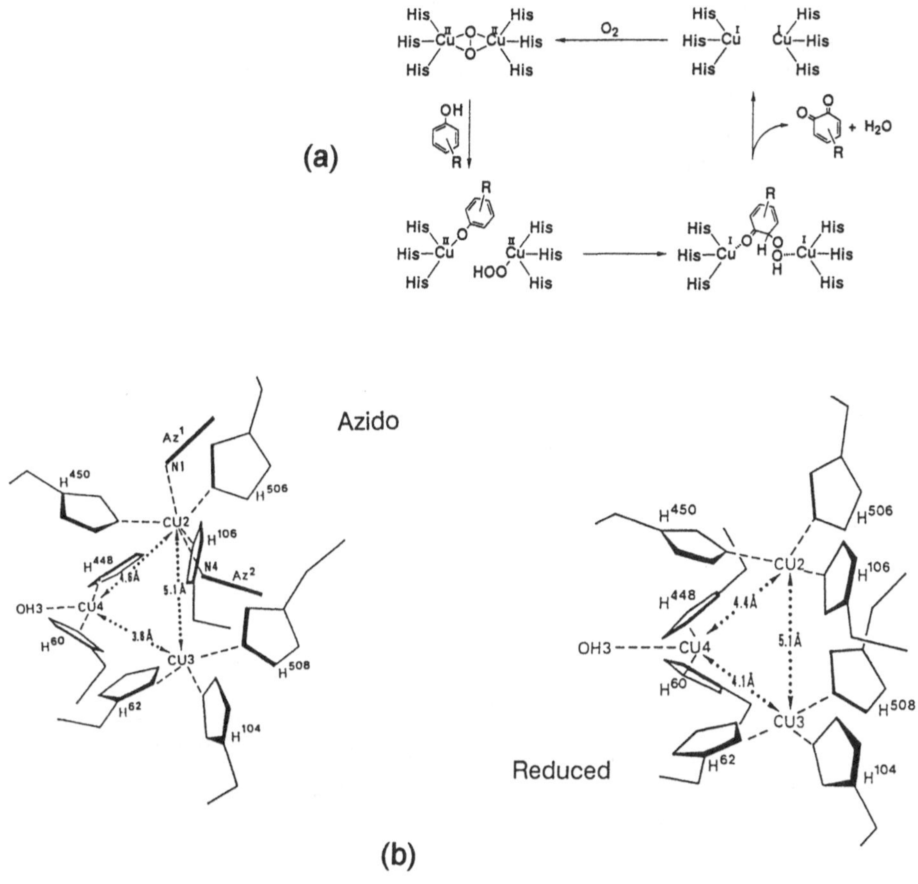

(a)

Azido

Reduced

(b)

Figure 16. Asymmetry in the active sites of (a) tyrosinase[24] and (b) ascorbate oxidase.[91]

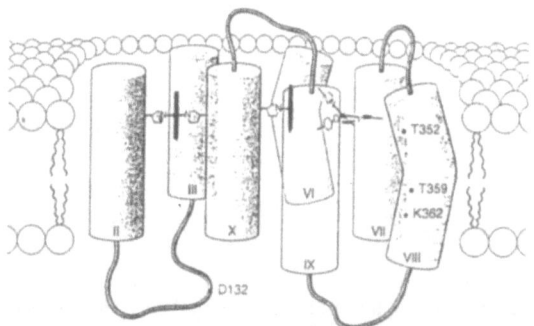

Figure 17. Current model for the active site of cytochrome c oxidase.[90a]

Asymmetry in Biological Iron-Sulfur Clusters. Most iron sulfur clusters are symmetric, ranging from the mononuclear Fe(SR)$_4$ species, where SR is a thiolato ligand from the amino acid cysteine, to the dinuclear [(RS)$_2$FeS]$_2$ to the tetranuclear [(RS)FeS]$_4$ species, illustrated in Figure 1(b).[12] An asymmetric variant of the dinuclear species occurs in Rieske centers, where the two thiolato ligands at one iron center are replaced by two histidine(imidazole) ligands. Asymmetric variants of the Fe$_4$S$_4$ distorted cubane-like motif exists, where one iron atom is missing, and where histidine supplements cysteine.[89] Higher order FeS clusters, as found in nitrogenase, are asymmetric, and include the eight-iron FeS cluster and the dinitrogen binding complex, the MoFe-cofactor, which contains a molybdenum atom. Many FeS and FeMoS clusters have been synthesized, but the clusters of nitrogenase have proved difficult to assemble in non-biological situations. These asymmetric species are illustrated in Figure 18.[12a]

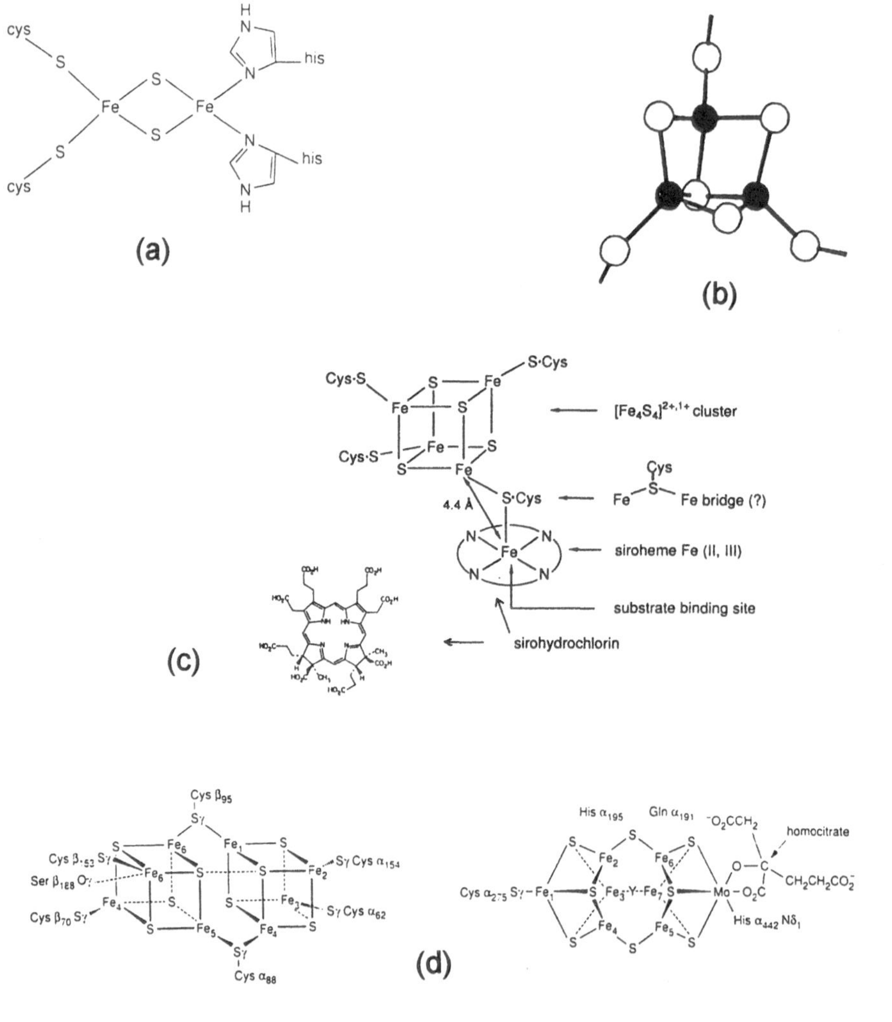

Figure 18. Asymmetric iron-sulfur clusters. (a) The less symmetrical (His)$_2$Fe(μ-S)$_2$Fe(Cys)$_2$ complex of Rieske centers[12] (b) The Fe$_3$S$_4$ cluster found in aconitase[12a] (c) Proposed active-site structure for sulfite reductase[12b] (d) Asymmetric Fe$_8$ and FeMoS clusters found in nitrogenase.[12a]

Asymmetry in Biological Oxo-Bridged Diiron Systems. The μ-oxo and μ-hydroxo diiron moieties occur in proteins of diverse functionality, including dioxygen transport (hemerythrin), mixed-function redox (methane monooxygenase and ribonucleotide reductase), and hydrolysis (purple acid phosphatase). The last activity, utilizing the very strong Lewis acid properties of high-spin iron(III) centers, is not found among the general functions of hemoproteins. The manifestations of the Fe-O-Fe motif in biological systems are illustrated in Figure 19.

The crystallographic elucidation of the active site of hemerythrin followed a tortuous path.[15,94] The distinctively strong antiferromagnetic coupling observed in *all* μ-oxo diiron(III) moieties characterized to date permitted the unambiguous identification of this moiety in met(oxidized)hemerythrin species long before crystallographic characterization was attempted.[22] The triply bridged, μ-oxo-bis(μ-carboxylato) motif, novel at the time of the first structural characterizations of hemerythrin (Figure 19a), took longer to recognize. μ-Oxo systems of only passing relevance to methemerythrin, as well as more explicit model systems were vital to characterization of the triply bridged core. Nonetheless, some apparent structural features of methemerythrin derivatives, specifically a highly unsymmetric placement of the bridging oxo ligand,[15a,95] remain at variance with expectations, and the mode of dioxygen coordination remains without crystallographically characterized precedent in a non-biological context.

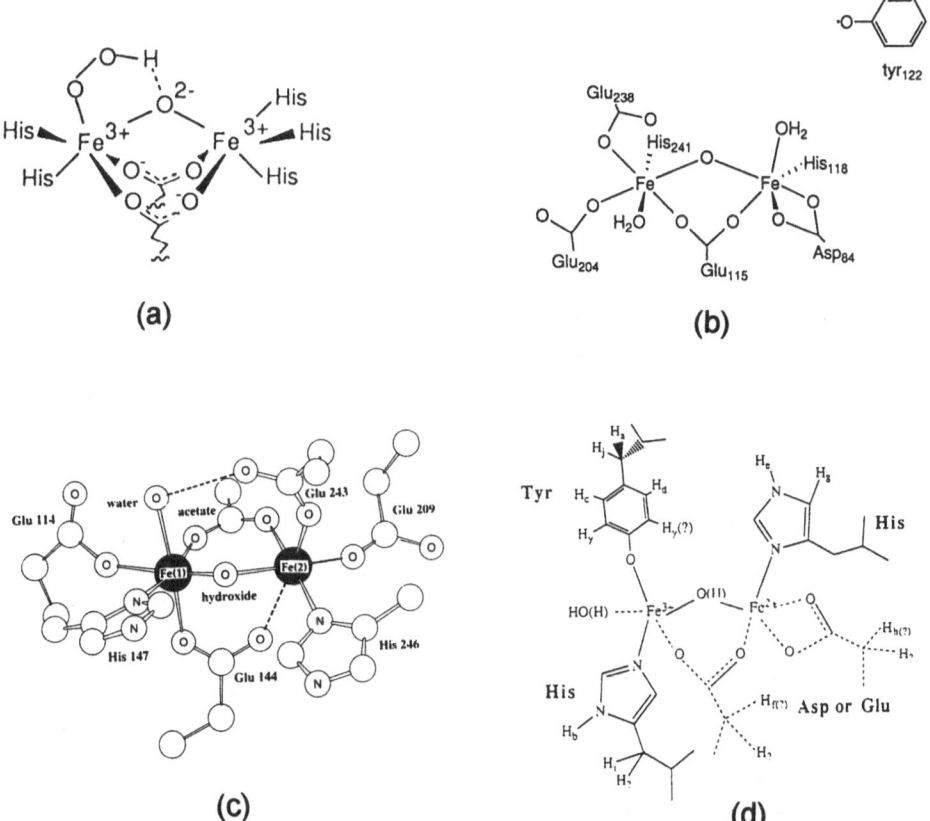

Figure 19. Asymmetry in the diiron complexes of some metalloproteins featuring oxobridged diiron complexes. (a) Oxyhemerythrin.[8] (b) Ribonucleotide reductase (oxidized).[8] (c) Methane monooxygenase (oxidized).[14] (d) Purple acid phosphatases (proposed).[17]

459

The structures of ribonucleotide reductase, in which a doubly bridged μ-oxo-μ-carboxylato core stabilizes a non-coordinated tyrosine radical (Figure 19c),[18] and methane monooxygenase (Figure 19b),[14,19] in which a μ-hydroxo species occurs, also feature asymmetric diiron complexes. In these cases, the complexes are richer in oxy anion-type ligands than hemerythrin, a feature which increases the basicity of the bridging oxo ligand so that at physiological pH it is protonated, at least in the case of methane monooxygenase. Asymmetry is not as strongly pronounced as in hemerythrin, at least in the species structurally characterized to date.

The active site of purple acid phosphatases, most studied as the purple acid phosphatase from bovine spleen and the uteroferrin from pig uterine fluid, is becoming clearer (Figure 19d).[17] The mechanism for stabilization of a mixed-valent $Fe^{III}Fe^{II}$ species, the physiologically more active form, remains unknown. The distinctive purple color is imparted by a phenoxo moiety from tyrosine that is terminally coordinated to the Fe^{III} site, giving pronounced asymmetry to the diiron species. Water is involved in the catalytic cycle of purple acid phosphatases, in which phosphodiesters are hydrolyzed, although the mechanism by which water is used remains unknown. The active site is an area of intense activity at present, but not all of the features deduced from spectroscopic studies to be present in the purple acid phosphatases have found expression in models systems yet.

Several other manifestations of the Fe^{III}-O-Fe^{III} motif are known.[14a,16] At least part of the active site of purple acid phosphatases resembles that for the mononuclear iron-tyrosinate proteins, such as the catechol dioxygenases,[96] and lactoferrin[97] and other members of the transferrin family.[98]

Strategies for Asymmetric Complexes

Several strategies exist for generating asymmetric complexes. Good luck or, more euphemistically, serendipitous self-assembly of unsymmetric species from symmetric building blocks, while intrinsically unreliable, is not to be abjured as a means of expanding horizons, especially since ligands intended to create obligately asymmetric dinuclear systems have often led to symmetric species, frustrating their creators' intentions.

Serendipitous Self-Assembly. The notion that the active sites of metalloproteins, especially non-heme proteins, might represent species that are thermodynamically stable and capable of an independent non-biological existence was articulated some time ago and termed "self assembly."[44] Thus, it is possible, even in aqueous solution, to generate $(RS)_4Fe_4S_4$ clusters,[99] and the triply bridged μ-oxo-bis(μ-carboxylato) diiron(III) core is, with hindsight, readily assembled from a a pair of iron(III) atoms, a couple of tridentate ligands, a couple of carboxylic acids, adventitious water and a bit of non-coordinating base.[16] While the latter recipe assembles symmetric complexes, some variants of this method lead to complexes that are unsymmetric in situations where symmetric species might have been expected -- serendipitous self-assembly. For example, the tripodal tetradentate ligand tris-(2-pyridyl-methyl)amine (TPA), has led to a number of unsymmetrical $\mu(O_2X)$-μ-oxo-diiron(III) species, where at one iron center, the amine is trans and at the other iron a pyridyl group is trans to the μ-oxo moiety, and where X = C-R, $P(OC_6H_5)_2$, etc., as illustrated in Figure 20(a).[100]

Figure 20. Ligands and accidentally unsymmetric complexes derived therefrom. (a) Tris(2-pyridylmethyl)-amine and an unsymmetric diiron(III) complex.[100] (b) 2,6-Bis(bis(2'-X)methylamine)-4-methyl-phenol and for X = benzimidazolylmethyl an unsymmetric dicopper(II) complex[101] and for X = benzimidazolyl- methyl[102a-c] or pyridyl mixed-valent ($Fe^{III}Fe^{II}$) diiron complexes.[102d] (c) 1-Bis(2'-benzimidazolylmethyl)-2-[(2'-benzimidazolyl-methyl)(2'-ethoxy)]diaminoethane, an unsymmetric μ-oxo diiron(III) complexes (X = Cℓ, Br).[87] (Facing page.)

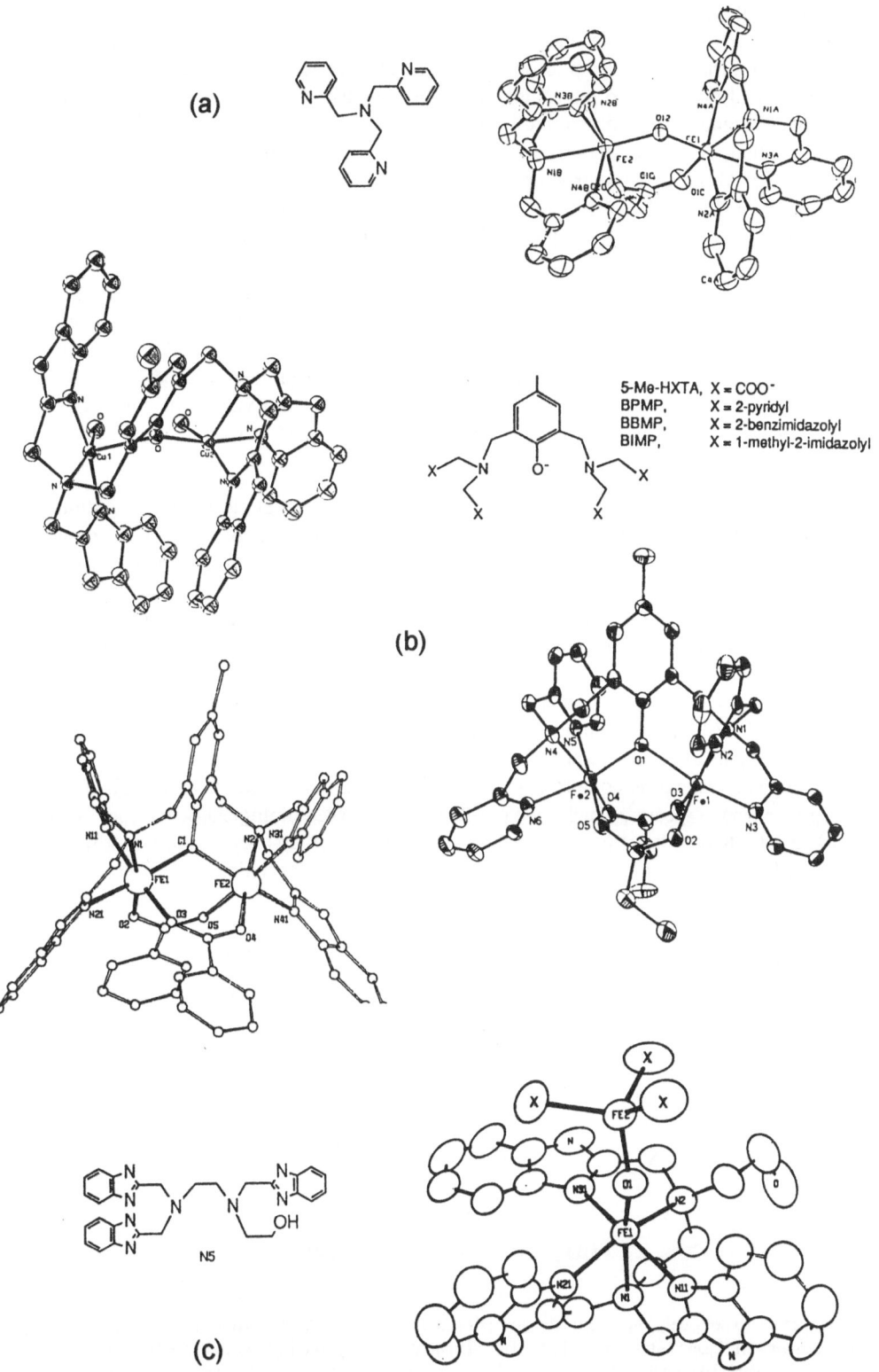

(a)

(b)

5-Me-HXTA, X = COO⁻
BPMP, X = 2-pyridyl
BBMP, X = 2-benzimidazolyl
BIMP, X = 1-methyl-2-imidazolyl

N5

(c)

In μ-phenoxodicopper(II) systems where two identical ligand-bearing arms are symmetrically attached to a central, bridging phenol moiety, substantially different copper(II) coordination environments and different Cu-O(phenoxo) separations result from the phenoxo bridge being axial to one copper(II) atoms and equatorial to the other.[101] A similar ligand that leads to the accidentally unsymmetric dicopper(II) species,[101] leads to slightly unsymmetric mixed-valent diiron(II,III) species,[102a-e] where in the solid state, but not in solution, the iron centers are distinguishable crystallographically and spectroscopically by means of Mössbauer spectroscopy.

An asymmetrically substituted pentadentate derivative of ethylenediamine, illustrated in Figure 20(c), that is potentially capable of bridging two iron centers, coordinates to one iron, leaving a single vacant site that can be occupied by alkoxo, halo and -O-FeX$_3$ (X = Cl, Br) moieties.[87] The μ-oxo derivatives could have in principle a single mirror plane in the plane of the Fe-O-Fe moiety, although in terms of coordination geometries at each iron -- approximately octahedral and tetrahedral -- the complexes are unequivocally unsymmetric.

In the realm of serendipitous self-assembly of unsymmetric species, as studies accumulate for the behavior of the ligand and metal ions in a variety of conditions, the ability improves to create a variety of products by manipulation of temperature, solvent and counter ions.

Asymmetric Ligands. An ostensibly more deliberate route to asymmetric dinuclear species, is *via* asymmetric ligand systems that lead to obligately asymmetric dinuclear complexes. Such ligand systems have proven synthetically challenging. A ligand bearing two pendant arms, but differing in one arm by the deletion of a methylene bridge between the amine and the central phenol ring (Figure 21a), leads to a remarkably symmetric-looking dicopper(II) complex, where both Cu-O(phenoxo) bonds are axial to each copper.[103a] Asymmetrically substituted 2,6-disubstituted-phenol molecules or aliphatic secondary alcohols offer routes into asymmetric dinuclear species. One such system, based upon Schiff base moieties,[103b] offers synthetic flexibility for its pendant arms and coordinates asymmetrically a pair of copper atoms. Congruence with any currently known dinuclear moiety in biological systems is compromised by the planar aromatic Schiff-base moiety and by bridging phenoxo species, which have presently no observed biological occurrence. Current manifestations of this ligand offer asymmetric complexes that are equally coordinatively saturated complexes at each copper center (Figure 21b). A variation of the ligand BPMP

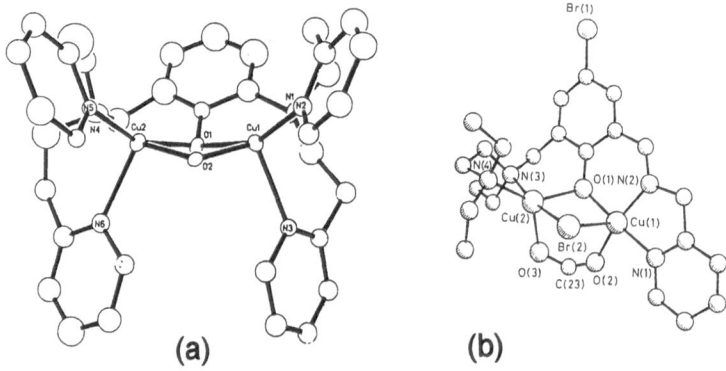

(a)   (b)

Figure 21. Two obligately asymmetric dicopper(II) complexes. (a) A slightly asymmetric dicopper(II) complex of the asymmetric ligand 2-(bis(2-pyridylmethyl)-methylamine-4-methyl-6-(bis(2-pyridylmethyl)-amine phenol.[103a] (b) The dicopper(II) complex of a Schiff base ligand offering asymmetric coordination.[103b]

(Figure 20b) where one pyridyl is replaced by a 2-phenol moiety has been used to prepare an asymmetric diiron(II,III) species, similar in overall stereochemistry to the diiron compounds illustrated in Figure 20b.[104]

Recently our group synthesized a new family of ligands capable of providing to a pair of metal atoms an asymmetric coordinatively-unsaturated environment that allows a variety of "exogenous" bridging and terminal ligands to bind.[105] The hexadentate ligand, abbreviated N3-O(H)-N2-R, is shown in Figure 22. This ligand is comprised of a bis-2-(benzimidazolyl)methylamine arm (denoted N3, and used previously to create symmetric triply bridged μ-oxo-bis(μ-carboxylato)diiron(III) complexes[106]), linked by a cresol moiety (denoted O(H)) to a 2-(benzimidazolyl)-methylamine arm (denoted N2-R). A variety of dicopper(II) complexes of general formula

$$[(X)Cu(II)(N3-\mu-O-N2-R)(\mu-Y)Cu(II)(Z)]$$

have been characterized, where X = ClO_4^- or nothing, Y = Cl^-, NO_3^-, CH_3COO^- and N_3^-, Z = C_2H_5OH or CH_3OH, and R = cyclohexyl or benzyl. The strategy for the synthesis of asymmetric ligand systems involves exploiting small differences in reactivity of secondary amines to assemble asymmetric ligands, counter to statistical expectations of a mixture of isomers, as illustrated in Figure 22. The general requirements for the ligand include:

(1) a different donor set in terms of number and types of ligands to each metal,
(2) benzimidazole moieties, as histidine analogues, primarily for the replaceable imidazole N-H proton,
(3) potential to have terminal phenoxo moieties, as tyrosine analogues, for application to purple acid phosphatase stereochemistry,
(4) a bridging group, phenoxo, to enforce dinuclearity,
(5) methylene spacers between donor groups; no Schiff bases or macrocycles
(6) ease of synthesis and of modification, especially replacement of a benzimidazole or the pendant R group by phenol.

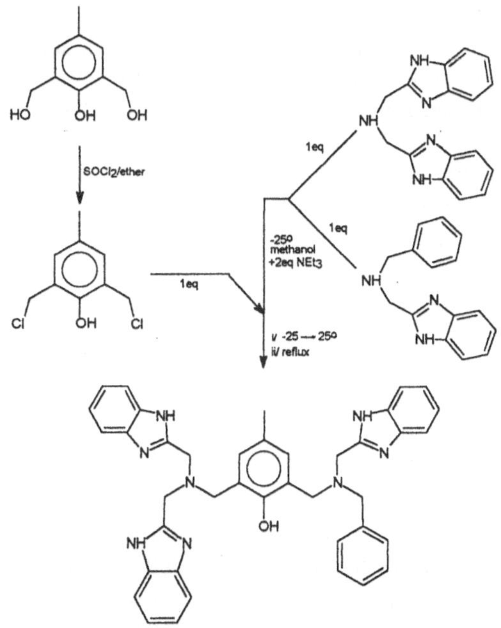

Figure 22. General synthetic route to asymmetric ligands offering different ligand suites and degree of coordinative unsaturation to a pair of metal ions.

This ligand is intended to offer a different suite of ligands to a pair of metal ions linked by the central phenol group. While congruence with any currently known metalloprotein is compromised by a bridging phenoxo group, in terms of steric demands and poor mediation of antiferromagnetic coupling between two metal centers, μ-phenoxo and μ-hydroxo moieties behave similarly. As probes of dicopper(I) chemistry and the stereochemical changes between dicopper(I) and dicopper(II) species, these ligands have limited utility.

In the next sections, the behavior of this ligand towards copper(II) and iron(III) atoms will be described. While the ligand readily bridges two copper(II) ions, it reluctantly, at this stage, dinucleates a pair of iron(III) atoms, instead preferring to bind in a 1:1 Fe:ligand ratio. The N3-μ-O-N2 ligand has insufficient ligands on the bidentate N2 branch for the chelate effect to entropically favor diiron(III) species, as iron(III) has a higher intrinsic coordination number than copper(II). The species derived do, however, have considerable bioinorganic interest, as described below, and indicate that there is still much to discover in the area of oxo-bridged diiron species. A similar ligand system, but with pendant arms bearing imidazole groups has been synthesized by a lengthy synthetic procedure.[107]

Other Strategies. The interaction of metal ions with single amino acids and dipeptides has been extensively studied, although little of relevance to dinuclear metal centers emerged. While early bioinorganic forays with simple amino acids and peptides as ligands have been of limited application bioinorganically, the now facile ability to synthesize oligopeptides has led to the reemergence of short polypeptide chains as ligands, especially in the area of models for metalloproteins with iron-sulfur clusters.[12b] The control of oligopeptide conformation and thence of metal nuclearity in resultant complexes remains problematic. This approach has the added difficulty that the resultant ligands would be substantially larger on an atoms per ligand donor atom basis than current entirely non-biological model systems, where, in most cases, the minimal number of atoms are used and chelate rings do not exceed 5 to 7 atoms (as opposed to more than 10 atoms for polypeptide-based ligand systems). Largely organic molecules in the 1000 dalton range also have proven difficult to crystallize, as they lack the nice, close-packing, solvent-poor characteristics of smaller molecules or the solvent-rich packing of protein molecules.

Direct study of proteins is increasingly more feasible, as cloning becomes more routine, and macroscopic quantities of protein can be synthesized. However, this leads to study, even with mutants, of a restricted range of conformational and stereochemical space and of a reduced variety of exogenous substrates. Relationships among structural, spectroscopic and magnetic properties established in one system may not be applicable to other systems or may prove constricting to the interpretation of data from other systems.

Unsymmetric Synthetic Dicopper Complexes of N3-O(H)-N2-R Ligands

Hosts of symmetrical alkoxo- and phenoxo-bridged dicopper(II) species[82] and a much smaller number of accidentally unsymmetric dicopper(II) complexes have been characterized,[101,108] but only a few obligately asymmetric species are known.[103-105,107] The ligand N3-O(H)-N2-R, illustrated in Figure 22, offers a tridentate branch (N3) to one copper center, a bidentate branch (N2) to second copper center, and the central phenol moiety bridges the two copper atoms. Coordination at one copper is completed by one or two more ligands and at the other by two or three more ligands; one of the additional ligands bridges the two copper ions. A chloro-bridged adduct is illustrated in Figure 23.[105] Acetato-, azido- and nitrato-bridged species have also been structurally characterized,[105b] and the coordination cores of these four compounds are illustrated in Figure 24. Although all complexes have structural features in common, the range of stereochemistry and resultant magnetic properties is wide and strongly dependent on the second bridging ligand. Selected stereochemical and spectroscopic data are summarized in Table 8.

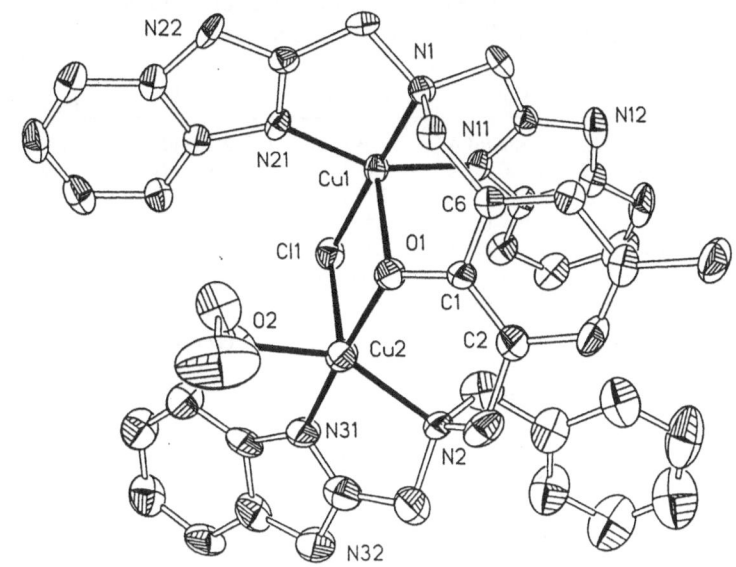

Figure 23. Structure of an asymmetric dicopper(II) complex, [Cu(N3-μ-O-N2-R)(μ-Cℓ)Cu-(C₂H₅OH)]²⁺.[105a]

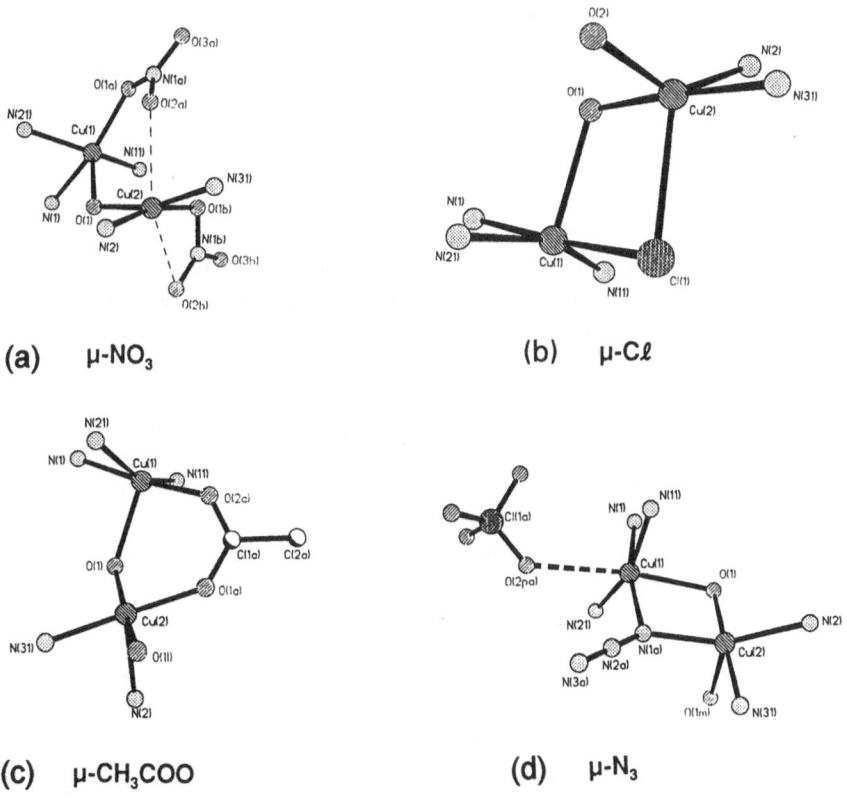

(a) μ-NO₃

(b) μ-Cℓ

(c) μ-CH₃COO

(d) μ-N₃

Figure 24. The core structures of dicopper(II) complexes of the dinucleating ligand N3-O(H)-N2-R, [(X)Cu(N3-μ-O-N2-R)(μ-Y)Cu(Z)]ⁿ⁺.[105] (a) X = nothing, Y = 1,2-NO₃⁻, Z = NO₃⁻-O,O'. (b) X = nothing, Y = Cℓ⁻, Z = C₂H₅OH. (c) X = nothing, Y = 1,2-CH₃COO⁻, Z = C₂H₅OH. (d) X = OCℓO₃⁻, Y = 1,1-N₃⁻, Z = CH₃OH.

Table 8. Stereochemical and spectroscopic properties of asymmetric dicopper(II) complexes of N3-O(H)-N2-R (R = cyclohexyl, benzyl).[a]

	μ-Chloro	μ-Nitrato	μ-Azido	μ-Acetato
Cu1-O1	2.352 (6) Å	2.193 (8)	2.266 (7)	2.165(11)
Cu2-O1	1.900 (6) Å	1.896 (8)	1.873 (7)	1.926(11)
Cu1-X	2.274 (2) Å	2.010 (9)[b]	1.999(10)	1.952(13)[b]
Cu2-X	2.649 (3) Å	2.592 (11)[b]	2.053 (8)	1.921(12)[b]
Cu..Cu	3.349 (2) Å	3.570 (3)	3.149 (2)	3.498 (4)
Cu1-O-Cu2	103.3 (3)°	121.4 (7)	98.6 (3)	117.4(5)
Cu1-X-Cu2	85.4 (1)°	N/A	102.1 (5)	N/A
Cu1(O1, ax)	sq py	sq py	oct (OCℓO₃)	sq py
Cu2(Y, ax)	sq py (μ-Cℓ)	oct (ONO₃'s)	sq py (MeOH)	sq py (N_Am)
Cu1-O-C-C	48.8°	53.8	51.1	39.7
Cu2-O-C-C	-3.6°	39.5	1.3	48.1
Magn. mom. (77 to 295 K)	2.50 to 2.45	2.15 to 2.37	2.95	2.75 to 2.70
2J (cm⁻¹)	-30	-90	400	40
	w antiferro	antiferro	s ferro	w ferro
EPR	g_y = 2.29 g_z = 2.11 g_x = 1.90	$g_{//}$ = 2.24 $g_⊥$ = 2.03	none	not meas.
Bridging-ligand orientations				
O(Ph)	ax-eq	ax-eq	ax-eq	ax-eq
X	eq-ax	eq-ax[a]	eq-eq	eq-eq[a]

[a] Abbreviations: sq py, square pyramidal; oct, distorted octahedral (six-coordinate); ax, axial ligand; eq, equatorial ligand. [b] Bidentate, μ-1,2 bridging ligand.

For the chloro-bridged species of N3-O(H)-N2-Cyclohexyl), extreme asymmetry in Cu-O(phenolato) and Cu-Cℓ bond distances is observed, and is correlated with essentially no coupling of electron spins of the two copper ions. A lesser degree of asymmetry in the Cu-O(phenolato) bonds is observed for the nitrato-bridged species -- the nitrato bridge is very unsymmetric; weak antiferromagnetic coupling is observed. For the azido-bridged species, the Cu-N(N₃⁻) separations are insignificantly different; strong ferromagnetic coupling is observed.

In the presence of copper(II) the dinucleating nature of the ligand is established. The basic [Cu(N3-μ-O-N2)Cu]³⁺ platform supports a variety of exogenous bridging and other ligands to raise the coordination to a minimum of five. To date, the bridging phenoxo ligand invariably adopts an axial position to the copper liganded by the N3 arm and an equatorial position to the copper liganded by the N2 arm, resulting in extreme asymmetry in the Cu-O(phenoxo) bond lengths (see Table 8).[105b] The N3 arm binds to copper(II) in a meridional fashion and leads to square-pyramidal (occasionally octahedral) complexes, with the N3 arm as part of the basal plane. Copper(II) complexes containing the N2 arm are usually highly distorted square-pyramidal species and in complexes of the N3-O(H)-N2-R ligand, the N2 and phenoxo moieties form part of the basal plane. The basal plane is variously completed by solvent or anionic bridging ligands.

The conformational flexibility of the [Cu(N3-μ-O-N2)Cu]³⁺ moiety allows other bridging ligands, whether μ-1,1 or μ-1,2 in mode, to be equatorial to the copper liganded by the N3 arm and either axial or equatorial to the copper liganded by the N2 arm (Cu2 in Table 8). The axial ligand, Y, at Cu2 is specified in Table 8, and is variously solvate, ligand nitrogen or bridging anion. An equatorial-equatorial conformation for the second bridging ligand leads to reduced asymmetry in the Cu-Y bond lengths of the second bridging ligand and facilitates ferromagnetic coupling (S = 1 ground state) between the two copper centers, as has been previously observed.[82b] For the μ-azido species, the ferromagnetic coupling at J ≈ 400 cm^{-1} is unusually strong, but comparable to that seen in a symmetric μ-hydroxo-μ-1,1-azido dicopper(II) species.[109] The weakly antiferromagnetically coupled systems are EPR active, contrary to simple notions that two (uncoupled) paramagnetic centers in close proximity will cause rapid spin relaxation and broaden into oblivion the EPR spectrum. In this regard, the observation of EPR signals for copper(II) in a structurally uncharacterized metalloprotein should not be construed as evidence for a single isolated copper(II) site.

While this family of asymmetric dicopper(II) complexes has only passing similarity with the active sites *currently* known for copper-containing metalloproteins, the stereochemistry and coordinative unsaturation of these complexes is amenable to studies of ligand substitution processes for copper, the wide range of spectroscopic and magnetic behavior associated with a particular motif has cautionary relevance to studies of copper-containing metalloproteins, and the obligate asymmetry should facilitate the deliberate assembly of mixed-metal complexes. Given the predilection for phenoxo moieties to bridge two metal atoms in non-biological systems, it seems likely that tyrosine may yet be found spanning two metals in a biological system, such as the possibly dinuclear EPR-active Cu$_A$ site of cytochrome c oxidase.

Towards Unsymmetric μ-Oxo Diiron Systems

Engendering asymmetry in diiron complexes has proved difficult, and almost 10 years after the first non-biological non-symmetric singly bridged μ-oxo complexes,[87,110] the family of unsymmetric synthetic diiron species has been extended by remarkably few new members. Unsymmetric species include most of the diiron(III) complexes of tris(2-pyridylmethyl)-amine (TPA),[100] most of the μ-phenoxo-bis(μ-carboxylato) mixed-valent species,[102,104] a diiron(II) tris(μ-carboxylato) species,[111] and several μ-alkoxo diiron(III) species.[112] With one exception, a mixed-valent, μ-phenoxo species featuring a terminal phenoxo moiety at only one iron center,[104] none of these species can be considered obligately asymmetric, and only a few are highly unsymmetric.[100e,104,111,112a] There is presently considerable activity directed towards diiron species with terminally coordinated phenoxo moieties[104,113,114] and coordinated phosphate or molybdate groups,[100b,114-116] as models for purple acid phosphatases (Figure 19d). In addition, species with two or more water-derived ligands[100d,100e,114,117,118] are relevant to the formation or hydrolysis of the Fe-O-Fe motif, as part of the mechanism of purple acid phosphatases. The processes by which two terminally coordinated water molecules may condense to form an oxo bridge are shown schematically in Figure 25. One route, featuring two bridges by water-derived species,

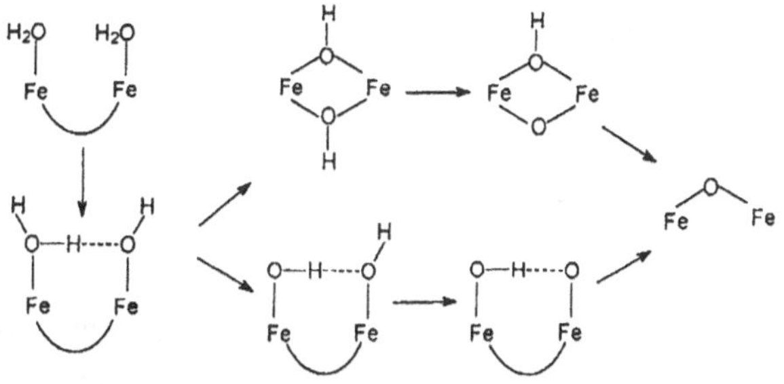

Figure 25. A possible mechanism for the formation of a μ-oxo bridge from adventitious water. The arc represents some other bridging ligand, such as a μ-1,2-carboxylato ligand.

involves expansion of the coordination sphere, or loss of one ligand at each iron. In the other route, the ligands at each metal are maintained. Several key species remain structurally uncharacterized. In addition, the perturbation of Fe-O bond lengths that may occur when the coordinated oxo or hydroxo moieties are hydrogen bonded to other species remains undefined.

Asymmetry and Perturbation of the Fe-O-Fe Moiety. The structures of several derivatives of octameric methemerythrin all share a surprising asymmetry in the Fe-O(μ-oxo) bond lengths,[95] an asymmetry that is not replicated in the structure of monomeric met-azidomyohemerythrin,[15b] nor in a variety of singly, doubly and triply bridged μ-oxodiiron(III) complexes. Although, this asymmetry was strongly qualified,[15a,95] hopes that it might be valid continue to be expressed.[100d] However, the structures of [N5FeOFeX$_3$]$^+$ (N5 is the functionally pentadentate ligand, 1-bis(2'-benzimidazolylmethyl)-2-[(2'-benzimidazolylmethyl)(2'-ethoxy)]diaminoethane; X = Cℓ, Br) show an octahedral iron(III) center linked to a tetrahedral iron(III) center by a single oxo bridge (Figure 20c). Notwithstanding the extreme asymmetry in coordination environments for the iron center, the Fe-O bond lengths differ by less than 0.07 Å (compared to the 0.24 Å reported for methemerythrin derivatives[15a]). The longest Fe-O(μ-oxo) bond reported to date is 1.83 Å for an unsymmetric μ-oxo-μ-(aquo..hydroxo) diiron(III) compound of TPA;[100d,e] the shortest is 1.72 Å for [N5FeOFeBr$_3$]$^+$.[87b] There are no data to support the conclusion that in methemerythrin derivatives pairs of Fe-O(oxo) bond lengths differ by more than 0.05 Å from each other or by more than 0.03 from 1.79 Å.

The consequences of asymmetry on the distinctive structural and spectroscopic properties of the μ-oxodiiron(III) moiety are of considerable importance. Currently, strong antiferromagnetic coupling (-J $\approx$ 100 cm^{-1}) between two iron(III) centers is uniquely mediated by an oxo bridge, independent of bridging angle, coordination number and nature of ligands -- at least for symmetric species.[87,100a-c] An hydroxo, alkoxo, phenoxo or carboxylato bridge in place of an oxo bridge all diminish the strength of coupling by nearly an order of magnitude.[8,9,16] For mixed-valent μ-phenoxodiiron(II,III) species[102,104] and for μ-hydroxodiiron(II) species magnetic coupling is weak,[119,120] and for μ-aquodiiron(II) species[120] possibly ferromagnetic.

For unsymmetric μ-oxodiiron(III) species, does the remarkably narrow range of magnetic behavior seen for symmetric species still obtain? Magnetic properties for [N5FeOFeX$_3$]$^+$ are perturbed relative to symmetric singly bridged species, but even in these cases of extreme asymmetry strong antiferromagnetic coupling still occurs.[87] Further, are other spectroscopic properties perturbed by asymmetry? Resonance Raman spectroscopy is a useful techniques for metalloproteins, as metal-ligand vibrational modes that are coupled with a metal chromophore can be selectively enhanced. The ratio of intensities for the asymmetric and symmetric Fe-O-Fe stretches, I_{as}/I_s, for μ-oxodiiron(III) species, are typically less than 0.1 in symmetric species, and become as large as ~0.3 in methemerythrin species.[88] For the highly unsymmetric [N5FeOFeX$_3$]$^+$, values of I_{as}/I_s greater than two are observed.[88]

It should be noted that Fe-O bond lengths of around 1.80 Å are *not* unique to Fe-O(μ-oxo) moieties. For example, the complex [N5Fe(OC$_2$H$_5$)]$^{2+}$ has an Fe-O separation of 1.793 (1) Å.[87d] Thus, in the application of EXAFS techniques to metalloproteins, a short Fe-O bond length is not diagnostic of an oxo bridge, unless accompanied by an Fe..Fe feature at ~ 3.0 to 3.6 Å, a feature that is now, but was not always, routinely discernible from EXAFS data. Complexes, such as [N5FeOFeX$_3$]$^+$, which are not explicit models for hemerythrin or any other oxo-bridged diiron metalloprotein, still provide useful insight into biological systems, both in assessing reliability of structural data and in establishing ranges of spectroscopic and magnetic behavior associated with the Fe^{III}-O-Fe^{III} motif.

Iron(III) Complexes of N3-O(H)-N2-R. The N3-O(H)-N2-R ligand that readily leads to crystalline asymmetric dicopper(II) complexes,[105] forms dinuclear iron(III) reluctantly at this stage. In protic environments two symmetric complexes with a 1:1 ratio of ligand-to-iron have been characterized very recently. They are of relevance to hydrolysis processes involving μ-oxo diiron(III) species and to the active site of purple acid phosphatases. A third complex is a symmetric dimer of an unsymmetric diiron(III) complex, where the ligand binds in the desired 1:2 ligand-to-iron ratio.

$[Fe_2(N3-O(^-)-N2H(^+)-R)_2(\mu-O_2P(OC_6H_5)_2)(\mu-OHO)]^{2+}$ **(I)**. This structure, illustrated in Figure 26, features a novel μ(oxo..hydroxo)diiron(III) moiety, one of the intermediate species of Figure 25.[114] Only the N3 arm of the N3-O(H)-N2-R ligand coordinates; the putatively central phenoxo coordinates in a terminal manner. The benzimidazole H-N of the N2 arm stabilizes the oxo..hydroxo bridge by hydrogen bonding. A bridging diphenyl-phosphate group completes coordination in an unusual manner that leads to an Fe..Fe separation of 5.164 (2) Å. This extremely long separation, the longest yet observed in a genuinely dinuclear diiron complex, is the result of steric clash of the non-coordinated benzimidazole groups of the N2 arm, and is not intrinsic to the oxo..hydroxo bridge, since in the hydroxo..aquo-bridged complex the Fe..Fe separation is 3.389 (2) Å.[100d,e] In terms of terminal coordination of a phenoxo ligand, coordination of imidazole groups, bridging coordination of diphenylphosphate and a second bridging ligand derived from water, this species shares stereochemical features proposed for the stereochemistry of phosphate derivative of oxidized purple acid phosphatase.[17]

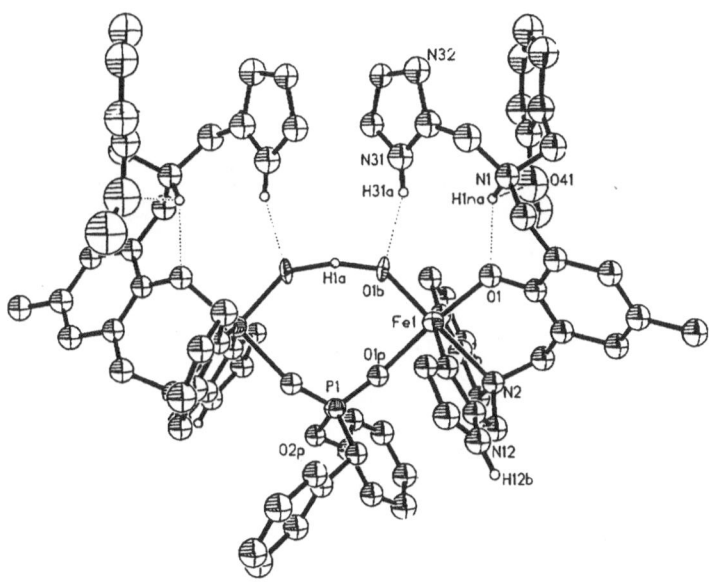

Figure 26. The compound $[Fe_2(N3-O(^-)-N2H(^+)-R)_2(\mu-O_2P(OC_6H_5)_2)(\mu-OHO)]^{2+}$.[114] Note the *syn* arrangement of the uncoordinated arms.

The distinctive color of purple acid phosphatases arises from an Fe^{III}-phenoxo charge transfer band. The Fe^{III}-phenoxo charge transfer band of **I** at 598 nm may be compared to those observed for the phosphate derivative of oxidized purple acid phosphatase at 550 nm[121] and for a μ-alkoxo-bis(μ-diphenylphosphato) species at 587 nm.[112b] However, the low intensity of this band for $\mathbf{I}$ (ϵ = 600 M^{-1} cm^{-1}), compared to other terminally coordinated Fe^{III}-phenoxo species (1200 to 4000 M^{-1} cm^{-1}), is attributed to the trans relationship between the phenoxo and bridging phosphato groups.

The protic binding pocket surrounding the μ-O..H..O moiety leads to nearly reversible redox behavior under cyclic voltammetry, with well-separated couples with waves of equal but opposite height, as shown in Figure 27. The first couple, corresponding to

$$Fe^{III}..Fe^{III} \rightleftharpoons Fe^{III}..Fe^{II}$$

occurs at 380 mV, potential similar to that for oxidized phosphate-free purple acid phosphatase at 306 mV at pH 6.0,[123] although the extreme pH sensitivity of the redox couple for purple acid phosphatase negates much of the value of the comparison. The value observed for $\mathbf{I}$ is at much more positive potential than those seen for alkoxo-, phenoxo and oxo-bridged compounds, consistent with the recently demonstrated absence of an oxo bridge for purple acid phosphatase.[17,124,125] The second couple occurs at -257 mV. Reversible redox behavior is rare for diiron species, outside biological systems, except for alkoxo- and phenoxo-bridged species,[102,113b] where the bridge and pendant ligands give entropic stability to iron in lower oxidation states. The partly reversible behavior observed for $\mathbf{I}$, [Fe_2(N3-O($^-$)-N2H($^+$)-R)$_2$(μ-O$_2$P(OC$_6$H$_5$)$_2$)(μ-OHO)]$^{2+}$, is attributed to the proximity of protons needed to stabilize the more basic complexes as charge decreases through reduction.

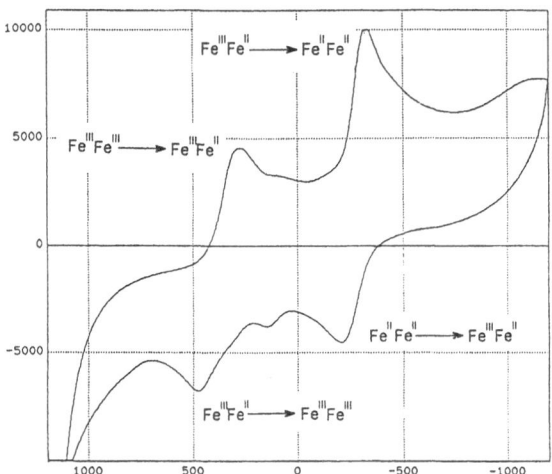

Figure 27. Cyclic voltammogram of [Fe_2(N3-O($^-$)-N2H($^+$)-R)$_2$(μ-O$_2$P(OC$_6$H$_5$)$_2$)(μ-OHO)]$^{2+}$.[114]

[Fe$_2$(N3-O($^-$)-N2-R)$_2$(μ-OH)$_2$]$^{2+}$ **(II)**.[118] In this bis-hydroxo bridged compound, illustrated in Figure 28, the bridging hydroxo moieties are hydrogen-bonded to water molecules, which in turn are hydrogen bonded to the uncoordinated benzimidazole moieties. As a result of the hydrogen bonding and the weak coordination of the amine part of the N3 ligand, the shorter Fe-O(hydroxo) bond length (1.901 (5) Å) is indistinguishable from that observed for a μ-oxo-μ-hydroxo species.[117] The long Fe-O(hydroxo) separation (2.064 (5) Å) is trans to a phenoxo moiety, as shown in Figure 28. By way of comparison, a bis(μ-hydroxodiiron(III) derivative of a Schiff-base ligand has Fe-μ-OH separations of 1.986 (5) Å and 2.055 (6) Å.[126]

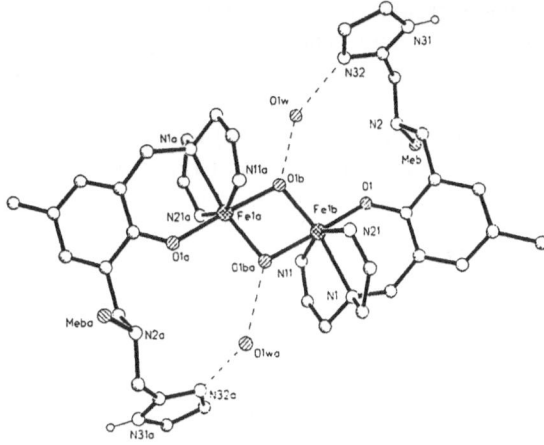

Figure 28. The compound [Fe$_2$(N3-O($^-$)-N2-R)$_2$(μ-OH)$_2$]$^{2+}$. Note the water molecules hydrogen bonded to bridging hydroxo groups and uncoordinated benzimidazole groups. Benzimidazole groups on the coordinating N3 arm are omitted for clarity.[125] Note the *anti* arrangement of the uncoordinated arm of the ligand

[Fe$_2$(N3-μ-O-N2-R)(μ-1,2-O$_2$P(OC$_6$H$_5$)$_2$)$_2$]$_2$O^{4+} (**III**). A μ-oxo dimer of an unsymmetric diiron(III) complex has been characterized (Figure 29).[116] The N3-O(H)-N2-R ligand *spans* two iron centers, as desired, and the central phenoxo group bridges the two metal centers. Coordination is completed by two bridging diphenylphosphate anions, and at the remaining vacant site an oxo bridge links the two asymmetric dinuclear complexes together. The structure is notable for the highly unsymmetric bridging of the phenoxo moiety. The asymmetry in Fe-O$_{phenoxo}$ bond lengths, which differ by 0.35 Å and which is unique in the stereochemistry of diiron(III) species, is attributed to the strong trans influence of the oxo bridge.

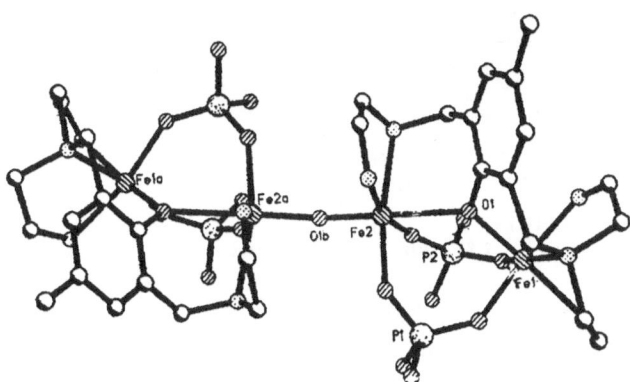

Figure 29. A μ-oxo dimer of an unsymmetric diiron complex of the ligand N3-O(H)-N2-R, [Fe$_2$(N3-μ-O-N2-R)-(μ-1,2-O$_2$P(OC$_6$H$_5$)$_2$)$_2$]$_2$O^{4+}. Benzimidazole groups are omitted for clarity.[123]

The three new diiron(III) structures of the potentially dinucleating ligand N3-O(H)-N2-R, **I**, **II** and **III**, reveal:

(1) a potential intermediate, the μ-O..H..O moiety, in the hydrolysis of water to an oxo bridge;

(2) the role of available protons for reversible redox behavior for compounds where two irons are not linked by a polydentate dinucleating ligand;

(3) the influence of hydrogen bonding on Fe-μ-OH separations in a bis(μ-hydroxo) structure;

(4) the extreme asymmetry in Fe-O(phenoxo) bond lengths possible for a bridging phenoxo group in a asymmetric dinuclear iron(III) compound;

(5) possible routes into asymmetric diiron complexes.

Stereochemistry of Aquo-, Hydroxo- and Oxo-Bridged Diiron Species.

Table 9 summarizes bond parameters for bonds between iron(III) or iron(II) and water-derived ligands. Legions of synthetic μ-oxodiiron(III) species are known,[8,9,16,21] but only a few μ-hydroxodiiron(III) species relevant to metalloproteins currently known to have oxo- or hydroxo-bridged diiron centers have been reported.[8,9,16,115a,127,128] Very recently several species relevant to the condensation/hydrolysis processes summarized in Figure 25 have been characterized. These include a μ-oxo-μ(aquo..hydroxo) species,[100d,e] and a disordered μ-oxo-μ-hydroxo species,[117] as well as compounds **I** and **II** described above.[114,118] An hydroxo-[119] and an aquo-bridged[120] diiron(II) species have been characterized, but the only mixed-valent species structurally characterized to date feature a phenoxo bridge. Some phenoxo- and alkoxo-bridged species are also included to highlight trans effects in diiron species. A single phenoxo-bridged diiron(II) species is known.[129] A more comprehensive listing is available in reference 8. Species characterized since then are highlighted in bold-face type. For unsymmetrically bridged compounds, both Fe-μ-O bond distances are entered; for symmetrically bridged compounds, the range of values observed are entered. Except for the mixed-valent μ-phenoxo compounds, the Fe-O separations for Fe^{III} and Fe^{II} species are from different compounds.

Table 9. Bond distances for Fe-O moieties.

Ligand	Fe^{III}		Fe^{II}	Reference
H$_2$O (terminal)	2.094(7)		-----	100e
OH (terminal)	~1.90[a]		-----	100d
O$_{phenoxo}$ (terminal)	1.85-1.95		-----	112b,114,104
μ(H$_2$O..OH)	**1.913(7)**	**2.040(9)**	-----	100d
	2.094(7)	2.140(7)	-----	100e
μ-OH-μ-O [b]	**1.91(1)**	**1.97(1)**	-----	117
μ(OHO)		**1.897 (10)**	-----	114
μ-OH$_2$	-----		**2.176(4)**	120a
(μ-OH)$_2$	1.986(5)	2.055(6)	-----	126
μ-OH-(μ-Y)$_2$	1.94-1.96		1.987(8)	119,127,128
μ-O-(μ-Y)$_2$	1.78-1.81		-----	8,9b,106,119,128
μ-O$_{phenoxo}$	2.02-2.06		2.057	113b,129,130
μ-O$_{phenoxo}$ (asymm)	**1.987(5)**	**2.340(5)**		116
	2.107(9)	**2.158(11)**[c]	-----	112b
μ-O$_{phenoxo}$ (Fe^{II}Fe^{III})	1.974 (3)		2.082 (4)	102c-e
	1.949(3)		2.073(3)	104[d]
μ-O$_{alkoxo}$	1.98- 2.06		___	112a,112b

[a] Taken from the μ(hydroxo.aquo) structure.[100d] [b] Disordered μ-oxo-μ-hydroxo bridge. [c] Terminal phenoxo trans to bridging phenoxo. [d] Terminal phenoxo cis to bridging phenoxo.

While Fe-μ-O bond lengths are nearly invariant to the number, type and disposition of other ligands, Fe-μ-OH and Fe-μ-O(phenoxo) and Fe-μ-O(alkoxo) are quite sensitive to trans influences of other ligands, especially oxo but also phosphato. Fe-μ-O(phenoxo) bonds are considerably elongated (by ~ 0.35 Å) when trans to an oxo bridge, and are somewhat elongated (0.15 Å) when trans to a phosphate oxygen atom. There are still many features of the stereochemistry of oxo-bridged diiron species that remain to be defined, with both direct and indirect relevance to the stereochemistry of iron in metalloproteins containing oxo-bridged diiron(III) species. As is readily apparent from Table 9, the stereochemistry of Fe^{II} in dinuclear complexes is incompletely characterized, and there are some puzzling inconsistencies, for example, the two independent and significantly different measurements for the μ(H$_2$O..OH) bridge both used the TPA ligand.

FINAL PERSPECTIVES

Site-directed mutagenesis and random mutants of myoglobin have produced a wealth of new data, including many structural data of stable species, partially liganded species, species trapped in alternative quaternary states, and even of a photodissociated intermediate. New insights into ligand binding processes have emerged, especially concerning the role of water in the ligand-binding pocket in facilitating dioxygen binding relative to carbon monoxide binding. Several features of ligand binding remain unresolved and are likely to remain unresolvable until more precise stereochemical data become available. Presently, the only source of very precise structural data is from small model systems. The stereochemistry of the Fe-CO moiety in model systems showing high and low affinity for carbon monoxide is well-characterized. However, in biological systems, while off-axis orientation of the CO moiety is unequivocal, even in mutants where few steric constraints on a linear Fe-CO geometry exist, the extent of bending *versus* tilting is not discernible because of the absence of atomic resolution in the diffraction data. The precision of the stereochemistry of the Fe-OO moiety remains low -- the precise determination of this stereochemistry is a high priority, especially as grist for theoretical mills. Interpretation of kinetic and thermodynamic data critically require structural data; such structural data remains scarce for model systems and frustratingly imprecise for biological systems. Data from synthetic metalloporphyrinato species, even those of only passing biological relevance, now permeate any analysis of ligand binding to the hemoglobins and hemoproteins in general. However, the role of model complexes in interpreting data from the hemoglobins themselves increasingly is becoming more indirect, with insight into the biological processes being gained as much from differences as from similarities between model and biological systems.

In the area of metalloproteins featuring multinuclear iron and copper complexes, there remains considerable scope for further studies into the range of structural, spectroscopic and magnetic properties associated with a particular structural motif. Part of this work is the creation of a reference library of structural features that may be associated with particular spectroscopic signatures, in order to identify reliably and efficiently the stereochemistry of newly characterized metalloproteins. In particular, there remains the challenge of preparing asymmetric complexes in order to define the contributions of asymmetry to structure, function and reactivity of metalloproteins. Functional model systems remain rare, so that the detail of ligand binding processes that is being addressed in hemoprotein systems remains inappropriate for many non-heme systems. However, it is in the area of functional models that insight into mechanism can be obtained: unstable intermediates in the biological system, may be rendered stable in a model system; the steric specificity of metalloenzymes for substrate means that electronic effects involved in substrate activation may be better studied *via* model systems, for which a wider range of substrates may be studied. Putative model systems that are at variance with data from biological systems are, in fact, essential to the elimination of possible structure, function and reactivity relationships for metalloproteins.

The development of a comprehensive understanding of structure, function and reactivity relationships in metalloproteins has proceeded synergistically from studies of protein and model systems. In addition to seeking to understand the mechanisms by which porphyrinato iron(II) species bind small molecules, both in biological and non-biological systems, several long-standing justifications for and applications of such research are within sight of being achieved. These include the development of both genetically engineered biological and non-biological substitutes for hemoglobin,[131] and the modification of functionality of a protein, in this case from an O_2 carrier into an oxygenase.

5. ACKNOWLEDGEMENTS

I thank the organizers of the symposium for the invitation to present a contribution, and to augment that into this article. The support of the National Institutes of Health for aspects of this work and of Georgetown University and Massey University is gratefully acknowledged. I thank mentors who introduced me to bioinorganic chemistry, and students at Georgetown University for their dedication and inspiration. And I am indebted to the efforts of graduate students Peter Kamaras and Miroslav Rapta for their recent work that has led to the obligately asymmetric dicopper(II) systems and to the novel diiron(III) systems. Finally, I thank Professor E.N. Baker and Dr. A.K. Burrell for helpful discussions and reading of this manuscript.

6. REFERENCES

1. (a) For a group of reviews on metal-dioxygen chemistry, see: *Chem. Rev.* 94(3): 1994.
    (b) For a recent textbook on aspects of bioinorganic chemistry, see for example: "Bioinorganic Chemistry," I. Bertini, H.B. Gray, S.J. Lippard, and J.S. Valentine, eds., University Science Books, Mill Valley, CA (1994).
    (c) For a recent textbook on principles of bioinorganic chemistry, see: J. Berg and S.J. Lippard, "Principles of Bioinorganic Chemistry," University Science Books, Mill Valley, CA (1994).
    (d) For a group of reviews concentrating on structure and function of metalloproteins, see for example: *Adv. Protein Chem.* 42: (1991).
2. G.B. Jameson and J.A. Ibers, Biological and synthetic dioxygen carriers, *in*: "Bioinorganic Chemistry," I. Bertini, H.B. Gray, S.J. Lippard, and J.S. Valentine, eds., University Science Books, Mill Valley, CA, pp 167-252 (1994).
3. (a) J.S. Valentine, Dioxygen reactions, *in*: "Bioinorganic Chemistry," I. Bertini, H.B. Gray, S.J. Lippard, and J.S. Valentine, eds., University Science Books, Mill Valley, CA, pp 253-313, (1994).
    (b) See also ref. 96.
4. M. Momenteau and C.A. Reed, *Chem. Rev.* 94: 659-698 (1994).
5. (a) B.A. Springer, S.S. Sligar, J.S. Olson, and G.N. Phillips Jr., *Chem. Rev.* 94: 699-714 (1994).
    (b) T. Li, M.L. Quillin, and G.N. Phillips Jr., *Biochemistry* 33: 1433-1446 (1994).
    (c) T.E. Carver, R.E. Brantley Jr., E.W. Singletoin, R.M. Arduini, M.L. Quilin, G.N. Phillips, and J.S. Olson, *J. Biol. Chem.* 267: 14443-14450.
6. (a) G.B. Jameson and J.A. Ibers, *Comments Inorg. Chem.* 2: 97-226 (1993).
    (b) G.B. Jameson, W.T. Robinson, and J.A. Ibers, Structural results for model compounds of significance in hemoglobin chemistry, *in*: "Hemoglobin and Oxygen Binding," C. Ho, ed., Elsevier North Holland Inc., Amsterdam, pp 25-35 (1982).
7. (a) K. Kim and J.A. Ibers,. *J. Am. Chem. Soc.* 113: 6077-6081(1991).
    (b) K. Kim, J. Fettinger, J.L. Sessler, M. Cyr, J. Hugdahl, J.P. Collman, and J.A. Ibers, *J. Am. Chem. Soc.* 111: 403-405 (1989).
8. L. Que Jr. and A.E. True, *Prog. Inorg. Chem.* 38: 98-200 (1990).
9. (a) J.B. Vincent, G.L. Olivier-Lilley, and B.A. Averill, *Chem. Rev.* 90: 1447-1467(1990).
    (b) D.M. Kurtz Jr. *Chem. Rev.* 90: 585-606 (1990).
10. (a) E.T. Adman, *Adv. Protein Chem.* 42: 144-197 (1991).

(b) K.D. Karlin and Y. Gultneh, *Prog. Inorg. Chem.* 35: 219-327 (1987).

(c) B.P. Murray, *Coord. Chem. Rev.* 124: 63-105 (1993).

(d) E.I. Solomon, F. Tuczek, D.E. Root, and C.A. Brown, *Chem. Rev.* 94: 827-856 (1994).

11. G.R. Moore and G.W. Pettigrew, "Cytochromes c: Evolutionary, Structural and Physicochemical Aspects," Springer-Verlag, Berlin (1990).

12. (a) E.I. Steifel and G.N. George, Ferrodoxins, hydrogenases and nitrogenases: metal-sulfide proteins, *in*: "Bioinorganic Chemistry," I. Bertini, H.B. Gray, S.J. Lippard, and J.S. Valentine, eds., University Science Books: Mill Valley, CA, pp 253-313 (1993).

(b) R.H. Holm, S. Ciurli, and J.A. Weugel, *Prog. Inorg. Chem.* 38: 1-74 (1990).

13. J.B. Howard and D.C. Rees, *Adv. Protein Chem.* 42: 199-280 (1991).

14. (a) A.L. Feig and S.J. Lippard, *Chem. Rev.* 94: 759-805 (1994).

(b) A.C. Rosenzweig and S.J. Lippard, *Acc. Chem. Res.* 27: 229-236 (1994).

15. (a) R.E. Stenkamp, *Chem. Rev.* 94: 715-726 (1994).

(b) S. Sheriff, W.A. Hendrickson, and J.L. Smith, *J. Mol. Biol.* 197: 273-296 (1987).

16. S.J. Lippard, *Angew. Chem., Int. Ed. Eng.* 27: 344-361 (1988).

17. Z. Wang, L.-J. Ming, L. Que Jr., J.B. Vincent, M.W. Crowder, and B.A. Averill, *Biochemistry* 31: 5263-5268 (1992).

18. (a) P. Nordlund and H. Eklund, *J. Mol. Biol.* 232: 123-164 (1993).

(b) P. Nordlund, B.-M. Sjöberg, and H. Eklund, *Nature* 345: 593-598 (1990).

19. A.C. Rosenzweig, C.A. Frederick, S.J. Lippard, and P. Nordlund, *Nature* 366: 537-543 (1993).

20. (a) J. Kuriyan, S. Wilz, M. Karplus, and G.A. Petsko, *J. Mol. Biol.* 192: 133-154 (1986).

(b) X. Cheng and B.P. Shoenborn, *J. Mol. Biol.* 220: 381-399 (1991).

(c) L. Powers, J.L. Sessler, G.L. Woolery, and B. Chance, *Biochemistry* 23: 5519-5523 (1984).

(d) A. Bianconi, A. Congiu-Castellano, P.J. Durham, S.S. Hasnain, and S. Phillips, *Nature*, 318: 685-687 (1985).

(e) Many biochemistry textbooks show Fe-CO moieties as bent as Fe-OO moieties (120°).

21. (a) K.S. Murray, *Coord. Chem. Rev.* 12: 1-35 (1974).

(b) J. Sanders-Loehr, Binuclear iron proteins, *in*: "Iron Carriers and Iron Proteins," T.M. Loehr, ed., VCH, New York, Vol. 5, pp 373-476 (1989).

22. I.M. Klotz, and D.M. Kurtz Jr., *Acc. Chem. Res.* 17: 16-22 (1984).

23. D.E. Wilcox, J.R. Long, and E.I. Solomon, *J. Am. Chem. Soc.* 106: 2186-2194 (1984).

24. (a) N. Kitajima and Y. Moro-oka, *Chem. Rev.* 94: 737-757 (1994).

(b) N. Kitajima, K. Fujisawa, C. Fujimoto, Y. Moro-oka, S. Hashimoto, T. Kitagawa, K. Toriumi, K. Tatsumi, and A. Nakamura, *J. Am. Chem. Soc.* 114: 1277-1291 (1992).

25. K.A. Magnus, H. Ton-Hat, and J.E. Carpenter, *Chem. Rev.* 94: 727-735 (1994).

26. G.B. Jameson, F.S. Molinaro, J.A. Ibers, J.P. Collman, J.I. Brauman, E. Rose, and K.S. Suslick, *J. Am. Chem. Soc.* 102: 3224-3237 (1980).

27. (a) J.H. Wang, *J. Am. Chem. Soc.* 80: 3168-3169 (1958).

(b) T.G. Traylor, *Acc. Chem. Res.* 14: 102-109 (1981).

28. (a) D.L. Anderson, D.J. Weschler, and F. Basolo, *J. Am. Chem. Soc.* 96: 5599-5600 (1974).

(b) G.C. Wagner and R.J. Kassner, *J. Am. Chem. Soc.* 96: 5593-5595 (1974).

(c) W.S. Brinigar, C.K. Chang, J. Geibel, and T.G. Traylor, *J. Am. Chem. Soc.* 96: 5597-5599 1974).

(d) J. Almog, J.E. Baldwin, R.L. Dyer, J. Huff, and C.J. Wilkerson, *J. Am. Chem. Soc.* 96: 5600-5601 (1974).

29. J.P. Collman, R.R. Gagne, T.R. Halbert, J.C. Marchon, and C.A. Reed, *J. Am. Chem. Soc.* 95: 7868-7870 (1973).

30. (a) J.B. Weiss, *Nature* 202: 83-84 (1964); *ibid.* 203: 183 (1964).

(b) L. Pauling, *Nature* 203: 182-183 (1964).

(c) H.B. Gray, *Adv. Chem. Ser.* 100: 365-389 (1971).

31. M.F. Perutz, *Nature* 228: 726-739 (1970).

32. B.M. Hoffman, and D.H. Petering, *Proc. Natl. Acad. Sci, USA* 67: 637-643 (1970).

33. E.C. Niederhoffer, J.H. Timmons, and A.E. Martell, *Chem. Rev.* 84: 137-203 (1984).

34. G.A. Rodley and W.T. Robinson, *Nature* 235: 438-439 (1972).

35. (a) J.P. Collman, R.R. Gagne, C.A. Reed, W.T. Robinson, and G.A. Rodley, *Proc. Natl. Acad. Sci, USA* 71: 1326-1329 (1974).

(b) G.B. Jameson, G.A. Rodley, W.T. Robinson, R.R. Gagne, C.A. Reed, and J.P. Collman, *Inorg. Chem.* 17: 850-857 (1978).

36. G.B. Jameson, F.S. Molinaro, J.A. Ibers, J.P. Collman, J.I. Brauman, E. Rose, and K.S. Suslick, *J. Am. Chem. Soc.* 100: 6769-6770 (1978).

37. S.E.V. Phillips, *Nature* 273: 247-248 (1978).

38. W. Steigemann and E. Weber, *J. Mol. Biol.* 127: 309-338 (1979).

39.	B. Shaanan, B. *J. Mol. Biol.* 171: 31-59 (1983).

40.	A Brzozowski, Z. Derewenda, E. Dodson, G. Dodson, M. Grabowski, R. Liddington, T. Skarżyński, and D. Vallely, *Nature* 307: 74-76 (1984).

41.	(a) A. Arnone, P. Rogers, N.V. Blough, J.L. McGourty, and B.M. Hoffman, *J. Mol. Biol.* 188: 693-706 (1986).

	(b) B. Luisi, B. Liddington, G. Fermi, and N. Shibayama, *J. Mol. Biol.* 214: 7-14 (1990).

	(c) R. Liddington, Z. Derenda, G. Dodson, and D. Harris, *Nature* 331: 725-728 (1988).

42.	(a) I. Schlichting, J. Berendzen, G.N. Phillips Jr., and R.M. Sweet, *Nature* 371: 808-812 (1994).

	(b) T.-Y. Teng, V. Šrajer, and K. Moffat, *Nature Struct. Biol.* 1: 701-705 (1994).

43.	S.-M. Peng and J.A. Ibers, *J. Am. Chem. Soc.* 98: 8032-8036 (1976).

44.	J.A. Ibers and R.H. Holm, *Science* 209: 223-235 (1980).

45.	S.G. Boxer, *Nature Struct. Biol.* 1: 226 (1994).

46.	W.R. Scheidt and Y.J. Lee, *Structure and Bonding*, 1987: 1-70 (1987).

47.	I. Bytheway and M.B. Hall, *Chem. Rev.* 94: 639-658 (1994).

48.	(a) J.P. Collman, R.R. Gagne, H.B. Gray, and J.W. Hare, *J. Am. Chem. Soc.* 96: 6522-6224 (1974).

	(b) A. Dedieu, M.-M. Rohmer, M. Benard, and A. Veillard, *J. Am. Chem. Soc.* 98: 3717-3718 (1976).

	(c) M. Cerdonio, A. Congiu-Castellano, F. Mogno, B. Pispisa, G.L. Romani, and S.Vitale, *Proc. Natl. Acad. Sci, USA* 74: 398-400 (1977).

	(d) L. Pauling and C.D. Coryell, *Proc. Natl. Acad. Sci, USA* 22: 210-216 (1936).

	(e) Z.S. Herman and G.H. Loew, *J. Am. Chem. Soc.* 102: 1815-1821 (1980).

	(f) A. Dedieu, M.-M. Rohmer, and A. Veillard, *in:* "Metal Ligand Interactions in Organic Chemistry and Biochemistry," Reidel, part 2 pp 101-130 (1977).

	(g) W.A. Goddard III and B.D. Olafson, *Ann. N.Y. Acad. Sci.* 367: 419-433 (1981).

	(h) L. Pauling, *Proc. Natl. Acad. Sci, USA* 74: 2612 (1977).

	(i) B. Boso, P.G.Debrunner, G.C. Wagner, and T. Inubushi, *Biochim. Biophys. Acta* 791: 244-251 (1984).

	(j) J.S. Philo, U. Dreyer, and T.M. Schuster, *Biochemistry* 23: 865-873 (1984).

	(k) J.P. Savicki, G. Lang, and M Ikeda-Saito, *Proc. Natl. Acad. Sci, USA* 81: 5417-5419 (1984).

	(l) M. Cerdonio, S. Morante, D. Torresani, S. Vitale, A. De Young, R.W. Noble, *Proc. Natl. Acad. Sci, USA* 82: 102-103 (1985).

49.	(a) J.O. Alben, W.H. Fuchsman, C.A. Beaudreau, and W.S. Caughey, *Biochemistry* 7: 624-635 1968).

	(b) G.S. Hammond and C.-S. Wu, *Adv. Chem. Ser.* 7: 186-207 (1968).

	(c) J.P. Collman, *Acc.Chem. Res.* 10: 265-272 (1977).

50.	(a) D.-H. Chin, G.N. La Mar, and A.L. Balch, *J. Am. Chem. Soc.* 102: 4344-4350 (1980).

	(b) I.R Paeng, H. Shiwaku, and K. Nakamoto, *J. Am. Chem. Soc.* 110: 1995-1996 (1988).

51.	R.E. Brantley Jr., S.J. Smerdon, A.J. Wilkinson, E.W. Singleton, and J.S. Olson, *J. Biol. Chem.* 268: 6995-7010 (1993).

52.	(a) Q.H. Gibson and M.H. Smith, *Proc. Roy. Soc. London, Ser B.* 163: 206-214 (1965).

	(b) T. Okazaki and J.B. Wittenberg, *Biochim. Biophys. Acta* 111: 503 (1965).

	(c) A.P. Klock, J. Yang, F.S. Mathews, and D.E. Goldberg, *J. Biol. Chem.* 268: 17669-17671 (1993).

53.	M. Ikeda-Saito, M.Brunori, and T. Yonetani, *Biochim. Biophys. Acta* 533: 173-180 (1978).

54.	(a) T. Imamura, A. Riggs, and Q.H. Gibson, *J. Biol. Chem.* 247: 521-526 (1972).

	(b) J.B.Wittenberg, C.A. Appleby, and B.A. Wittenberg, *J. Biol. Chem.* 247: 527-531 (1972).

	(c) C.A. Appleby, *Biochim. Biophys. Acta* 60: 226(1962).

	(d) J.B. Wittenberg, F.J. Bergersen, C.A. Appleby, and G.L. Turner, *J. Biol. Chem.* 249: 4057-4066 (1974).

55.	(a) N. Alberding, R.H. Austin, K.W. Beeson, S.S. Chan, L. Eisenstein, H. Frauenfelder, and T.M. Nordlund, *Science* 192:1002-1004 (1976).

	(b) S. Dasgupta and T.G. Spiro, *Biochemistry* 25: 5941-5948 (1986).

	(c) M.R. Chance, J.L. Parkhurst, G.L. Woolery, and B. Chance, *J. Biol. Chem.* 261: 5689-5692 (1986).

56.	G.B. Jameson and R.S. Drago, *J. Am. Chem. Soc.* 107: 3017-3020 (1985).

57.	D. Lavalette, C. Tétreau, M. Mispelter, M. Momenteau, and J.-M. Lhoste, *Eur. J. Biochem.* 145: 555-565 (1984).

58.	(a) M. Momenteau and D. Lavalette, *J. Chem. Soc., Chem. Commun.* 341-343 (1982)

	(b) J. Mispelter, M. Momenteau, D. Lavalette, and J.-M. Lhoste, *J. Am. Chem. Soc.* 105: 5165-5166 (1983).

	(c) I.P. Gerothanassis, M. Momenteau, and B. Loock, *J. Am. Chem. Soc.* 111: 7006-7012 (1989).

59.	(a) K. Nagai, B. Luisi, D. Shih, G. Miyazaki, K. Imai, C. Poyart, A. De Young, L. Kwiatkowsky, R.W. Noble, S.-H. Lin, and N.-T. Yu, *Nature* 329: 858-860 (1987).

(b) J. S. Olson, A. J. Mathews, R. J. Rohlfs, B. A. Springer, K. D. Egeberg, S. G. Sligar, J. Tame, J.-P. Renaud, and K. Nagai, *Nature* 336: 265-266 (1988).

60. M. Bolognesi, A. Coda, F. Frigerio, G. Gatti, P. Ascenzi, and M. Brunori, *J. Mol. Biol.* 213:621-625 (1990).

61. (a) J.P. Collman, J.I. Brauman, B.L. Iveson, J.L. Sessler, J.M. Morris, and Q.H. Gibson, *J. Am. Chem. Soc.* 105: 3052-3064 (1983).

(b) J.P. Collman, Brauman, T.R. Halbert, and K.S. Suslick, *Proc. Natl. Acad. Sci., USA* 73: 3333-3337 (1976).

62. See reference 4 for a comprehensive and up-to-date compilation of sterically hindered and non-hindered model systems.

63. (a) M.L. Quillin, R.M. Arduini, J.S. Olson, and G.N. Phillips Jr., *J. Mol. Biol.* 234: 140-155 (1993).

(b) T. Li, M.L. Quillin, G.N. Phillips Jr., and J.S. Olson, *Biochemistry* 33: 1433-1446 (1994).

64. M.C.M. Chung and H.D. Ellerton, *Progr. Biophys. Mol. Biol.* 35: 53-102 (1979).

65. W.E. Royer Jr., *J. Mol. Biol.* 235: 657-681 (1994).

66. P.R. Kolatkar, M.L. Hackert, and A.F. Riggs, *J. Mol. Biol.* 237: 87-97 (1994).

67. (a) E. Antonini and M. Brunori, "Hemoglobin and Myoglobin in Their Reactions with Ligands," North Holland, 1971.

(b) K. Imai, "Allosteric Effects in Hemoglobin," Cambridge University Press (1982).

68. T. Ochiai, S. Hoshina, and I. Usuki, *Biochim. Biophys. Acta* 1203: 310-314 (1993).

69. J. Monod, J. Wyman, and J.-P.Changeux, *J. Mol. Biol.* 12: 88-118 (1965).

70. M.F. Perutz, G. Fermi, B. Luisi, B.Shaanan, and R.C. Liddington, *Acc. Chem. Res.* 20: 307-321 (1987).

71. J.Baldwin and C. Chotia *J. Mol. Biol.* 129: 175-195 (1979).

72 (a) G.E.O. Borgstahl, P.H. Rogers, and A. Arnone, *J. Mol. Biol.* 236: 817-830 (1994).

(b) G.E.O. Borgstahl, P.H. Rogers, and A. Arnone, *J. Mol. Biol.* 236: 831-843 (1994).

(c) There is a report in the review literature[70] for an R-state deoxyhemoglobin structure.

73 (a) G.K. Ackers, and F.R. Smith, *Ann. Rev. Biophys. Biophys.Chem.* 16: 583-609 (1987).

(b) G.K. Ackers, *Biophys. Chem.* 37: 371-382 (1990).

(c) V.J. LiCata, P.M. Dalessio, and G.K. Ackers, *Proteins: Struct. Funct. Genet.* 17: 279-296 (1993).

74 M.M. Silva, P.H. Rogers, and A. Arnone, *J. Biol. Chem.* 267: 17248-17256 (1992).

75. M.L. Doyle, G. Lew, G.J. Turner, D. Rucknagel, and G.K. Ackers, *Proteins: Struct. Funct. Genet.* 14: 351-362 (1992).

76. (a) F.R. Smith, E.E. Lattman, and C.W. Carter Jr. *Proteins* 10: 81-91 (1991).

(b) J. Janin and S.J. Wodak, *Proteins: Struct. Funct. Genet.* 15: 1-4 (1993).

77. G.J Turner, F. Galacteros, M.L.Doyle, B. Hedlund, D.W. Pettigrew, B.W. Turner, F.R. Smith, W. Moo-Penn, D.L. Rucknagel, and G.K. Ackers, *Proteins: Struct. Funct. Genet.* 14: 333-350 (1992).

78. S. Balasubramanian, D.G. Lambright, J.H. Simmons, S.J. Gill, and S.G. Boxer, *Biochemistry* 33: 8355-8360 (1994).

79. T.B. Freedman, J.S. Loehr, and T.M. Loehr, *J. Am. Chem. Soc* 98: 2809-2815 (1976).

80. (a) K.D. Karlin and Z. Tyeklar, *Adv. Inorg. Biochem.* 9: 123 (1993).

(b) K.D. Karlin and Y. Gultneh, *Progr. Inorg. Chem.* 35: 219-327 (1987).

81. (a) D.C. Bradley, J.S. Ghotra, F.A. Hart, M.B. Hursthouse, and P.R. Raithby, *J. Chem. Soc. Dalton Trans.* 1166-1172 (1977).

(b) R. Haegele and J.C.A. Boeyens, *J. Chem. Soc. Dalton Trans.* 648-650 (1977).

82. (a) N. Kitajima, *Adv. Inorg. Chem.* 39: 1-77 (1992).

(b) T.N. Sorrell, *Tetrahedron* 45: 3-68 (1989).

(c) Z. Zanello, S. Tamburini, P.A. Vigato, and G.A. Mazzocchin, *Coord. Chem. Rev.* 77: 165-273 (1987).

83. K.D. Karlin, American Chemical Sociey Meeting, Washington, DC, August 21-25, Division of Inorganic Chemistry, paper 310 (1994).

84. (a) See reference 25 for a discussion on the active site of molluscan hemocyanin and tyrosinase.

(b) E.T. Adman, *Adv. Protein Chem.* 42: 144-157 (1991).

85. (a) W.H. Armstrong and S.J. Lippard, *J. Am. Chem. Soc.* 105: 4837-4838 (1983).

(b) K. Wieghardt, J. Pohl, and W. Gebert, *Angew. Chem., Int. Ed. Engl.* 22: 727 (1983).

86. (a) D.H. Busch and N.W. Alcock, *Chem.Rev.* 94: 585-623 (1994).

(b) M.H. Dickman and M.T. Pope, *Chem.Rev.* 94: 569-584 (1994).

87. (a) P. Gomez-Romero, G.C. De Fotis, and G.B. Jameson, *J. Am. Chem. Soc.* 108: 851-853 (1986).

(b) P. Gómez-Romero, E.H. Witten, W.M. Reiff, G. Backes, J. Sanders-Loehr, and G.B. Jameson, *J. Am. Chem. Soc.* 111: 9039-9047 (1989).

(c) P. Gómez-Romero, E.H. Witten, W.M. Reiff, and G.B. Jameson, G. B. *Inorg. Chem.* 29: 5211-

5217 (1990).

(d) P. Gómez-Romero, Ph.D. Thesis, Georgetown University, Washington, DC (1987).

88. J. Sanders-Loehr, W.D. Wheeler, A.K. Shiemke, B.A. Averill, and T.M. Loehr, *J. Am. Chem. Soc.* 111: 8084-8093 (1989).

89. M.J. Maroney, American Chemical Sociey Meeting, Washington, DC, August 21-25, Division of Inorganic Chemistry, paper 335 (1994).

90. (a) M.W. Calhoun, J.W. Thomas, and R.B. Gennis, *Trends Biochem. Sci.* 19: 325-330 (1994).

(b) For an issue devoted to cytochrome *c* oxidase, see: *J. Bioenerg. Biomembr.* 25(2): (1993).

(c) B.G. Malmström, *Acc. Chem. Res.* 26: 332-337 (1993).

(d) T. Ogura, S. Takahashui, S.Hirota, K. Shinzawa-Itoh, S. Yoshikawa, E.H. Appleman, and T.Kitagawa, *J. Am. Chem. Soc.* 115: 8527-8536 (1993).

91. (a) A. Messerschmidt, *in*: "Bioinorganic Chemistry of Copper," K.D. Karlin and Z. Tyeklar, eds., Chapman & Hall, New York, pp478-484 (1993).

(b) A. Messerschmidt, H. Leucke, and R. Huber, *J. Mol. Biol.* 230: 997-1012 (1993).

(c) A. Messerschmidt and R. Huber, *Eur. J. Biochem.* 341-347 (1990).

92. J.C. Severns and D.R. McMillin, *Biochemistry* 29: 8592-8597 (1990).

93. (a) See reference 82(a).

(b) A. Nanthakumar, S. Fox, N.N. urthy, K.D. Karlin, N. Ravi, B.H. Huynh, E.P. Day, K.S. Hagen, and N.J. Blackburn, *J. Am. Chem. Soc.* 115: 8513-8514 (1993).

(c) S. Lee and R.H. Holm, *J. Am. Chem. Soc.* 115: 11789-11798 (1993).

94. R.E. Stenkamp and L.H. Jensen, *Adv. Inorg. Biochem.* 1: 219 (1979).

95. (a) R.E. Stenkamp, L.C. Sieker, and L.H. Jensen, *J. Am. Chem. Soc.* 106: 618-622 (1984).

(b) M.A. Holmes and R.E. Stenkamp, *J. Mol. Biol.* 220: 723-737 (1991).

(c) M.A. Holmes, I. Le Trong, S. Turley, L.C. Sieker, and R.E. Stenkamp, *J. Mol. Biol.* 218: 583-593 (1991).

96. (a) D.H. Ohlendorf, J.D. Lipscomb, P.C. Weber, *Nature* 336: 403-405 (1988).

(b) L. Que Jr., The catechol dioxygenases, *in*: "Iron Carriers and Iron Proteins," T.M. Loehr, ed., VCH, New York, Vol. 5, pp 467-524 (1989).

97. E.N. Baker, B.F. Anderson, H.M.Baker, M. Haridas, G.B. Jameson, G.E. Norris, S.V. Rumball, and C.A. Smith, *Int. J. Biol. Macromol.* 13: 122-129 (1991).

98. (a) D.C. Harris and P. Aisen, Physical biochemistry of the transferrins, *in*: "Iron Carriers and Iron Proteins," T.M. Loehr, ed., VCH, New York, Vol. 5, pp 239-351 (1989).

(b) P. Aisen, Physical biochemistry of the transferrins: update, 1984-1988, *in*: "Iron Carriers and Iron Proteins," T.M. Loehr, ed., VCH, New York, Vol. 5, pp 239-351 (1989).

99. D.N.Kutrtz Jr. and W.C. Stevens, *J. Am. Chem. Soc.* 106: 1523-1524 (1984).

100. (a) S. Yan, D.D. Cox, L.L. Pearce, C. Juarez-Garcia, L. Que Jr., J.H. Zhang, and C.J. O'Connor, *Inorg. Chem.* 28: 2507-2509 (1989).

(b) R.C. Holz, T.E. Elgren, L.L. Pearce, J.H. Zhang, C.J. O'Connor, and L. Que Jr., *Inorg. Chem.* 32: 5844-5850 (1993).

(c) R.E. Norman, R.C. Holz, J.H. Zhang, C.J. O'Connor, S. Menage, and L. Que Jr. *Inorg. Chem.* 29: 4629-4637 (1990).

(d) A. Hazell, K.B. Jensen, C.J. McKenzie, and H. Toftlund, *Inorg. Chem.* 33: 3127-3134 (1994).

(e) E.C. Wilkinson, Y. Dong, and L. Que Jr., *J. Am. Chem. Soc.* 116: 8394-8395 (1994).

101. (a) H.P. Berends and D.W. Stephan, *Inorg. Chem,* 26: 749-754 (1987).

(b) H.P. Berends and D.W. Stephan, Inorg. Chim Acta, 99: L53-L54 (1987).

102 (a) M. Suzuki, A. Uehara, and K. Endo, *Inorg. Chim. Acta* 123: L9-L10 (1986).

(b) M. Suzuki, H. Osho, A. Uehara, K. Endo, M. Yanaga, S. Kida, and K. Saito, *Bull. Chem. Soc. Japan* 61: 3907-3913 (1988).

(c) A. Ben-Hussein, N.L. Morris, G.J. Long, P. Gomez-Romero, and G.B. Jameson, unpublished structures of various salts of compounds described in reference 102(a).

(d) A.S. Borovik, V. Papaefthymiou, L.F. Taylor, O.P. Anderson, and L. Que Jr., *J. Am. Chem. Soc.* 111: 6183-6195 (1989).

(e) M.S. Mashuta, R.J. Webb, J.K. McKusker, E.A. Schmitt, K.J. Oberhausen, and J.F. Richardson, R.M. Buchanan, and D.N. Hendrickson, *J. Am.Chem. Soc.* 114: 3815-3827 (1992).

103. (a) M.S. Nasir, K.D. Karlin, D. McGowty, and J. Zubieta, *J. Am. Chem. Soc.* 113: 698-700 (1991).

(b) J.D. Crane, D.E. Fenton, J.-M. Latour, and A. Smith, *J. Chem. Soc., Dalton Trans.* 2979-2987 (1991).

104. E. Bernard, W. Moneta, J. Laugier, S. Chardon-Noblat, A. Deronzier, J.-P. Tuchagues, and J.-P. Latour, *Angew Chem, Int. Ed. Engl.* 33: 887-889 (1994).

105. (a) P. Kamaras, M.C. Cajulis, M. Rapta, G.A. Brewer, and G.B. Jameson, *J. Am. Chem. Soc.* 116: 10334-10335 (1994).

(b) P. Kamaras, Ph. D. Dissertation, Georgetown University, Washington, DC (1994).

106. P. Gómez-Romero, N. Casan-Pastor, A. Ben Hussein, and G.B. Jameson, *J. Am. Chem. Soc.* 110: 1988-1990 (1988).

107. R.M. Buchanan, American Chemical Sociey Meeting, Washington, DC, August 21-25, Division of Inorganic Chemistry, paper 213 (1994).

108. (a) H. Adams, G. Candeland, J.D. Crane, D.E. Fenton, and A. Smith, *J. Chem. Soc., Chem. Commun.* 93-95 (1990).

(b) A. Bencini, D Gatteschi, C. Zanchii, O. Kahn, M. Verdaguer, and M. Julve, *Inorg. Chem.* 25: 3181-3183 (1986).

(c) M. Julve, M. Verdaguer, A. Gleizes, M. Piloche-Levisalles, and O. Kahn, *Inorg. Chem.* 23: 3808-3818 (1984).

(d) G. Cros, J.-P. Laurent, and F. Dahan, *Inorg. Chem.* 26: 596-599 (1987).

109. O. Kahn, S. Sikorav, J. Gunterou, Y. Jeannin, and J. Jeannin, *Inorg. Chem.* 22: 2577-2578 (1983).

110. I. Collamati, G. Dessy, and V. Fares, *Inorg. Chim. Acta* 111: 149-155 (1986).

111. W.B. Tolman, S. Liu, J.G. Bentsen, and S.J. Lippard, *J. Am. Chem. Soc.* 113: 152-164 (1991).

112. (a) B. Bremer, K. Schepers, P. Fleischhauer, W. Haase, G. Henkel, and B. Krebs, *J. Chem. Soc. Chem. Comun.* 510-511 (1991).

(b) B. Krebs, K. Schepers, B. Bremer, G. Henkel, E. Althaus, W. Müller-Warmuth, K. Griesar, and W. Haase, *Inorg. Chem.* 33: 1907-1914 (1994).

113. (a) V.D. Campbell, E.J. Parsons, and W.T. Pennington, *Inorg. Chem.* 32: 1773-1778 (1993).

(b) A. Neves, M.A. de Brito, I. Vencato, V. Drago, K. Griesar, W. Haase, and Y.P. Mascarenhas, *Inorg. Chim. Acta* 214: 5-8 (1993).

114. M. Rapta, P. Kamaras, G.A. Brewer, and G.B. Jameson, *J. Am. Chem. Soc.*, submitted.

115. (a) P.N. Turowski, W.H. Armstrong, S. Liu, S.N. Brown, and S.J. Lippard, *Inorg. Chem.* 33: 636-645 (1994).

(b) P.N. Turowski, W.H. Armstrong. M.E. Roth, and S.J. Lippard, *J. Am. Chem. Soc.* 112: 681-690 (1990).

(c) R.E. Norman, S. Yan, L. Que Jr., G. Backes, J. Ling, J. Sanders-Loehr, and J.H. Zhang, and C.J. O'Connor, *J. Am. Chem. Soc.* 112: 1554-1562 (1990).

116. M.Rapta and G.B. Jameson, unpublished results.

117. Y. Zang, G. Pen, L. Que Jr., B.G. Fox, and E Münck, *J. Am. Chem. Soc.* 106: 3653-3654 (1984).

118. M. Rapta, P. Kamaras, J.A. Cooley, and G.B. Jameson, American Chemical Sociey Meeting, Washington, DC, August 21-25, Division of Inorganic Chemistry, paper 276 (1994).

119. J.R. Hartman, R.L. Rardin, P. Chaudhuri, K. Pohl, K. Wieghardt, B. Nuber, J. Weiss, G.C. Papaefthymiou, R.B. Frankel, and S.J. Lippard, *J. A. Chem. Soc.* 109: 7387-7396 (1987).

120. (a) K.S. Hagen and R. Lachiotte, *J. Am. Chem. Soc.* 114: 8741-8742 (1992).

(b) R. Lachiotte, A.Kitaygarodskiy, and K.S. Hagen, *J. Am. Chem. Soc.* 115: 8883-8884 (1993).

121. H.D. Campbell, D.A. Dionysus, D.T. Keough, B.F. Wilson, J. de Jersey, and B. Zerner, *Biochem. Biophys. Res. Commun.* 82: 615-620 (1978).

122. S. Yan, L. Que, L. Jr., L.F. Taylor, and O.P. Anderson, *J. Am. Chem. Soc.* 110: 5222-5224 (1988).

123. D.L. Wang, R.C. Holz, S.S. David, L. Que Jr., and M.T. Stankovich, *Biochemistry* 30: 8187-8194 (1991).

124. A.E. True, R.C. Scarrow, C.R. Randall, R.C. Holz, and L. Que Jr., *J. Am.Chem. Soc.* 115: 4246-4254 (1993).

125. (a) S. Gehring, P. Fleischhauer, W. Haase, M. Dietrich, and H. Witzel, *Biol. Chem. Hoppe-Seyler* 371: 786 (1990). This report contradicts earlier reports (125(b) and (c)) of strong antiferromagnetic coupling for oxidized purple acid phosphatase.

(b) E. Sinn, C.J. O'Connor, J. de Jersey, B. Zerner, *Inorg.Chim. Acta* 78: L13-L15 (1983).

(c) J.C. Davis and B.A. Averill, *Proc. Natl. Acad. Sci., USA* 79: 4623-4627 (1982).

126. (a) L. Borer, L. Thalken, C. Ceccarelli, M. Glick, J.H. Zhang, and W.M. Reiff, *Inorg.Chem.*22: 1719-1724 (1983).

(b) J.A. Thich, C.-C. Ou, D. Powers, B. Vasilious, D. Mastropaolo, J.A. Potenza, and H.J. Schugar, *J. Am. Chem. Soc.* 98: 1425-1432 (1976).

(c) C.-C Ou, R.A. Lanlancette, J. A. Potenza, and H.J. Schugar, *J. Am. Chem. Soc.* 100: 2053-2057 (1978).

127. W.H. Armstrong and S.J. Lippard, *J. Am. Chem. Soc.* 106: 4632-4633 (1984).

128. A. Ben-Hussein, P. Gómez-Romero, N.L. Morris, O. Zafarullah,W.M. Reiff, and G.B. Jameson, unpublished structure of $[N3Fe(O_2CCH_3)]_2OH^{3+}$.

129. A.S. Brovik and L. Que Jr. *J. Am Chem. Soc.* 110: 2345-2347 (1988).

130. B.P. Murch, F.C.Bradley, and L. Que Jr., *J. Am Chem. Soc.* 110: 2345-2347 (1988).

131. E. Tschuchida, American Chemical Sociey Meeting, Washington, DC, August 21-25, Division of Polymeric Materials: Science and Engineering Inc., International Symposium on Metal-Containing Polymeric Materials: Bioinorganic Polymers, paper 362 (1994).

INCORPORATION OF SQUARE-PLANAR METAL BINDING SITES INTO PROTEIN POLYMERIC STRUCTURES

Eric C. Long, Paula D. Eason, and Daniel F. Shullenberger

Department of Chemistry
Indiana University-Purdue University Indianapolis
402 North Blackford Street
Indianapolis, Indiana 46202-3274

INTRODUCTION

The ability to specifically incorporate artificial transition metal binding sites into proteins has led to the development of novel biomolecules with unique chemical and physical properties.[1] Recent experiments demonstrate that such "engineered" sites of metal ion binding can, for example, assist in the conformational stabilization and assembly of polypeptides[2] or serve to regulate enzymatic activity.[3] In addition, the specific incorporation of *redox-active* metal ions, capable of generating oxidizing equivalents competent to cleave the polymeric backbone of proteins or nucleic acids, has facilitated investigations of protein three-dimensional structure,[4] folding,[5] and involvement in macromolecular binding interactions through affinity cleavage experiments (e.g., with nucleic acids).[6] A growing appreciation of the utility of artificial metal binding domains has therefore accelerated the development of new methodologies which permit their efficient and specific installation within the tertiary structures of proteins.

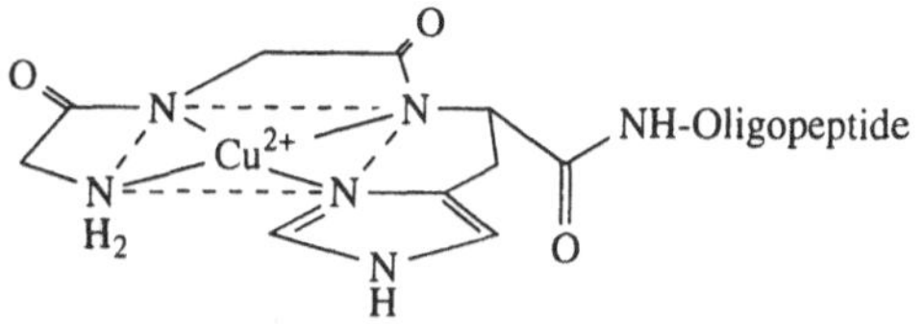

Figure 1. The structure of amino-terminal $Cu^{2+} \cdot$Gly-Gly-His metallopeptides.

Among recently developed strategies employed in the modification of proteins for affinity cleavage experiments, the addition of the tripeptide NH_2-Gly-Gly-His (GGH) to the amino-terminus of proteins[7] or oligopeptides[8] has shown considerable promise due to its relative ease of synthesis and potential for biosynthetic installation.[9] This simple peptide sequence mimics the square-planar metal-chelating domain found at the amino-terminus of serum albumins.[10] At physiological pH, GGH forms a 1:1 Cu^{2+} ($K_D \sim 10^{-16}$ to 10^{-17}) or Ni^{2+} complex[11] through the histidine imidazole nitrogen, two deprotonated amide nitrogens, and terminal α-amine. Importantly, upon chemical activation, this metal-ligand system produces oxidizing equivalents that induce DNA[7-9,12] and protein[13] cleavage adjacent to the position of the appended metal complex. Unfortunately, while the reactivity of metallo-GGH is highly beneficial in pinpointing the exact location and orientation of a protein bound to another biological macromolecule, the central role played by the terminal α-amine in the coordination of metal ions limits incorporation of GGH to the amino-terminus of polypeptides. To circumvent this problem, we have sought to design a similar peptide sequence that *preserves the metal binding, electronic, and catalytic properties of GGH while also permitting its incorporation at any site along a polypeptide chain.* Described herein is the redesign and synthesis of a novel metal chelating domain that fulfills these requirements through the use of (δ)-Orn-Gly-His.

RESULTS

In order to incorporate a metal binding domain similar to GGH at either peptide termini or *within* an oligopeptide chain, a strategy was developed[14] involving (δ)-Orn-Gly-His that employs solid phase peptide synthesis.[15] As illustrated in Figure 2, model peptides **1** and **2** represent a *carboxy-terminal* and *interior* metal binding domain, respectively, using the (δ)-Orn-Gly-His sequence. This specific amino acid connection simultaneously allows the peptide chain to be extended through the δ-amino group of ornithine while leaving the α-amino group of ornithine free to participate in metal complexation in a fashion analogous to that of the terminal glycine in NH_2-Gly-Gly-His.[10]

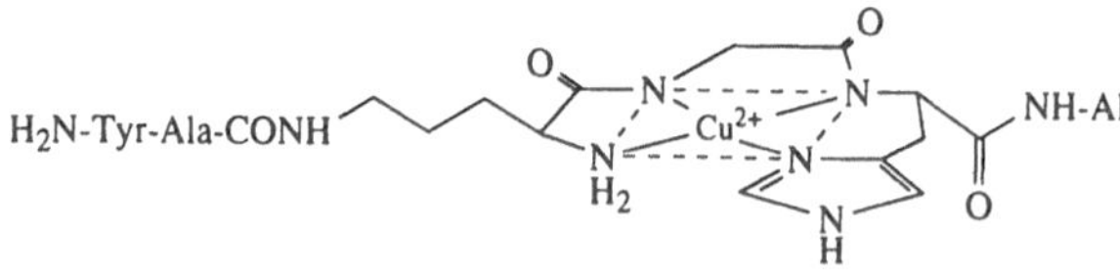

1

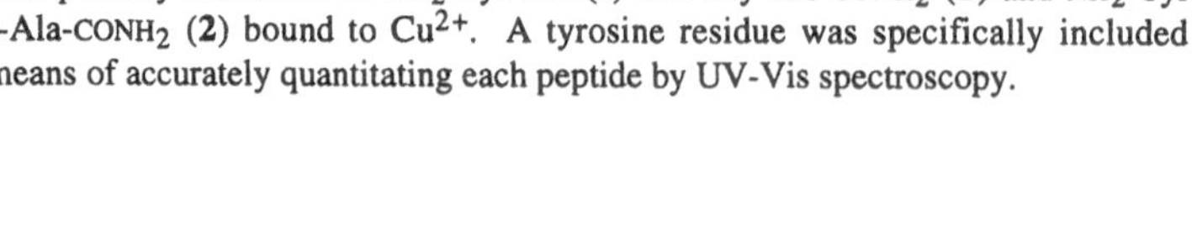

2

Figure 2. The primary structures of NH_2-Tyr-Ala-(δ)-Orn-Gly-His-$CONH_2$ (**1**) and NH_2-Tyr-Ala-(δ)-Orn-Gly-His-Ala-Ala-$CONH_2$ (**2**) bound to Cu^{2+}. A tyrosine residue was specifically included to provide a convenient means of accurately quantitating each peptide by UV-Vis spectroscopy.

As illustrated in Figure 3, the synthesis of (δ)-Orn-Gly-His peptides is easily accomplished by conventional automated or manual solid-phase techniques[15] using N-δ-Boc-N-α-Fmoc-ornithine. After purification by reversed-phase HPLC and structure

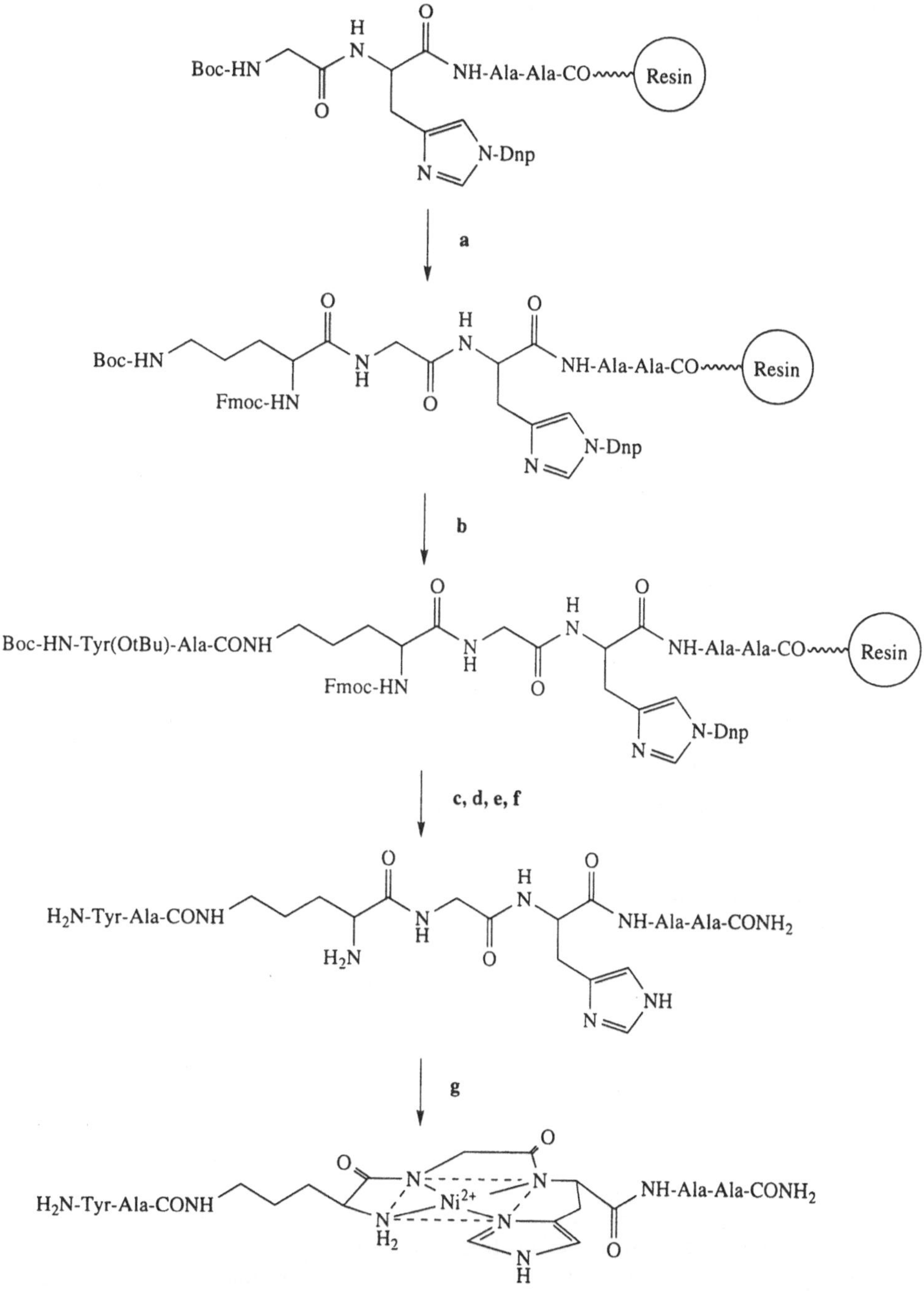

Figure 3. Solid-phase peptide synthesis (SPPS) of ornithine-based metal binding oligopeptides. **a:** coupling of Boc-Orn(Fmoc) via Boc-benzyl SPPS method; **b:** Boc-benzyl SPPS (two cycles); **c:** thiophenol/DMF; **d:** piperidine/DMF; **e:** TFA/CH$_2$Cl$_2$; **f:** NH$_3$/CH$_3$OH; **g:** Ni(OAc)$_2$/10 mM sodium cacodylate.

verification [sequencing analysis & FAB-MS: 560.3 and 702.4 for **1** and **2**, respectively], the ability of **1** and **2** to bind Cu^{2+} (Figure 4) or Ni^{2+} in the intended fashion was demonstrated by UV-Vis spectroscopy; each complex yielded a characteristic absorbance at λ_{max}= 520-530 nm and λ_{max}= 425 nm exactly analogous to that observed with $Cu^{2+}\cdot$GGH and $Ni^{2+}\cdot$GGH, respectively, and similar metallopeptides which utilize a terminal α-amino functionality in metal chelation.[10,11,16] In parallel, control experiments were also performed with a corresponding GGH-containing peptide (NH$_2$-Gly-Gly-His-Ala-Tyr-CONH$_2$); at pH 7.5 (25 mM sodium cacodylate buffer) identical molar extinction coefficients (ε = 108 cm^{-1} M^{-1}; λ_{max} = 525-527 nm; broad, flattened absorbance) for **1**, **2**, and GGHAY were calculated. Additionally, titration studies of each peptide with Cu^{2+} indicate the formation of 1:1 peptide:Cu^{2+} complexes with affinities for metal similar to those previously described ($K_D\sim10^{-16}$).[10,16] Along with optical absorbance spectroscopy, an ESR spectrum (Figure 5) identical to those of previously characterized $Cu^{2+}\cdot$GGH complexes[16] (g_o = 2.090, superhyperfine splitting constant = 12.8 G) provided additional evidence for the ability of **1** and **2** to bind Cu^{2+} as depicted in Figure 2.

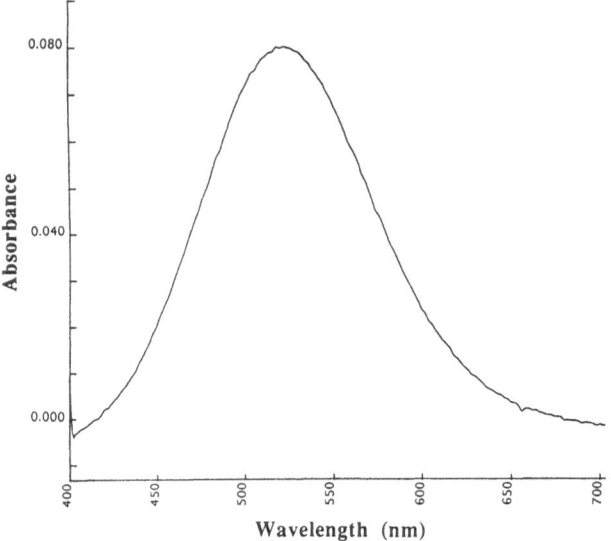

Figure 4. UV-Vis absorbance spectrum of the Cu^{2+} complex of **1** [1.0 mM **1**, 0.8 mM Cu(OAc)$_2$] in 25 mM sodium cacodylate buffer (pH 7.5). An identical spectrum was obtained for **2** and GGH bound to Cu^{2+}.

The above data provides strong evidence that in spite of its ornithine side chain, (δ)-Orn-Gly-His, when installed at either the interior or carboxy-terminus of oligopeptides, provides a high affinity binding site for metal ions such as Cu^{2+} or Ni^{2+} in a fashion identical to amino-terminal Gly-Gly-His metallopeptides. These observations are also supported by previous studies of Gly-Gly-His metallopeptides which indicated that amino acid substitutions with side chains bulkier than glycine within the parent tripeptide sequence have little effect on metal complexation at or above physiological pH values, indicating that the primary determinants of metal chelation are the histidine imidazole, two intervening deprotonated amides and the free α-amine.[16]

In light of the spectroscopic evidence supporting their similarities to previously characterized Gly-Gly-His metallopeptides, molecular models were constructed of each metal-bound peptide (using the previously characterized crystal structure of $Cu^{2+}\cdot$Gly-Gly-His as a basis for model building[10a]) to provide further insight into their structure. As shown in Figure 6, metal ions can be easily chelated by the (δ)-Orn-Gly-His sequence

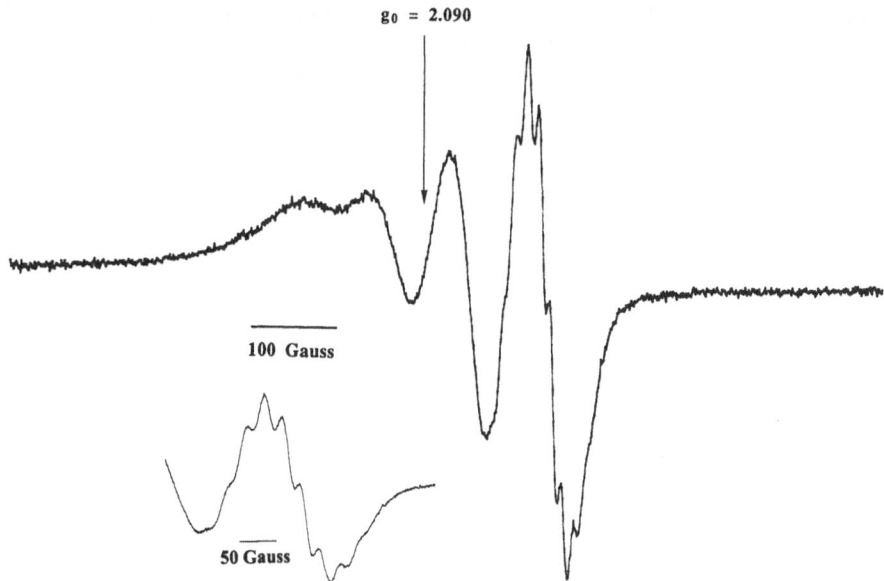

Figure 5. Room temperature ESR analysis of the Cu^{2+} complex of **1** (inset: expansion of region showing superhyperfine splitting due to the coordination of four nitrogen ligands). The spectrum was obtained on a 4.9 mM solution of metallopeptide in 25 mM sodium cacodylate buffer, pH 7.5 ($v = 9.165$ MHz; modulation amplitude, 2 G; scan range, 1000 G; microwave power, 50 mW; gain setting, 2 x 10^4). The $g_0$value and the superhyperfine splitting constant measured from this spectrum were 2.090 and 12.8 G, respectively, as compared to $g_0 = 2.090$ and 12.6 G for published room temperature spectra of Cu^{2+} complexes using and α-amino functionality in metal chelation.[16]

regardless of its position in the primary sequence of an oligopeptide chain. Inspection of these molecular models indicated that the complex formed is actually relatively small in relation to the overall peptide structure, an important consideration in the design of modified protein or oligopeptide agents. Additionally, given the positioning of the imidazole ring of the histidine residue and the α-amino functionality that completes the four-coordinate ligation of the metal, the main chain of the oligopeptide was actually found to be minimally constrained (i.e., the main chain is not required to "wrap around" the central metal ion) suggesting that this metal binding domain may be incorporated at a variety of intervening peptide segments (e.g., loop regions) spanning elements of defined secondary and tertiary structure within proteins.

Along with our examination of models of all L-amino acid containing peptides, it became apparent that inverting the chirality at selected positions within the metal binding domain could lead to changes in the orientation of the incoming and outgoing oligopeptide chains. As illustrated in Figure 6, substitution of D-His for L-His within the central (δ)-Orn-Gly-His region produces a predominantly linear oligopeptide while the L-His containing structure orients the amino- and carboxy-termini of the metal binding region at distinct angles from one another. This finding suggests that through a judicious substitution of residues, the (δ)-Orn-Gly-His metal binding domain could be customized to suit a particular structural environment within a folded protein. Further, the ability to control the subtle positioning of this metal binding domain (within a folded protein) through amino acid substitutions could permit the oxidizing equivalents that it produces to be "aimed" at a particular portion of the substrate molecule, assisting in understanding the precise location and orientation of a bound ligand through affinity cleavage.

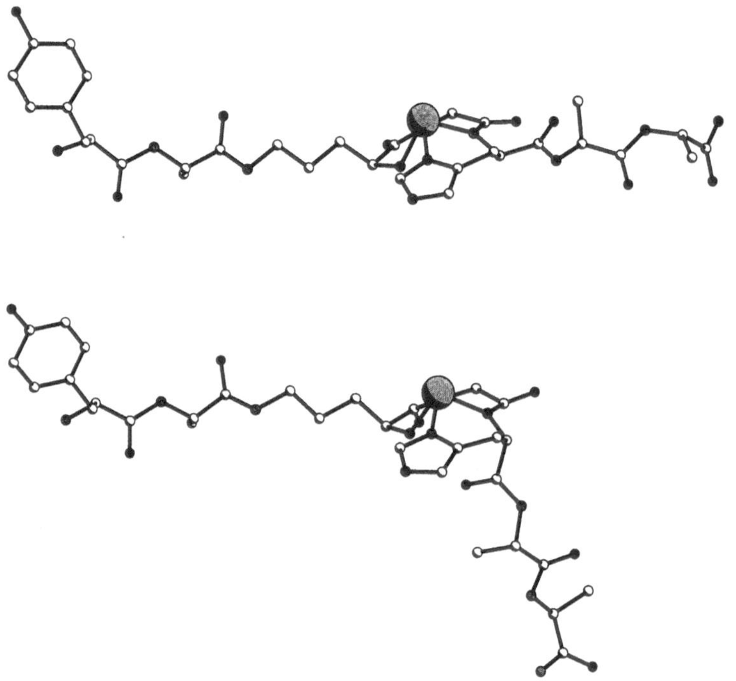

Figure 6. Molecular models of metallopeptide **2** containing the sequence (δ)-Orn-Gly-D-His (Top) and (δ)-Orn-Gly-L-His (Bottom).

In conjuction with the spectroscopic and structural studies discussed above, experiments were also conducted to determine if the Cu^{2+} and Ni^{2+} complexes of **1** and **2** possessed redox properties capable of effecting DNA cleavage like that of GGH metal complexes.[6-9,12] The assay of activity employed in this experiment involved monitoring the topological changes that occur upon cleavage of a supercoiled plasmid DNA (Form I) which can be relaxed to nicked, closed circular DNA (Form II) or linearized (Form III) upon breakage of the DNA backbone. This assay provides a relatively sensitive means of evaluating the DNA cleavage potential of a molecule by comparing the amounts of each form of DNA that appear in a reaction mixture through their separation by agarose gel electrophoresis. The experiments conducted indicate that preincubation of the Cu^{2+} or Ni^{2+} complexes of **1** and **2** with ΦX174 RF plasmid followed by activation with ascorbate/H_2O_2 or oxone,[17] respectively, resulted in facile conversion of the intact Form I DNA into Forms II & III (Figure 7); essentially identical results were obtained with the control peptide NH_2-Gly-Gly-His.

Additionally, analysis of the DNA cleavage induced by Ni^{2+}·(δ)-Orn-Gly-His complexes in the presence of ^{32}P-end labeled restriction fragments provided further information pertaining to their mechanism of action. Cleavage of this substrate DNA produced isolated, single nucleotide sites of cleavage at A, T, G, or C residues that were enhanced upon treatment with alkali. These observations suggest: (1) the operation of a non-diffusible oxidizing equivalent; (2) cleavage that is not nucleobase-dependent; and (3) the formation, in part, of DNA lesions that require a subsequent chemical modification to yield polymeric strand scission.

Overall, the DNA cleavage results presented above indicate that the (δ)-Orn-Gly-His metallopeptides not only provide a structural and electronic environment that is similar to GGH metallopeptides, but also one that provides the wherewithal to produce "activated" intermediates capable of inducing the strand scission of a DNA substrate. Moreover, the activated species generated by the Ni^{2+}·(δ)-Orn-Gly-His complexes appears to be unique in comparison to previously characterized Ni^{2+} complexes that predominantly induce the modification of G residues.[17]

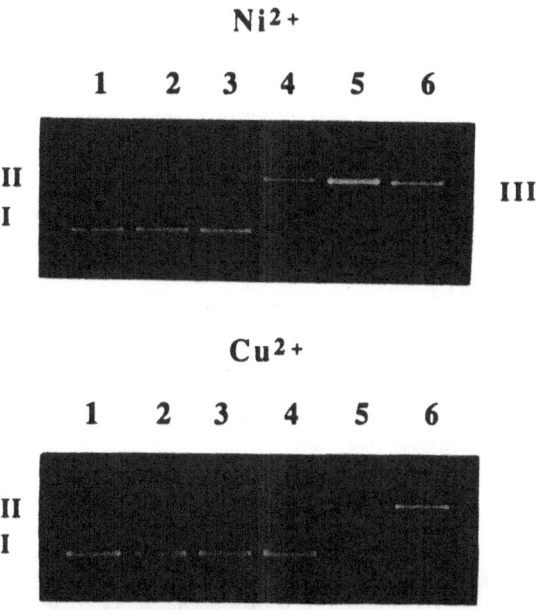

Figure 7. Cleavage of ΦX174 RF DNA (11 μM in base pairs) by Ni^{2+}·1 and Cu^{2+}·1. Reactions with Ni^{2+} were initiated with oxone and quenched after 1.0 min with EDTA (120 mM) loading buffer. Reactions with Cu^{2+} were initiated by the simultaneous addition of sodium ascorbate and H_2O_2 to preformed Cu^{2+} complexes and quenched after 1.5 min with EDTA buffer. All reactions were performed in 10 mM sodium cacodylate buffer, pH 7.5. Ni^{2+}·1 (top) lane 1: reaction control, DNA alone; lane 2: reaction control [100 μM **1**, 80 μM Ni^{2+}]; lane 3: reaction control [80 μM Ni^{2+} 100 μM oxone]; lane 4: cleavage reaction [1 μM **1**, 80 μM Ni^{2+}, 100 μM oxone]; lane 5: cleavage reaction [5 μM **1**, 80 μM Ni^{2+}, 100 μM oxone]; lane 6: cleavage reaction [25 μM **1**, 80 μM Ni^{2+}, 100 μM oxone]. Cu^{2+}·1 (bottom) lane 1: reaction control, DNA alone; lane 2: reaction control [150 μM **1**, 120 μM Cu^{2+}, 150 μM sodium ascorbate]; lane 3: reaction control [150 μM **1**, 120 μM Cu^{2+}, 150 μM H_2O_2]; lane 4: reaction control [150 μM **1**, 120 μM Cu^{2+}]; lane 5: cleavage reaction [25 μM **1**, 20 μM Cu^{2+}, 25 μM sodium ascorbate, 25 μM H_2O_2]; lane 6: cleavage reaction [150 μM **1**, 120 μM Cu^{2+}, 150 μM sodium ascorbate, 150 μM H_2O_2]. (Reprinted from ref. 14)

DISCUSSION AND CONCLUSIONS

The preceding observations demonstrate that (δ)-Orn-Gly-His provides an environment for metal ion binding that is similar electronically and structurally to that presented by NH_2-Gly-Gly-His. Additionally, metal complexes of (δ)-Orn-Gly-His are capable of inducing DNA strand scission comparable to the parent tripeptide. This new peptide sequence thus creates a versatile metal binding motif that can be synthetically or semi-synthetically[18] incorporated *within* a polypeptide chain at sterically-permissible locations and, theoretically, at several locations simultaneously. In this regard, (δ)-Orn-Gly-His could have applications in mapping protein-nucleic acid interactions through the creation of a series of metal binding derivatives that systematically change the location and orientation of the (δ)-Orn-Gly-His unit. Also, such derivatives may not be limited to Boc-Orn(Fmoc) as the provider of the α-amino group involved in metal coordination; a similar incorporation of analogously protected amino acids with side chain substitutions that are shorter or longer than ornithine should also be capable of providing the proper environment necessary for metal chelation.

Relative to other designs, therefore, (δ)-Orn-Gly-His demonstrates several unique properties including: (1) the formation of a complete, high affinity ($K_D \sim 10^{-16}$) metal binding site within three *contiguous* amino acids; (2) the use of commercially-accessible chemistries that can be easily incorporated into automated peptide synthesis and semi-synthesis protocols (and are thus readily accessible to researchers); (3) the elimination of the need for external ligands to fill empty coordination sites; and (4) the production of non-diffusible oxidizing equivalents that result in focused sites of substrate modification. Current research efforts in our laboratory seek to determine the utility of this unique peptide sequence and to further explore means of effecting its installation into much larger, structured oligopeptides.

ACKNOWLEDGEMENTS

The authors would like to acknowledge the Petroleum Research Fund, administered by the American Chemical Society and the National Science Foundation for molecular modeling equipment. In addition, one of us (P.D.E.) wishes to thank the Department of Education for fellowship support (GAANN).

REFERENCES

1.   Arnold, F. H.; Haymore, B. L. *Science* **1991**, *252*, 1796.
2.   (a) Hellinga, H. M.; Caradonna, J. P.; Richards, F. M. *J. Mol. Biol.* **1991**, *222*, 787. (b) Ghadiri, M. R.; Choi, C. *J. Am. Chem. Soc.* **1990**, *112*, 1630. (c) Ruan, F.; Chen, Y.; Hopkins, P. B. *J. Am. Chem. Soc.* **1990**, *112*, 9403. (d) Lieberman, M.; Sasaki, T. *J. Am. Chem. Soc.* **1991**, *113*, 1470. (e) Ghadiri, M. R.; Soares, C.; Choi, C. *J. Am. Chem. Soc.* **1992**, *114*, 4000. (f) Imperiali, B.; Fisher, S. L. *J. Am. Chem. Soc.* **1991**, *113*, 8527. (g) Merkle, D. L.; Schmidt, M. H.; Berg, J. M. *J. Am. Chem. Soc.* **1991**, *113*, 5450; (h) Muheim, A.; Todd, R. J.; Casimiro, D. R.; Gray, H. B.; Arnold, F. H. *J. Am. Chem. Soc.* **1993**, *115*, 5312. (i) Handel, T.; DeGrado, W. F. *J. Am. Chem. Soc.* **1990**, *112*, 6710.
3.   Higaki, J. N.; Haymore, B. L.; Chen, S.; Fletterick, R. J.; Craik, C. S. *Biochemistry* **1990**, *29*, 8582.
4.   Rana, T. M.; Meares, C. F. *Proc. Natl. Acad. Sci. USA* **1991**, *88*, 10578. (b) Rana, T. M.; Meares, C. F. *J. Am. Chem. Soc.* **1990**, *112*, 2457. (c) Rana, T. M.; Meares, C. F. *J. Am. Chem. Soc.* **1991**, *113*, 1859.
5.   Ermacora, M. R.; Delfino, J. M.; Cuenoud, B.; Schepartz, A.; Fox, R. O. *Proc. Natl. Acad. Sci. USA* **1992**, *89*, 6383. (b) Platis, I. E.; Ermacora, M. R.; Fox, R. O. *Biochemistry* **1993**, *32*, 12761.
6.   Dervan, P. B. *Methods Enzymol.* **1991**, *208*, 497.

7.   (a) Mack, D. P.; Iverson, B. L.; Dervan, P. D. *J. Am. Chem. Soc.* **1988**, *110*, 7572. (b) Mack, D. P.; Dervan, P. B. *J. Am. Chem. Soc.* **1990**, *112*, 4604. (c) Mack, D. P.; Dervan, P. B. *Biochemistry* **1992**, *31*, 9399.

8.   Shullenberger, D. F.; Long, E. C. *Bioorg. Med. Chem. Lett.* **1993**, *3*, 333.

9.   Nagaoka, M.; Hagihara, M.; Kuwahara, J.; Sugiura, Y. *J. Am. Chem. Soc.* **1994**, *116*, 4085.

10.   (a) Camerman, N.; Camerman, A.; Sarkar, B. *Can. J. Chem.* **1976**, *54*, 1309. (b) Lau, S.-J.; Kruck, T. P. A.; Sarkar, B. *J. Biol. Chem.* **1974**, *249*, 5878.

11.   (a) Bossu, F. P.; Margerum, D. W. *Inorg. Chem.* **1977**, *16*, 1210. (b) Bannister, C. E.; Raycheba, J. M. T.; Margerum, D. W. *Inorg. Chem.* **1982**, *21*, 1106. (c) Sakurai, T.; Nakahara, A. *Inorg. Chem.* **1980**, *19*, 847.

12.   (a) Chiou, S.-H.; Chang, W.-C.; Jou, Y.-S.; Chung, H.-M. M.; Lo, T.-B. *J. Biochem.* **1985**, *98*, 1723. (b) Chiou, S.-H. *J. Biochem.* **1983**, *94*, 1259.

13.   Cuenoud, B.; Tarasow, T. M.; Schepartz, A. *Tetrahedron Lett.* **1992**, *33*, 895.

14.   Shullenberger, D. F.; Eason, P. D.; Long, E. C. *J. Am. Chem. Soc.* **1993**, *115*, 11038.

15.   Stewart, J. M.; Young, J. D. *Solid-Phase Peptide Synthesis*; Pierce Chemical Co.; Rockford Il, 1984 .

16.   (a) Lau, S.-J.; Laussac, J.-P.; Sarkar, B. *Biochem. J.* **1989**, *257*, 745. (b) Iyer, K. S.; Lau, S.-J.; Laurie, S. H.; Sarkar, B. *Biochem. J.* **1978**, *169*, 61. (c) Rakhit, G.; Sarkar, B. *J. Inorg. Biochem.* **1981**, *15*, 233.

17.   Chen, X.; Rokita, S. E.; Burrows, C. J. *J. Am. Chem. Soc.* **1991**, *113*, 5884.

18.   Wuttke, D. S.; Gray, H. B.; Fisher, S. L.; Imperiali, B. *J. Am. Chem. Soc.* **1993**, *115*, 8455.

DYNAMIC ASPECTS OF ELECTRON-TRANSFER REACTIONS IN METALLOPROTEIN COMPLEXES

Nenad M. Kostić

Department of Chemistry
Gilman Hall
Iowa State University
Ames, IA 50011

INTRODUCTION

Various metalloproteins act as electron carriers and redox enzymes in photosynthesis, respiration, and other biological processes. Because these processes involve oxidation and reduction, that is, electron transfer, mechanisms of electron-transfer reactions involving metalloproteins are very important. Despite much recent research, molecular mechanisms of these reactions are only beginning to be understood.[1-4] Because the protein enveloping metal complexes (often termed metal sites or metal centers) is an organic polymer, mechanisms of electron transfer through proteins may interest polymer scientists and other materials scientists, especially those who study electroconductive materials. If one understands how electrons are transferred over 10-20 Å or even longer distances between metal complexes in metalloproteins, one may be more successful in designing a synthetic electroconductive polymer.

Most of the research in oxidoreduction (also called redox) reactions of metalloproteins has dealt with the following reaction partners: metalloproteins and redox-active organic compounds; metalloproteins and redox-active metal complexes; metalloproteins and electrodes; redox enzymes made up of multiple subunits; and metalloproteins with one another. The author's research is of the last type, concerning reactions between two metalloproteins. In these reactions electrons move (tunnel) from the reducing agent (reductant, electron donor) inside one protein molecule through the organic matter and across the protein-protein interface, to the oxidizing agent (oxidant, electron acceptor) inside another protein molecule. Previous studies have dealt mostly with kinetics (rates) of the reactions under various conditions in solution, such as pH and ionic strength. The author's research group and several other research groups are doing kinetic studies but are interested mostly in the dynamic processes, because protein mobility affects the overall electron-transfer reactions between metalloproteins.

The Proteins

All the metalloproteins discussed here are electron carriers in respiratory (in animals) and photosynthetic (in green plants) electron-transport chains. Cytochrome c, designated c, is a prototypical heme protein;[5] heme is a complex of iron with a particular type of porphyrin. An edge of the heme is slightly exposed on the protein surface and is surrounded by a ring of

Metal-Containing Polymeric Materials
Edited by C.U. Pittman, Jr., *et al.*, Plenum Press, New York, 1996

positively-charged amino-acid residues. Much evidence indicates that this exposed part of the metal complex is the place through which an electron can be picked up by the iron(III) form and released by the iron(II) form of cytochrome c. Likewise, the positively-charged (basic) patch around the exposed edge of the heme is involved in docking with physiological and other redox partners and in molecular recognition.

Cytochrome f is another heme protein. Again, the oxidation states of iron are II and III, so that this protein, too, is an electron carrier. It was long considered to be structurally similar to cytochrome c, but a very recent crystallographic analysis revealed an unexpected structure. The heme and a positively-charged site on the surface are far apart in this protein, not close to each other as in cytochrome c.

Plastocyanin, designated pc, is a prototypical blue copper protein.[6] The copper atom is proximate to an electroneutral (hydrophobic) patch and remote from a negatively-charged (acidic) patch on the protein surface. Plastocyanin seems to use both of these patches for interactions with its physiological and other redox partners, but causes and consequences of choosing one or the other remain unclear. Copper can exist in two oxidation states, I and II, and plastocyanin therefore acts as an electron carrier.

Cytochrome b_5 is yet another heme protein, in which the iron atom can adopt the oxidation states II and III. Unlike cytochrome c, this protein has a negatively-charged (acidic) patch on the surface around the exposed heme edge. In this respect cytochrome b_5 resembles plastocyanin, but the two differ in one important respect. Cytochrome b_5 has only this one surface patch for recognition and docking, whereas plastocyanin has also the hydrophobic patch.

All of the aforementioned proteins act as redox agents in their *ground* electronic states. When, however, the iron(II) ion in cytochrome c is replaced by the zinc(II) ion, the reconstituted protein, zinc cytochrome c, can easily be excited into a long-lived triplet state, designated ^{3}Znc. This excited state is a strong reducing agent, which readily donates an electron from the porphyrin (the heme).[7] The zinc(II) ion does not change its oxidation state, and in this respect the reconstituted protein differs from its native parent, in which oxidoreduction occurs at the iron ion itself. Reconstitution of cytochrome c with zinc(II) does not significantly perturb the protein structure, and it makes possible the study of photoinduced electron-transfer reactions.

The main properties of the proteins are summarized in Table 1. The first three are used as electron donors (reductants), and the last two are used as electron acceptors (oxidants).

Table 1. Properties of the Metalloproteins

Protein	Symbols	Redox-active group	Net charge at pH 7.0	Surface patch(es) and charge	Reduction potential, $E°$, in V vs. NHE[a]
Cytochrome c	c(II), c(III)	Fe(II), Fe(III)	+	Basic (+)	0.26
Cytochrome f	f(II), f(III)	Fe(II), Fe(III)	–	Basic (+)	0.36
Zinc cytochrome c	Znc, Znc$^+$	Heme, heme$^+$	+	Basic (+)	–0.88[a]
Plastocyanin	pc(I), pc(II)	Cu(I), Cu(II)	–	Acidic (–) and hydrophobic (o)	0.36
Cytochrome b_5	b$_5$(II),b$_5$(III)	Fe(II), Fe(III)	–	Acidic (–)	0.00

[a]All the potentials except one are for ground-state reactions involving the oxidation states shown for the redox-active group. This one potential is for the reaction Znc$^+$ + e$^-$ → ^{3}Znc, involving the triplet excited state.

The Protein Complexes

Association between proteins in aqueous solutions is governed more by local than by overall charges. The first three proteins in Table 1, which have a basic patch each, form electrostatic complexes with the last two proteins in this table, which have an acidic patch each. The two protein pairs that will be discussed the most in this article are shown in eqs. 1 and 2.

$$c + pc \rightleftharpoons c/pc \qquad (1)$$

$$Znc + pc \rightleftharpoons Znc/pc \qquad (2)$$

The slant represents association, whether electrostatic or covalent. The former occurs simply upon mixing the proteins in solution; the latter requires cross-linking and will be discussed below. The lower the ionic strength of aqueous solution, the stronger the electrostatic interactions. For example, the binding constant (association constant) for the processes shown in eqs. 1 and 2 is 6×10^5 M^{-1} at the ionic strength of 10 mM. There is ample evidence that the basic patch of cytochrome c abuts the acidic patch of plastocyanin.[8-10] Replacement of iron with zinc in the former protein does not perturb the protein surface and does not alter this protein-protein association.[7]

Protein-Protein Orientation

We use two experimental methods to learn about the spatial arrangement (configuration) of metalloproteins in the electron-transfer event. First, an electrostatic complex can be reinforced by formation of direct covalent bonds between basic and acidic side chains, as shown in eq. 3. This cross-linking can be accomplished with various carbodiimides, without

$$\sim NH_3^+ \ + \ ^-O-\overset{\overset{\textstyle O}{\|}}{C}\sim \ \xrightarrow{\ RN=C=NR'\ } \ \sim NH-\overset{\overset{\textstyle O}{\|}}{C}\sim \ + \ H_2O \qquad (3)$$

detectable perturbation of the protein structures.[10] Comparative kinetic studies of an electrostatic diprotein complex, which is flexible, and of its covalent analog, which is rigid, can show whether the proteins react in the same configuration in which they dock. If cross-links affect the kinetics little or not at all, we can conclude that the docking configuration is reactive. But if the covalent and electrostatic complexes of the same two proteins differ greatly in their reactivity, the only conclusion is that the reactive and the docking configurations are completely different. This finding tells little, if anything, about the actual reactive configuration.

The second method was developed by van Leeuwen, and it gives positive information about the reactive orientation.[11] A detailed analysis of electrostatic interactions between macromolecules such as proteins must include not only net charges but also dipole moments. Analyses along the familiar lines of Debye-Hückel theory usually fail because proteins can not be approximated as point charges (monopoles). Asymmetric distribution of charged atoms and groups gives rise to a large dipole moment. For example, the values for cytochrome c and plastocyanin are 280 and 360 D, respectively. Whereas monopole-monopole interactions are isotropic, monopole-dipole and dipole-dipole interactions are anisotropic, i.e., they depend on the protein-protein orientation. Experimentally, rate constants for the (bimolecular) electron-transfer reaction between unassociated protein molecules are determined in solutions spanning a wide range of ionic strength. Fitting of the rate dependence on ionic strength to the van Leeuwen theory yields structural information about the transient diprotein complex formed by the two proteins when they collide to exchange an electron.

RESULTS AND DISCUSSION

Ground-State or Thermal Reactions

Driving force for an oxidoreduction reaction is approximately equal to the difference between the reduction potentials of the electron acceptor and the electron donor. As Table 1 shows, the driving force for the intracomplex reaction in eq. 4 is small, only 0.10 eV. We

$$c(II)/pc(II) \longrightarrow c(III)/pc(I) \qquad (4)$$

studied this reaction by different experimental methods and in both electrostatic and covalent complexes c/pc.[12,13] All of these experiments consistently showed that the reaction within the electrostatic complex is relatively fast (the rate constant of 1300 s^{-1}), whereas the same reaction within the covalent complex is undetectable (the rate constant of < 0.2 s^{-1}). Thorough spectroscopic, electrochemical, and kinetic experiments showed that the electrostatic and covalent complexes c/pc are structurally similar and that the cross-linked proteins remain redox-active toward external redox agents. They just would not exchange an electron with each other! We concluded that covalent cross-links impede some rearrangement that is coupled with the electron-transfer step. Evidently, cytochrome c does not inject an electron into plastocyanin from the initial binding site, which is located in the broad acidic patch on the plastocyanin surface. Some rearrangement is required, but of what kind? There are two general possibilities for surface diffusion (sd), depicted in Scheme 1. Either cytochrome c has to migrate far away from the initial binding site before electron transfer can occur, or the two proteins fluctuate in the same general docking orientation and this fluctuation is coupled to the electron-transfer step. Further evidence disfavors a large migration and favors configurational fluctuation, as will be explained below.

Scheme 1

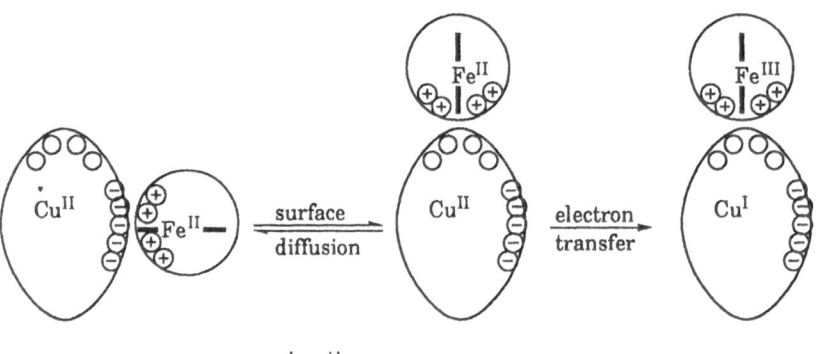

migration

configurational
fluctuation

But is this finding a mere curiosity? Cytochrome c, after all, is not a physiological partner of plastocyanin; cytochrome f is. Perhaps the true partners have evolved so as to dock and react with each other in the same configuration, without rearrangement?

To test this hypothesis, we studied the reaction in eq. 5, again within electrostatic and covalent complexes.[14,15] Since cytochrome f and plastocyanin have equal reduction potentials,the driving force for this reaction is nil. Therefore the reaction is reversible, and

$$f(II)/pc(II) \rightleftharpoons f(III)/pc(I) \qquad (5)$$

hence the double arrows in eq. 5. Again, kinetic studies by different experimental methods showed that the electron-transfer reaction occurs within the electrostatic complex but not within the covalent one. The respective rate constants are 2800 s^{-1} and < 0.1 s^{-1}. Even the complex between the physiological redox partners seems to rearrange for electron transfer. This finding shows the importance of dynamic processes in oxidoreduction reactions involving proteins and perhaps other polymers as well.

Excited-State or Photoinduced Reactions and Gating

In the study of ground-state reactions discussed above, electron transfer between the protein molecules was initiated by injecting an electron into the cytochrome (c or f) molecule by an external reducing agent. This external reaction, which was not shown in eqs. 4 and 5, complicates the kinetic analysis. Fortunately, the complication can be avoided if zinc cytochrome c is used. A laser flash excites the heme group into a triplet state (eq. 6), which is

$$Znc \xrightarrow[\text{decay}]{\text{flash}} {}^3Znc \qquad (6)$$

$$\text{ground} \qquad\qquad \text{excited}$$
$$\text{state} \qquad\qquad \text{state}$$

a potent reducing agent and can react according to Scheme 2.[16] Symbol P stands for another protein, which can accept an electron from the triplet state, ^3Znc, and become reduced to P$^-$. If P stands for pc(II) or b$_5$(III), then P$^-$ stands for pc(I) or b$_5$(II).

The degree of electrostatic association between the proteins Znc and P can be finely controlled by adjusting ionic strength. At high ionic strength (the left side of Scheme 2) the reactants are separate and have to collide with each other, i.e., to form a transient encounter complex, in order to react. At low ionic strength (the right side of Scheme 2) the proteins Znc and P are mostly associated in the so-called preformed complex. (In the covalent complex, cross-links make this association complete and permanent.) At intermediate ionic strengths Znc and P exist both free and associated with each other, but the corresponding bimolecular and unimolecular electron-transfer reactions can still be analyzed.[17]

Laser flash on the reaction mixture creates the electron donor, ^3Znc, in situ. If it is already associated with the acceptor P (the right side), the so-called forward reaction occurs and results in the oxidation of the heme into Znc$^+$ and reduction of the acceptor into P$^-$. If ^3Znc and P are separate, they have to encounter each other before the forward reaction can occur. Because Znc$^+$ lacks an electron, it is a good acceptor (oxidizing agent), which takes back an electron from P$^-$, to restore the situation that existed before the flash. What was done in the forward reaction, involving the excited state ^3Znc, is immediately undone in the back reaction, involving both proteins in their ground electronic states. Fortunately, the forward and the back reactions can be monitored separately, on a microsecond time-scale, with suitably short laser pulses.

The corresponding reaction intermediates on the two sides of Scheme 2 have identical compositions, and therefore identical formulas. Although these intermediates may be truly identical, we allowed for the possibility that they are not. Because these reaction intermediates may have different electron-transfer properties, the subscripts for the corresponding rate

495

constants are printed differently: f and F for the forward reaction, b and B for the back reaction.

Scheme 2

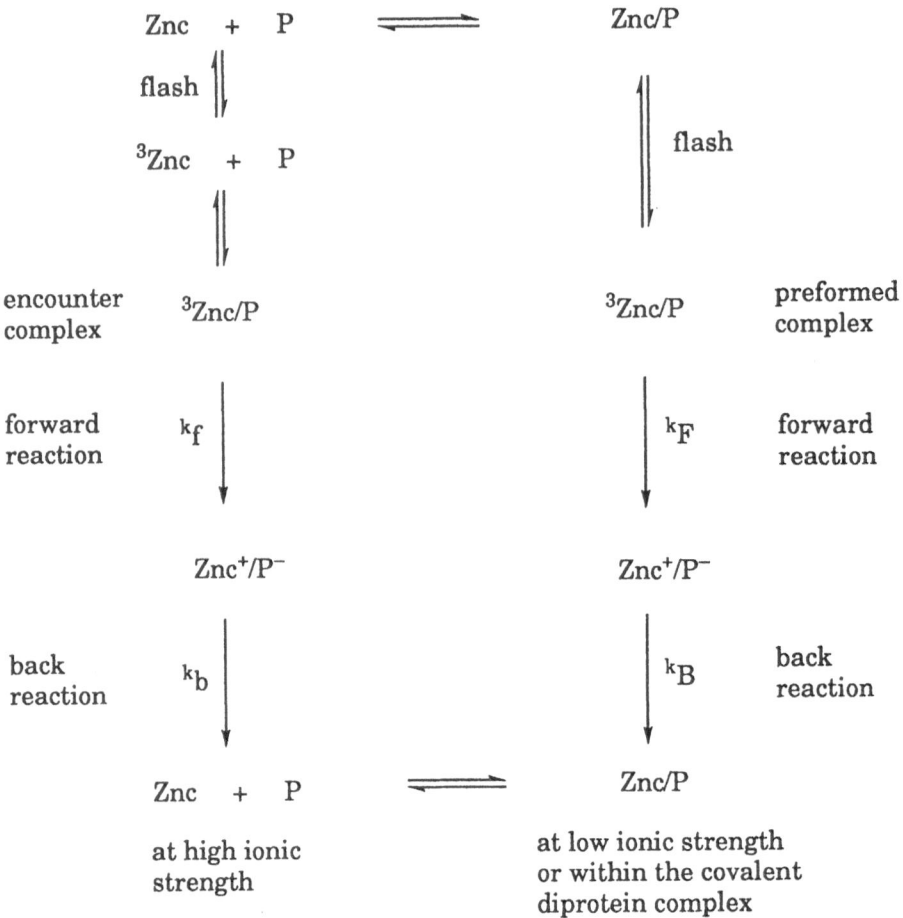

The kinetic experiments gave unexpected results.[17] First, $k_f = k_F = 2.5 \times 10^5$ s^{-1}. Either the two diprotein complexes are identical, or one or both of them rearrange(s) rapidly prior to the forward reaction so that the reactive form of both complexes is the same. Second, the covalent complex Znc/pc still undergoes the intracomplex reactions k_F and k_B, but these rate constants in the covalent complex are approximately one-tenth of the corresponding rate constants within the electrostatic Znc/pc complex. Whereas covalent cross-links abolished the thermal reaction in eq. 4, they only slowed down the photoinduced reactions in Scheme 2. Evidently, the native protein c(II) cannot, whereas the reconstituted protein ^{3}Znc can, reduce plastocyanin from the initial binding site on the plastocyanin surface. Indeed, analysis of protein-protein orientation by van Leeuwen theory (see above) showed that both the forward (eq. 7) and back (eq. 8) reactions occur with Znc (the excited triplet state or the cation) at the acidic patch of plastocyanin.[18,19]

$$^3Znc \quad + \quad pc(II) \quad \longrightarrow \quad Znc^+ \quad + \quad pc(I) \qquad (7)$$

$$Znc^+ \quad + \quad pc(I) \quad \longrightarrow \quad Znc \quad + \quad pc(II) \qquad (8)$$

The question remains, why does cross-linking of the proteins lower the rate constant k_F tenfold, from 2.5×10^5 s^{-1} in the electrostatic complex ^{3}Znc/pc(II) to 2×10^4 s^{-1} in the covalent complex ^{3}Znc/pc(II)? To answer this question, we studied the effect of solution viscosity on the rate constant k_F for the reaction in eq. 9 within the electrostatic and covalent

$$^3\text{Znc/pc(II)} \quad \xrightarrow{\quad k_F \quad} \quad \text{Znc}^+/\text{pc(I)} \tag{9}$$

complexes.[20,21] (This photoinduced reaction was already shown in Scheme 2 and discussed above.) The results, shown in Figure 1, are revealing. Increasing viscosity lowers the rate constant in the electrostatic complex, which is flexible, but does not affect the rate constant in the covalent complex, which is rigid. Clearly, high viscosity dampens, and cross-links abolish, the dynamic process that is coupled with electron transfer. Moreover, the two plots converge at high viscosity. Evidently, covalent cross-links reinforce the same diprotein configuration (orientation) that is trapped by the viscous solvent.

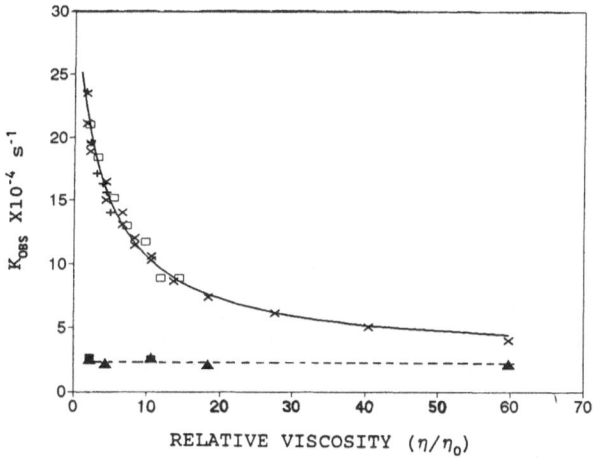

Figure 1. Rate constant k_F (the same as k_{obs} shown) for the electron-transfer reaction within the electrostatic (curve) and covalent (horizontal line) complexes ^{3}Znc/pc(II) at the ionic strength of 2.5 mM and pH 7. The viscosity was adjusted with glycerol, ethylene glycol, or D-glucose. Reproduced, with permission of American Chemical Society, from ref. 21.

The curved plot in Figure 1 can be explained most simply in terms of two configurations of the electrostatic complex Znc/pc(II) — the initial one that optimizes protein-protein docking but not the electron-transfer reaction, and a rearranged one that optimizes the reaction.[20,21] Before the laser flash, there can be no reaction and the equilibrium greatly favors the former configuration. When a flash creates ^{3}Znc/pc(II) and the reaction becomes possible, the equilibrium shifts in favor of the more reactive (rearranged) configuration and the reaction proceeds. The electron-transfer reaction in the rearranged configuration of ^{3}Znc/pc(II) is very fast, and the rate-limiting step is the rearrangement, which we view as surface diffusion (sd) of the protein molecules on each other. Fitting of the curved plot in Figure 1 to this mechanism yields the rate constant $k_{sd} = 2.5 \times 10^5$ s^{-1}, exactly the same value found for k_f and k_F before! Evidently, the electron-transfer reaction in eq. 9 (and in Scheme 2) is gated, i.e., controlled by some other, non-redox process. In this case, this process is rearrangement of the diprotein complex.

Effects of ionic strength on electron-transfer reactions within diprotein complexes are often explained imprecisely in terms of gating. For example, an increase of the rate constant as the ionic strength increases is attributed to "loosening of the complex" as the electrostatic attraction between the protein molecules weakens, so that the "more flexible" complex rearranges faster. Although such an explanation is intuitively plausible, it is incorrect. Ionic strength of a solution governs the equilibrium between the free and bound protein molecules, that is, ionic strength controls the extent of association. But ionic strength does not control the interactions within the diprotein complex. Our experiments with the reaction in eq. 9 within the electrostatic diprotein complex prove this point. The viscosity dependence at five different values of ionic strength (2.5, 5.0, 10, 15, and 20 mM) is the same as in Figure 1 and coincides with the curved plot in Figure 1. The marked dependence of the rate constant k_F on viscosity remains, as a direct proof that the system Znc/pc(II) is dynamic. Nevertheless, k_F does not depend on ionic strength. As the ionic strength increases, from 2.5 to 20 mM, the *concentration* of the electrostatic complex Znc/pc(II) decreases and the amplitude of the kinetic trace for the reaction in eq. 9 becomes smaller and smaller. The rate constant k_F, however, remains the same.

The rearrangement rate ($k_{sd} = 2.5 \times 10^5 \text{ s}^{-1}$) is the same for the c(II)/pc(II) and Znc/pc(II) complexes because replacement of Fe(II) by Zn(II) in the interior of cytochrome c does not significantly alter the topography and electrostatic properties of this protein's surface. The ground-state reaction in eq. 4 has a small driving force (ca. 0.10 eV). Because its rate constant (1300 s^{-1}) is lower than k_{sd}, the observed rate constant represents true electron transfer. The excited-state reaction in eq. 9 has a large driving force (ca. 1.2 eV), which greatly enhances the electron-transfer process but, of course, has no effect on the mechanical rearrangement. Now the true rate constant for electron transfer is higher than k_{sd}. The rearrangement becomes the step which limits the overall rate and which is observed in kinetic experiments. The increase in the driving force within essentially the same protein pair converts an electron-transfer system that is not gated (eq. 4) to another system that is gated (eq. 9 and Scheme 2).

Nature of the Rearrangement

Aforementioned studies of the bimolecular reaction in eq. 7 showed that ^{3}Znc reduces plastocyanin from the acidic patch on the surface of the latter.[18,19] Studies of the analogous unimolecular (intracomplex) reaction in eq. 9 showed that this electron-transfer event is modulated by a dynamical process.[20,21] This combined evidence disfavors large migration and favors local configurational fluctuations; these two processes are depicted in Scheme 1 in reference to a ground-state reaction (eqs. 4 and 5), but the same concepts apply also to the photoinduced reaction (eq. 9) under discussion here.

Like plastocyanin, cytochrome b_5 has a large acidic patch for docking with cytochrome c or other proteins. But unlike plastocyanin, which has also the hydrophobic patch, cytochrome b_5 lacks other surface domains for interactions with redox partners. The rate constant for the reaction in eq. 10 depends on viscosity just like the reaction in eq. 9, and the results are the

$$^3\text{Znc/b}_5(\text{III}) \longrightarrow \text{Znc}^+/\text{b}_5(\text{II}) \tag{10}$$

same as those shown in Figure 1.[22] Moreover, the fitted rate constant for rearrangement of Znc/b$_5$(III) is the same as that given above for Znc/pc(II), $k_{sd} = 2.5 \times 10^{-5}$ s^{-1}. Clearly, dynamic behavior and dependence of the intracomplex rate constant on viscosity are caused by configurational fluctuations of the electrostatic complex, such that the basic patch of zinc cytochrome c remains within the broad acidic patch of plastocyanin or cytochrome b_5. See Scheme 3.

Scheme 3

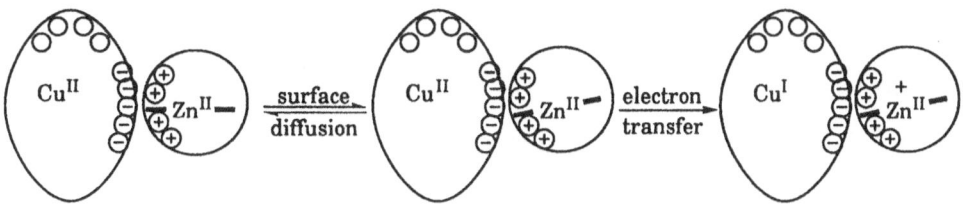

configurational
fluctuation

Pathways for Electron Tunneling

What does the diprotein complex, say Znc/pc(II), gain by configurational fluctuations? We applied the approximate quantum-chemical method of Beratan and Onuchic[23,24] to analyze electron-tunneling paths through covalent bonds, hydrogen bonds, and interatomic spaces, in various configuration of the complex c(II)/pc(II).[25] Although this electrostatic complex contains native cytochrome c, findings of this theoretical analysis should pertain also to the complex Znc/pc(II), with which most of our experimental studies have been done. We compared five configurations of the cyt(II)/pc(II) complex, all of which are stabilized by electrostatic attraction between the two proteins. We considered water molecules inside the proteins, we simulated hydration near the interface, and we recognized the anisotropy of electronic coupling of the copper(II) atom and its four ligands in plastocyanin. It is important to realize that different ligands contribute to different extents, or not at all, to that molecular orbital of the blue copper site which eventually accepts the electron. We critically compared the results obtained with different sets of electronic-structure parameters and with different degrees of hydration. These calculations consistently showed that the configuration that optimizes protein-protein *interactions* does not optimize the conditions for the electron-transfer *reaction* between them. Configurational fluctuation moves cytochrome c away from the initial binding site on the plastocyanin surface but does not entirely remove it from the acidic patch. This fluctuation maximizes coupling between the electron donor, iron(II) or the excited state, on the one hand and the electron acceptor, copper(II), on the other.

Current and Future Research

We are now collaborating with biochemists and molecular biologists who can replace specific amino acids in blue copper proteins with other amino acids and thus alter the electrostatic and steric properties of the protein surface. Comparisons among the native plastocyanin and its various mutant forms will allow us to analyze protein-protein recognition, electron-tunneling paths through proteins, and the interplay between dynamical rearrangement and electron-transfer reactions.

CONCLUSIONS

This author may be a typical chemist in his belief that design of new materials requires thorough knowledge of scientific fundamentals on which this design rests. Study at a molecular level of electron transfer through well-characterized metalloproteins may help in the study of electrical conductance by synthetic polymers.

ACKNOWLEDGMENTS

This research has been supported by the National Science Foundation through grants CHE-8858387 and MCB-9222741. The original research has ably been carried out by the author's coauthors in the publications that are cited below. The author is grateful to all of them, especially to Dr. Jian Zhou, Dr. Ling Qin, and Matthias Ullmann, for their dedication and excellent work.

REFERENCES

1. G. McLendon and R. Hake, *Chem. Rev.* 92:481 (1992).
2. J. R. Winkler and H. B. Gray, *Chem. Rev.* 92:369 (1992).
3. B. M. Hoffman, M. J. Natan, J. M. Nocek, and S. A. Wallin, *Struct. Bonding* 75:86 (1991).
4. N. M. Kostić, *Metal Ions Biol. Syst.* 27:129 (1991).
5. G. R. Moore and G. W. Pettigrew, *"Cytochromes c: Evolutionary, Structural and Physicochemical Aspects"*, Springer-Verlag: Berlin (1990).
6. (a) A. G. Sykes, *Struct. Bonding* 75:177 (1991). (b) A. G. Sykes, *Adv. Inorg. Chem.* 36:377 (1991).
7. J. M. Vanderkooi, F. Adar, and M. Erecińska, *Eur. J. Biochem.* 64:381 (1976).
8. S. Bagby, P. C. Driscoll, K. G. Goodall, C. Redfield, and H. A. O. Hill, *Eur. J. Biochem.* 188:413 (1990).
9. V. A. Roberts, H. C. Freeman, E. D. Getzoff, A. J. Olson, and J. A. Tainer, *J. Biol. Chem.* 266:13431 (1991).
10. J. S. Zhou, H. M. Brothers II, J. P. Neddersen, T. M. Cotton, L. M. Peerey, and N. M. Kostić *Bioconjugate Chem.* 3:382 (1992).
11. J. W. van Leeuwen, *Biochim. Biphys. Acta* 743:408 (1983).
12. L. M. Peerey and N. M. Kostić, *Biochemistry* 28:1861 (1989).
13. L. M. Peerey, H. M. Brothers II, J. T. Hazzard, G. Tollin, and N. M. Kostić, *Biochemistry* 30:9297 (1991).
14. L. Qin and N. M. Kostić, *Biochemistry* 31:5145 (1992).
15. L. Qin and N. M. Kostić, *Biochemistry* 32:6073 (1993).
16. J. S. Zhou and N. M. Kostić, *The Spectrum* 5, no 2:1 (1992).
17. J. S. Zhou and N. M. Kostić, *J. Am. Chem. Soc.* 113:6067 (1991).
18. J. S. Zhou & N. M. Kostić, *J. Am. Chem. Soc.* 113:7040 (1991).
19. J. S. Zhou and N. M. Kostić, *Biochemistry* 32:4539 (1993).
20. J. S. Zhou and N. M. Kostić, *J. Am. Chem. Soc.* 114:3562 (1992).
21. J. S. Zhou and N. M. Kostić, *J. Am. Chem. Soc.* 115:10796 (1993).
22. L. Qin and N. M. Kostić, *Biochemistry* 33:12592 (1994).
23. D. N. Beratan, J. N. Onuchic, and H. B. Gray, *Metal Ions Biol. Syst.* 27:97 (1991).
24. J. J. Regan, S. M. Risser, D. N. Beratan, and J. N. Onuchic, *J. Phys. Chem.* 97:13083 (1993).
25. G. M. Ullmann and N. M. Kostić, *J. Am. Chem. Soc.* 117:issue of May 10 (1995).

CONTRIBUTORS
(Includes addresses of principal authors*)

Ronald D. Archer,* Huiyong Chen, Jon A. Cronin
    and Sharon M. Palmer
Department of Chemistry
University of Massachusetts
Amherst, MA 01003-4150
USA

Anjali Bajpai,* Milind Khandwe and
    Udai D. N. Bajpai
Department of Chemistry
Government Autonomous Science College
Jabalpur 482001
INDIA

Udai D. N. Bajpai,* Sandeep Rai and Anjali
    Bajpai
Polymer Research Laboratory
Department of Post-Graduate Studies and
    Research in Chemistry
Rani Durgavati University
Jabalpur 482001
INDIA

Natalia M. Bravaya,* Anatolii. D. Pomogailo,
    V. A. Maksakov and V. P. Kirin
Institute of Chemical Physics in Chernogolovka
Russian Academy of Sciences
Chernogolovka 142432
Moscow Region
RUSSIA

Charles E. Carraher, Jr.,* J. William Louda,
    Dorothy Sterling, Earl Baker, Alberto
    Rivalta and Qingmao Zhang
Florida Atlantic University
Boca Raton, FL 33431
USA

Charles E. Carraher, Jr.* and Qingmao
    Zhang
Florida Atlantic University
Boca Raton, FL 33431
USA

G. I. Dzhardimalieva* and Anatolii D.
    Pomogailo
Institute of Chemical Physics in
    Chernogolovka
Russian Academy of Sciences
Chernogolovka 142432
Moscow Region
RUSSIA

Aiaz A. Efendiev,* Junshud J. Orujev, Elmar
    B. Amanov, and Yusif M. Sultanov
Azerbaijan Academy of Sciences
Institute of Polymer Materials
Samed Vurgun str. 124
Sumgait 373204
AZERBAIJAN REPUBLIC

Osamu Funayama,* Tomoko Aoki and Takeshi
    Isoda
Tonen Corporation
Corporate Research & Development Laboratory
1-3-1 Nishi-Tsurugaoka Ohi-machi Iruma-gun
Saitama 356
JAPAN

Georges Gelbard,* David C. Sherrington,
    Francois Breton, Mohamed
    Benelmoudeni, Marie-Thérèse Charreyre
    and Doan Dong
Institut de Recherdes sur la Catalyse-CNRS
2 avenue Albert Einstein
69626 Villuerbanne Cedex
FRANCE

Takahiro Gunji* and Yoshimoto Abe
    Department of Industrial Chemistry
    Faculty of Science and Technology
    Science University of Tokyo
2641 Yamazaki, Noda, Chiba 278
JAPAN

Wolfgang Haase,* K. Griesar, E. A.
    Soto-Bustamante and Yu. G.
    Galyametdinov
Technische Hochschnule Darmstadt
Institut für Physikalische Chemie
Petersenstr. 20
64287 Darmstadt
GERMANY

Michael Hanack
Institut für Organische Chemie
Lehrstuhl für Organische Chemie II der
    Universität Tübingen
Auf der Morganstelle 18
D-72076 Tübingen
GERMANY

Richard G. Jones,* Anthony C. Swain, S. J.
    Holder, A. J. Wiseman, M. J. Went and
    R. E. Benfield
Centre for Materials Research, Chemical Lab.
University of Kent, Canterbury
Kent CT2 7NH
GREAT BRITAIN

Nenad M. Kostić
Department of Chemistry
Gilman Hall
Iowa State University
Amers, IA 50011
USA

Nina E. Kotelnikova,* Deitrich Fengel, and
    Valery P. Kotelnikov
Russian Academy of Sciences
Institute of Macromolecular Compounds
Bolshoi pr. 31
St. Petersburg 199004
RUSSIA

Geoffrey B. Jameson
Department of Chemistry and Biochemistry
Massey University
Palmerston North
NEW ZEALAND

Eric C. Long,* Paula D. Eason and Daniel F.
    Shullenberger
Department of Chemistry
Indiana University-Purdue University
    Indianapolis
402 North Blackford Street
Indianapolis, IN 46202-3274
USA

Fred B. McCormick,* Bradford B. Wright and
    Jerry W. Williams
3M Corporate Rresearch Labs
230-2G-14, 3M Center
St. Paul, MN 55144-1000
USA

Leonid S. Molochnikov,* Elena G.
    Kovalyova, Igor A. Grigor'ev and
    Vladimir A. Reznikov
Ural Wood Technology Academy
Siberian Highway 37
Ekaterinburg, 620032
RUSSIA

Yoshiyuki Okamoto* and Hannia
    Luján-Upton
Department of Chemistry and the Polymer
    Research Institute
Polytechnic University
6 Metrotech Center
Brooklyn, NY 11201
USA

Charles U. Pittman, Jr.* and Ruicheng Ran
Department of Chemistry
University/Industry Chemical Research Ctr.
Mississippi State University
Mississippi State, MS 39762
USA

Anatolii D. Pomogailo,* A. S. Rozenberg and
    G. I. Dzhardimalieva
Institute of Chemical Physics in Chernogolovka
Russian Academy of Sciences
Chernogolovka 142432
Moscow Region
RUSSIA

Olga I. Shchegolikhina,* Inessa V. Blagodatskikh,
    Yulia A. Pozdnyakova and Alexandre A.
    Zhdanov
Institute of Organo-Element Compounds
Russian Academy of Sciences
Vavilova str 28
Moscow 117813
RUSSIA

John E. Sheats,* Charles E. Carraher, Jr.,
    Charles U. Pittman, Jr., Martel Zeldin
    and Bill M. Culbertson
Department of Chemistry
Rider University
Lawrenceville, NJ 08648
USA

Duward F. Shriver* and Glenn C. Rawksy
Department of Chemistry and Materials
    Research Center
Northwestern University
Evanston, IL 60201-3113
USA

Warren P. Steckle, Jr.,* J. R. Duke, Jr. and
    B. S. Jorgensen
Los Alamos National Labs
MST-7, MS E-549
Los Alamos, NM 87545
USA

Herbert H. Stewart,* Charles E. Carraher,
        Jr., Winn J. Soldani, Lisa Reckleben,
        Jose de la Torre and Shi Li Miao
Departments of Biological Sciences and
        Chemistry
Florida Atlantic University
Boca Raton, FL 33431
USA

Claire Tessier,* Gan Ouyang and Richard
        Simons
Department of Chemistry
University of Akron
Akron, OH 44325-3601
USA

Larry T. Taylor,* Adley F. Rubira and James D.
        Rancourt
107 Davison Hall
Department of Chemistry
Virginia Tech
Blacksburg, VA 24061-0212
USA

David W. Thompson,* Robin E. Southward,
        D. Scott Thompson and Anne K. St.
        Clair
Langley Research Center
NASA, MS 227
USA

David W. Thompson,* Robin E. Southward,
        D. Scott Thompson, Maggie L. Caplan
        and Anne K. St. Clair
Langley Research Center
NASA, MS 227
Hampton, VA 23665-5225
USA

Eishun Tsuchida* and Kimihisa Yamamoto
Department of Polymer Chemistry
Waseda University
Tokyo 169
JAPAN

Eishun Tsuchida,* Kimihisa Yamamoto and
        Kenichi Oyaizu
Department of Polymer Chemistry
Waseda University
Tokyo 169
JAPAN

Omar M. Yaghi
Department of Chemistry & Biochemistry
Arizona State University
Box 871604
Tempe, AZ 85287-1604
USA

Luping Yu* and Zhenan Bao
Department of Chemistry
University of Chicago
5735 S. Ellis Ave.
Chicago, IL 60637
USA

INDEX

Acetate
  erbium, 339, 343, 344
  europium, 341, 343, 344
  thulium, 339, 345
Acetic acid, 268, 272
Acetyl CoA, 95, 96
Acetylacetonate, 172
tris(Acetylacetonato)lanthanide(III), 338
tris(Acetylacetonato)thulium(III), 345, 346
Acid red 88, 416
Acidic salts, 269
Acrylic acid, 64, 320
  Ba(III) acetate, 324
  Cu(II) acetate, 324
  Y(III) nitrate, 324
Acrylic acid with divinyl benzene (KB-2), 395
  copolymerization, 395
Albo Lacinatus, 93
AlCl3, 272
Alkali ion cellulose complexes, 403
Alkali metal reduction, 189, 190, 194
Alkene, 265, 267, 274
Alkenyl alcohols, 269
μ-Alkoxo diiron(III), 467
Alkylation, 268
Allosteric effectors, 456
Allyl alcohol, 255–257, 260–263, 269
  hydrogenation of, 255, 256
α,ω-bis(Allyloxy)poly(ethylene glycol), 387
Al/Ti ratio, effect of
  on butadiene polymerization, 247, 248
  on isobutylene polymerization, 248, 249
  on ethylene polymerization, 250
  on isobutylene/isoprene copolymerization, 251
Aluminosilicate, 386
Aluminosilicate/poly(ehtylene glycol) copolymers, 383
Amberlite IRC-50, 399, 400
Amberlite resins, 277
  crown ethers sorbed, 277
Amide, 272, 482
Amidification, 272
Amine, 266, 269, 274
α-Amine, 482, 484, 485, 488
Amino, 161
Amino acid, 482, 486, 488
Amino terminated oligomers of hexamethylene tere-
    phthalamide, 119

1,3-bis(3-Aminophenoxy)benzene (APB), 337, 338,
    358
2,2'-bis(4-Aminophenyl)hexafluoropropane, 337
2-Aminopyridine, 268
a-Amylase, 98, 100
b-Amylase, 98
Animation, 174
Anion-exchange resins, 266
Anionic polymerization, 20, 24, 25
Anisotropic electron flow, 109
Anthracenocyanine, 331
Antiferromagnetic 1-D-Heisenberg behaviors, 287
Antiferromagnetically coupled Fe-(μ-O)-Fe-units,
    297, 298
Aryl halides, 168
Arylether, 161
Ascorbate, 486
Ascorbate oxidase, 456, 457
Asymmetric
  ligands, 462
  dicopper(II) complexes, 467
Atomic absorption, 173
Autocatalytic reduction, 74
Autooxidation control of porphyrinatoiron(II), 431,
    432

Bahr effect, 452
Ballistic heating, 413
Basic, 270
"Basket handle" porphyrins, 445, 446
1,3,5-Benzenetricarboxylic acid, 220
1-bis(2'Benzimidazoylmethyl)-2-[(2'-benzi-mida-
    zolylmethyl)(2'-ethoxy)]-diamino-ethane,
    460, 468
3,3',4,4'-Benzophenonetetracarboxylic dianhydride
    (BTDA), 338, 349, 357
  derived films, 367
3,3',4,4'-Benzophenonetetracarboxylic acid dianhy-
    dride-4,4'-oxydianiline anhydride (BTDA-
    ODA), 360, 362, 363, 365–367
Betaines, 268
Betainic, 269
BF3•OEt2, 172
Bidentate coordination of RCOO—, 66, 69
N,O-Bidentate pendant groups, 288
Bifunctional acrylic monomers, 76
Bimolecular electron transfer reaction, 493, 498

505

510